BIOGEOCHEMICAL, HEALTH, AND ECOTOXICOLOGICAL PERSPECTIVES ON GOLD AND GOLD MINING

BIOGEOCHEMICAL, HEALTH, AND ECOTOXICOLOGICAL PERSPECTIVES ON GOLD AND GOLD MINING

Ronald Eisler, Ph.D.

CRC Press
Taylor & Francis Group
Boca Raton London New York

CRC Press is an imprint of the
Taylor & Francis Group, an **informa** business

CRC Press
Taylor & Francis Group
6000 Broken Sound Parkway NW, Suite 300
Boca Raton, FL 33487-2742

First issued in paperback 2019

© Taylor & Francis Group, LLC
CRC Press is an imprint of Taylor & Francis Group, an Informa business

No claim to original U.S. Government works

ISBN-13: 978-0-8493-2898-5 (hbk)
ISBN-13: 978-0-367-39369-4 (pbk)

Library of Congress Card Number 2004051932

Library of Congress Cataloging-in-Publication Data

Eisler, Ronald, 1932–
    Biogeochemical, health, and ecotoxicological perspectives on gold and gold mining. Ronald Eisler.
      p. cm.
    Includes bibliographical references and index.
    ISBN 0-8493-2898-5 (alk. paper)
    1. Gold mines and mining—Health aspects. 2. Gold mines and mining—Environmental aspects. 3. Gold—Toxicology. 4. Gold—Physiological effect. I. Title.

RC965.M48E35 2004
615.9′02—dc22   2004051932

Visit the Taylor & Francis Web site at
http://www.taylorandfrancis.com

and the CRC Press Web site at
http://www.crcpress.com

Dedication

To Jeannette, Renée, David, Charles, Julie, and Eb

Preface

Over the past several decades I have been tasked by environmental specialists of the U.S. Fish and Wildlife Service with the preparation of risk assessment documents of chemical and biological hazards of various compounds to wildlife. For the most part, these documents involved analysis of measurable risks associated with agricultural, industrial, municipal, military, and industrial chemicals and their wastes. Risk assessment — still an inexact science — depends heavily on well-documented databases that include the compound's source and use; its physical, chemical, and metabolic properties; concentrations in field collections of abiotic materials, plants, and animals; lethal and sublethal effects, including effects on survival, growth, reproduction, metabolism, mutagenicity, carcinogenicity, and teratogenicity; proposed regulatory criteria for the protection of human health and sensitive natural resources; and recommendations for additional research when databases are incomplete. However, this approach was only partially successful in attempting to evaluate gold and gold mining practices because none of the preceding reports — unlike the present account on gold — relied significantly on social, political, economic, medical, or psychological variables in assessing risk to the biosphere. For the past several years — through literature analysis, visits to operating gold mines, and consultations with colleagues — I have tried to evaluate critically the enormous effort expended by society in producing gold mainly for coinage, bullion, and personal jewelry, regardless of environmental damage. This book is the result; however, all interpretations are my own and do not necessarily reflect those of the U.S. Geological Survey or any other federal agency. Moreover, mention of trade names or commercial products is not an endorsement or recommendation for use by the U.S. government.

Ronald Eisler, Ph.D.
Senior Research Biologist
U.S. Geological Survey
Patuxent Wildlife Research Center
Laurel, Maryland

Acknowledgments

I owe a special debt of gratitude to Judd A. Howell, the Director of the Patuxent Wildlife Research Center (PWRC), and Harry N. Coulombe, Chief of Research at PWRC, for their encouragement and financial support during the course of this five-year effort. I thank Wanda Manning and Lynda J. Garrett for providing essential library services, and my colleagues at PWRC for discussions and technical support, specifically, Peter H. Albers, Thomas W. Custer, Gary H. Heinz, David J. Hoffman, T. Peter Lowe, Glenn H. Olsen, Oliver H. Pattee, Matthew C. Perry, Barnett A. Rattner, Graham W. Smith, and Nimish Vyas. Computer assistance was kindly provided by PWRC specialists Kinard Boone and Henry C. Bourne. I am also obligated to others for their insights on gold mining aspects, including David N. Weissman (National Institute of Occupational Safety and Health), Stanley N. Wiemeyer (U.S. Fish and Wildlife Service), Rory E. Lamp (State of Nevada Department of Wildlife), Roger D. Congdon and David J. Vandenberg (U.S. Bureau of Land Management), David Gaskin (State of Nevada Bureau of Mining Regulation and Reclamation), Michael L. Strobel (U.S. Geological Survey), Tom Jeffers (U.S. Forest Service), Jeff White (Newmont Mining Corporation), and Ron A. Espell and Peter G. Bodily (Barrick Goldstrike Mines, Inc.). Finally, I thank CRC staffers Randi Cohen, Gail Renard, and Kathy Johnson for their professionalism in expediting publication of this volume.

About the Author*

Ronald Eisler received his B.A. degree from New York University in biology and chemistry, and his M.S. and Ph.D. degrees from the University of Washington in aquatic sciences and radioecology, respectively.

Eisler has been a senior research biologist since 1984 at the U.S. Geological Survey, Patuxent Wildlife Research Center in Laurel, Maryland. Prior to 1984, he held, in order, the following positions: bioscience advisor, U.S. Fish and Wildlife Service, Washington, D.C.; research aquatic toxicologist, U.S. Environmental Protection Agency, Narragansett, Rhode Island; fishery research biologist, U.S. Fish and Wildlife Service, Highlands, New Jersey; radiochemist, University of Washington Laboratory of Radiation Ecology; aquatic biologist, New York State Department of Environmental Conservation, Raybrook, New York. During the Korean War, he served in the U.S. Army Medical Service Corps.

He has held a number of special assignments and teaching appointments including senior science advisor to the American Fisheries Society, adjunct professor of zoology at American University in Washington, D.C., adjunct professor of oceanography at the University of Rhode Island, and visiting professor of marine biology and resident director of the Marine Biological Laboratory of Hebrew University in Eilat, Israel. Since 1955, he has authored more than 135 technical articles and books on contaminant hazards to living organisms, mainly on physiological and toxicological effects of trace metals, as well as agricultural chemicals, municipal wastes, crude oils and oil dispersants, and military and industrial wastes.

He resides in Potomac, Maryland, with his wife Jeannette, a teacher of French and Spanish.

* Dr. Eisler retired in July 2004 after 45 years of federal service.

Books by Ronald Eisler

Handbook of Chemical Risk Assessment: Health Hazards to Humans, Plants, and Animals. Volume 1. Metals; Volume 2. Organics; Volume 3. Metalloids, Radiation, Cumulative Index to Chemicals and Species. Lewis Publishers, Boca Raton, Florida, 2000.

Trace Metal Concentrations in Marine Organisms. Pergamon Press, New York, 1981.

List of Tables

2.1 Gold production in Canada in 1975, 1985, and 1992 by source 13

2.2 Total gold production in the United States, 1799–1965 16

2.3 Total placer gold production in the United States, 1792–1969 17

2.4 U.S. gold production by state: 1995 vs. 2000 18

4.1 Oxidation states of gold, examples, and stability in water 41

5.1 Gold concentrations in selected abiotic materials 52

5.2 Gold concentrations in selected plants and animals 57

6.1 Distribution of $^{198}Au^+$ in anesthetized Sprague-Dawley rats 79

9.1 Gold thiolate compounds used medicinally 133

10.1 Metal concentrations (in µg/L) in stream waters at Goldenville Gold Mine, Nova Scotia 176

10.2 Average concentrations of Cd, Cu, Pb, and Zn in waters, soils, and crops near Korean gold mining activities 176

10.3 Acute toxicity of aged gold mill effluent to marine fishes and crustaceans 181

10.4 Tissue metal burdens of juvenile tanner crabs, *Chionoecetes bairdi* 181

10.5 U.S. Food and Drug Administration guidance for arsenic, cadmium, lead, and nickel in shellfish 182

10.6 Drinking water limits and soil threshold values for protection of human health 184

11.1 Cyanide and metals concentrations in water and sediments downstream of Portovela-Zaruma cyanide-gold mining area, Ecuador; dry season, 1988 192

11.2 Single oral dose toxicity of sodium cyanide (mg NaCN/kg body weight) fatal to 50% of selected birds and mammals 200

12.1 Arsenic concentrations in biota and abiotic materials collected near gold mining and processing facilities 226

12.2 Lethal and sublethal effects of various arsenicals on humans and selected species of plants and animals 231

12.3 Proposed arsenic criteria for the protection of human health and selected natural resources 242

13.1 Total mercury concentrations in abiotic materials, plants, and animals near active Brazilian gold mining and refining sites 259

13.2 Total mercury concentrations in abiotic materials, plants, and animals near historic gold mining and refining sites in the United States 271

13.3 Lethal effects of mercury to sensitive species of aquatic organisms, birds, and mammals 278

13.4 Proposed mercury criteria for the protection of selected natural resources and human health 287

14.1 Metals and arsenic in tailings, soils, rice, and groundwater near an abandoned gold-silver-copper-zinc mine, Dongil, Korea, 2000–2001 311

Contents

Part 1   Gold and Gold Compounds 1

Chapter 1   Introduction 3
Literature Cited 5

Chapter 2   Geology, Sources, and Production 7
2.1   Geology 8
2.2   Sources and Production 10
    2.2.1   Asia and Environs 12
    2.2.2   Canada 12
    2.2.3   Europe 13
    2.2.4   Republic of South Africa (RSA) 14
    2.2.5   South America 14
    2.2.6   United States 14
2.3   Summary 19
Literature Cited 19

Chapter 3   Uses 23
3.1   Jewelry 23
3.2   Coinage 24
3.3   Electronics 26
3.4   Radiogold 26
3.5   Medicine 28
3.6   Dentistry 31
3.7   Delivery Vehicle 32
3.8   Electron Microscopy 32
3.9   Other 33
3.10   Summary 33
Literature Cited 33

Chapter 4   Properties 39
4.1   Physical Properties 39
4.2   Chemical Properties 41
4.3   Biochemical Properties 43
4.4   Summary 47
Literature Cited 48

Chapter 5   Gold Concentrations in Field Collections 51
5.1   Abiotic Materials 51
5.2   Plants 56
5.3   Animals 59
5.4   Summary 60
Literature Cited 61

Chapter 6　　The Effects of Gold on Plants and Animals 65
6.1　Aquatic Organisms 65
　　6.1.1　Monovalent Gold 65
　　6.1.2　Trivalent Gold 66
6.2　Accumulation 67
　　6.2.1　Microorganisms, Fungi, and Higher Plants 68
　　6.2.2　Aquatic Macrofauna 72
　　6.2.3　Animal Fibrous Proteins 73
6.3　Laboratory Mammals 73
　　6.3.1　Metallic Gold 73
　　6.3.2　Monovalent Gold: Obese Mouse Model 74
　　6.3.3　Monovalent Gold: Other 75
　　6.3.4　Trivalent Gold 80
6.4　Summary 81
Literature Cited 82

Part 2　　Human Health Impacts 89

Chapter 7　　Health Risks of Gold Miners 91
7.1　Historical Background 91
7.2　Health Risks: Underground Miners 93
　　7.2.1　Australia 93
　　7.2.2　North America 94
　　7.2.3　South America 95
　　7.2.4　Europe 96
　　7.2.5　Africa 96
7.3　Health Risks: Surface Miners Who Use Mercury 100
　　7.3.1　Case Histories 101
　　7.3.2　Mercury in Tissues 102
　　7.3.3　Mercury in Air and in Fish Diet 104
7.4　Summary 105
Literature Cited 106

Chapter 8　　Human Sensitivity to Gold 113
8.1　History 113
8.2　Adverse Reactions 115
　　8.2.1　Suicide Attempt 115
　　8.2.2　Teratogenicity and Carcinogenicity 115
　　8.2.3　Hypersensitivity 115
8.3　Case Histories 118
　　8.3.1　Hypersensitivity 118
　　8.3.2　Goldschlager Syndrome 121
　　8.3.3　Prostheses 122
　　8.3.4　Protective Effect of Gold Rings 123

8.4   Dental Aspects.. 123
     8.4.1   Allergic Reactions and Sensitization................................ 123
     8.4.2   Case Histories 125
8.5   Summary 126
Literature Cited... 126

Chapter 9   Chrysotherapy 131
9.1   History.. 131
9.2   Proposed Modes of Action 134
     9.2.1   Au$^+$ and Au$^+$ Metabolites 135
     9.2.2   Immunomodulatory Activity... 139
     9.2.3   Tumor Necrosis Factor... 141
     9.2.4   Bone Resorption... 142
     9.2.5   Leukocyte Infiltration.. 142
     9.2.6   Lysosomal Enzymes.. 142
     9.2.7   Macrophages 143
     9.2.8   Polymorphic Neutrophils 143
     9.2.9   Sulfhydryl Binding Sites 143
     9.2.10 Superoxide Ion 144
     9.2.11 Thioredoxin Reductase 144
9.3   Treatment Regimes, Case Histories, and Adverse Effects........................ 144
     9.3.1   Treatment Regimes 145
     9.3.2   Case Histories 146
     9.3.3   Adverse Effects 148
9.4   Summary 154
Literature Cited... 155

Part 3   Effects of Gold Extraction on Ecosystems.................... 161

Chapter 10   Gold Mine Wastes: History, Acid Mine Drainage, and Tailings
                 Disposal... 163
10.1   Overview 163
     10.1.1 Lode Mining... 165
     10.1.2 Placer Mining... 166
10.2   Acid Mine Drainage... 168
     10.2.1 Effects... 169
     10.2.2 Mitigation... 170
10.3   Tailings 172
     10.3.1 Freshwater Disposal 172
     10.3.2 Marine Disposal 178
     10.3.3 Terrestrial Storage 183
10.4   Waste Rock.. 183
10.5   Summary 183
Literature Cited... 184

Chapter 11   Cyanide Hazards to Plants and Animals from Gold Mining
and Related Water Issues...189
11.1 History of Cyanide Use in Gold Mining....................................189
11.2 Cyanide Hazards 195
       11.2.1 Aquatic Ecosystems 195
       11.2.2 Birds 199
       11.2.3 Mammals...201
       11.2.4 Terrestrial Flora...203
11.3 Cyanide Mitigation and Research Needs 203
11.4 Water Management Issues 206
       11.4.1 Affected Resources 207
       11.4.2 Pit Lakes...211
11.5 Water Quality and Management Research Needs........................213
11.6 Summary 214
Literature Cited...215

Chapter 12   Arsenic Hazards from Gold Mining for Humans, Plants, and
Animals..221
12.1 Arsenic Sources to the Biosphere from Gold Mining 221
12.2 Arsenic Risks to Human Health...223
12.3 Arsenic Concentrations in Abiotic Materials and Biota near Gold
Extraction Facilities 225
12.4 Arsenic Effects on Sensitive Species...230
12.5 Proposed Arsenic Criteria 241
12.6 Summary 244
Literature Cited...245

Chapter 13   Mercury Hazards from Gold Mining for Humans, Plants, and
Animals..251
13.1 History of Mercury in Gold Mining...251
13.2 Ecotoxicological Aspects of Amalgamation...............................255
       13.2.1 Brazil 255
       13.2.2 South America Other than Brazil 265
       13.2.3 Africa...267
       13.2.4 People's Republic of China 268
       13.2.5 The Philippines 268
       13.2.6 Siberia..269
       13.2.7 Canada..269
       13.2.8 The United States...270
13.3 Lethal and Sublethal Effects of Mercury 276
       13.3.1 Aquatic Organisms..277
       13.3.2 Birds 281
       13.3.3 Mammals..284
       13.3.4 Other Groups..286
13.4 Proposed Mercury Criteria..286

13.5 Summary 292
Literature Cited..293

Chapter 14 Abandoned Underground Gold Mines.............................307
14.1 Habitat for Biota 307
14.2 Land Development 309
14.3 Effects on Water Quality...310
14.4 Science Site Potential..312
14.5 Summary 312
Literature Cited..312

Part 4 Mining Legislation, Concluding Remarks, and
Indices 315

Chapter15 Selected Mining Legislation...317
15.1 United States 317
    15.1.1 Federal Laws 317
    15.1.2 Mining Public Lands in the Western United States 323
    15.1.3 State Laws 327
    15.1.4 Mining Law Reform 328
15.2 Foreign...328
15.3 Summary 330
Literature Cited..331

Chapter 16 Concluding Remarks 333

General Index..337

Species Index..351

וְנָהָר יֹצֵא מֵעֵדֶן לְהַשְׁקוֹת אֶת־הַגָּן וּמִשָּׁם יִפָּרֵד וְהָיָה
לְאַרְבָּעָה רָאשִׁים: שֵׁם הָאֶחָד פִּישׁוֹן הוּא הַסֹּבֵב אֵת
כָּל־אֶרֶץ הַחֲוִילָה אֲשֶׁר־שָׁם הַזָּהָב: וּזֲהַב הָאָרֶץ הַהִוא
טוֹב

And a river goes out from Eden to water the garden. And from there it divides and becomes four branches. The name of the first is Pishon and is the one that winds through all the land of Havilah,* where there is gold. And the gold of that land is good ....

Hebrew Bible, Genesis II, 10-12**

---

* Havilah is in northern Arabia on the Persian Gulf. Arabia was famed in antiquity for its gold.
** Translation by R. Eisler.

PART 1

Gold and Gold Compounds

CHAPTER 1

Introduction

Since antiquity, gold has been valued for its scarcity, beauty, and resistance to corrosion. Gold is the best known of all native elements and the most likely to be found in a metallic state (Pough 1991). It is the universal standard of value and the common medium of exchange in world commerce (Koschmann and Bergendahl 1968). Gold is almost everywhere considered to be the symbol of everything precious and of enduring value because of the effort required to extract it from nature, and because of its scarcity relative to other metals (Petralia 1996; Merchant 1998). Gold was known and highly valued by the earliest civilizations: Egyptian, Minoan, Assyrian, and Etruscan. Ornaments of great beauty and workmanship have survived from these periods, many of them as perfect as when they were first made, several thousands of years ago (Rose 1948). The earliest mining work of which traces remain was on gold ores in Egypt; gold washing is depicted on Egyptian monuments of the fourth dynasty, about 2000 BCE. The legend of the Golden Fleece may actually describe an expedition around 1200 BCE to seize gold washed out from the river sands by Armenians using sheepskins (Rose 1948). Financial investors regard gold as an excellent hedge against inflation (Greer 1993). In recent years, the net buyers of gold were central banks and government monetary agencies. In Asia, investment demand for gold has risen sharply. In 1988, about 75% of new gold output went to Taiwan, Japan, South Korea, Hong Kong, and Singapore as a store of wealth and as adornment (Greer 1993).

Gold has produced deep political and social changes through the ages on entire continents (Morteani 1999). Examples include the Spanish conquistadors who destroyed indigenous cultures in Central and South America in their search for "El Dorado"; the discovery of rich gold mines in Bohemia, Silesia, the northern Carpathians, and northern Romania in the 12th and 13th centuries, which produced a large immigration of miners; the settling of the American West, started mainly by the California gold rush in 1849; the Australian gold rush of 1851, resulting in a doubling of that population in seven years to one million; and the Alaskan gold rush that started in 1897 (Morteani 1999). Unfortunately, the human costs of mining, extraction, refining, marketing, and accumulation of gold include war, slavery, conscripted and convict labor, unhealthy and shortened life span, and degraded living conditions (Anikin 1983). Gold as an investment is difficult to justify (Greer 1993). The intangibility of investor anxieties, the volatility of gold's international market price, and the infectious nature of the desire to hoard, all strongly indicate that complex forces are involved in gold as an investment. Greer (1993) states that it is difficult to justify the production of gold in view of society's minimal need for it and the shortfalls in, for example, food and timber production. However, the appeal of gold has not diminished over the past century, nor has the damage caused by its extraction (Greer 1993).

Long believed to be relatively benign, it is now known that gold is, in fact, a relatively common allergen that induces dermatitis about the face and eyelids and at other sites of direct skin contact (Ehrlich and Belsito 2000). In patch tests worldwide the prevalence of gold allergy might be as high as 13%, with 9.5% the most recent estimate in North America (Fowler 2001). The main exposure sources of gold contact dermatitis are personal jewelry and dental alloys (Ahnlide et al. 2000; Suarez et al. 2000; Vamnes et al. 2000; Tsuruta et al. 2001). Occupations most frequently causative of contact dermatitis due to gold include photography, chinaware or glass decorating, manufacturing of dental alloys, and crafting of jewelry (Suarez et al. 2000). In humans, contact allergy to gold may be lifelong, or at least extend for years, unlike in some strains of rodents that recover quickly when the gold stimulus is removed (Lee and Maibach 2001). The growing evidence of human health concerns related to gold, coupled with known adverse ecotoxicological aspects of various gold extraction techniques, has prompted a critical review of gold in the environment.

This book reviews environmental aspects of gold, gold salts, and gold mining, and is divided into four sections. The first section deals with the uses of gold and gold salts, with emphasis on jewelry, bullion and coinage, medicine, dentistry, and electronics; properties, including geological, physical, chemical, and biological; sources and production; concentrations in field collections worldwide of abiotic materials, flora, and fauna; and lethal and sublethal effects to plants and animals, emphasizing effects on aquatic organisms and laboratory mammals. The second section is devoted to human health aspects, including health risks to gold miners; medical aspects of gold drugs, with special attention to the anti-inflammatory action of gold thiol drugs in the treatment of rheumatoid arthritis; and human sensitivity to gold, including contact dermatitis reactions and other concerns. The third section evaluates ecotoxicological aspects of various gold extraction technologies, with emphasis on mercury amalgamation, cyanidation technologies, and related water management issues; environmental effects of gold mining wastes discharged into the biosphere, including arsenic-containing tailings and acid mine drainage; and impact of abandoned underground gold mines on human and natural resources. The last section includes selected legislation governing disposal of gold mining wastes, concluding remarks, and comprehensive indexes by subject and by biological species.

LITERATURE CITED

Ahnlide, I., B. Bjorkner, M. Bruze, and H. Moller. 2000. Exposure to metallic gold in patients with contact allergy to gold sodium thiosulfate, *Contact Dermatitis*, 43, 344–350.

Anikin, A.V. 1983. Gold — the yellow devil. International Publ., New York, 244 pp.

Ehrlich, A. and D.V. Belsito. 2000. Allergic contact dermatitis to gold, *Cutis*, 65, 323–326.

Fowler, J.F., Jr. 2001. Gold, *Amer. Jour. Contact Dermatitis*, 12, 1–2.

Greer, J. 1993. The price of gold: environmental costs of the new gold rush, *The Ecologist*, 23 (3), 91–96.

Koschmann, A.H. and M.H. Bergendahl. 1968. Principal gold-producing districts of the United States, U.S. Geol. Surv. Prof. Paper 610, 283 pp.

Lee, E.E. and H.L. Maibach. 2001. Is contact allergy in man lifelong? An overview of patch test follow-ups, *Contact Dermatitis*, 44, 137–139.

Merchant, B. 1998. Gold, the noble metal and the paradoxes of its toxicology, *Biologicals*, 26, 49–59.

Morteani, G. 1999. History, economics and geology of gold, in *Gold: Progress in Chemistry, Biochemistry and Technology*, H. Schmidbaur, (Ed.), John Wiley & Sons, New York, 39–63.

Petralia, J.F. 1996. *Gold! Gold! A Beginner's Handbook and Recreational Guide: How & Where to Prospect for Gold!* Sierra Outdoor Products Co., San Francisco, 143 pp.

Pough, F.H. 1991. *Peterson First Guide to Rocks and Minerals*. Houghton Mifflin, Boston, 128 pp.

Rose, T.K. 1948. Gold, *Encyclopaedia Britannica*, 10, 479-485.

Suarez, I., M. Ginarte, V. Fernandez-Redondo, and J. Toribio. 2000. Occupational contact dermatitis due to gold, *Contact Dermatitis*, 43, 367–368.

Tsuruta, K., K. Matsunaga, K. Suzuki, R. Suzuki, H. Akita, Y. Washimi, A. Tomitaka, and H. Ueda. 2001. Female predominance of gold allergy, *Contact Dermatitis*, 44, 55–56.

Vamnes, J.S., T. Morken, S. Helland, and N.R. Gjerdet. 2000. Dental gold alloys and contact hypersensitivity, *Contact Dermatitis*, 42, 128–133.

CHAPTER 2

Geology, Sources, and Production

Since before recorded time, gold has been mined, collected from alluvial deposits, or separated from the ores of silver, copper, and other metals (Merchant 1998). Gold is the first metal mentioned in the Old Testament in Genesis 2:11 (Petralia 1996). One gold mine in Saudi Arabia has been mined for more than 3000 years (Kirkemo et al. 2001). Artisans of ancient civilizations used gold lavishly in decorating tombs and temples, and gold objects made more than 5000 years ago have been found in Egypt (Kirkemo et al. 2001). Among the most productive gold fields in ancient times were those in Egypt, where in the deep mines the slave laborers were maltreated, and in Asia Minor near the River Pactolus, the source of Croesus' wealth. The Romans obtained much of their gold from Transylvania (Rose 1948). The gold in the Aztec and Inca treasuries of Mexico and Peru was plundered by the Conquistadors during their explorations of the New World, melted and cast into coins and bars, destroying priceless artifacts of Indian culture (Kirkemo et al. 2001). Slaves were used to mine gold in Brazil from 1690 to 1850 (Lacerda 1997). From 1850 to 1860, gold production in the United States and Australia was at its peak. In the 1890s, the placers (gold-bearing gravels or sand which has eroded from the surrounding mountains) of the Canadian Klondike and Alaska were prominent gold producers. By 1927, the Transvaal (Republic of South Africa) had been the richest gold field in the world for many years, although there were important gold fields on every continent and in most countries (Rose 1948). Major population shifts resulted as gold discoveries were documented in Chile in 1545; in Brazil between 1696 and the 1970s; in Siberia between 1744 and 1866; in the United States in 1799 (North Carolina), 1847 (California), 1858 (Colorado), 1859 (Nevada), 1862 (Idaho), 1864 (Montana), and 1884 (Alaska); in Canada between 1857 and 1896; in Australia between 1850 and 1893; in New Zealand from 1862 to 1865; and in South Africa between 1873 and 1886 (Nriagu and Wong 1997).

This chapter briefly summarizes geological characteristics of gold-bearing deposits, sources, and production of gold with emphasis on Asia, the Republic of South Africa, Canada, Europe, South America, and the United States. Reliable data on these subjects were scarce and difficult to obtain; therefore, interpretations should be treated with caution.

2.1 GEOLOGY

Gold originates at considerable depth and is carried upward by hot fluids and magma that force their way into rock fractures (Cvancara 1995). Crystallization, most often in quartz veins, occurs as the fluids cool and pressures diminish. As lode deposits break up by weathering, such materials are carried downslope and accumulate as gold-bearing sand or gravel in streams, along beaches, in front of melting glaciers, and in sand dunes. Placer gold in sand and gravel is the most readily available source of gold for the recreational prospector (West 1971; Cvancara 1995). Geological events such as uplift and subsidence may cause prolonged and repeated cycles of erosion and concentration, and where these processes occur, deposits may be enriched (West 1971). The greater specific gravity of gold (19.3) compared with residual rock (about 2.7) leads to gold settling out while lighter rocks are washed away (Puddephatt 1978). Alluvial gold, once discovered, is often easy to extract as nuggets or grains by simple gravity concentration, and this fact gave impetus to the gold rushes of California in 1848, Australia in 1850, and the Yukon in 1896. Much of the gold in the Ural Mountains of the former Soviet Union is alluvial (Puddephatt 1978).

Gold is widespread in the environment. It occurs in minute quantities in almost all rocks, especially igneous and metamorphic rocks. Gold is usually obtained from quartz lodes or veins, or from deposits derived from them by denudation into river gravel (Rose 1948). Gold occurs in about 40 minerals, but only native gold (Au) and electrum (Au-Ag) are common (Gasparrini 1993). Gold generally occurs in native form, and also in combination with tellurium as the ore calaverite ($AuTe_2$) and with silver and tellurium as the ore sylvanite [$(Au,Ag)Te_2$] (Krause 1996). The mineral most commonly found with gold in lodes is iron pyrites, a yellow sulfide of iron (Rose 1948). Others are copper pyrites, arsenical pyrites, and other metal sulfides. Although no mineral is an infallible guide to gold, limonite, a yellow oxide of iron, is considered a reliable indicator of lode gold on ground surfaces. Magnetite (black iron sand) is useful as an indicator of placer deposits (Rose 1948).

The main concerns of the miner or prospector interested in a lode deposit of gold are to determine the gold content (tenor) per ton of mineralized rock, and the size of the deposit. Usually, a fire assay is used to report results as grams of gold per metric ton or troy ounces of gold per short avoirdupois ton (Kirkemo et al. 2001). Gold can be profitably extracted from ores containing 3.8 to 6.4 g/t or 0.1 to 0.2 oz/t (Gasparrini 1993). Placer deposits represent concentrations of gold derived from lode deposits by erosion, disintegration, or decomposition of the enclosing rock, and subsequent concentration by gravity (Kirkemo et al. 2001). Prospectors look for gold where black sands have concentrated and settled with the gold. Magnetite is the most common mineral in black sands, but others include cassiterite, monazite, ilmenite, chromite, platinum-group metals, and sometimes gem stones. The content of recoverable free gold in placer deposits is determined by mercury amalgamation of the gold-bearing concentrate and expressed as grams per cubic meter.

Gold veins fill bedrock cracks or fissures (Krause 1996). They can range from about one centimeter in thickness to hundreds of meters thick or long. Lodes are ore deposits of veins that are close to each other. The "Mother Lode" on the western slopes of the Sierra Nevada Mountains in California is the area where the 1848–1849

Gold Rush originated (Krause 1996). Gold occurs as thin veins in quartz rocks in the gold fields of South Africa. This gold, known as vein or reef gold, is present as microscopic particles so that extraction is comparatively difficult. In general, reef gold is found in quartz or albite rocks, often together with iron pyrites. Deposits with silver and other metals often occur in volcanic regions controlled by major fault zones where concentration of gold takes place by thermal metamorphism from basic rocks and deposition in sedimentary rocks (Puddephatt 1978).

Certain locations in Canadian stream beds and drainage basins favor preferential accumulation of gold and other heavy minerals, and here many models have been formulated to locate placer formations (Day and Fletcher 1991). One model was based on the process of erosion and redeposition of the bed during annual flood events that effected changes in equilibrium transport rates. The model postulates that increases in bed roughness (diameter of materials in the stream) results in preferential accumulation of heavy minerals, with enrichment increasing with increasing density, as is the case for gold. Decreasing channel slope also results in enrichment of heavy minerals. Distribution of gold predicted by the model agreed with field data from a 5-km stretch along a gravel bed stream in British Columbia (Day and Fletcher 1991).

In northeastern Nevada, high-grade ores containing as much as 24.7 g of gold per metric ton, with an estimated total gold endowment of 2200 tons, were first discovered in 1962 in an areal extent of 8.5 km by 2 km (Bettles 2002). This deposit — the Carlin gold deposit — is one of the largest hydrothermal disseminated replacement deposits discovered in North America, with four zones of gold mineralization, mostly in the upper 250 m (Radtke 1985). It was formed in the late Tertiary, about 70 million years ago, over a period of at least 100,000 years, as a result of high angle faulting, igneous and hydrothermal activity, and other processes. Ores contained, in mg/kg, about 8 Au, 21 Hg, 222 to 409 As, and 52 to 106 Sb (Radtke 1985). Gold, in unoxidized form was found mainly within arsenian pyrite and associated with mercury, antimony, and thallium (Bettles 2002). Characteristics of the host rocks that may enhance their favorability to gold deposition include the presence of reactive carbonate, porosity, permeability, and the presence of iron, which can be sulfidized to form auriferouspyrite (Bettles 2002). The mining districts of north-central Nevada are localized by major structural features, including the Roberts Mountains thrust fault on which clastic and volcanic rocks of early and middle Paleozoic age have ridden eastward over correlative carbonate rocks (Roberts 1960). Genesis of gold deposits in northern Nevada is not fully understood and is subject to conflicting models (Williams and Rodriguez 2000). A consensus of these models is that regional structures control the spatial distribution of deposits. Measurement of the earth's magnetic and electrical fields through magnetotelluric soundings demonstrates resistivity structure and fracture zones associated with subsurface gold deposits (Williams and Rodriguez 2000) and shows promise in locating future subsurface gold deposits.

Distribution of gold deposits in Russia's Far East was mapped using information about gold discoveries, as well as geological, gravity, and magnetic data (Eirish and Moiseenko 1999). Factors responsible for gold localization and the erosion of ore bodies were evaluated using mineral composition, geologic formation boundaries, lithosphere structure, and others. It is now established that the main source of gold is the mantle and its derivatives, and that the mobilization and transport mechanisms are granitic magmas and fluids occurring at various depths. Gold is generally mobilized in the fluid phase in the form of chlorides, hydrosulfides, and complex compounds, and deposited in sediments with high carbon, sulfide, and iron content. However, it can also be mobilized by hydrothermal solutions and granitic intrusion; deep crustal faults are important in the accumulation and transport of ore-bearing solutions (Eirish and Moiseenko 1999).

The Circle Mining District in Alaska is a major gold repository (Yeend 1991). The area once contained granite, quartzite, quartzite schist, and mafic schist overlain by colluvium gravel, fan deposits, silt, organic material, and several ages of gold-bearing gravel. Mafic schist seems to be the bedrock source of the gold. The Tintina fault zone in the northeast section of the Circle District is a dominant structure in the area. The zone contains at least three ages of superimposed fan gravel: late Tertiary, late Pleistocene, and Holocene, the last being the most gold-enriched and with 16.1% silver; all samples contained antimony, which distinguishes this area from other gold-bearing areas of Alaska (Yeend 1991).

Most aspects and characteristics of porphyry and epithermal systems containing at least 200 tons of gold are similar to those that typify their smaller and lower-grade counterparts (Sillitoe 1997). Nevertheless, hypothetical mechanisms operating in the mantle, in high-level magma chambers, during exsolution of magmatic fluids and sites of gold deposition are considered to be particularly favorable for either the liberation, concentration, transport, or precipitation of gold and, hence, for the formation of large deposits. When considering the 25 largest gold deposits over 200 tons in the circum-Pacific region, a number of criteria are identified as favorable indicators for large gold accumulations. Both gold-rich porphyry copper and various epithermal gold deposits seem to be more common in association with igneous rocks containing high concentrations of potassium. The porphyry deposits are characterized by high hydrothermal magnetite contents and impermeable host rocks, especially limestones. Epithermal gold deposits, in contrast, are controlled by marked lithologic differences and proximity to volcanic settings (Sillitoe 1997).

2.2 SOURCES AND PRODUCTION

Gold was recovered from the rocky desert of Egypt between the Nile and the Red Sea, according to the first known mining map dated 1100 BCE. Hardrock gold mining on a large scale with thousands of workers was exercised here for the first time, with production estimated as high as 10 tons annually (Bachmann 1999). Gold was mined in what is now Saudi Arabia during the reign of King Solomon in 961 to 922 BCE (Kirkemo et al. 2001). Between 500 and 200 BCE, Gaul became a center of gold production. Egyptian mines were still in operation in 60 BCE using time-tested methods, namely, gold recovered mechanically, aided by fire, and separation from the rock slurry by gravity (Bachmann 1999). Later, Imperial Rome was capable of similar efforts. Between 0 and 200 CE, the Old World's gold reserves were concentrated in Rome; primary production, as well as plunder and tribute, were all significant. More important than hardrock gold mining was gold recovery from placers. Stream and river gravel as well as sands can be easily panned for gold. The content of gold in mined hardrock and placer deposits varied from 0.5 to 50 g/t. It is assumed that in antiquity deposits with 2 g or less of gold per ton or less have been worked (Bachmann 1999).

Total world production of gold is estimated at 3.4 billion troy ounces of which more than 67% was mined in the past 50 years and with 45% of the total world production coming from the Witwatersrand district of South Africa (Kirkemo et al. 2001). World gold production in 1979 was about 39 million ounces, with the Republic of South Africa (RSA) producing 58% of the global output (Elevatorski 1981). The Soviet Union ranked second and Canada third. About 87% of the gold production came from primary sources, such as lodes and placers, and the remainder as a by-product from the refining of copper and base metal ores. The United States was fourth in 1979 with an output of 964,000 ounces, or 2.5% of the world's production; however, 57% of the output was primary and 43% by-product. In addition to the RSA, the Soviet Union, Canada, and the United States, significant (>300,000 ounces) gold production in 1979 was also documented in the Dominican Republic, Brazil, Ghana, Zimbabwe, Papua New Guinea, the Philippines, and Australia (Elevatorski 1981). During 1979, the three largest gold-producing states in the United States, with 75% of the U.S. total, were Utah, South Dakota, and Nevada. The Utah production was almost entirely a by-product of copper mining. Gold was also produced in quantity in Alaska and in 11 western states (Elevatorski 1981). Total world production in 1994 was estimated at 220 tons or 60 million ounces (Korte et al. 2000), but this needs verification.

In developing countries, gold mining activities increased substantially in the late 1960s following the end of the 1944 Breton Woods agreement, which limited the price of gold to US $35 per troy ounce (Meech et al. 1998). The price rose gradually during the 1970s, leading to reworking of ores previously considered too low grade for gold extraction. In May 1985, representatives from 20 countries at a World Bank meeting concluded that legal title to gold-mined lands was the major concern, superseding lack of technology, environmental impact, and financial support. As of 1998, about 3 million people were directly involved in gold mining throughout the developing world. Mercury pollution problems, as a result of gold mining, are particularly severe in Brazil, Peru, Ecuador, Colombia, Bolivia, Venezuela, Indonesia, and the Philippines (Meech et al. 1998).

In 1986, the largest gold producers were the RSA with 38.5% of the world's production, the Soviet Union (18.8%), the United States (7.0%), Canada (6.6%), Australia (5.5%), China (4.1%), and Brazil (4.1%) (Gasparrini 1993). By 1988, the surge in Amazonian gold mining placed Brazil second in worldwide production: behind RSA (621 tons), and ahead of the United States (205 tons), Australia (152 tons), and Canada (116 tons; de Lacerda and Salomons 1998). In 1990, gold production in the United States exceeded that of the Soviet Union for the first time in 50 years (Gasparrini 1993). Between 1980 and 1990, annual production of new gold increased from 962 to 1734 metric tons, with production significantly increased in the United States, Canada, and Australia (Greer 1993). These three nations controlled about 33% of the new gold's market share in 1990, up from 12% in 1980. New gold output from less industrialized nations, such as Brazil, the Philippines, Papua New Guinea, Ghana, Zimbabwe, and Ecuador, rose to 26% in 1990, with production increases being greatest in Brazil and Papua New Guinea. In industrialized countries, it was corporate mining interests that propelled this growth. However, in Latin America, Asia, and Africa (excluding RSA), most of the new gold was produced by small-scale or informal sector mining operations, employing several million people worldwide and accounting for about 25% of global gold output (Greer 1993). The market price of gold per troy ounce has moved from about $20 in 1873 to $35 in 1934, $100 in 1972, $200 in July 1978, $600 in late 1978, a record $850 in January 1980, $474 in March 1980, $750 in September 1980, $526 in December 1980, and $275 in December 2001 (Gasparrini 1993). The incentive to produce gold is high; at $500 an ounce, a cube of gold 30.5 cm on a side (one foot) weighs about 550 kg and is valued at approximately 9 million dollars (Petralia 1996). Fluctuations in the price of gold are considered irrational and based on fear, greed, and anxiety rather than on industrial production and associated costs (Gasparrini 1993).

Details follow on the specific geographical areas associated with gold production, including Canada, the United States, the Republic of South Africa, South America, Asia, and Europe.

2.2.1   Asia and Environs

In 1995, the People's Republic of China produced 105 tons of gold, about one-third from small-scale mines using mercury amalgamation extraction techniques (Lin et al. 1997). The Bo Hai Sea area of northeast China is adjacent to onshore occurrence of gold lodes that have made the region the largest gold-producing area in China (Clark and Li 1991). The best provenances for the placer gold materials that occur in the Bo Hai Sea are the Liadong-Koren and Shangdong peninsulas, which are rich in epithermal gold deposits. The depositional sandy coast is considered the most favorable area for the formation of placer gold due to a large and consistent source of detrital minerals and a suitable concentrating environment. Submarine morphologic features, such as sand bars and shoals, sea floor platforms, and submerged rivers, are the most favorable features for formation of placer gold deposits within the Bo Hai Sea (Clark and Li 1991).

Since ancient times, gold has been produced by territories in the Urals, Caucasus, Kazakhstan, and Central Asia, with 60% deriving from placer deposits (Gasparrini 1993). By the 1970s, the Soviet Union was considered to be the second largest producer of gold in the world, although the precise amounts mined were not published (Puddephatt 1978).

In recent years, India produced about 2 tons of gold annually using the cyanidation process (Agate 1996).

2.2.2   Canada

Gold has been discovered and mined in Alberta, British Columbia, Manitoba, Nova Scotia, Quebec, and the Yukon Territory. The earliest recorded notice of gold in Canada was of placer findings in Quebec in 1823, but little effort was expended

Table 2.1   Gold Production in Canada in 1975, 1985, and 1992 by Source

Year	Total Mined, in kg	Auriferous-Quartz Deposits	Placer Operations	Base Metal Deposits
1975	51,433	73.0%	0.6%	26.4%
1985	87,562	76.8%	4.0%	19.2%
1992	160,351	88.5%	2.2%	9.3%

Source: Modified from Ripley et al. 1996.

during the next 25 years (Ransom 1975). Lode gold was first mined in Nova Scotia in the 1850s (Ripley et al. 1996). The gold rush to California in 1849 and to Australia in 1851 stimulated the gold prospecting fever in Canadians. In 1858, placer gold was discovered in northern British Columbia. In 1862, lode deposits were opened in Nova Scotia (Ransom 1975). In 1866, lode gold was discovered in southeastern Ontario (Ripley et al. 1996). In 1896, the discovery of rich placer gravel along the Klondike River in the Yukon Territory stimulated the great gold rush of 1898 that served as the springboard for intensive prospecting in Alaska (Ransom 1975). By 1913, Yukon gold production was about 11 tons (Ripley et al. 1996). In 1909, important gold discoveries were made in Ontario and Manitoba (Ransom 1975).

Between 1858 and 1972, Canada produced 195.7 million troy ounces of gold, with peak production in 1941 (Ransom 1975). Ontario was the leading producer, followed by Quebec, with most of the gold coming from quartz lode mines or as a by-product of base-metal refining. About 90% of the Canadian gold production is from the northern and eastern half of Canada, all of it from lode mines and placers. In 1975, Canada ranked second among the world's gold producing nations, first being the Republic of South Africa (Ransom 1975). In 1992, Canada ranked fifth in world production of gold, mainly from auriferous-quartz mines, followed by base metal mine and placer mining operations (Ripley et al. 1996; Table 2.1).

For more than 100 years, exploration and extraction of gold have taken place in the coastal regions of the western Canadian margin (Barrie and Emory-Moore 1994). There is a potential for significant marine placer deposits (up to 4 g Au/t) over an extensive area (60 km), in the nearshore and shelf regions off western Canada, with thicknesses of up to one meter, especially in the Queen Charlotte Islands of British Columbia. The major source of the gold and minable minerals, mainly titanium, is the early Holocene beach deposits that form the coastal bluffs of the area. Deposits in any particular area are ephemeral, the result of wave action in a macrotidal setting along an eroding and unconsolidated coast (Barrie and Emory-Moore 1994).

2.2.3   Europe

Native gold and secondary minerals were identified in the quartz veins from Bastogne, Belgium, and the extended area. Native gold occurs as metallic flakes up to 5 mm in diameter and is associated with pyrrhotite and chalcopyrite (Hatert et al. 2000). Gold mineralization was found at the site of abandoned kaolin quarries in southern Sardinia, Italy (Cidu et al. 1997). Tailings from this quarry were of low pH and contained elevated concentrations of arsenic, zinc, copper, and other potentially harmful metals. If open-pit mining of gold is initiated using the cyanidation process, similar tailings problems are expected. An environmental impact study was recommended (Cidu et al. 1997), but results are not available.

2.2.4   Republic of South Africa (RSA)

Gold production in all of Africa for the period 1976 to 1985 was about 230 million troy ounces, with RSA producing about 96% of the total; Zimbabwe, Ghana, and Zaire producing about 3.8%; and the rest coming from Ethiopia, Mali, Liberia, Gabon, Congo, Central African Republic, Rwanda, Tanzania, Sudan, Burundi, Cameroon, Kenya, and the Malagasy Republic (Ikingura et al. 1997). More than 95% of all gold produced in the RSA has its source in the Witwatersrand area, and almost all of it is lode gold (Gasparrini 1993). The largest producing gold mine in Witwatersrand is 1000 to 3000 m beneath the surface (Korte et al. 2000). By the 1970s, the gold fields of RSA had yielded more than 31 million kg of gold, or more than half the gold reserves of the world (Puddephatt 1978). The genesis and tectonic setting of the Witwatersrand gold deposits are somewhat similar in hydrothermal lode equivalents to other large deposits in eastern Russia, but differ in their sedimentary–hydrothermal metamorphic origin (Shcheglov 1997).

2.2.5   South America

The most recent gold rush in South America was triggered in 1980 by the discovery of a large gold mine in Serra Pelada, Amazon Region, Brazil. From the Amazon, the fever spread to neighboring countries including Venezuela, Guyana, French Guiana, and Suriname (Mol et al. 2001). Most of the gold produced in Brazil is from individuals operating in remote areas. Production figures are difficult to obtain; however, Rojas et al. (2001) aver that Brazil produces more gold than any nation except RSA, with about 90% coming from informal mining in Amazonia using mercury amalgamation. This will be discussed in detail later.

Small-scale mining in Suriname started in 1876 (Mol et al. 2001). At its height in 1907, the industry produced about 1200 kg of gold yearly, but after 1910, production declined. Between 1930 and 1970 annual production was about 200 kg. However, gold mining revived in the 1990s due to the rise in the price of gold, the deterioration of the economy in Suriname, the immigration of Brazilian miners, and the presence of foreign prospecting companies. About 35,000 workers are currently involved in uncontrolled small-scale gold mining in Suriname, or almost 9% of the total population of 400,000. The estimated production of gold increased from 10,000 kg in 1995 to 20,000 kg/yr in 1998 and 1999. About 30% of the gold production is registered in official figures, the rest being sold illegally or smuggled out of the country (Mol et al. 2001).

2.2.6   United States

Except for small recoveries by Indians and Spanish explorers, gold was produced largely in southern California as early as 1775, in the southern Appalachian region of the eastern United States in 1792 (Kirkemo et al. 2001), and in North Carolina in 1799 (Koschmann and Bergendahl 1968; Da Rosa and Lyon 1997). New discoveries were made in other Appalachian states in the 1820s and 1830s. Appalachian gold was mined by panning and dredging of stream gravels (placer mining). There were also a few underground mines in North Carolina, Virginia, and Georgia that used slave labor. At that time, gold was typically concentrated using the mercury amalgamation process (Da Rosa and Lyon 1997). These states produced significant amounts of gold until the Civil War (1861 to 1865). After the discovery of gold in California in 1848, California and other western states contributed the bulk of domestic gold production (Koschmann and Bergendahl 1968). In general, gold was derived from three types of ore: ore in which gold is the main metal of value, base-metal ore which yields gold as a by-product, and placers. In the early years, most of the gold was mined from placers, but after 1873, production came mainly from lode deposits. By 1905, gold deposits had been discovered in Alaska and Nevada, and gold production in the United States exceeded 4 million troy ounces annually — a level maintained until 1917; during World War I (1914 to 1918) and for some years afterward, annual domestic gold production declined to 2 million ounces (Kirkemo et al. 2001). Since the 1930s, by-product gold has become a measurable fraction of the annual domestic gold output (Koschmann and Bergendahl 1968). The largest single source of by-product gold in the United States is the porphyry deposit at Bingham Canyon, Utah, which has produced about 18 million troy ounces of gold since 1906 (Kirkemo et al. 2001).

In 1934, President Franklin Delano Roosevelt issued an order limiting the amount of gold per individual to a maximum of 200 ounces (Ransom 1975). In 1942, the War Production Board issued Order L-208 closing all gold mines for the duration of World War II. Only a few mines reopened after the war ended. On March 17, 1968, the U.S. Treasury discontinued buying and selling domestically mined gold and ceased to control its price. When U.S. Treasury buying ended, the official fixed price of gold was set at $35 a troy ounce, 1.000 fine — the price set in 1934 (Ransom 1975). Since 1968, miners can sell gold to any willing purchaser at prices determined by supply and demand. The restriction on the 200-ounce limit of possession ended on December 31, 1974 (Ransom 1975). With government controls lifted, the price of gold began rising until by 1975 it had risen from $35/oz to almost $600/oz (Ransom 1975). This sharp price increase resulted in increased gold prospecting and mining worldwide, although in 2002 the price of gold was less than $300/oz. The industrial break-even cost of producing an ounce of gold in 1995 in the United States was estimated at $256 (Dobra 1997). Noncash costs and profitability added another $51, for a total of $307. This did not include production costs, such as exploration and development. One estimate for producing a sustainable supply of gold in the United States is $370 to $400/troy ounce (Dobra 1997).

From the end of World War II through 1983, domestic mine production of gold did not exceed 2 million ounces per year (Kirkemo et al. 2001). Since 1985, annual production has risen about 1 million ounces every year, reaching about 9 million troy ounces in 1990, and — for the first time — exceeding domestic consumption, estimated at about 3.5 million ounces per annum (Kirkemo et al. 2001).

The most abundant areas for gold in the conterminous United States are located in the Sierra Nevada and the Rocky Mountain ranges (Petralia 1996). Of the 50 states,

Table 2.2   Total Gold Production in the United States, 1799–1965

State	Production, in Millions of Troy Ounces	Percent of Total
California	106.0	34.5
Colorado	40.8	13.3
South Dakota	31.3	10.2
Alaska	29.9	9.7
Nevada	27.5	9.0
Utah	17.8	5.8
Montana	17.8	5.8
Arizona	13.3	4.3
Idaho	8.3	2.7
Oregon	5.8	1.9
Washington	3.7	1.2
New Mexico	2.3	0.7
Others (MI, PA, NC, GA, SC, VA, AL, TN, WY)	2.8	0.9
Total USA	307.3	100.0

Source: From Koschmann and Bergendahl 1968.

32 have reported the existence of gold in sufficient showings to interest the casual collector (Ransom 1975). At least 21 states produced more than 10,000 troy ounces of placer and lode gold through 1959; however, more than 77% of the 307.1 million ounces of gold mined between 1799 and 1965 came from only five states: Alaska, California, Colorado, Nevada, and South Dakota (Koschmann and Bergendahl 1968; Table 2.2). It is likely that numerous early prospectors and miners, and later amateur and semi-professional gold seekers, never recorded their finds, gleaning thousands of ounces of gold annually from abandoned mining districts (Ransom 1975). There are more than 500 districts in the United States that have produced at least 10,000 ounces of gold, and 45 have produced more than 1 million ounces. Four districts have produced more than 10 million ounces: Lead, South Dakota; Cripple Creek, Colorado; Grass Valley, California; and Bingham, Utah (Koschmann and Bergendahl 1968). The largest gold mine in the United States is the Homestake mine at Lead, which has accounted for almost 10% of total domestic gold production since it opened in 1876, with combined production and reserves of about 40 million troy ounces (Kirkemo et al. 2001). Open-pit mines in Nevada, which started production in 1965, now account for about 1.5 million troy ounces annually (Kirkemo et al. 2001). Gold mining returned to the South Carolina Piedmont in the 1980s and 1990s, and renewed gold exploration is under way in parts of Appalachia (Da Rosa and Lyon 1997).

It was placer gold prospecting that opened the frontier West and Alaska to settlement (West 1971; Ransom 1975; Table 2.3). When the known alluvial gold deposits were exhausted, miners began searching for the original sources of gold buried nearby within the earth, usually in quartz veins. Thus, following the era of placer mining, came the era of lode-gold mining. Placer gold is still abundant. In 1973, a nugget weighing 25 pounds (11.3 kg) was found in Sierra County, California,

Table 2.3   Total Placer Gold Production in the United States, 1792–1969

State	Production, in 1000s of Troy Ounces	Placer Share of All Gold Produced in State (%)	Placer Share of All Gold Produced in USA (%)[a]
California	68,470	64.3	21.7
Alaska	21,130	70.2	6.7
Montana	9001	50.8	2.9
Idaho	5625	67.7	1.8
Oregon	3500	60.2	1.1
Nevada	1901	7.1	0.6
Colorado	1798	4.4	0.6
Georgia	600	68.8	0.2
New Mexico	505	22.0	0.2
Arizona	500	3.7	0.2
South Dakota	351	1.1	0.1
Washington	275	7.0	0.1
North Carolina	245	20.5	0.1
Utah	75	0.4	<0.1
South Carolina	52	16.3	<0.1
Virginia	50	29.8	<0.1
Wyoming	43	97.6	<0.1
Alabama	18	30.0	<0.1

[a] Based on total gold production of 316.77 million ounces, including 4.66 million ounces from undesignated sources. Placer share of all gold was about 37%.

Source: Modified from West 1971.

possibly the largest nugget ever found in California (Ransom 1975). Recreational placer gold mining is also popular in California and elsewhere (West 1971).

Valuable gold-bearing coastal sediments have been reported in Alaska (Koschmann and Bergendahl 1968; Bronston 1990; Eyles 1990; Yeend 1991; Garnett 1997). Between the first discovery of gold in Alaska in 1848 and 1965, a total of 29.8 million ounces have been mined (Koschmann and Bergendahl 1968). Coastal sediments in the Gulf of Alaska between Cape Yakataga and Icy Bay — a distance of about 50 km — produced about 15,000 ounces (0.46 metric tons) of gold between 1898 and 1913 (Eyles 1990). Gold was recovered from sands exposed at low tide along the high-energy, storm-dominated coastline. The White River Glacier and the Yakataga Glacier drainage basins were the main sources of gold along the Yakataga coastline (Eyles 1990). Up to 1986, the total onshore placer gold production from the immediate vicinity of Nome was about 155.5 metric tons, or 5 million troy ounces (Garnett 1997). Between 1986 and 1988, offshore placer mining near Nome produced more than 2.18 metric tons of gold from depths up to 50 meters below sea level, using drilling techniques suitable for all seasons; collected ores were processed onshore (Bronston 1990). The offshore Nome gold resource was parallel to the coast, about 25 km in length, extending to about 4.8 km offshore (Garnett 1997). In deeper waters, the highest gold grades were found in the upper 2 meters of the sediments in log gravel. Closer inshore, payable grades extended to about 20 meters below the seabed. Of practical importance were silt (seawater turbidity), clay (interfered with gold recovery onshore), and cobbles and boulders that resisted

Table 2.4   U.S. Gold Production by State: 1995 vs. 2000

State	Troy Ounces	Percent of Total U.S. Production
Alaska	141,800 vs. 546,000	1.4 vs. 4.8
California	783,000 vs. 447,000	7.7 vs. 3.9
Idaho	96,500 vs. 72,000	1.0 vs. 0.6
Montana	437,200 vs. 212,000	4.3 vs. 1.9
Nevada	6,765,000 vs. 8,585,000	66.5 vs. 75.6
South Dakota	559,100 vs. 171,000	5.5 vs. 1.5
Utah	729,700 vs. 700,000	7.2 vs. 6.1
Others	660,400 vs. 616,100	4.4 vs. 5.4
Total	10,172,700[a] vs. 11,349,100[b]	

[a] Worth $3.906 billion @ US$384/troy ounce
[b] Worth $4.358 billion

Source: Modified from Dobra 1997, 2002.

excavation, caused circuit blockages, increased mechanical damage and wear to the dredge and plant parts, and reduced the throughput rate. The dredging process was complicated by high winds (15 to 32 km/h in summer, up to 90 km/h in other seasons) for periods up to 55 hours, average wave heights of 6.5 m for periods up to 12 hours, summer storms, and extreme temperatures ranging from 30°C in summer to –41°C in winter. The dredging operation was discontinued because the equipment was not suitable for prevailing geological and climatic conditions (Garnett 1997). Gold placers of the Circle Mining District in east-central Alaska produced more than a million ounces of gold since its discovery in 1893 using comparatively primitive techniques (Yeend 1991). Most placer gold in the Circle District was recovered at the gravel–bedrock contact. The lowermost meter of gravel and the uppermost 0.5 m of bedrock contained 80 to 90% of the gold that was ultimately recovered. Current mining in the Circle District requires the use of expensive, sophisticated equipment (Yeend 1991).

In 1995, the United States became the second largest producer of gold in the world, behind the Republic of South Africa (Dobra 1997). Although gold was mined from hundreds of lode mines and placer sites across the United States, almost 40% (3.9 million ounces) came from the two largest producers operating in north-central Nevada. In fact, the top four producers accounted for almost 55% of U.S. production in 1995 (Dobra 1997). About 90,000 people were directly involved in the production of precious metals in the United States, which includes gold and silver, with 51,457 jobs in Nevada alone (Dobra 1997). About 79% of the ore was processed by heap leach methods to produce 32% of the gold, and 21% of the ore was processed using various milling techniques to produce 68% of the gold. There was also a shift of production from near-surface, low-grade oxide ores, to deeper, higher-grade refractory ores (Dobra 1997). Nevada produced about 6.8 million ounces in 1995, or 67% of total U.S. production, followed by California with 0.8 million ounces, and Utah with 0.7 million ounces (Table 2.4). By the year 2000, Nevada produced 8.58 million ounces of gold and accounted for 75.6% of all U.S. production (Table 2.4; Dobra

2002). In 2001, Nevada produced 8.125 million ounces of gold, the fourth year in a row above the 8 million ounce mark (Driesner and Coyner 2002). In 2001, Nevada became the third largest producer of gold in the world — producing about 21% of all mined gold — ranking behind the Republic of South Africa and Australia; underground operations contributed about 20% of the gold production in Nevada in 2001 (Driesner and Coyner 2002).

A large part of the gold stock of the United States is stored in the vault of the Fort Knox Bullion Depository (Kirkemo et al. 2001). Gold in the form of bricks ($7 \times 3.6 \times 1.75$ inches) that weigh about 27.5 lb each (about 400 troy ounces) is stored without wrapping in the depository vaults.

2.3 SUMMARY

Gold is ubiquitous in the environment, although usually occurring in minute amounts. Geological characteristics of commercial gold-bearing strata are related, in part, to proximity to volcanic settings, mobilization and transport rates of granitic magmas and fluids, pyrites of iron and other metals, and igneous rocks containing potassium. Accurate production figures for new gold are difficult to obtain, but probably exceed 39 million troy ounces annually (1209 metric tons). The major commercial producer of gold is the Republic of South Africa; others include the former Soviet Union, the United States, Canada, Australia, China, and Brazil. In 1995, the United States ranked second in global gold production, with most of its gold produced in Nevada.

LITERATURE CITED

Agate, A.D. 1996. Recent advances in microbial mining, *World Jour. Microbiol. Biotechnol.*, 12, 487–495.

Bachmann, H.G. 1999. Gold for coinage: history and metallurgy, in *Gold: Progress in Chemistry, Biochemistry and Technology*, H. Schmidbaur, (Ed.), John Wiley & Sons, New York, 3–37.

Barrie, J.V. and M. Emory-Moore. 1994. Development of marine placers, northeastern Queen Charlotte Islands, British Columbia, Canada, *Marine Georesour. Geotechnol.*, 12, 143–158.

Bettles, K. 2002. Exploration and geology, 1962 to 2002, at the Goldstrike property, Carlin Trend, Nevada, *Soc. Econ. Geol.*, Spec. Publ. 9, 275–298.

Bronston, M.A. 1990. Offshore placer drilling technology: a case study from Nome, Alaska, *Mining Engin.*, 42(1), 26–31.

Cidu, R., R. Caboi, L. Fanfani, and F. Frau. 1997. Acid drainage from sulfides hosting gold mineralization (Furtei, Sardinia), *Environ. Geol.*, 30, 231–237.

Clark, A. and C. Li. 1991. Gold placer deposits in the Bo Hai Sea, *Mar. Mining*, 10, 195–214.

Cvancara, A.M. 1995. *A Field Manual for the Amateur Geologist*. John Wiley & Sons, New York, 335 pp.

Da Rosa, C.D. and J.S. Lyon (Eds.). 1997. *Golden Dreams, Poisoned Streams*, Mineral Policy Center, Washington, D.C., 269 pp.

Day, S.J. and W.K. Fletcher. 1991. Concentration of magnetite and gold at bar and reach scales in a gravel-bed stream, British Columbia, Canada. *Jour. Sediment. Petrol.*, 61, 871–882.

de Lacerda, L.D. and W. Salomons. 1998. *Mercury from Gold and Silver Mining: A Chemical Time Bomb?* Springer, Berlin, 146 pp.

Dobra, J.L. 1997. The U.S. gold industry 1996, Nevada Bur. Mines Geol., Spec. Publ. 21. 32 pp.

Dobra, J.L. 2002. The U.S. gold industry 2001, Nevada Bur. Mines Geol., Spec. Publ. 32, 40 pp.

Driesner, D. and A. Coyner. 2002. Major mines of Nevada 2001, Nevada Bur. Mines Geol., Spec. Publ. P-13, 28 pp.

Eirish, L.V. and V.G. Moiseenko. 1999. Distribution of gold deposits in Russia's Far East, *Geol. Pacific Ocean*, 12, 327–344.

Elevatorski, E.A. 1981. *Gold Mines of the World.* Minobras, Dana Point, CA, 107 pp.

Eyles, N. 1990. Glacially derived shallow-marine gold placers of the Cape Yakataga district, Gulf of Alaska, *Sediment Geol.*, 68, 171–185.

Garnett, R.H.T. 1997. Problems with dredging in offshore Alaska, *Mining Engin.*, 49(3), 27–33.

Gasparrini, C. 1993. *Gold and Other Precious Metals. From Ore to Market.* Springer-Verlag, Berlin, 336 pp.

Greer, J. 1993. The price of gold: environmental costs of the new gold rush, *The Ecologist*, 23 (3), 91–96.

Hatert, F., M. Deliens, M. Houssa, and F. Coune. 2000. Native gold, native silver, and secondary minerals in the quartz veins from Bastogne, Belgium, *Bull. Sci. Terre Aardwetenschappen,* 70, 223–229.

Ikingura, J.R., M.K.D. Mutakyahwa, and J.M.J. Kahatano. 1997. Mercury and mining in Africa with special reference to Tanzania, *Water Air Soil Pollut.*, 97, 223 –232.

Kirkemo, H., W.L. Newman, and R.P. Ashley. 2001. *Gold,* U.S. Geological Survey, Denver, 23 pp.

Korte, F., M. Spiteller, and F. Coulston. 2000. The cyanide leaching gold recovery process is a nonsustainable technology with unacceptable impacts on ecosystems and humans: the disaster in Romania, *Ecotoxicol. Environ. Safety*, 46, 241–245.

Koschmann, A.H. and M.H. Bergendahl. 1968. Principal gold-producing districts of the United States, U.S. Geol. Surv. Prof. Paper 610, 283 pp.

Krause, B. 1996. *Mineral Collector's Handbook.* Sterling, New York, 192 pp.

Lacerda, L.D. 1997. Evolution of mercury contamination in Brazil, *Water Air Soil Pollut.*, 97, 209–221.

Lin, Y., M. Guo, and W. Gan. 1997. Mercury pollution from small gold mines in China, *Water Air Soil Pollut.*, 97, 233–239.

Meech, J.A., M.M. Veiga, and D. Tromans. 1998. Reactivity of mercury from gold mining activities in darkwater ecosystems, *Ambio*, 27, 92–98.

Merchant, B. 1998. Gold, the noble metal and the paradoxes of its toxicology, *Biologicals*, 26, 49–59.

Mol, J.H., J.S. Ramlal, C. Lietar, and M. Verloo. 2001. Mercury contamination in freshwater, estuarine, and marine fishes in relation to small-scale gold mining in Suriname, South America, *Environ. Res.*, 86A, 183–197.

Nriagu, J. and H.K.T. Wong. 1997. Gold rushes and mercury pollution, in *Mercury and Its Effects on Environment and Biology*, A. Sigal and H. Sigal, (Eds.), Marcel Dekker, New York, 131–160.

Petralia, J.F. 1996. *Gold! Gold! A Beginner's Handbook and Recreational Guide: How & Where to Prospect for Gold!* Sierra Outdoor Products, San Francisco, 143 pp.

Puddephatt, R.J. 1978. *The Chemistry of Gold.* Elsevier, Amsterdam, 274 pp.

Radtke, A.S. 1985. Geology of the Carlin gold deposit, Nevada, U.S. Geol. Surv. Prof. Paper 1267, 124 pp.

Ransom, J.E. 1975. *The Gold Hunter's Field Book.* Harper & Row, New York, 367 pp.

Ripley, E.A., R.E. Redmann, and A.A. Crowder. 1996. *Environmental Effects of Mining.* St. Lucie Press, Delray Beach, FL, 356 pp.

Roberts, R.J. 1960. Alinement of mining districts in north-central Nevada, U.S. Geol. Surv. Prof. Paper 400-B, 17–19.

Rojas, M., P.L. Drake, and S.M. Roberts. 2001. Assessing mercury health effects in gold workers near El Callao, Venezuela, *Jour. Occup. Environ. Med.*, 43, 158–165.

Rose, T.K. 1948. Gold, *Encyclopaedia Britannica*, 10, 479–485.

Shcheglov, A.D. 1997. Some specific features of the Witwatersrand gold deposits and their equivalents in eastern Russia, *Geol. Pacific Ocean*, 13, 443–452.

Sillitoe, R.H. 1997. Characteristics and controls of the largest porphyry copper-gold and epithermal gold deposits in the circum-Pacific region, *Austral. Jour. Earth Sci.*, 44, 373–388.

West, J.M. 1971. How to mine and prospect for placer gold, U.S. Dept. Interior, Bur. Mines Inform. Circ. 8517.

Williams, J.M. and B.D. Rodriguez. 2000. Deep electrical geophysical measurements across the Carlin trend, Nevada, U.S. Geol. Surv., Open-File Rept. 00-419, 11 pp.

Yeend, W. 1991. Gold placers of the Circle district, Alaska — past, present, and future, *U.S. Geol. Surv. Bull.*, 1943, 42 pp.

CHAPTER 3

Uses

The most common uses of gold are in personal jewelry, coinage, and bullion. Because of its unique properties, gold is also used in electronics, human medicine, dentistry, physiology, and immunology. Radiogold isotopes have been used to tag various species of wildlife and to treat human carcinomas. Secondary uses of gold are expected to increase significantly.

3.1 JEWELRY

The ancient Egyptians used Nubian gold to produce jewelry and to decorate statues, tombs, crypts, and sarcophagi. They also used it extensively in cosmetics to adorn their own bodies, a practice which continues in some parts of the world today (Merchant 1998). Skilled Etruscan artisans produced numerous gold ornaments 3300 years ago, including ear pendants, armbands, garment clasps with animal motifs, necklaces, dental prostheses, and vessels decorated with figurines (Loevy and Kowitz 1997). For the past 7000 years, gold jewelry has been sought by consumers for traditional, social, and stylistic reasons (Raub 1999). As a result, jewelry fabrication has accounted for the largest share of global gold production (Gasparrini 1993; Korte and Coulston 1995; Raub 1999). In 1991, for example, gold production including the use of scrap amounted to 2111.1 tons, and more than 95% was used in jewelry manufacture (Raub 1999). Gold alloys, especially those containing silver and copper, have been preferred by goldsmiths for the past 5000 years. Alloys used in jewelry today also contain zinc and nickel with the result that gold can appear yellow, white, green, or red depending on the mixture. But it is only since 1960 that the International Gold Council has decided to support research and investigations on metallurgy of gold alloys. This is clearly connected with the fact that jewelry accounts for the largest use of gold in every country of the world (Raub 1999).

The gold concentration of alloys can be expressed either in terms of caratage, 24 carats representing pure gold, or in fineness, which is the weight fraction expressed in 1000ths (Raub 1999). In Europe, 8, 14, and 18 carat jewelry is most common (0.333, 0.585, and 0.750 fine, respectively). Low carat alloys are inferior to the widely used 14 and 18 carat alloys with respect to tarnish and superficial corrosion resistance. In general, white gold alloys show high nickel and silver and low copper and gold. Blue or purple gold, an aluminum alloy ($AuAl_2$), being 0.785 fine, is rather brittle and not corrosion resistant. A recent addition to jewelry alloys is the so-called Spangold, an Au–Cu–Al alloy, about 18 carat, in yellow (76% Au, 19% Cu, 5% Al) and pink (76% Au, 10% Cu, 6% Al). The most recent alloy in jewelry is gold-titanium at 23.5 carat, which has unusual hardness and wear resistance (Raub 1999).

Gold is also used to prepare ruby glass and to decorate pottery and earthenware (Rose 1948). When a solution of auric chloride is precipitated with a solution of stannous chloride, a reddish-purplish precipitate is formed containing both metallic gold and tin hydroxide. This product is used mainly in the preparation of ruby glass. Liquid gold is used mainly in the decoration of pottery. It is a sulfo-resinate of gold dissolved in oils, together with small quantities of bismuth, rhodium, and sometimes other metals. The liquid gold is applied to the surface of the glaze. After drying, it is fired at 700 to 800°C, producing a brilliant film of metallic gold on the surface of the ware (Rose 1948).

3.2 COINAGE

Gold was the first metal used in currency and has probably been used for this purpose for more than 5000 years (Bachmann 1999). The pre-monetary currency of gold, the *shat* (7 g of gold) is known from Egypt around 2620–2061 BCE. Reference to gold used as payment for the purchase of tin was mentioned in sources from the dynasty of Akkod in 2050–1950 BCE (Bachmann 1999).

As early as 1091 BCE, gold circulated in China as small cubes that people understood by their weight and size to be worth a certain amount (Ransom 1975; Morteani 1999). The first gold coins were made around 640 BCE in Asia Minor from placer gold with a high content of silver (Bachmann 1999). Persian gold coins circulated through the Middle East from 500 to 300 BCE. The Babylonian heavy gold *shekel* weighed 262.67 grains (one grain = 0.0648 g), a light one weighed 126.5 grains; the heavy gold *mina* was 12,630 grains and the light one 6315 grains. The largest Babylonian monetary unit, the heavy *talent*, weighed 758,000 grains; the light talent weighed 379,000 grains. Therefore, a heavy shekel, mina, and talent weighed almost 16.4, 818.4, and 48,967 g, respectively, equivalent to 0.035, 1.8, and 107.8 lb, respectively (Ransom 1975). The first coins of pure gold were probably struck in 561 BCE by King Croesus. He was also the first to fix the relation of silver to gold in coinage at 13 to 1, a relation not far from that of the terrestrial abundance of the metals which is 0.08 mg/kg for silver and 0.004 mg/kg for gold, or 20 to 1 (Morteani 1999). By 477–350 BCE, coins contained about 40% gold, 50% silver, and 10% copper (Bachmann 1999).

Gold coinage in the Middle Ages was recorded in Hungary, Silesia, Siberia, India, and Japan (Bachmann 1999). Throughout the Middle Ages, various measures for the valuation of gold were used. From Florence came the *florin* weighing 48 grains (3.1 g). The *tower pound* — 5400 grains, or about 350 g — was the standard of England until it was replaced in 1527 by the troy system, which originated in Troyes, France (Ransom 1975). In the 14th century, the west coast of Africa became known as the Gold Coast. Iran, Iraq, and Egypt were major importers and producers of gold. Excluding the Orient, the total amount of gold produced during medieval times (a period of about 500 years) is estimated at 2000 tons. In 1344, Edward III of England introduced the *noble* as a coin suitable for international trade. This coin weighed 8 to 9 g with a fineness of 0.995, viz., 23.875 carat. Nobles were copied in the Rhineland and the Netherlands. In Spain, Charles I (1516–1556) created the *escudo d'oro* with a weight of 3.375 g and an initial fineness of 0.917, later reduced to 0.875. It remained in circulation until 1873, being popular in Europe and the most common gold coin in Spanish America. Numerous gold coins were introduced before and after the *escudo d'oro* by various countries. From 1837 until 1939, the United States minted 0.900 fine $10 and $20 coins. The $20 piece weighed 33.4363 g corresponding to 30.0926 g of 24 carat (1.000 fine) gold. In 1932, the currency of 31 countries was based on gold 1.000 fine, nine countries on gold 0.91667 fine, and one country on gold 0.875 fine (Bachmann 1999). In 1934, the United States fixed the price of gold at $35 per troy ounce (31.1035 g), which was accepted as the international standard (Pain 1987). The price stayed constant even though by the 1950s most of Alaska's gold mines were in decline. In 1967, controls were lifted and the price of gold soared, with subsequent reopening of many mines. In 1980, the price of gold reached $850 per troy ounce, but fell to between $350 and $400 in the mid-1980s. With improved technology for working the deposits, miners found it cost effective to work poorer deposits and to reopen old mines (Pain 1987).

Gold has become available in larger quantities only in the past 100 years. Before the great gold discoveries, the metal was in relatively short supply (Morteani 1999). In the 100 years between 1800 and 1900, more gold was mined than in the preceding 5000 years. World gold production in 1895 was about 250 tons and in 1992 about 2200 tons (Morteani 1999). Today the troy system is currently used by jewelers and pharmacists: 1 troy ounce = 1.097 avoirdupois ounces = 31.103 g = 480 grains; 1 grain = 0.0648 g (Ransom 1975). In more recent times, gold in the form of coins, bullion, and jewelry has been used as a hedge against inflation, particularly in countries outside the United States during periods of adverse world conditions and economic uncertainty (Petralia 1996).

There are three qualities of fine gold used in most applications (Dahne 1999): (1) For the international market, gold bars of about 12.5 kg (400 troy ounces) with a purity of at least 99.5% Au are prescribed. Each bar must be marked with the purity in four digits, the trademark of the refinery, and a serial number. The mark and the dimensions of the bar are registered by the London Bullion Market. The bars can then be traded worldwide without further control. Most of the official hoardings of national banks are in these bars. (2) Small bars of 1 g to 1 kg of 99.99% purity are produced. The bars need to contain less than 100 mg Ag/kg, <20 mg Cu/kg, and <30 mg total of other base metals/kg. These bars are also stamped with the purity, the refiner's trademark, and a serial number. These bars are mostly used for private investment purposes. (3) For some highly technical purposes, mainly in electronic systems (see Section 3.3), gold is required at 99.999% purity (0.99999 fine). These bars must contain, in mg/kg, <5 platinum, <5 palladium, <3 silver, <3 iron, <2 lead, <2 bismuth, <0.5 copper, and <0.5 nickel (Dahne 1999).

Gold coins are no longer issued as official currency. The types minted today are but another form of gold bars (Bachmann 1999). The greatest hoard of gold known is the 30,000 tons of bullion owned by about 80 nations and now residing in the vaults of the Federal Reserve Bank of New York (Morteani 1999).

3.3   ELECTRONICS

At ordinary temperatures, the electrical conductivity of gold is about 75% that of pure silver; however, the electrical resistance of gold, which is the reverse of conductivity, decreases with decreasing temperature, and at about 5° Absolute, resistance has practically disappeared, at which point gold is a near perfect conductor of electricity (Rose 1948). Gold is now used as a soldering component in high technology electronics, electrical contacts, plating materials, wear-resistant contacts in rockets, submarines, computers, and signaling devices (Pethkar and Paknikar 1998). In 1991, 145,000 kg of gold was used in electronics (Puddephatt 1999). Gold is the metal most resistant to corrosion, has low electrical resistivity, and excellent malleability, ductility, and softness. All of these properties are useful in making very thin wires and other connectors, and in allowing joining of components by thermo-compression binding (Puddephatt 1999). In 1994, 165,700 kg of gold was used in the field of electrical engineering and electronics, mostly in electromechanical devices such as connectors, switches, or relays in which gold is the electrical contact material (Grossmann et al. 1999). This is largely due to the high electrical conductivity and the high corrosion resistance of gold and many of its alloys. Gold and gold-rich alloys are the contact materials of choice for electromechanical devices operating at low electrical loads and with low contact forces, especially if high reliability is required and aggressive atmospheres are encountered (Grossmann et al. 1999).

3.4   RADIOGOLD

Radiogold isotopes have been used to treat mucosal carcinomas of the mouth (Takeda et al. 1996) and adenocarcinomas of the prostate (Hochstetler et al. 1995); to tag various species of wildlife, including mice (Kaye 1959, 1961), bats (Cope et al. 1960), and lizards (O'Brien et al 1965); to measure radiation exposures from nuclear accidents (Iwatani et al. 1994; Komura et al. 2000); and as a chemical label for water-soluble gold-compound pharmaceuticals (Berning et al. 1998).

Early attempts to destroy unwanted tissues with colloidal radiogold-198 (physical half-life of 2.7 days) were discontinued because [198]Au was not sufficiently site-specific (Sadler 1976). More recently, [198]Au was used successfully to treat mucosal carcinomas of the mouth using a specially constructed applicator in the form of an oral mold that contained [198]Au (Takeda et al. 1996). Oral-mold therapy is the treatment of choice for tumors of the hard palate in the lower or upper alveoli where the mucosa is too thin to hold an implant. In one study, radiogold-198 mold therapy was given to 27 patients with 29 oral cancers. Initial tumor control was obtained in all lesions, with a 5-year survival rate of 82% (Takeda et al. 1996). Patients (N = 157) with adenocarcinoma of the prostate were treated with transcutaneous radiogold-198 seeds, receiving a median dose of 164 mCi (Hochstetler et al. 1995). Cancer-specific survival at 5 years was 100% for early stages, and 76% for advanced stage C cancer. The survival rates of patients treated with [198]Au seed implementation for localized cancer were equivalent or superior when compared with historical data of patients treated with iodine-125 implantation, external beam radiotherapy, a combination of [198]Au seed implantation plus external radiation, or radical prostate-ctomy (Hochstetler et al. 1995).

Movement of various wildlife species was accomplished through the use of radiogold-198 tags. Eastern harvest mice (*Reithrodontomys humulis*) were tagged subcutaneously with 20-gauge wires 10 mm in length, containing radiogold-198. During the first two days, mice could be detected in the field with a portable Geiger counter at a distance of about 3 m; by day seven, this was about 0.5 m (Kaye 1959, 1961). Movements of a northern fence lizard (*Sceloporus undulatus hyacinthus*) was studied for 17 days after external tagging with a gold wire containing one mCi (37 million Bq) of [198]Au. The tag was detectable at a distance of 3.7 m at the start of the study. By day seven, the tag was not detectable at distances >1.5 m, and a new tag was attached to the lizard at that time (O'Brien et al. 1965). Radiogold-198 in the form of a liquid was applied to the inner surface of wing bands of bats (Cope et al. 1960). The bands were attached to bats some distance from their home roost. Bands could be detected with a Geiger counter from a distance of one mile over six days, and were effective in locating bats behind brick walls, slate, tin, shingles, and rafters.

Neutron transport, as judged in part by yields of [198]Au activation from a [252]Cf fission neutron source, was useful in estimating radiation exposures from nuclear weapons releases (Iwatani et al. 1994). The [198]Au induced by an accident at a Japanese uranium processing plant on September 30, 1999, in Tokaimura was detected up to 1.4 km from ground zero through analysis of gold samples — mainly jewelry and coins — within 2.7 km of the site (Komura et al. 2000).

Development of water-soluble gold compounds continues to be an important area of chemical research because of their increasing use in chemotherapeutic appli-cations in the control of tumor activity, leukemia, and cytotoxicity (Berning et al. 1998). Radiogold-199 (with a physical half-life of 3.2 days) provides sufficient time for distribution and makes it suitable for radiolabeled drugs that are efficiently cleared from the blood and nontarget tissue. Also important is the fact that [199]Au can be readily produced indirectly as a no-carrier-added product in reactors. However, the development of [199]Au has been largely unexplored because of the lack of suitable ligand systems that produce [199]Au-labeled complexes with good stability under *in vivo* conditions. Radiogold-198 complexes with water-soluble phosphines were prepared and evaluated as models for potential radiogold-199 pharmaceuticals. The [198]Au (HMPB)$_2^+$ complex ([198]Au 1,2-bis[bis(hydroxymethyl)phosphino]benzene) was found to exhibit good *in vitro* stability over wide pH ranges and temperatures and was the only compound tested to show satisfactory *in vivo* stability (Berning et al. 1998).

3.5 MEDICINE

Gold has a medicinal history that can be traced through the written history of every culture and far into pre-history by means of archaeological records (Sadler 1976; Shaw 1999a). The biological use of gold can be traced to the Chinese in 2600 BCE (Sadler 1976; Brown and Smith 1980; Asperger and Cetina-Cizmek 1999). The Chinese used gold, probably as dust or flakes of its metallic form (Au^0), as a medicinal agent as early as 2500 BCE (Merchant 1998), although attempts to use colloidal gold (Au^0) as a drug in experimental animals have been unsuccessful to date (Sadler 1976). Examination of some Chinese recipes from 600 BCE indicated that gold was dissolved by an oxidation procedure that included KNO_3 and IO_3^- (Asperger and Cetina-Cizmek 1999). The reduction of IO_3^- with organic material, or with $FeSO_4$, yielded soluble gold dissolved as an $[AuI_2]^-$ anion, which was first used in the treatment of leprosy (Asperger and Cetina-Cizmek 1999).

Up to the eighth century, metallic gold was considered a panacea for every known disease (Sadler 1976). By the 1400s, auric chloride (prepared from metallic gold and aqua regia, followed by neutralization with chalk) was used in the treatment of leprosy. In India, calcined gold preparations, colloidal gold, and gold bromide were alleged to exhibit antiepileptic properties (Vohora and Vohora 1999). Gold drugs were used frequently during the 1700s and 1800s, with claims of great success. It was Koch's observation in 1890 that gold cyanide (AuCN) inhibited the growth of the bacteria that caused tuberculosis and represented the beginning of systematic gold molecular pharmacology (Sadler 1976; Brown and Smith 1980; Merchant 1998; Asperger and Cetina-Cizmek 1999). Koch's experiments showed that tubercle bacilli were killed at 0.5 mg AuCN *in vitro*, but met with limited success when introduced into the blood serum of infected animals (Sadler 1976). The "gold decade" between 1925 and 1935 marked extensive use of gold compounds in treatment of tuberculosis and syphilis, although toxic side effects remained a problem (Sadler 1976).

About 1913 to 1927 marked a period of intense searching for Au^+ compounds of lower toxicity. Until the 1930s, Au^+ salt therapy was extended to the treatment of rheumatoid arthritis and lupus erythematosus because of the (mistaken) belief that these diseases were atypical forms of TB (Merchant 1998). Some Au^+ complexes have antimicrobial and antifungal activity (Savvaidis et al. 1998).

Monovalent organogold salts have been used, with mixed results, in the treatment of pemphigus (skin and mucous membrane blisters), lupus erythematosus, ulcerative colitis, Crohn's disease, various types of arthritis, bronchial asthma, bullous skin conditions, discoid lupus, several forms of rheumatism, ankylosing spondylitis affecting peripheral joints, kala-azar (caused by the flagellate protozoan *Leishmania donovani* transmitted by the bite of sand flies), tuberculosis, and malaria (Singh et al. 1989; Suzuki et al. 1995; Jones and Brooks 1996; Hostynek 1997; Navarro et al. 1997; Tomioka and King 1997; Wang et al. 1997; Merchant 1998; Pandya and Dyke 1998; Ueda 1998; Lacaille et al. 2000). In malaria, for example, monovalent gold compounds show promise in treating a disease currently affecting about 400 million people and threatening another billion worldwide (Navarro et al. 1997). Chloroquine is the preferred treatment against non-resistant strains of *Plasmodium falciparum*, the parasite responsible for malaria. Tests with chloroquine-resistant strains of *Plasmodium* and a monovalent gold–chloroquine complex show promise in controlling *Plasmodium*. The gold–chloroquine complex displayed high *in vitro* activity — markedly higher than other metal–chloroquine complexes tested — against the blood stage of two chloroquine-resistant strains of *P. falciparum*. It was also active *in vitro* and *in vivo* against *Plasmodium berghei* (rodent malaria), showing that incorporation of Au^+ produced a marked enhancement in chloroquine efficacy (Navarro et al. 1997).

Gold compounds have been used to control various microbial pathogens. The concept of gold as an antibacterial therapy is traced back to the eighth century (Elsome et al. 1996). In 1890, it was firmly established that gold cyanide was inhibitory to TB bacilli *in vitro*, but not *in vivo*. The introduction of gold compounds for the treatment of rheumatoid arthritis was based on the assumption that bacteria were responsible for this condition. In one study, *in vivo* tests with mouse skin demonstrated that a monovalent gold thiocyanate compound applied topically controlled antibiotic-resistant strains of bacteria and the yeast *Candida albicans* (Shaw 1999a; Elsome et al. 1996). *In vitro* tests with monovalent organogold treatments used in rheumatoid arthritis therapy showed antibacterial activity against *Helicobacter pylori*, a bacterium associated with gastritis and peptic ulcer (Paimela et al. 1995). Antimicrobial activities of two isomeric Au^+-triphenylphosphine complexes with nitrogen-containing heterocycles were documented against two species of Gram-positive bacteria (*Bacillus subtilis*, *Staphylococcus aureus*) and *Candida albicans*, and were more effective against these organisms than were corresponding silver (Ag^+) complexes; however, they were not effective against tested species of Gram-negative bacteria and molds (Nomiya et al. 2000).

Antitumor properties of monovalent and trivalent gold complexes are documented (Cagnoli et al. 1998; Kamei et al. 1999; Messori et al. 2000; Tiekink 2002, 2003). Intramuscular injections of sodium aurothiomalate (Au^+) were successful in controlling tumor-associated antigens in patients with tongue carcinoma or pulmonary cancer (Kamei et al. 1999). Selected Au^{+3} complexes are reasonably stable under physiological conditions and show cytotoxic properties when tested *in vitro* on the human ovarian tumor cell line A2780. The cell-killing properties of stable Au^{+3} complexes were attributed to the Au^{+3} center with ligands of ethylenediamine, diethylenetriamine, or cyclam (Messori et al. 2000). Four monovalent gold compounds coordinated with different phosphines showed varying degrees of cytotoxicity when tested *in vitro* against three human ovarian cancer cell lines, and one — 1,2-bis(diphenylphosphino)ethane bis[Au^+ lupinylsulfide] dihydrochloride — was more effective than conventional antitumor platinum compounds (Cagnoli et al. 1998). More research on the antitumor properties of gold compounds seems warranted.

Oral gold treatment with a monovalent organogold compound has been used successfully to treat psoriatic arthritis in a patient simultaneously infected with HIV (Shapiro and Masci 1996). The use of gold compounds in HIV-associated inflammatory arthritis and the possible antiretroviral effects of gold merit additional study (Shaw 1999b). And reports of anti-HIV activity for cyanide and thioglucose derivatives of gold compounds may be even more significant (Shaw 1999a).

Gold needles are now used by acupuncturists, and gold prostheses by ophthalmologists and otologists. Japanese acupuncturists use gold needles in patients with rheumatoid arthritis to relieve severe pain and to slow the progression of joint destruction (Oba et al. 1999). Deficient eyelid closure is a major visual threat to patients with unresolved facial nerve palsy. Gold weight implants assisted closure in patients with incomplete paralysis of the orbicularis oculi, ameliorating complaints of dry eye, excessive tearing, and corneal epithelial breakdown. Upper lid loading with a gold implant is a currently preferred surgical option to treat the problem, with a success rate between 43 and 88%, and no cases of postoperative infection or implant rejection (Chuke et al. 1996; Abell et al. 1998; Choo et al. 2000; Misra et al. 2000). For patients with prolonged or permanent facial paralysis, the combination of a gold weight implant for the upper lid with tightening of the lower lid provides the most efficacious surgical solution (Julian et al. 1997). The custom-shaped gold implants typically weigh between 0.8 and 1.2 g, but some weigh as much as 2.8 g (Jobe 2000). Gold has also been used successfully in synthetic middle ear prostheses (Gjuric and Schagerl 1998). In one study, 62 hearing-impaired patients afflicted with otosclerosis experienced significant auditory restoration when the stapes of the inner ear was replaced with a 0.999 fine gold prosthesis. There were no postoperative infections, possibly due to gold's bacteriostatic action, or other postoperative complications (Tange et al. 1998).

Sporadic use of gold drugs in the treatment of tuberculosis led to the use of sodium gold thiomalate in the treatment of rheumatoid arthritis (Brown and Smith 1980). Most gold drugs in current medical use are monovalent gold thiol complexes that are stable in air and water, and usually administered by injection. Some gold phosphine complexes, however, can be administered orally. The medical community views gold with mixed emotions: optimism, because gold drugs can cause remission of rheumatoid arthritis and other diseases; and concern, because of the high frequency of toxic side effects (Brown and Smith 1980). In current medical practice, chrysotherapy — the treatment of rheumatoid arthritis (RA) with gold-based drugs — is well established (Sadler 1976; Shaw 1999a). It derives its name from Chryseis, a golden-haired daughter of a priest of Apollo in Greek mythology. Five monovalent organogold complexes are now widely used throughout the world in these treatments. Thiomalatogold (myochrisin or gold sodium thiomalate), thioglucose gold (solganol), and thiopropanosulfonate gold (allochrysin) are oligomeric complexes that contain linear Au^+ ions connected by bridging thiolate ligands. Bis(thiosulfate) gold (sanochrysin) contains gold bound to the terminal sulfur donor atoms of $S_2O_3^{2-}$. The newest drug, auranofin, licensed in the mid-1980s, contains coordinated triethylphosphine and 2,3,4,6-tetra-O-acetyl-B-1-D-thioglucose ligands. Auranofin has the advantage of being orally absorbed but is considered less effective than the injectable gold thiolates (Sadler 1976; Shaw 1999a). After more than 75 years of use in the management of RA, chrysotherapy with injectable monovalent organogold salt formulations is now routinely used in the treatment of RA and other diseases having an autoimmune or inflammatory component to their pathogenesis, despite a significant incidence of adverse side effects (ten Wolde et al. 1995; Merchant 1998; Ueda 1998). The Empire Rheumatism Council in 1961 confirmed the effectiveness of gold salt treatment for RA, and today injectable gold salts are considered a major line of treatment of progressive RA (Ueda 1998). As recently as 1985, injectable monovalent organogold compounds were the initial drugs of choice for Canadian patients with RA; by 1992, gold and methotrexate were equally preferable in moderate RA, and methotrexate was preferable to gold in aggressive RA (Maetzel et al. 1998). Between 1985 and 1994, parenteral gold was the most frequently prescribed treatment overall for RA patients by rheumatologists in Edmonton, Alberta, Canada; however, there was widespread variation in use of these drugs (Galindo-Rodriguez et al. 1997). It is noteworthy that the mechanism of action for chrysotherapy is not known with certainty despite almost 75 years of use and major advances in bioinorganic chemistry (Shaw 1999b); however, there is no lack of theories, and several are discussed in detail in Chapter 9.

Gold thioglucose ($C_6H_{11}O_5SAu$), or GTG, was initially developed and marketed as a therapeutic agent for the treatment of arthritis and rheumatism. However, a single intraperitoneal or subcutaneous injection in mice destroyed the appetite center located in the hypothalamus region of the brain producing hypophagia (uncontrolled eating) and, within a few weeks, many of the characteristics of human obesity (Blair et al. 1996). The degree of obesity induced by GTG is dependent on the dose administered and the strain of the mouse. This topic is discussed in greater detail in Section 6.3.2.

3.6 DENTISTRY

Gold dental prostheses comprised of gold, silver, and copper were used extensively by the Etruscans, a society that dominated the Italian Mediterranean area about 2800 years ago (Loevy and Kowitz 1997; Rothaut 1999). The Etruscan prostheses used gold bands soldered with copper into rings instead of the gold wires as used in Egyptian and Phoenician cultures during the same period. These bands were used to hold substitute teeth of human or animal origin and also to fix in place mobile teeth caused by periodontal diseases. Many prostheses were easily removable and were probably used only on social occasions. By 300 BCE, the Etruscan city states were absorbed by the Romans (Loevy and Kowitz 1997; Rothaut 1999).

Gold is now used in restorative dentistry mainly as the base element in precious metal alloys for gold inlays, crowns, and bridges. High gold dental alloys are distinguished by their superior corrosion resistance when compared with other metal alloys (Hostynek 1997; Merchant 1998; Rothaut 1999). Gold is usually the major component of dental alloys containing silver, copper, and small amounts of platinum and lead; these alloys can be heat-treated to develop strengths as great as 150,000 psi (Merchant 1998). Most dental gold alloys containing platinum or palladium show conspicuous age hardening characteristics due to the formation of a metastable gold–copper complex. The American Dental Association recommends that dental casting gold alloys should contain more than 75% gold and other platinum group metals (Kim et al. 2001). Titanium is now under consideration to replace other metals used in gold alloys, including platinum and palladium, because of superior mechanical, thermal, and corrosion properties (Fischer 2000).

High noble alloys currently used in dental prostheses contain at least 40% gold and 60% (total) gold, palladium, and platinum. Argedent 52 (manufactured by the

Argen Corporation), for example, is a commonly used alloy in the state of Maryland, and is composed of 52.5% gold, 26.9% palladium, 16.0% silver, 2.5% indium, 2.0% tin, and 0.1% ruthenium (www.identalloy.org.html).

Gold-based dental solders may contain tin, cadmium, indium, zinc, gallium, and other metals to improve their flow properties and melting range; however, cadmium, zinc, and gallium have high risks for release from gold-based dental alloys (Wataha et al. 1999). Since gold-based dental solders are in long-term intimate contact with oral tissues, and since the biological properties of dental solders are almost unknown under these conditions, it is recommended that cytotoxicity of gold-based solders be assessed using a relevant substrate. At present, the cytotoxicity of most solders tested with a gold–palladium substrate alloy was acceptably low based in comparisons with a variety of dental casting alloys in common use (Wataha et al. 1999).

Amalgam (Au–Hg) fillings are a mainstay of restorative dental care for the population of many countries, but these are considered to be the main source of human mercury intake next to methylmercury contained in food (Halbach et al. 1998).

3.7 DELIVERY VEHICLE

Studies using minute metallic gold particles as delivery vehicles for introduction of exogenous proteins and antibodies, and for gene therapy are documented. Proteins, such as serum albumin, prevent flocculation of colloidal gold sols; labeled antibodies adsorbed onto colloidal gold particles retained full activity, with major implications for immunization (Sadler 1976). DNA-coated gold beads were effective at introducing DNA into cells without serious cell injury (Webster et al. 1994). Immunization of ferrets, for example, with a plasmid DNA expressing influenza virus hemagglutinin, provided complete protection from influenza virus. Delivery of DNA-coated gold beads by gene gun to the epidermis was significantly more efficient at inducing immunity than was an intramuscular injection of DNA in aqueous solution (Webster et al. 1994). In another study, animals were successfully immunized with biotin by using colloidal gold particles as carrier, with implications for other immunoassay procedures (Dykman et al. 1996). Bombardment of cells with DNA on gold microparticles was developed as a safe method of gene transfer into target cells and tissues; naked DNA, including defined gene sequences, was adsorbed to the surface of gold particles and delivered via helium pulse to cells of the inferior epidermis (Merchant 1998). Repeating polypeptides able to bind to metallic gold beads were found to retain their binding properties when freed from the protein used to select them (Brown 1997). In another study, proteins affixed to colloidal gold particles were successfully incorporated through phagocytotic action into absorptive cells of embryos of a goodeid fish *Ameca splendens* (Schindler and Greven 1992).

3.8 ELECTRON MICROSCOPY

Colloidal gold was first made in the 1600s for use in microscopy (Hainfeld and Powell 2000). Colloidal gold particles are excellent markers in electron microscopy because they are electron-dense, spherical in shape, and can be prepared in sizes from 1 to 25 nm. Colloidal gold particles also make useful tags because of their ability to bind to immunoglobulins, lectins, enzymes, and other compounds (Sadler 1976; Bendayan 2000; Gardea-Torresdey et al. 2000). Colloidal gold (10 nm) is a useful tracer to obtain information about the processes involved in the immunological response (Glazyrin et al. 1995). Colloidal (10 nm) gold-tagged ligands aided in the identification of the ultrastructural site of acrosome-inducing substance of sperm from starfish, *Asterias amurensis* (Longo et al. 1995). Particles can adsorb antibodies for use in immunoreactive studies.

Gold clusters are used extensively as heavy atom labels for proteins and other biological substances (Jahn 1999). Gold clusters used as labels in structural biology are organometallic compounds formed by a tightly packed core of gold atoms surrounded by a shell of phosphine ligands. This configuration allows the localization of specific sites by electron microscopy. Gold clusters are usually stable in water at physiological pH, insensitive to oxygen, and are easily prepared (Jahn 1999). Gold clusters can be made water soluble by altering the organic group and derivatized to link to proteins. Chemical crosslinking of gold particles to biologically active molecules has produced new probes, including gold–lipid, gold–ATP, and others (Hainfeld et al. 1999; Hainfeld and Powell 2000). For example, gold is used for labeling enzymes and proteins for x-ray diffraction studies, the usual labeling agents being $Au(CN)_2^-$, $AuCl_4^-$, and AuI_4^- (Savvaidis et al. 1998).

3.9   OTHER

A minute portion of the world's inhabitants is exposed to powdered gold (Au^0) as a decoration for pastries, in chocolates, or in alcoholic beverages (Merchant 1998). A rare instance of severe lichen planus in response to metallic gold dust included in a liquor is discussed in Chapter 8.

3.10   SUMMARY

The most common use for privately owned gold is jewelry and personal ornaments. Coinage and bullion rank second, especially gold coins, minted by many nations, which are attractive to small investors. Secondary — and increasing — uses of gold include the disciplines of electronics, dentistry, physiology, immunology, electron microscopy, and human medicine.

LITERATURE CITED

Abell, K.M., R.S. Baker, D.E. Cowan, and J.D. Porter. 1998. Efficacy of gold weight implants in facial nerve palsy: quantitative alterations in blinking, *Vision Res.*, 38, 3019–3023.

Asperger, S. and B. Cetina-Cizmek. 1999. Metal complexes in tumour therapy, *Acta Pharmaceut.*, 49, 225–236.

Bachmann, H.G. 1999. Gold for coinage: history and metallurgy, in *Gold: Progress in Chemistry, Biochemistry and Technology*, H. Schmidbaur, (Ed.), John Wiley & Sons, New York, 3–37.

Bendayan, M. 2000. A review of the potential and versatility of colloidal gold cytochemical labeling for molecular morphology, *Biotech. Histochem.*, 75, 203–242.

Berning, D.E., K.V. Katti, W.A. Volkert, C.J. Higginbotham, and A.R. Ketring. 1998. [198]Au-labeled hydroxymethyl phosphines as models for potential therapeutic pharmaceuticals, *Nucl. Med. Biol.*, 25, 577–583.

Blair, S.C., I.D. Caterson, and G.J. Cooney. 1996. Glucose and lipid metabolism in the gold thioglucose injected mouse model of diabesity, in *Lessons from Animal Diabetes VI*, E. Shafrir, (Ed.), Birkhauser, Boston, 237–265.

Brown, D.H. and W.E. Smith. 1980. The chemistry of the gold drugs used in the treatment of rheumatoid arthritis, *Chem. Soc. Rev.*, 9, 217 –240.

Brown, S. 1997. Metal-recognition by repeating polypeptides, *Nature Biotechnol.*, 15, 269–272.

Cagnoli, M., A. Alama, F. Barbieri, F. Novelli, C. Bruzzo, and F. Sparatore. 1998. Synthesis and biological activity of gold and tin compounds in ovarian cancer cells. *Anti-Cancer Drugs*, 9, 603–610.

Choo, P.H., S.R. Carter, and S.R. Seiff. 2000. Upper eyelid gold weight implantation in the Asian patient with facial paralysis, *Plastic Reconstruct. Surg.*, 3, 855–859.

Chuke, J.C., R.S. Baker, and J.D. Porter. 1996. Bell's palsy-associated blepharospasm relieved by aiding eyelid closure, *Ann. Neurol.*, 39, 263–268.

Cope, J.B., E. Churchwell, and K. Koontz. 1960. A method of tagging bats with radioactive gold-198 in homing experiments, *Proc. Indiana Acad. Sci.*, 70, 267–269.

Dahne, W. 1999. Gold refining and recycling, in *Gold: Progress in Chemistry, Biochemistry, and Technology*, H. Schmidbaur, (Ed.), John Wiley & Sons, New York, 120–141.

Dykman, L.A., L.Y. Matora, and V.A. Bogatyrev. 1996. Use of colloidal gold to obtain antibiotin antibodies, *Jour. Microbiol. Meth.*, 24, 247–248.

Elsome, A.M., J.M.T. Hamilton-Miller, W. Brumfitt, and W.C. Noble. 1996. Antimicrobial activities *in vitro* and *in vivo* of transition element complexes containing gold (I) and osmium (VI), *Jour. Antimicrobiol. Chemotherapy*, 37, 911–918.

Fischer, J. 2000. Mechanical, thermal, and chemical analyses of the binary system Au-Ti in the development of a dental alloy, *Jour. Biomed. Mater. Res.*, 52, 678–686.

Galindo-Rodriguez, G., J.A. Avina-Zubieta, A. Fitzgerald, S.A. LeClerq, A.S. Russell, and M.E. Suarez-Almazor. 1997. Variations and trends in the prescription of initial second line therapy for patients with rheumatoid arthritis, *Jour. Rheumatol.*, 24, 633–638.

Gardea-Torresdey, J.L., K.J. Tiemann, G. Gamez, K. Dokken, I. Cano-Aguilera, L.R. Furenlid, and M.W. Renner. 2000. Reduction and accumulation of gold(III) by *Medicago sativa* alfalfa biomass: X-ray absorption, spectroscopy, pH, and temperature dependence, *Environ. Sci. Technol.*, 34, 4392–4396.

Gasparrini, C. 1993. *Gold and Other Precious Metals. From Ore to Market*. Springer-Verlag, Berlin, 336 pp.

Gjuric, M. and S. Schagerl. 1998. Gold prostheses for ossiculoplasty, *Amer. Jour. Otol.*, 19, 273–276.

Glazyrin, A.L., S.I. Kolesnikov, G.N. Dragun, E.L. Zelentsov, K.V. Zolotarev, Y.A. Sorin, G.N. Kulipanov, V.N. Gorchakov, and I.P. Dolbnya. 1995. Distribution of colloidal gold tracer within rat parasternal lymph nodes after intrapleural injection, *Anat. Rec.*, 241, 175–180.

Grossmann, H., K.E. Saeger, and E. Vinaricky. 1999. Gold and gold alloys in electrical engineering, in *Gold: Progress in Chemistry, Biochemistry, and Technology*, H. Schmidbaur, (Ed.), John Wiley & Sons, New York, 200–236.

Hainfeld, J.F., W. Liu, and M. Barcena. 1999. Gold-ATP, *Jour. Struct. Biol.*, 127, 120–134.

Hainfeld, J.F. and R.D. Powell. 2000. New frontiers in gold labeling, *Jour. Histochem. Cytochem.*, 48, 471–480.

Halbach, S., L. Kremers, H. Willrath, A. Mehl, G. Welzl, F.X. Wack, R. Hickel, and H. Greim. 1998. Systemic transfer of mercury from amalgam fillings before and after cessation of emission, *Environ. Res.*, 77A, 115–123.

Hochstetler, J.A., K.J. Kreder, C.K. Brown, and S.A. Loening. 1995. Survival of patients with localized prostate cancer treated with percutaneous transperineal placement of radioactive gold seeds: stages A2, B, and C, *Prostate*, 26, 316–324.

Hostynek, J.J. 1997. Gold: an allergen of growing significance, *Food Chem. Toxicol.*, 35, 839–844.

Iwatani, K., M. Hoshi, K. Shizuma, M. Hiraoka, N. Hayakawa, T. Oka, and H. Hasai. 1994. Benchmark test of neutron transport calculations: indium, nickel, gold, europium, and cobalt activation with and without energy moderated fission neutrons by iron simulating the Hiroshima atomic bomb casing, *Health Physics*, 67, 354–362.

Jahn, W. 1999. Review: chemical aspects of the use of gold clusters in structural biology, *Jour. Structural Biol.*, 127, 106–112.

Jobe, R. 2000. Lid loading with gold for upper lid paralysis. *Plastic Reconstruct. Surg.*, 106, 735–736.

Jones, G. and P.M. Brooks. 1996. Injectable gold compounds: an overview. *Brit. Jour. Rheum.*, 35, 1154–1158.

Julian, G.G., J.F. Hoffman, and C. Shelton. 1997. Surgical rehabilitation of facial nerve paralysis, *Otolaryngol. Clinics North Amer.*, 30, 701–726.

Kamei, H., T. Koide, T. Koijima, Y. Hashimoto, and M. Hasegawa. 1999. Effect of gold on tumor-associated antigens, *Cancer Biother. Radiopharmaceut.*, 14, 403–406.

Kaye, S.V. 1959. Gold-198 wires used to study movements of small mammals, *Science*, 131, 824.

Kaye, S.V. 1961. Movements of harvest mice tagged with gold-198, *Jour. Mammalogy*, 42, 323–337.

Kim, H-I., Y-K. Kim, M-I. Jang, K. Hisatsune, and A.A.E.S. Sakrana. 2001. Age-hardening reactions in a type III dental gold alloy, *Biomaterials*, 22, 1433–1438.

Komura, K., A.M. Yousef, Y. Murata, T. Mitsugashira, R. Seki, and T. Imanaka. 2000. Activation of gold by the neutrons from the JCO accident, *Jour. Environ. Radioact.*, 50, 77–82.

Korte, F. and F. Coulston. 1995. From single-substance evaluation to ecological process concept: the dilemma of processing gold with cyanide, *Ecotoxicol. Environ. Safety*, 32, 96–101.

Lacaille, D., H.B. Stein, J. Raboud, and A.V. Klinkhoff. 2000. Longterm therapy of psoriatic arthritis: intramuscular gold or methotrexate? *Jour. Rheum.*, 27, 1922–1927.

Loevy, H.T. and A.E. Kowitz. 1997. The dawn of dentistry: dentistry among the Etruscans. *Inter. Dental Jour.*, 47, 279–284.

Longo, F.J., A. Ushiyama, K. Chiba, and M. Hoshi. 1995. Ultrastructural localization of acrosome reactin-inducing substance (ARIS) on sperm of the starfish (*Asterias amurensis*), *Molec. Reprod. Develop.*, 41, 91–99.

Maetzel, A., C. Bombardier, V. Strand, P. Tugwell, and G. Wells. 1998. How Canadian and US rheumatologists treat moderate or aggressive rheumatoid arthritis: a survey, *Jour. Rheum.*, 25, 2331–2338.

Merchant, B. 1998. Gold, the noble metal and the paradoxes of its toxicology, *Biologicals*, 26, 49–59.

Messori, L., F. Abbate, G. Marcon, P. Orioli, M. Fontani, E. Mini, T. Mazzei, S. Carotti, T. O'Connell, and P. Zanello. 2000. Gold(III) complexes as potential antitumor agents: solution chemistry and cytotoxic properties of some selected gold(III) compounds, *Jour. Med. Chem.*, 43, 3541–3548.

Misra, A., R. Grover, S. Withey, A.O. Grobbelaar, and D. H. Harrison. 2000. Reducing postoperative morbidity after the insertion of gold weights to treat lagophthalmos, *Ann. Plastic Surg.*, 45, 623–628.

Morteani, G. 1999. History, economics and geology of gold, in *Gold: Progress in Chemistry, Biochemistry, and Technology*, H. Schmidbaur, (Ed.), John Wiley & Sons, New York, 39–63.

Navarro, M., H. Perez, and R.A. Sanchez-Delgado. 1997. Toward a novel metal-based chemotherapy against tropical diseases. 3. Synthesis and antimalarial activity *in vitro* and *in vivo* of the new gold-chloroquinone complex [Au(PPh$_3$)(CQ)]PF$_6$, *Jour. Med. Chem.*, 40, 1937–1939.

Nomiya, K., R. Noguchi, K. Ohsawa, K. Tsuda, and M. Oda. 2000. Synthesis, crystal structure and antimicrobial activities of two isomeric gold(I) complexes with nitrogen-containing heterocycle and triphenylphosphine ligands, [Au(L)(PPh$_3$] (HL = pyrazole and imidazole), *Jour. Inorg. Biochem.*, 78, 363–370.

Oba, T., T. Ishikawa, and M. Yamaguchi. 1999. Different effects of two gold compounds on muscle contraction, membrane potential and ryanodine receptor, *Europ. Jour. Pharmacol.*, 374, 477–487.

O'Brien, G.P., H.K. Smith, and J.R. Meyer. 1965. An activity study of a radioisotope tagged lizard, *Sceloporus undulatus hyacinthinus* (Sauria: Iguanidae), *Southwest. Natural.*, 10, 179–187.

Paimela, L., M. Leirisalo-Repo, and T.U. Kosunen. 1995. Effect of long term intramuscular gold therapy on the seroprevalence of *Helicobacter pylori* in patients with early rheumatoid arthritis, *Ann. Rheum. Dis.*, 54, 437.

Pain, S. 1987. After the goldrush, *New Scientist*, 115 (1574), 36–40.

Pandya, A.G. and C. Dyke. 1998. Treatment of pemphigus with gold, *Arch. Dermatol.*, 134, 1104–1107.

Pethkar, A.V. and K.M. Paknikar. 1998. Recovery of gold from solutions using *Cladosporium cladosporoides* biomass beads, *Jour. Biotechnol.*, 63, 121–136.

Petralia, J.F. 1996. *Gold! Gold! A Beginner's Handbook and Recreational Guide: How & Where to Prospect for Gold!* Sierra Outdoors Product Co., San Francisco, 143 pp.

Puddephatt, R.J. 1999. Gold metal and gold alloys in electronics and thin film technology, in *Gold: Progress in Chemistry, Biochemistry, and Technology*, H. Schmidbaur, (Ed.), John Wiley & Sons, New York, 237–256.

Ransom, J.E. 1975. *The Gold Hunter's Field Book*. Harper & Row, New York, 367 pp.

Raub, C.J. 1999. Gold metal and gold alloys in jewellery, in *Gold: Progress in Chemistry, Biochemistry, and Technology*, H. Schmidbaur, (Ed.), John Wiley & Sons, New York, 105–118.

Rose, T.K. 1948. Gold, *Encyclopaedia Britannica*, 10, 479–485.

Rothaut, J. 1999. Gold and its alloys in dentistry, in *Gold: Progress in Chemistry, Biochemistry, and Technology*, H. Schmidbaur, (Ed.), John Wiley & Sons, New York, 173–198.

Sadler, P.J. 1976. The biological chemistry of gold: a metallo-drug and heavy-atom label with variable valency, *Structure Bonding*, 29,171–215.

Savvaidis, Y., V.I. Karamushka, H. Lee, and J.T. Trevors. 1998. Micro-organism-gold interactions, *Biometals*, 11, 69–78.

Schindler, J.F. and H. Greven. 1992. Protein-gold transport in the endocytic complex of trophotaenial absorptive cells in the embryos of a goodeid teleost, *Anat. Rec.*, 233, 387–398.

Shapiro, D.L. and J.R. Masci. 1996. Treatment of HIV associated psoriatic arthritis with oral gold, *Jour. Rheumatol.*, 23, 1818–1820.

Shaw, C.F., III. 1999a. Gold complexes with anti-arthritic, anti-tumour and anti-HIV activity, in *Uses of Inorganic Chemistry in Medicine*, N.C. Farrell, (Ed.), Royal Society of Chemistry, Cambridge, 26–57.

Shaw, C.F., III. 1999b. The biochemistry of gold, in *Gold: Progress in Chemistry, Biochemistry, and Technology*, H. Schmidbaur, (Ed.), John Wiley & Sons, New York, 260–308.

Singh, M.P., M. Mishra, A.B. Khan, S.L. Ramdas, and S. Panjiyar. 1989. Gold treatment for kala-azar, *Brit. Med. Jour.*, 299, 1318.

Suzuki, S., M. Okubo, S. Kaise, M. Ohara, and R. Kasukawa. 1995. Gold sodium thiomalate selectivity inhibits interleukin-5-mediated eosinophil survival, *Jour. Allergy Clin. Immunol.*, 96, 251–256.

Takeda, M., H. Shibuya, and T. Inoue. 1996. The efficacy of gold-198 grain mold therapy for mucosal carcinomas of the oral cavity, *Acta Oncol.*, 35, 463–467.

Tange, R.A., A.J.G. de Bruijn, and W. Grolman. 1998. Experience with a new pure gold piston in stapedotomy for cases of otosclerosis, *Auris Nasus Larynx*, 25, 249–253.

ten Wolde, S., B.A.C. Dijkmans, J.J. van Rood, F.H.J. Claas, R.R.P. de Vries, J.M.W. Hazes, P.L.C.M. van Riel, A. van Gestel, and F.C. Breedveld. 1995. Human leucocyte antigen phenotypes and gold-induced remissions in patients with rheumatoid arthritis, *Brit. Jour. Rheum.*, 34, 343–346.

Tiekink, E.R.T. 2002. Gold derivatives for the treatment of cancer, *Crit. Rev. Oncol. Hematol.*, 42, 225–248.

Tiekink, E.R.T. 2003. Gold compounds in medicine: potential anti-tumour agents, *Gold Bull.*, 36/4, 117–124.

Tomioka, H. and T.E. King, Jr. 1997. Gold-induced pulmonary disease: clinical features, outcome, and differentiation from rheumatoid arthritis, *Amer. Jour. Respir. Crit. Care Med.*, 155, 1011–1020.

Ueda, S. 1998. Nephrotoxicity of gold salts, D-penicillamine, and allopurinol, in *Clinical Nephrotoxins: Renal Injury from Drugs and Chemicals*, M.E. De Broe, G.A. Porter, W.M. Bennett, and G.A. Verpooten, (Eds.), Kluwer, Dordrecht, 223–238.

Vohora, D., and S.B. Vohora. 1999. Elemental basis of epilepsy, *Trace Elem. Electro.*, 16, 109–123.

Wang, Q., N. Janzen, C. Ramachandran, and F. Jirik. 1997. Mechanism of inhibition of protein-tyrosine phosphatases by disodium aurothiomalate, *Biochem. Pharmacol.*, 54, 703–711.

Wataha, J.C., P.E. Lockwood, M.N. Vuilleme, and M.-H. Zurcher. 1999. Cytotoxicity of Au-based dental solders alone and on a substrate alloy. *Jour. Biomed. Mater. Res.*, 48, 786–790.

Webster, R.G., E.F. Fynan, J.C. Santoro, and H. Robinson. 1994. Protection of ferrets against influenza challenge with a DNA vaccine to the haemagglutinin, *Vaccine*, 12, 1495–1498.

CHAPTER 4

Properties

Gold is a complex and surprisingly reactive element, with unique physical, chemical, and biochemical properties. Some of these properties are listed and discussed below.

4.1 PHYSICAL PROPERTIES

Gold is a comparatively rare native metallic element, ranking fiftieth in abundance in the earth's crust. The chemical symbol for gold is Au, from the Latin *aurum* for gold. Metallic gold is an exceptionally stable form of the element and most deposits occur in this form. The main elements with which gold is admixed in nature include silver, tellurium, copper, nickel, iron, bismuth, mercury, palladium, platinum, indium, osmium, iridium, ruthenium, and rhodium. The native gold–silver alloys have a color range from pale yellow to pure white, depending on the amount of silver present. Finely divided gold is black, like most other metallic powders, while colloidally suspended gold varies in color from deep ruby red to purple. Gold occurs as metallic gold (Au^0) and also as Au^+ and Au^{+3}, so that it occurs in combination with tellurium as calaverite ($AuTe_2$) and sylvanite ($AuAgTe_4$), and also with tellurium, lead, antimony, and sulfur as nagyagite, $Pb_5Au(Te,Sb)_4S_{5-8}$ (Rose 1948; Ransom 1975; Sadler 1976; Puddephatt 1978; Krause 1996).

Gold is characterized by an atomic weight of 196.967, atomic number of 79, a melting point of 1063°C, and a boiling point of about 2700°C. In the massive form, gold is a soft yellow metal with the highest malleability and ductility of any element. A single troy ounce of gold can be drawn into a wire over 66 km in length without breaking, or beaten to a film covering approximately 100 m^2. Traces of other metals interfere with gold's malleability and ductility, especially lead, but also cadmium, tin, bismuth, antimony, arsenic, tellurium, and zinc. It is extremely dense, being 19.32 times heavier than water at 20°C. A cube of gold 30 cm (12 in.) on a side weighs about 544 kg (1197 lb). Gold has high thermal and electrical conductivity, properties that make it useful in electronics. It is extremely resistant to the effects of oxygen and will not corrode, tarnish, or rust. Pure (100%) gold is 1.000 fine, equivalent to 24 carats. Gold is usually measured in troy ounces, wherein 1 troy ounce equals 31.1 g vs. 28.37 g in an ounce avoirdupois (Rose 1948; Ransom 1975; Sadler 1976; Puddephatt 1978; Elevatorski 1981; Gasparrini 1993; Cvancara 1995; Krause 1996; Petralia 1996; Merchant 1998).

Gold has 30 known isotopes, but only one, [197]Au, is stable. The nucleus of [197]Au contains 79 protons and 118 neutrons. Isotopes of mass numbers 177 to 183 are all α emitters and all have a physical half-life of <1 min. Isotopes of mass numbers 185 to 196 decay by electron capture accompanied by radiation and in some cases by positron emission. The only long-lived isotope is [195]Au with a half life of 183 days. The neutron-heavy isotopes of 198 to 204 all decay by emission accompanied by radiation. The isotope [198]Au is widely used in radiotherapy, in medical diagnosis, and for tracer studies (Puddephatt 1978; Windholz 1983).

The color of gold alloys depends on the metal mixture. Red gold is comprised of 95.41% Au and 4.59% copper (Cu); yellow gold of 80% gold and 20% silver (Ag); and white gold of 50% Au and 50% Ag. The white gold commonly used in jewelry contains 75 to 85% Au, 8 to 10% nickel, and 2 to 9% zinc, while more expensive white alloys include palladium (90% Au to 10% Pd) and platinum (60% Au, 40% Pt). Colloidally suspended gold varies in color from deep ruby red to purple, and is used in the manufacture of ruby glass. Gold–silver–copper alloys are frequently used in coinage and gold wares. A purple alloy results with 80% Au and 20% aluminum, but this compound is too brittle to be made into jewelry. Gold forms alloys with many other metals, but most of these are also brittle. As little as 0.02% of tellurium, bismuth, or lead makes gold brittle (Rose 1948; Ransom 1975; Puddephatt 1978).

Analytical methodologies to measure gold in biological samples and abiotic materials rely heavily on its physical properties. These methodologies include x-ray fluorescence (Borjesson et al. 1993; Messerschmidt et al. 2000), adsorptive stripping voltammetry (Lack et al. 1999), bacteria-modified carbon paste electrodes (Hu et al. 1999), inductively-coupled plasma mass spectrometry [ICP-MS] (Higashiura et al. 1995; Perry et al. 1995; Barefoot and Van Loon 1996; Christodoulou et al. 1996; Barefoot 1998; Barbante et al. 1999), atomic absorption spectrometry [AAS] (Brown and Smith 1980; Kehoe et al. 1988; Niskavaara and Kontas 1990; Ohta et al. 1995; Begerow et al. 1997), fire assay (Gasparrini 1993), and neutron activation [NA] and spectrometry (Shiskina et al. 1990). Analyses of gold based upon gravimetric, volumetric, and UV/visible spectrophotometric techniques have been largely displaced by instrumental methods, such as NA, AAS, and more recently ICP-MS and ICP-AAS. In ICP-MS, for example, detection limits of gold after preconcentration of samples were as low as 0.04 ng/g ash in vegetation, 0.1 to 0.8 ng/L in water and urine, and 0.1 ng/g in soils and sediments (Perry et al. 1995; Barefoot and Van Loon 1996; Barefoot 1998). It is noteworthy that the fire assay method to analyze ore samples for gold content is the most convenient and least expensive method used throughout the world, despite interferences from copper, nickel, lead, bismuth, and especially tellurium and selenium (Gasparrini 1993). The fire assay, known to metalworkers for at least 3000 years, involves a weighed sample of the pulverized rock melted at 1000°C in a flux containing lead oxide, a measured amount of silver, soda, borax, silica, and potassium nitrate (Kirkemo et al. 2001). The lead fraction contains the gold and added silver and settles to cool as a button, which is subsequently remelted, oxidized to remove the lead oxide, leaving behind a bead consisting of precious metals. The bead is dissolved in acid and usually analyzed by AAS.

4.2 CHEMICAL PROPERTIES

The chemistry of gold is complex. Gold can exist in seven oxidation states: -1, 0, $+1$, $+2$, $+3$, $+4$, and $+5$. Apart from Au^0 in the colloidal and elemental forms, only Au^+ and Au^{+3} are known to form compounds that are stable in aqueous media and important in medical applications (Table 4.1; Puddephatt 1978; Shaw 1999a, 1999b). The remaining oxidation states of -1, $+2$, $+4$, and $+5$ are not presently known to play a role in biochemical processes related to therapeutic uses of gold (Shaw 1999b). Neither Au^+ or Au^{+3} forms a stable aquated ion ($[Au(OH_2)_{2-4}{}^+]$ or $[Au(OH_2)_4{}^{3+}]$, respectively) analogous to those found for many transition metal and main group cations. Both are thermodynamically unstable with respect to elemental gold and can be readily reduced. The gold-based anti-arthritic agents are considered pro-drugs that undergo rapid metabolism to form new metabolites (Shaw 1999a), a phenomenon that will be discussed in detail later. In complexes containing a single gold atom, the oxidation states $+1$, $+2$, $+3$, and $+5$ are well established (Puddephatt 1978). Divalent gold (Au^{+2}) is rare, usually being formed as a transient intermediate in redox reactions between the stable oxidation states Au^+ and Au^{+3}. The first Au^{+5} complex containing the ion $AuF_6{}^-$ was reported in 1972. The compound AuF_5 can also be prepared. Both are powerful oxidizing agents. Gold also forms many complexes with metal–metal bonds in which it is difficult to assign formal oxidation states. Additional information on stereochemistry, stability of complexes, oxidation–reduction potentials, current theories, and other aspects of gold chemistry is presented in detail by Sadler (1976), Puddephatt (1978), Merchant (1998), and Schmidbaur (1999).

Table 4.1 Oxidation States of Gold, Examples, and Stability in Water

Oxidation State	Example	Stable
−1	CsAu, ammoniacal Au^-	No
0	Metallic and colloidal gold	Yes
+1	$Au(CN)_2{}^-$, aurothiomalate	Yes
+2	$Au_2(CH_2PMe_2CH_2)_2Cl_2$	No
+3	$AuCl_4{}^-$, $Au(CN)^-$	Yes
+4	$Au(S_2C_6H_4)_2$	No
+5	$AuF_6{}^-$	No

Source: Data from Puddephatt 1978; Shaw 1999a, 1999b.

Metallic gold (Au^0) is comparatively inert chemically. Gold is resistant to tarnishing and corrosion during lengthy underground storage or immersion in seawater. It does not oxidize or burn in air even when heated. However, gold reacts with tellurium at high temperatures to yield AuTe$_2$ and reacts with all the halogens. Bromine is the most reactive halogen and, at room temperatures, reacts with gold powder to produce Au$_2$Br$_6$. At temperatures below 130°C, chlorine is adsorbed onto the gold, forming surface compounds; at 130 to 200°C, further reactions occur but the rate is limited by the diffusion rate of chlorine through the surface layer of gold chlorides; at >200°C, a high reaction rate occurs as the gold chlorides sublime, continually exposing a gold surface. Atomic gold is considerably more reactive than the massive metal (Puddephatt 1978). Evaporation of gold at high temperatures under vacuum followed by cocondensation of the vapor with a suitable reagent onto an inert noble-gas matrix at liquid helium temperature produces Au(O$_2$), Au(C$_2$H$_4$), Au(CO), and Au(CO$_2$). Cocondensation of atomic gold with carbon monoxide and dioxygen gives the complex Au(CO)$_2$O$_2$; all of these gold compounds decompose on warming the matrix (Puddephatt 1978). When auric oxide is treated with strong ammonia, a black powder is formed called fulminating gold (AuN$_2$H$_3$, 3H$_2$O). Dried, it is a powerful explosive as it detonates by either friction or on heating to about 145°C. Caution is advised when handling this compound (Rose 1948).

Halogen compounds of gold are well known, especially aurous chloride (AuCl) and auric chloride (AuCl$_3$; Rose 1948). Aurous chloride is a yellowish-white solid that is insoluble in cold water, but it undergoes slow decomposition into Au^0 and AuCl$_3$. Auric chloride takes the form of a reddish brown powder or ruby red crystals. The auric chloride of commerce is aurichloric or chloroauric acid (HAuCl$_4$·3H$_2$O), a brown deliquescent substance that is soluble in water or ether. Aurichloric acid forms a series of salts called aurichlorides or chloroaurates. Aurichlorides of Li, K, and Na are very soluble in water, and those of Rb and Cs much less soluble. The sodium salt, NaAuCl$_4$·2H$_2$O, is sold as sodio-gold chloride and, unlike aurichloric acid, is not deliquescent. Two gold bromides are known, AuBr and AuBr$_3$, corresponding to their chlorine counterparts. Auric iodide (AuI$_3$) is unstable and decomposes into aurous iodide (AuI) and free iodine. Iodine in aqueous-alcoholic solutions combines with metallic gold to form aurous iodide, a white or lemon-yellow powder that is insoluble in water (Rose 1948).

Gold is inert to strong alkalis and virtually all acids, except aqua regia — a mixture of concentrated nitric acid (1 part) and hydrochloric acid (3 parts). The nitric and hydrochloric acids interact forming nitrosylchloride (NOCl) together with free chlorine, which reacts with gold. In aqua regia, gold forms tetrachloroauric acid, HAuCl$_4$, which is the source of gold chloride. Gold is also soluble in hot selenic acid forming gold selenate, and in aqueous solutions of alkaline sulfides and thiosulfates (Rose 1948; Krause 1996; Merchant 1998). Gold will dissolve in hydrochloric acid in the presence of hypochlorite or ferric iron (Fe^{+3}) as oxidant. The dissolution of gold in cyanide solutions with air or hydrogen peroxide as oxidant is another example of this effect (Ransom 1975; Puddephatt 1978). The reaction with oxygen as oxidizing agent apparently takes place by adsorption of oxygen onto the gold surface, followed by reaction of this surface layer to yield AuCN, followed by the complex Au(CN)$_2^-$, which passes into solution (Rose 1948; Puddephatt 1978).

Gold is also soluble in liquid mercury and in dilute solutions of sodium or calcium cyanide. The cyanide solvent was used in Australia in 1897 where it was used to remove finely disseminated gold from pulverized rock. The cyanide process is the only known method of profitably treating massive low-grade gold ores. Using the cyanide process, auriferous rocks containing as little as 1 part gold in 300,000 parts of worthless materials can be treated successfully (Ransom 1975; Cvancara 1995).

Gold is readily dissolved by halide or sulfide ions in the presence of oxidizing agents to yield Au^{+3} or Au^+ complexes (Puddephatt 1978). It is probably in this way that gold is dissolved when hot volcanic rock is buried, or when a hot granite intrusion rises near the surface of the earth's crust. As the solution cools to 300 to 400°C, concentrations of oxygen and hydrochloric acid decrease sharply, and gold is redeposited. Hydrothermal transfer of gold as the complex ion $[Au(SH)_2]^-$ may occur in some cases. Dissolution and redeposition of gold in stream beds may also be responsible for the formation of large crystals of alluvial gold (Puddephatt 1978).

Solutions containing gold complexes, such as $AuCl_4^-$, are easily reduced to Au^0 and under controlled conditions colloidal gold may be formed. Colloids of gold — first reported in the 18th century — may be red, blue, or violet depending on the mean particle size and shape (Puddephatt 1978). Various reducing agents can be used for preparing colloidal gold including tannin, phosphorus, formaldehyde, and hydrazine hydrate. The "purple of Cassius" is a mixed colloid of hydrated Sn^{+4} oxide and gold formed by reducing $AuCl_4^-$ with Sn^{+2} chloride, A purple or ruby-red precipitate is formed on heating the solution. A sensitive test for gold is based on this process, that is, a purple color is formed if a 10^{-8} M solution of $AuCl_4^-$ is added to a saturated solution of $SnCl_2$ (Puddephatt 1978).

The chlorination process, introduced in 1867, remains one of the most important refining processes for raw gold (Dahne 1999). Chlorination makes use of the fact that silver, copper, and base metals in raw gold react with chlorine at about 1100°C to form stable chlorides while gold and platinum chlorides are unstable at >400°C. At 1100°C, silver chloride and copper chloride are molten, and base metal chlorides are volatile. The silver and copper chlorides are removed by skimming. Chlorination — which is usually completed within a few hours — is usually stopped at 99% gold so that gold losses by vaporization are avoided. Other refining processes for gold include electrolysis, and wet-chemical separation of gold from silver and base metals. In electrolysis, the anode plates are a mixture of tetrachloroauric acid, hydrochloric acid, and raw gold, and the cathodes are thin titanium (Dahne 1999).

4.3 BIOCHEMICAL PROPERTIES

Gold is not an essential element for living systems (Brown and Smith 1980). Indeed, the administration of gold to patients has been more similar to that of toxic elements, such as mercury, than to that of biologically utilized transition elements such as copper and iron. Gold distributes widely in the body and the number of possible reactions and reaction sites is large. Most of the *in vivo* gold chemistry is concerned with the reaction of gold species with thiols. Within mammalian systems subjected to Au^0, Au^+, or Au^{+3}, gold metabolism resulted in both monomeric and polymeric species. Most gold complexes administered orally or parenterally were absorbed, but rate and extent of accumulation were highly variable among gold compounds. Gold circulated in blood mainly by way of the serum proteins, especially albumin. Gold was deposited in many tissues and was dependent on dose and compound administered. Likely storage forms included colloidal Au^0, insoluble Au^+ deposits, and possibly Au^{+3} polymers. Accumulated gold containing sulfur was documented. There is no suitable animal model available for testing mechanisms of action of gold compounds used in human medicine (Brown and Smith 1980).

Gold has a unique biochemical behavior (Sadler 1976). Biochemical behaviors of heavy metal ions show some similarities, particularly in their affinity for polarizable ligands. But they also show important differences. Gold, for example, has a comparatively low affinity for amino and carboxylate groups, a stable higher oxidation state in water, and proven anti-inflammatory activity of selected Au^+ organic salts (Sadler 1976). The biochemistry of gold has developed mainly in response to prolonged use of gold compounds in treating rheumatoid arthritis and in response to efforts to develop complexes with anti-tumor and anti-HIV activity (Shaw 1999b). [See Chapter 9 for additional details.]

Chemical reactions of gold drugs exposed to body fluids and proteins are mainly ligand exchange reactions that preserve the Au^+ oxidation state (Shaw 1999a). Aurosomes (lysosomes that accumulate large amounts of gold and undergo morphological changes) taken from gold-treated rats contain mainly Au^+, even when Au^{+3} has been administered. However, the potential for oxidizing Au^+ to Au^{+3} *in vivo* exists. Monovalent gold drugs can be activated *in vivo* to an Au^{+3} metabolite that is responsible for some of the immunological side effects observed in chrysotherapy. For example, treatment of rodents and humans with anti-arthritic monovalent gold drugs generates T-cells that react to Au^{+3} but not to the parent compound (Shaw 1999a).

Although metallic gold (Au^0) is arguably the least corrosive and most biologically inert of all metals, it can be gradually dissolved by thiol-containing molecules such as cysteine, penicillamine, and glutathione to yield Au^+ complexes (Merchant 1998). Metallic gold reacted with cysteine in aqueous or saline solution in the presence of oxygen to produce an Au^+–cysteine complex; Au^+ and cysteine formed a 1:1 Au^+–cysteine complex; L-cysteine reduced most Au^{+3} compounds in solution to produce the Au^+–L-cysteine complex (Brown and Smith 1980). With D-penicillamine, Au^0 formed a Au^0–penicillamine complex; Au^+ under a nitrogen environment formed a R_3PAu^+–penicillamine complex; and Au^{+3} formed a bis complex with penicillamine. With glutathione, Au^+ formed a stable 1:1 complex in solution; Au^{+3} oxidized glutathione to sulfoxide, the gold being reduced to Au^+, which was stabilized by complexing with unreacted glutathione (Brown and Smith 1980). These processes were amplified at alkaline pH, significantly at pH 7.2, and perceptibly in acidic environments having pH values as low as 1.2 (Merchant 1998). The rate of the reaction was controlled by the concentrations of thiol-containing molecules and by the pH; reactions might take place within cells and inside lysosomes. Under favorable conditions, reactions occurred at low rates on skin surfaces. Skin samples taken from beneath gold wedding bands of normal individuals averaged 0.8 mg/kg dry weight skin. *In vitro* studies designed to simulate conditions inside phagocytic lysosomes showed substantial dissolution of Au^0 in the presence of hydrogen peroxide and amino acids such as histidine and glycine. There are reported instances of rheumatoid arthritis patients who, on initiation of gold drug treatment (chrysotherapy), have promptly produced rashes in the skin areas that have had regular contact with gold jewelry. Gold jewelry, if in close contact with skin, could be slowly dissolved by sweat. Thus, the thinning of gold rings over time, thought to be due mainly to abrasion, could also be due, in part, to dissolution (Merchant 1998).

Colloidal gold is readily accumulated by macrophages (Sadler 1976). The gold particles are taken into small vesicles, which form by surface invagination, and into vesicles fusing to form vacuoles with subsequent transport to the centrosomic region. The part played by the surface of the Au^0 particle may be due to Au^+ ions on the surface, which promote uptake. A soluble gold-uptake stimulating factor of MW <100,000 is reportedly secreted by lymphocytes and acts upon the macrophages (Sadler 1976).

$Gold^+$ drugs were metabolized rapidly in vivo (Shaw 1999a). The half life for gold excretion in dogs was 20 days, but major metabolites had half-life times of 8 to 16 hours. Within 20 minutes of administration, gold was protein-bound mainly in the serum. Injectable $gold^+$ drugs were not readily taken up by most cells, but bound to cell surface thiols where they affected cell metabolism. The high affinity of Au^+ for sulfur and selenium ligands suggested that proteins, including enzymes and transport proteins, were critical in vivo targets. It was clear that extracellular gold in the blood was primarily protein bound, suggesting protein-mediated transport of gold during therapy (Shaw 1999a). Metallothioneins play an important role in metal homeostasis and in protection against metal poisoning in animals (Eisler 2000). Metallothioneins are cysteine-rich (>20%), low-molecular-weight proteins with a comparatively high affinity for gold, copper, silver, zinc, cadmium, and mercury. These heat-stable metal-binding proteins were found in all vertebrate tissues and were readily induced by a variety of agents — including gold — to which they bind through thiolate linkages. The role of metallothioneins in maintaining low intracellular gold concentrations needs to be resolved.

Following a chrysotherapy-type regimen with gold disodium thiomalate in mice, Au^{+3} generation was analyzed with a lymph node assay system using T-lymphocytes sensitized to Au^{+3} (Merchant 1998). The findings were consistent with three separate anti-inflammatory mechanisms:

1. Generation of Au^{+3} from Au^+ scavenges reactive oxygen species, such as hypochloric acid.
2. Au^{+3} is a highly reactive chemical that irreversibly denatures proteins, including those lysosomal enzymes that nonspecifically enhance inflammation when they are released from cells at an inflammatory focus.
3. Au^{+3} may interfere with lysosomal enzymes involved in antigen processing or may directly alter molecules along the lysosomal–endosomal pathway, resulting in reduced production of arthritogenic peptides (Merchant 1998).

If all of these activities occurred within a redox system in phagocytic cells, then the anti-inflammatory actions of Au^+/Au^{+3} could be effective for protracted periods, and explain, in part, both the anti-inflammatory and the adverse effects of antirheumatic

Au[+] drugs. Deviation of proteins could also contribute to the rare instances of auto-immunity reported in association with chrysotherapy (Merchant 1998).

Knowledge of Au[+] binding sites on large molecules, such as proteins, is limited to a few studies using $Au(CN)_2^-$ (Sadler 1976). Although $Au(CN)_2^-$ is one of the most stable gold ions in solution, it is considered too toxic for clinical use. The simple Au[+] cation does not appear to exist in solution, and most Au[+] compounds are insoluble or unstable in water. Mercaptides stabilize Au[+] in water, and sodium gold thiomalate is now in widespread use as an anti-inflammatory drug. Ionic Au[+] seems to enter many cells but localize within the lysosomes of the phagocytic cells called macrophages. Here they may inhibit enzymes important in inflammation. Studies with sodium gold thiomalate suggest that anti-tumor mechanisms, like inflammation, are also macro-phage-mediated (Sadler 1976).

Canumalla et al. (2001) report on two recent advances in understanding gold metabolism *in vivo*. In one finding, gold[+] drugs and their metabolites react *in vivo* with cyanide, forming dicyanoaurate[+], $(Au^+(CN)_2)^-$; this ion has been identified as a common metabolite of Au[+] drugs in blood and urine of chrysotherapy patients. Second, Au[+] is the primary oxidation state found *in vivo* although there is increasing evidence for the generation of Au[+3] metabolites. Biomimetic studies indicate that the oxidation of sodium gold[+] thiomalate and sodium gold[+] thioglucose by hypochlo-rite ion (OCl)[-], released when cells are induced to undergo the oxidative burst at inflamed sites, is rapid and thermodynamically feasible in the formation of Au[+3] species. The OCl[-] ion is involved in both the generation of $Au(CN)_2^-$ and the formation of Au[+3] species *in vivo* (Canumalla et al. 2001).

The potential anti-tumor activity of gold complexes is driven by three rationales: (1) analogy to immunomodulatory properties underlying the benefit from Au[+] com-plexes in treating rheumatoid arthritis; (2) the structural analogy of square-planar Au[+3] to platinum[+2] complexes, which are potent anti-tumor agents; and (3) complex-ation of Au[+] or Au[+3] with other active anti-tumor agents in order to enhance the activity and alter the biological distribution of Au[+3] (Shaw 1999a). For example, the rate of hydrolysis of $AuCl_4^-$ in water is 375 times greater than that of $PtCl_4^-$ (Sadler 1976). There is potential for developing new cytotoxic gold complexes that have anti-tumor properties, and this requires robust, new ligand structures that can move gold through cell membranes and into the cytoplasm, and perhaps into the cell nucleus (Shaw 1999a). Trivalent gold (Au[+3]) compounds are potential anticancer agents (Calamai et al. 1997). These compounds are soluble in organic solvents, such as methanol or DMSO, but poorly soluble in water. In water, $AuCl_3$ undergoes hydrolysis of the bound chloride without loss of the heterocycle ligand. When Au[+3] compounds react with proteins, like albumin or transferrin, Au[+3] is easily reduced to Au[+]. Cytotoxicity studies with tumorous cells showed marked anticancer activity of Au[+3] complexes, probably mediated by a direct interaction with DNA. However, rapid hydrolysis of Au[+3] to Au[+] under physiological conditions may severely restrict their use. More studies are needed to understand the biological mechanisms of gold complexes, including extent of cell penetration and biodistribution (Calamai et al. 1997).

Anti-HIV activity of monovalent gold compounds were associated with inhibition of reverse transcriptase (RT), an enzyme that converts RNA into DNA in the host cell (Shaw 1999a). Other reports indicate that Au^+ inhibits the infection of cells by HIV strains without inhibiting the RT activity, with the critical target site tentatively identified as a glycoprotein of the viral envelope. Other reports show that $Au(CN)_2^-$ at concentrations as low as 20 μg/L is incorporated into a T-cell line susceptible to HIV infection, and retards the proliferation of HIV in these cells. This concentration is well tolerated in patients with rheumatoid arthritis, suggesting that $Au(CN)_2^-$ may have promise for existing HIV patients (Shaw 1999a).

When Au^{+3} compounds were used as labels for crystalline proteins, the nature of the bound species was uncertain (Sadler 1976). Labelling with AuI_4^- has been claimed, but this ion appears unstable in aqueous solution. In addition, Au^{+3} compounds often have strong oxidizing properties. With a careful choice of ligands for Au^{+3}, a range of antitumor drugs may emerge because Au^{+3} has a high affinity for polynucleotides and may interfere with cell division properties (Sadler 1976).

4.4 SUMMARY

Elemental gold is a soft yellow metal with the highest malleability and ductility of any known element. It is dense, being 19.32 times heavier than water at 20°C; a cube of gold 30 cm on a side weighs about 544 kg. Metallic gold is inert to strong alkalis and virtually all acids; however, solubility is documented for aqua regia, hot selenic acid, aqueous solutions of alkaline sulfides and thiosulfates, cyanide solutions, and liquid mercury. Sensitive analytical methodologies developed to measure gold in biological samples and abiotic materials relied heavily on its physical properties. Gold has 30 known isotopes and exists in seven oxidation states. Apart from Au^0 in the colloidal and elemental forms, only Au^+ and Au^{+3} are known to form compounds that are stable in aqueous media and important in medical applications. The remaining oxidation states of -1, $+2$, $+4$, and $+5$ are not presently known to play a role in biochemical processes related to the therapeutic uses of gold. Gold has a unique biochemical behavior, characterized by a comparatively low affinity for amino and carboxylate groups, a stable higher oxidation state in water, and proven anti-inflammatory activity of selected Au^+ organic salts. The biochemistry of gold has developed mainly in response to prolonged use of gold compounds in treating rheumatoid arthritis and in response to efforts to develop complexes with anti-tumor and anti-HIV activity.

Most of the *in vivo* gold chemistry is concerned with the reaction of gold species with thiols, especially Au^+. Gold is not considered essential to life, although it distributes widely in the body and the number of possible reactions and reaction sites is large. Monovalent organogold drugs were metabolized rapidly *in vivo*, usually within 20 minutes of administration; however, half-time excretion rates ranged between 8 hours and 20 days, depending on the metabolite.

LITERATURE CITED

Barbante, C., G. Cozzi, G. Capodaglio, K. van de Velde, C. Ferrari, C. Boutron, and P. Cescon. 1999. Trace element determination in alpine snow and ice by double focusing inductively coupled plasma mass spectrometry with microconcentric nebulization, *Jour. Anal. Atomic Spectr.*, 14, 1433–1438.

Barefoot, R.R. 1998. Determination of the precious metals in geological materials by inductively coupled plasma mass spectrometry, *Jour. Anal. Atom. Spectrom.*, 13, 1077–1084.

Barefoot, R.R. and J.C. Van Loon. 1996. Determination of platinum and gold in anticancer and antiarthritic drugs and metabolites, *Anal. Chim. Acta*, 334, 5–14.

Begerow, J., M. Turfeld, and L. Dunemann. 1997. Determination of physiological noble metals in human urine using liquid-liquid extraction and Zeeman electrothermal atomic absorption spectrometry, *Anal. Chim. Acta*, 340, 277–283.

Borjesson, J., M. Alpstein, S. Huang, R. Jonson, S. Mattsson, and C. Thornberg. 1993. *In vivo* X-ray fluorescence analysis with applications to platinum, gold and mercury in man — experiments, improvements, and patient measurements, in *Human Body Composition*, K.J. Ellis and J.D. Eastman, (Eds.), Plenum, New York, 275–280.

Brown, D.H. and W.E. Smith. 1980. The chemistry of the gold drugs used in the treatment of rheumatoid arthritis, *Chem. Soc. Rev.*, 9, 217–240.

Calamai, P., S. Carotti. A. Guerri, L. Messori, E. Mini, P. Orioli, and G.P. Speroni. 1997. Biological properties of two gold(III) complexes: $AuCl_3$ (Hpm) and $AuCl_2$ (pm), *Jour. Inorg. Biochem.*, 66, 103–109.

Canumalla, A.J., N. Al-Zamil, M. Phillips, A.A. Isab, and C.F. Shaw III. 2001. Redox and ligand exchange reactions of potential gold(I) and gold (III)-cyanide metabolites under biomimetic conditions, *Jour. Inorg. Biochem.*, 85, 67–76.

Christodoulou, J., M. Kashani, B.M. Keohane, and P.J. Sadler. 1996. Determination of gold and platinum in the presence of blood plasma proteins using inductively coupled plasma mass spectrometry with direct injection nebulization, *Jour. Anal. Atomic Spectrom.*, 11, 1031–1035.

Cvancara, A.M. 1995. *A Field Manual for the Amateur Geologist*. John Wiley & Sons, New York, 335 pp.

Dahne, W. 1999. Gold refining and recycling, in *Gold: Progress in Chemistry, Biochemistry and Technology*, H. Schmidbaur, (Ed.), John Wiley & Sons, New York, 120–141.

Eisler, R. 2000. Zinc, in *Handbook of Chemical Risk Assessment: Health Hazards to Humans, Plants, and Animals, Volume 1. Metals.* Lewis Publishers, Boca Raton, FL, 605–714.

Elevatorski, E.A. 1981. *Gold Mines of the World*. Minobras, Dana Point, CA, 107 pp.

Gasparrini, C. 1993. *Gold and Other Precious Metals. From Ore to Market.* Springer-Verlag, Berlin, 336 pp.

Higashiura, M., H. Uchida, T. Uchida, and H. Wada. 1995. Inductively coupled plasma mass spectrometric determination of gold in serum: comparison with flame and furnace atomic absorption spectrometry, *Anal. Chim. Acta*, 304, 317–321.

Hu, R., W. Zhang, Y. Liu, and J. Fu. 1999. Determination of trace amounts of gold (III) by cathodic stripping voltammetry using a bacteria-modified carbon paste electrode, *Anal. Commun.* (Roy. Soc. Chem.), 36, 147–148.

Kehoe, D.F., D.M. Sullivan, and R.L. Smith. 1988. Determination of gold in animal tissue by graphite furnace atomic absorption spectrophotometry, *Jour. Assoc. Off. Anal. Chem.*, 71, 1153–1155.

Kirkemo, H., W.L. Newman, and R.P. Ashley. 2001. *Gold.* U.S. Geological Survey, Denver, CO, 23 pp.

Krause, B. 1996. *Mineral Collector's Handbook*. Sterling, New York., 192 pp.

Lack, B., J. Duncan, and T. Nyokong. 1999. Adsorptive cathodic stripping voltammetric determination of gold (III) in presence of yeast mannan, *Anal. Chim. Acta*, 385, 393–399.

Merchant, B. 1998. Gold, the noble metal and the paradoxes of its toxicology, *Biologicals*, 26, 49–59.

Messerschmidt, J., A. von Bohlen, F. Alt, and R. Klockenkamper. 2000. Separation and enrichment of palladium and gold in biological and environmental samples, adapted to the determination by total reflection X-ray fluorescence, *Analyst*, 125, 397–399.

Niskavaara, H., and E. Kontas. 1990. Reductive coprecipitation as a separation method for the determination of gold, palladium, platinum, rhodium, silver, selenium and tellurium in geological samples by graphite furnace atomic absorption spectrometry, *Anal. Chim. Acta*, 231, 273–283.

Ohta, K., T. Isiyama, M. Yokoyama, and T. Mizuno. 1995. Determination of gold in biological materials by electrothermal atomic absorption spectrometry with a molybdenum tube atomizer, *Talanta*, 42, 263–267.

Perry, B.J., R.R. Barefoot, and J.C. Van Loon. 1995. Inductively coupled plasma mass spectrometry for the determination of platinum group elements and gold, *Trends Anal. Chem.*, 14, 388–397.

Petralia, J.F. 1996. *Gold! Gold! A Beginner's Handbook and Recreational Guide: How & Where to Prospect for Gold!* Sierra Outdoor Products Co., San Francisco, 143 pp.

Puddephatt, R.J. 1978. *The Chemistry of Gold*, Elsevier, Amsterdam, 274 pp.

Ransom, J.E. 1975. *The Gold Hunter's Field Book*, Harper & Row, New York, 367 pp.

Rose, T.K. 1948. Gold, *Encyclopaedia Britannica*, 10, 479–485.

Sadler, P.J. 1976. The biological chemistry of gold: a metallo-drug and heavy-atom label with variable valency, *Structure Bonding*, 29, 171–215.

Schmidbaur, H. (Ed.). 1999. *Gold: Progress in Chemistry, Biochemistry, and Technology*, John Wiley & Sons, New York, 894 pp.

Shaw, C.F., III. 1999a. Gold complexes with anti-arthritic, anti-tumour and anti-HIV activity, in *Uses of Inorganic Chemistry in Medicine*, N.C. Farrell, (Ed.), Royal Society of Chemistry, Cambridge, UK, 26-57.

Shaw, C.F., III. 1999b. The biochemistry of gold, in *Gold: Progress in Chemistry, Biochemistry, and Technology*, H. Schmidbaur, (Ed.), John Wiley & Sons, New York, 260–308.

Shishkina, T.V., S.N. Dmitriev, and S.V. Shishkin. 1990. Determination of gold in natural waters by neutron activation and -spectrometry after preconcentration with tributyl phosphate as solid extractant, *Anal. Chim. Acta*, 236, 483–486.

Windholz, M. (Ed.). 1983. *The Merck Index, 10th edition*. Merck & Co., Rahway, NJ, 1463 pp.

CHAPTER 5

Gold Concentrations in Field Collections

Gold concentrations in various abiotic materials collected worldwide (rainwater, seawater, lakewater, atmospheric dust, soils, snow, sewage sludge, sediments) are listed and discussed in this chapter, as well as similar data for terrestrial and aquatic plants, terrestrial and aquatic invertebrates, fishes, and humans (Eisler 2004).

5.1 ABIOTIC MATERIALS

Gold concentrations in air, the earth's crust, freshwater, rainwater, seawater, sediments, sewage sludge, snow, soil, and volcanic rock are summarized in Table 5.1.

Most gold in ocean surface waters comes from fallout of atmospheric dust. Riverine sources of gold into seas and oceanic coastal waters are minor, as judged by studies of manganese transport (Gordeyev et al. 1997). Dissolved gold was discovered in seawater in 1872, and many unsuccessful attempts to recover the gold commercially from seawater have since been made (Puddephatt 1978). The most famous attempt was made by German scientists in the years 1920 to 1927, with the intention of paying off the German war debt incurred during World War I. The method was based on reduction to metallic gold using sodium polysulfide (Puddephatt 1978). Unfortunately, the German calculations of 0.004 µg Au/L were 100 to 400 times higher than the recently calculated range for dissolved oceanic gold of 0.00001 to 0.00004 µg/L (Gordeyev et al. 1997). At these low concentrations it was not possible to directly determine what gold species were present. However, based on redox potentials of gold compounds and seawater composition, it is probable that $AuCl_2^-$ predominates, with smaller amounts of $AuClBr^-$, as well as bromo-, iodo-, and hydroxy complexes of Au^+ (Puddephatt 1978) in oxidation states of Au^0, Au^+, and Au^{+3} (Karamushka and Gadd 1999). Dissolved gold may be usable as a tracer of hydrothermal influence on bottom waters near vents. Concentration of gold in bottom water samples of the mid-Atlantic ridge in 1988, near hydrothermal vents, was 0.0015 µg/L vs. 0.0007 µg/L at a reference site; hydrothermal vent samples also had elevated concentrations for manganese and turbidity (Gordeyev et al. 1991).

Table 5.1   Gold Concentrations in Selected Abiotic Materials*

Material	Concentration	Reference[a]
Air		
Dust near high-traffic road in Frankfurt/Main, Germany	440 DW	17
Earth's Crust	4 to 5 DW	18, 22
Freshwater		
Canada; Murray Brook, New Brunswick; active gold mining site 1989-92; dissolved gold in adjacent stream		
Prior to mining (1988)	Not detectable	1
Post mining (1997)		
Near mine site	Max. 19 FW	1
3 km downstream	Max. 3 FW	1
Poland and Czech Republic; near former gold mining site	Not detectable (<0.22 FW)	19
Ultrapure	0.00007 FW	2
Rainwater		
Uzbekistan, single rain event		
Solid phase	0.00046 DW	3
Soluble phase	0.001 FW	3
Seawater		
Global average	0.004 FW	4
Global average	1.0 DW	18
Global average	0.00001–0.00004 FW	5
Atlantic Ocean		
Mid-Atlantic ridge, 1988, near hydrothermal vents vs. reference site	0.00153 FW vs. 0.0007 FW	6
Northeastern Atlantic Ocean, 1989, surface waters 0.5–1.0 m	0.0002–0.0007 FW	5
Sediments		
Canada; Murray Brook, New Brunswick; stream sediments receiving leachate from oxidized pyrites tailings pile from gold mining activities between 1989 and 1992 using a cyanide vat leach process		
Prior to mining (1988)	<5 DW	1
Post mining (1997)		
Near tailings	Max. 256,000 DW	1
3 km downstream	Max. 6000 DW	1
Japan Sea; 1990; coastal sediments from <100–1500 m		
Near Niigata Prefecture	3.8 (0.3–35.0) DW	7
Near Sado Island	>10 DW	7
Mid-Atlantic ridge and northeast Pacific Ocean		
Gold-rich sulfides	800–5000 DW	8
Pyritic assemblages	>1000 DW	8

Table 5.1 (continued)   Gold Concentrations in Selected Abiotic Materials*

Material	Concentration	Reference[a]
Sphalerite with sulforates	Max. 18,000 DW	8
Sphalerite	Max. 5700 DW	8
New Zealand, North Island; base metal mine closed in 1974; reexamined in 1999	Max. 163 DW	21
Pacific Ocean		
Near Japan, terrigenous origin	2.4 DW	9
Central Pacific, pelagic origin	1.4 DW	9
Papua New Guinea, 1995, Manus Basin	Mean 3 DW, Max. 15 DW	9
Southwest Pacific Ocean; polymetallic sulfides recovered from hydrothermal vents	Mean 3100 DW, Max. 28,700 DW	10
Sewage Sludge		
Southeastern Australia		
Industrialized areas	430–1260 DW	11
Rural areas	180–2350 DW	11
Germany	500–4500 DW	11
USA	500–3000 DW	11
Snow		
France; Italian Alps; 4250 m elevation; 140 m core representing 200-year period; analysis based on [197]Au content of particulate matter		
All samples	0.07–0.35 DW	12
1778	0.20 DW	12
1887	0.14 DW	12
1918	0.17 DW	12
1945	0.18 DW	12
1961	0.30 DW	12
1971	0.28 DW	12
1981	0.35 DW	12
1991	0.07 DW	12
Italy; eastern Alps; 1997–1998		
Surface	1.0 FW	2
Alpine snow		
Mont Blanc	0.00021 (0.0001–0.0004) FW	2
Monte Rosa	0.00013 (0.00006–0.0003) FW	2
Russia; Kola Peninsula; April 1996; 1-year surface deposition; near ore roasting and smelter facilities		
Near ore roasting plant	330–340 DW	13
Near smelter		
Copper-nickel complex fraction	2530 DW	13
Copper concentrate fraction	630 DW	13
Soil		
Egypt, Aswan; agricultural soil; 10–60 cm depth	150–180 DW	20
Nevada; Sixmile Canyon; alluvial fan soil		
Premining (before 1859)	13 (5–29) DW	14
Postmining		

Table 5.1 (continued)   Gold Concentrations in Selected Abiotic Materials*

Material	Concentration	Reference[a]
Fan deposits	473 (80–843) DW	14
Modern channel	166 (15–424) DW	14
New York; Cornell University orchard site; sludge applied in 1978 containing 350 µg Au/kg DW to depth of 15 cm; sampled 15 years later in 1993		
Surface soil	43 DW	15
Subsoil (15–35 cm)	4.4 DW	15
Volcanic Rock		
Papua New Guinea; 1995; recovered by deep-sea submersible	Max. 15 DW	16

* Values are in µg/L or µg/kg fresh weight (FW) or dry weight (DW).
[a] 1, Leybourne et al. 2000; 2, Barbante et al. 1999; 3, Kist 1994; 4, Puddephat 1978; 5, Gordeyev et al. 1997; 6, Gordeyev et al. 1991; 7, Terashima et al. 1991; 8, Hannington et al. 1991; 9, Terashima et al. 1995; 10, Herzig et al. 1993; 11, Lottermoser 1995; 12, Van de Velde et al. 2000; 13, Gregurek et al. 1999; 14, Miller et al. 1996; 15, McBride et al. 1997; 16, Moss et al. 1997; 17, Messerschmidt et al. 2000; 18, Sadler 1976; 19, Samecka-Cymerman and Kempers 1998; 20, Rashed and Awadallah 1998; 21, Sabti et al. 2000; 22, Korte et al. 2000.

Known gold-rich sea-floor deposits in the southwest Pacific Ocean occur along the axis of a major gold belt extending from Japan through the Philippines, New Guinea, Fiji, Tonga, and New Zealand (Herzig et al. 1993). Polymetallic sulfides recovered from the sea-floor hydrothermal systems of this region contain up to 28.7 mg Au/kg (about 1 ounce per ton) with an average of 3.1 mg Au/kg. These samples are among the most gold-rich hydrothermal precipitates reported from the sea floor. The gold is generally of high purity, containing less than 10% silver. In one hydrothermal vent field, gold concentrations averaged 30 mg/kg and visible gold was seen in the sulfide chimney. Gold concentrations decreased sharply to <0.02 mg/kg when the temperature dropped from about 280 to 300°C in the center of the chimney to about 200°C at its outer margin. Subsea-floor boiling and precipitation of sulfides is important in separating gold from base metals in the ascending hydrothermal fluids. Gold seemed to be precipitated largely from aqueous sulfur complexes [$Au(HS)_2^-$] as a result of the combined effects of conductive cooling, mixing with seawater, and oxidation of H_2S. Sulfide deposits in this basin and elsewhere in the southwest Pacific Ocean are similar to some gold-rich massive sulfides on land (Herzig et al. 1993). Gold enrichment in high-sulfide marine sediments is usually — but not always — associated with elevated concentrations of silver, arsenic, antimony, lead, zinc, and various sulfosalts, especially iron-poor sphalerite (zinc sulfide); in contrast, gold is typically depleted in samples with high levels of cobalt, selenium, or molybdenum (Hannington et al. 1991). In one study, high gold concentrations in marine sediments were associated with elevated arsenic (1100 to 6600 mg/kg), antimony (85 to 280 mg/kg), and lead, but the correlations between these elements and gold were variable (Herzig et al. 1993). Moss et al. (1997) showed no significant correlation between gold and other trace metals measured or with silicon, iron, and magnesium.

Abnormally high gold concentrations (>10 µg Au/kg) found in the sediments around Sado Island in the Sea of Japan were attributed to auriferous mineralization of the island and anthropogenic mining activities (Terashima et al. 1991, 1995). Gold is probably supplied to marine sediments in dissolved form through rivers and seawater and, to a lesser extent, as discrete minerals. Gold distribution in coastal sediments of the Sea of Japan is controlled by geologic characteristics of the catchment area of rivers, the grain size of the sediments, redox potential, water depths of the sampling locations, and dissolved oxygen. For example, gold is more abundant in the finer fraction sediments than in coarse ones. In cases where there is a clear negative correlation between gold content and redox potential of the sediments, the gold occurs mostly in the dissolved form; if the correlation is not significant, the gold occurs in metallic form. Dissolved gold is converted by reduction to Au^0 in oxygen-depleted environments. The suspended gold particles are subsequently adsorbed on mineral surfaces or precipitated as hydroxide or sulfide (Terashima et al. 1991, 1995).

Freshwater sediments in Murray Brook, New Brunswick, Canada, received gold between 1989 to 1992 from a vat leach cyanidation process used to separate gold from ores (Leybourne et al. 2000). The gossan (oxidized pyrites) tailings pile in Murray Brook leached gold into the adjacent freshwater stream sediments from complexation of gold to $Au(CN)_2^-$ by residual cyanide within the tailings. The elevated gold concentrations (up to 256 mg Au/kg) in stream sediments close to the headwaters of the creek near the tailings suggest that $Au(CN)_2^-$ is degraded and the gold removed from solution via reduction of Au^+ by Fe^{2+}. Gold is converted from a complexed form to a colloidal form with increasing distance downstream, consistent with dissolved nitrate contents, which decreased from 5.2 mg/L near the headwaters to 1.4 mg/L at the lower end of the stream (Leybourne et al. 2000).

Worldwide accumulation of gold in sewage is about 360 tons each year (Lottermoser 1995). Sewage is commonly dumped on land or at sea. Discharge of excessive sewage into coastal areas poses a threat to human health and coastal fisheries, diminishes the recreational use of the littoral zone, and may result in the formation of anthropogenic labile-metal deposits. Sewage solids from a southeastern Australian community with a gold mining history of more than 100 years contained 0.18 to 2.35 mg Au/kg DW. These concentrations are similar to those of ore deposits currently mined for gold (Lottermoser 1995). Gold in sewage sludge containing 0.35 mg Au/kg DW applied to agricultural surface soils migrates downwards; after 15 years, about 60% of the gold was found in subsurface soils (McBride et al. 1997).

Gold concentrations in different strata of snow/ice cores from the French-Italian Alps deposited over a period of 200 years were consistently low (0.07 to 0.35 µg/kg fresh weight [FW], detection limit of 0.03 µg/kg), except for minor increases resulting from atmospheric deposition from nearby smelters (Van de Velde et al. 2000). In northwestern Russia, however, gold concentrations in the annual winter snow cover of 1995 to 1996 were greatly elevated (>350 µg/kg DW; Gregurek et al. 1999). Dust and smokestack emissions from the local ore roasting and metal smelters were the sources. Concentrations of gold in snow increased with proximity to these industrial sources. The high concentrations of gold and other precious metals (rhodium, platinum, palladium) deposited on snow during a single winter season suggest that modernization of the industrial plants to recover these metals would result in substantial economic benefits (Gregurek et al. 1999).

5.2 PLANTS

Gold levels in selected terrestrial and aquatic vegetation are summarized in Table 5.2.

Gold accumulator plants, such as *Artemisia persia*, *Prangos popularia*, and *Stripa* spp. grasses, routinely contain >0.1 mg Au/kg DW and may contain as much as 100 g of gold per metric ton or 100 mg Au/kg (Sadler 1976). Microorganisms in the plant roots may be responsible for solubilizing the gold, allowing ready uptake by these species. Some strains of *Bacillus megaterium*, for example, secrete amino acids, aspartic acid, histidine, serine, alanine, and glycine to aid in gold dissolution (Sadler 1976). Bioaccumulation of gold from metals-contaminated soils was documented in stems and needles of Corsican pine trees (*Pinus laricio*) from the Mount Olympus area of the island of Cyprus (Pyatt 1999), and plants grown in soils containing 1 to 25 µg Au/kg DW soil had comparatively high concentrations of gold in seeds and pericarp, but low concentrations in pods, leaves, and stems (Awadallah et al. 1995). In a recent study, faba beans (*Vicia* sp.) were shown to contain about the same amount of gold in their leaves as did the soils in which they were grown (170 µg/kg DW vs. 150 to 180 µg/kg DW; Rashed and Awadallah 1998); however, leaves, sugar, and juice of sugarcane (*Saccharum officinarum*) grown in Egypt contained 17 to 130 times less gold than did the soil of their sugarcane fields (Mohamed 1999).

Gold was detected in aquatic macrophytes from streams draining abandoned base-metal mines, suggesting use of these plants in biorecovery (Sabti et al. 2000). Bryophytes collected downstream of a gold mine in Wales had slightly higher concentrations of gold than did upstream samples, with a maximum value of 37 µg Au/kg DW (Samecka-Cymerman and Kempers 1998). In Poland and the Czech Republic, aquatic bryophytes reflected increased amounts of gold in a biotype with high arsenic mineralization; highest values recorded were in *Fontinalis antypyretica* (18.8 µg Au/kg DW) and *Chiloscyphus pallescens* (20.2 µg Au/kg DW) from areas of former gold mining (Samecka-Cymerman and Kempers 1998).

In the gold mining communities of Sri Lanka, peat and algal mats have been found to contain elevated concentrations of gold (Table 5.2). In peat, gold is positively correlated with increasing depth as well as with increasing concentrations of iron, manganese, cobalt, zirconium, sodium, magnesium, and potassium (Dissanayake and Kritsotakis 1984). In euryhaline algal mats, gold concentrations increase in a seaward direction, suggesting a greater geochemical mobility of dissolved gold with increasing concentrations of chloride ions.

Gold exploration in tropical or subtropical countries has indirectly accelerated efforts to understand the behavior of gold within lateritic formations (Davies 1997). Gold uptake by vegetation is a significant mechanism for mobilizing gold in tropical forests more than 100,000 years old. Pure gold dissolves only under organic conditions.

Table 5.2 Gold Concentrations in Selected Plants and Animals*

Taxonomic Group, Species, and Other Variables	Concentration	Reference[a]
Plants		
Brazil		
Vegetation; normal vs. near gold mining operations	<5 DW vs. 3–19 DW	1
Egypt		
Faba bean, *Vicia faba*; Aswan area; grown in soil containing 150–180 µg Au/kg DW		
Leaves	170 DW	2
Stems	50 DW	2
Pods	40 DW	2
Pericarp	36 DW	2
Testa, seeds, cotyledon	<7 DW	2
Germany		
Poplar, *Populus* sp. roots; hydroponic cultivation	2–28 DW	3
Coniferous trees; various; barks and twigs	nondetectable (<10 DW)	4
Japan		
Seaweeds; *Porphyra* sp. vs. *Ulva* sp.; maximum values	21 DW vs. 35 DW	5
New Zealand		
North Island; near gold mine closed in 1974, reexamined in 1999; aquatic macrophyte, *Egeria densa* from sediments containing up to 163 µg Au/kg DW	302–672 DW	6
Poland and Czech Republic		
Aquatic bryophytes; 5 species; collected spring–summer		
10 locations draining an area with high arsenic mineralization	3.4 DW	7
2 locations as above in areas of former gold mining activities	19.4 DW	7
22 reference sites	0.8 DW	7
All locations	0.4–20.2 DW	7
Reference standard; orchard leaves	2 DW	8
Sri Lanka		
Peat, *Muther agawela*	159–882 DW	9
Algal mats	486–1065 DW	9
United Kingdom		
Wales; aquatic bryophytes; downstream from gold mine	Max. 37 DW	7
Various locations		
Gold accumulator plants; *Artemisia* sp.; *Prangos* sp.; *Stripa* sp.	>100 to Max. 100,000 DW	10
Invertebrates		
Marine molluscs; soft parts		
Common mussel, *Mytilus edulis*	2–38 DW	5
Clam, *Tapes* sp.	5.7 DW	5
Crustacean; shrimp, *Pandalus* sp.; soft parts	0.28 DW	5

Table 5.2 (continued)   Gold Concentrations in Selected Plants and Animals*

Taxonomic Group, Species, and Other Variables	Concentration	Reference[a]
Fish		
Mackerel, *Pneumatophorous japonicus*, muscle	0.12 DW	5
Nonhuman Mammals		
Reference standards; bovine liver vs. nonfat milk	5–9 DW vs. 11–25 DW	8
Humans		
Blood, whole		
Uzbekistan	49–110 DW	11
Normal (from literature)	0.2–2.0 DW	11
From rheumatoid arthritis patients given sodium gold thiomalate chrysotherapy	2390 FW	12
Breast milk		
Recent mothers (N = 27); mean (range) vs. 50% quartile	0.29 (0.10-2.06) FW vs. 0.18–0.46 FW	13
From healthy mothers who had successfully given birth to mature babies after uneventful pregnancies; Grosz, Austria; 1995–1996	0.1–2.1 FW	14
Fingernails; normal children; Nigeria	20 (8–39) DW	15
Hair, scalp		
Nigeria; normal adults	47 (6–880) DW	15
Italian goldsmiths (N = 73) vs. controls (N = 22)	1440 DW vs. 670 DW	16
Infant milk formula		
Normal	<0.27 FW	13
Purchased from local Austrian supermarkets; 4 formulas	0.05–0.20 FW	14
Kidney		
Rheumatoid arthritis patients (N = 11) receiving gold+ drugs		
Time, in months, since last treatment		
<1 (3 patients)	60,000–233,000 FW (total gold of 6650–10,480 mg)	18
1–4 (4 patients)	24,000–19,000 FW (total gold of 2630–6320 mg)	19
9–21 (3 patients)	<25,000–31,000 FW (total gold of 4500–8000 mg)	18
140 (1 patient)	<42,000 FW (total gold of 260 mg)	18
Urine		
Healthy	<0.002–0.02 FW	3
Healthy (N = 43)	0.03–0.85 FW	17
Healthy (N = 21)	0.01–0.31 FW	17
Students	0.02 (0.01–0.04) FW	17
Construction workers	0.03 (0.01–0.11) FW	17

Table 5.2 (continued)   Gold Concentrations in Selected Plants and Animals*

Taxonomic Group, Species, and Other Variables	Concentration	Reference[a]
Dental technicians	0.19 (0.01–1.11) FW	17
Whole body, healthy adult	35.0 FW (total of 2.45 mg in 70-kg person)	19

* Values are in µg/L or µg/kg fresh weight (FW) or dry weight (DW).
[a] 1, Davies, 1997; 2, Rashed and Awadallah 1998; 3, Messerschmidt et al. 2000; 4, Weber et al. 1997; 5, Eisler 1981; 6, Sabti et al. 2000; 7, Samecka-Cymerman and Kempers 1998; 8, Ohta et al. 1995; 9, Dissanayake and Kritsotakis 1984; 10, Sadler 1976; 11, Zhuk et al. 1994; 12, Hirohata 1996; 13, Krachler et al. 2000; 14, Prohaska et al. 2000; 15, Oluwole et al. 1994; 16, Caroli et al. 1998; 17, Begerow et al. 1999; 18, Shakeshaft et al. 1993; 19, Merchant 1998.

The three primary gold complexes of mobilized gold are: $[Au(OH)_3 \cdot H_2O]^0$, $AuClOH^-$, and $Au(OH)_2FA^-$, where FA indicates fulvic acid from soil organic matter. These gold complexes are believed to be stable under surficial equatorial rain forest conditions, but they could be leached from soils to rivers (Davies 1997).

5.3   ANIMALS

Gold concentrations found in selected invertebrates, fish, and humans are listed in Table 5.2.

In one study, gold concentrations in soft tissues of marine invertebrates ranged between 0.3 and 38 µg Au/kg DW; for fish muscle the mean concentrations were 0.12 µg/kg on a dry weight basis and 2.6 µg/kg on an ash weight basis (Eisler 1981). Insect galls induced by egg deposition of the chalcid wasp *Hemadas nubilpennis* on shoots of the lowbush blueberry *Vaccinum angustifoloium* had elevated levels of gold and other metals in epidermal tissues, especially near the stomata (Bagatto and Shorthouse 1994). Gold comprised up to 5.4% of the total weight of gall periderm and epiderm, but was not detectable in nutritive cells or other tissues. Emissions from the nearby Sudbury, Ontario, site of the largest nickel producer in the world may have confounded the results of this study (Bagatto and Shorthouse 1994).

In humans, gold concentrations in breast milk ranged from 0.1 to 2.1 µg/L; it is speculated that the highest concentrations were due to gold dental fillings and jewelry of the mothers (Krachler et al. 2000). In dental technicians, concentrations of gold in urine were found to be significantly higher than in urine from other groups tested, i.e., students and road construction workers (Begerow et al. 1999). Dental technicians also had elevated urinary concentrations of platinum and palladium when compared with students and laborers. The comparatively high gold excretion rates of dental technicians were due to the greater number of noble-containing artificial dentures worn by that group (Begerow et al. 1999).

Gold in scalp hair of Italian goldsmiths, when compared to controls, was significantly higher (1440 µg/kg DW vs. 670 µg/kg DW). Hair from goldsmiths also contained significantly higher concentrations, in µg/kg DW, of silver (1290 vs. 400), copper (13,300 vs. 11,100), and indium (0.0016 vs. 0.0008); there were no significant differences found for cadmium, cobalt, chromium, mercury, nickel, lead, platinum, or zinc (Caroli et al. 1998). In Nigeria, gold concentrations in hair of normal adults were low (6 to 880 µg/kg DW), and there were significant positive correlations of gold with concentrations of arsenic, lanthanum, and cobalt (Oluwole et al. 1994).

Gold in whole blood of Uzbekistan residents was elevated — following a single storm event — when compared with the rest of the world (Table 5.2). Based on regional medical statistics of Uzbekistan, blood gold concentrations were positively correlated strongly with hypertension and anemia. These findings may be useful in future human health screenings (Zhuk et al. 1994). Rheumatoid arthritis patients undergoing chrysotherapy had grossly elevated concentrations of gold in blood (up to 2.4 mg/L; Hirohata 1996), and kidney (up to 233.0 mg/kg FW; Shakeshaft et al. 1993). Chrysotherapy is discussed in detail later.

5.4 SUMMARY

Maximum gold concentrations documented in abiotic materials were 0.001 µg/L in rainwater, 0.0015 µg/L in seawater near hydrothermal vents, 5.0 µg/kg dry weight (DW) in the earth's crust, 19.0 µg/L in a freshwater stream near a gold mining site, 440 µg/kg DW in atmospheric dust near a high-traffic road, 843 µg/kg DW in alluvial soil near a Nevada gold mine, 2.53 mg/kg DW in snow near a Russian smelter, 4.5 mg/kg DW in sewage sludge, 28.7 mg/kg DW in polymetallic sulfides from the ocean floor, and 256.0 mg/kg DW in freshwater sediments near a gold mine tailings pile. In plants, elevated concentrations of gold were reported in terrestrial vegetation near gold mining operations (19 µg/kg DW), in aquatic bryophytes downstream from a gold mine (37 µg/kg DW), in leaves of beans grown in soil containing 150 µg Au/kg (170 µg/kg DW), in algal mats of rivers receiving gold mine wastes (up to 1.06 mg/kg DW), and in selected gold accumulator plants (0.1 to 100 mg/kg DW). Fish and aquatic invertebrates contained 0.1 to 38.0 µg Au/kg DW. In humans, gold concentrations of 1.1 µg/L in urine of dental technicians were documented vs. 0.002 to 0.85 µg/L in urine of reference populations, 2.1 µg/L in breast milk, 1.4 mg/kg DW in hair of goldsmiths vs. a normal range of 6 to 880 µg/kg DW, 2.39 mg/L in whole blood of rheumatoid arthritis patients receiving gold thiol drug therapy (chryso-therapy) vs. a normal range of 0.2 to 2.0 µg/L blood; and 60.0 to 233.0 mg/kg fresh weight (FW) in kidneys of rheumatoid arthritis patients undergoing active chryso-therapy vs. <42.0 mg/kg FW kidney in these same patients 140 months posttreatment.

The significance of gold concentrations in various environmental compartments, gold's mode of action, and mechanisms governing its uptake, retention, and trans-location are not known with certainty. To more fully evaluate the role of gold in the biosphere, systematic measurements of gold levels is recommended in abiotic mate-rials and organisms comprising diverse multitrophic food chains using sensitive analytical methodologies. Samples should also be analyzed for various metals, metalloids, and compounds known to modify ecological and toxicological properties of gold.

LITERATURE CITED

Ahnlide, I., B. Bjorkner, M. Bruze, and H. Moller. 2000. Exposure to metallic gold in patients with contact allergy to gold sodium thiosulfate, *Contact Dermatitis*, 43, 344–350.

Awadallah, R.M., A.E, Mohamed, M.H. Abou-El-Wafa, and M.N. Rashed. 1995. Assessment of trace element concentrations in fenugreek and lupin planted in the experimental farm, High Dam Lake Development Authority, Gerf Hussein beach locality, Egypt, *Pakistan Jour. Sci. Indus. Res.*, 38, 51–60.

Bagatto, G. and J.D. Shorthouse. 1994. Mineral concentrations within cells of galls induced by *Hemadas nubilpennis* (Hymenoptera: Pteromalidae) on lowbush blueberry: evidence from cryoanalytical scanning electron microscopy, *Canad. Jour. Bot.*, 72, 1387–1390.

Barbante, C., G. Cozzi, G. Capodaglio, K. van de Velde, C. Ferrari, C. Boutron, and P. Cescon. 1999. Trace element determination in alpine snow and ice by double focusing inductively coupled plasma mass spectrometry with microconcentric nebulization, *Jour. Anal. Atomic Spectr., 14*, 1433–1438.

Begerow, J., U. Sensen, G.A. Wiesmuller, and L. Dunemann. 1999. Internal platinum, palladium, and gold exposure in environmentally and occupationally exposed persons, *Zbl. Hyg. Umweltmed.*, 202, 411–424.

Caroli, S., O. Senofonte, N. Violante, S. D'Ilio, S. Caimi, F. Chiodo, and A. Menditto. 1998. Diagnostic potential of hair analysis as applied to the goldsmith sector, *Microchem. Jour.*, 59, 32–44.

Davies, B.E. 1997. Deficiencies and toxicities of trace elements and micronutrients in tropical soils: limitations of knowledge and future research needs, *Environ. Toxicol. Chem.*, 16, 75–83.

Dissanayake, A.B. and K. Kritsotakis. 1984. The geochemistry of Au and Pt in peat and algal mats — a case study from Sri Lanka, *Chem. Geol.*, 42, 61–76.

Ehrlich, A. and D.V. Belsito. 2000. Allergic contact dermatitis to gold, *Cutis*, 65, 323–326.

Eisler, R. 1981. *Trace Metal Concentrations in Marine Organisms*. Pergamon, New York, 687 pp.

Eisler, R. 2004. Gold concentrations in abiotic materials, plants, and animals: a synoptic review, *Environ. Monitor. Assess.*, 90, 73–88.

Eisler, R., D.R. Clark Jr., S.N. Wiemeyer, and C.J. Henny. 1999. Sodium cyanide hazards to fish and other wildlife from gold mining operations, in *Environmental Impacts of Mining Activities: Emphasis on Mitigation and Remedial Measures,* J.M. Azcue, (Ed.), Springer-Verlag, Berlin, 55–67.

Elevatorski, E.A. 1981. *Gold Mines of the World*. Minobras, Dana Point, CA, 107 pp.

Fowler, J.F., Jr. 2001. Gold, *Amer. Jour. Contact Dermatitis*, 12, 1–2.

Gordeyev, V.V., A.S. Yegorov, A.P. Lisitsyn, V.S. Letokhov, D.Y. Pakhomov, and V.M. Gulevich. 1997. Dissolved gold in surface waters of the northeastern Atlantic, *Geochem. Int.,* 35, 1007–1015.

Gordeyev, V.V., A.S. Yegorov, V.N. Radayev, and I.V. Zubow. 1991. Gold as a possible tracer of hydrothermal influence on bottom waters, *Oceanology*, 31, 178–182.

Greer, J. 1993. The price of gold: environmental costs of the new gold rush, *The Ecologist*, 23 (3), 91–96.

Gregurek, D., F. Melcher, H. Niskavaara, V.A. Pavlov, C. Reimann, and E.F. Stumpfl. 1999. Platinum-group elements (Rh, Pt, Pd) and Au distribution in snow samples from the Kola Peninsula, NW Russia, *Atmospher. Environ.*, 33, 3281–3290.

Hannington, M., P. Herzig, S. Scott, G. Thompson, and P. Rona. 1991. Comparative miner-
alogy and geochemistry of gold-bearing sulfide deposits on the mid-ocean ridges,
Mar. Geol., 101, 217–248.

Herzig, P.M., M.D. Hannington, Y. Fouquet, U.V. Stackelberg, and S. Petersen. 1993. Gold-
rich polymetallic sulfides from the Lau Back Arc and implications for the geochemistry of gold in sea-floor hydrothermal systems of the southwest Pacific, *Econ. Geol.*,
88, 2182–2209.

Hirohata, S. 1996. Inhibition of human B cell activation by gold compounds, *Clin. Immunol.
Immunopathol.*, 81, 175–181.

Karamushka, V.I. and G.M. Gadd. 1999. Interaction of *Saccharomyces cerevisiae* with gold:
toxicity and accumulation, *Biometals*, 12, 289–294.

Kirkemo, H., W.L. Newman, and R.P. Ashley. 2001, *Gold*. U.S. Geological Survey, Denver,
23 pp.

Kist, A.N. 1994. Investigation of element speciation in atmosphere, *Biol. Trace Elem. Res.*,
43-54, 259–266.

Korte, F., M. Spiteller, and F. Coulston. 2000. The cyanide leaching gold recovery process is
a nonsustainable technology with unacceptable impacts on ecosystems and humans:
the disaster in Romania, *Ecotoxicol. Environ. Safety*, 46, 241–245.

Krachler, M., T. Prohaska, G. Koellensperger, E. Rossipal, and G. Stingeder. 2000. Concentrations of selected trace elements in human milk and in infant formulas determined
by magnetic sector field inductively coupled plasma-mass spectrometry, *Biol. Trace
Elem. Res.*, 76, 97–112.

Lee, E.E. and H.L. Maibach. 2001. Is contact allergy in man lifelong? An overview of patch
test follow-ups, *Contact Dermatitis*, 44, 137–139.

Leybourne, M.I., W.D. Goodfellow, D.R. Boyle, and G.E.M. Hall. 2000. Form and distribution
of gold mobilized into surface waters and sediments form a gossan tailings pile,
Murray Brook massive sulphide deposit, New Brunswick, Canada, *Appl. Geochem.*,
15, 629–646.

Lottermoser, B.G. 1995. Noble metals in municipal sewage sludges of southeastern Australia,
Ambio, 24, 354–357.

McBride, M.B., B.K. Richards, T. Steenhuis, J.J. Russo, and S. Sauve. 1997. Mobility and
solubility of toxic metals and nutrients in soil fifteen years after sludge application,
Soil Sci., 162, 487–500.

Merchant, B. 1998. Gold, the noble metal and the paradoxes of its toxicology, *Biologicals*,
26, 49–59.

Messerschmidt, J., A. von Bohlen, F. Alt, and R. Klockenkamper. 2000. Separation and
enrichment of palladium and gold in biological and environmental samples, adapted
to the determination by total reflection X-ray fluorescence, *Analyst*, 125, 397–399.

Miller, J.R., J. Rowland, P.J. Lechler, M. Desilets, and L.C. Hsu. 1996. Dispersal of mercury-
contaminated sediments by geomorphic processes, Sixmile Canyon, Nevada, USA:
implications to site characterization and remediation of fluvial environments, *Water
Air Soil Pollut.*, 86, 373–388.

Mohamed, A. 1999. Environmental variations of trace element concentrations in Egyptian
cane sugar and soil samples (Edfu factories), *Food Chem.*, 65, 503–507.

Moss, R., S. Scott, and R.A. Binns. 1997. Concentrations of gold and other ore metals in
volcanics hosting the Pacmanus seafloor sulfide deposit, *JAMSTEC Jour. Deep Sea
Res.*, 13, 257–267.

Ohta, K., T. Isiyama, M. Yokoyama, and T. Mizuno. 1995. Determination of gold in biological
materials by electrothermal atomic absorption spectrometry with a molybdenum tube
atomizer, *Talanta*, 42, 263–267.

Oluwole, A.F., J.O. Ojo, M.A. Durosinmi, O.I. Asubiojo, O.A. Akanle, N.M. Spyrou, and R.H. Filby. 1994. Elemental composition of head hair and fingernails of some Nigerian subjects, *Biol. Trace Elem. Res.*, 43-45, 443–452.

Prohaska, T., G. Kollensperger, M. Krachler, K. De Winne, G. Stingeder, and L. Moens. 2000. Determination of trace elements in human milk by inductively coupled plasma sector field mass spectrometry (ICP-SFMS), *Jour. Anal. Atomic Spectr.*, 15, 335–340.

Puddephatt, R.J. 1978. *The Chemistry of Gold*. Elsevier, Amsterdam, 274 pp.

Pyatt, F.B. 1999. Comparison of foliar and stem bioaccumulation of heavy metals by Corsican pines in the Mount Olympus area of Cyprus, *Ecotoxicol. Environ. Safety*, 42, 57–61.

Rashed, M.N. and R.M. Awadallah. 1998. Trace elements in faba bean (*Vicia faba* L) plant and soil as determined by atomic absorption spectroscopy and ion selective electrodes, *Jour. Sci. Food Agric.*, 77, 18–24.

Sabti, H., M.M. Hossain, R.R. Brooks, and R.B. Stewart. 2000. The current environmental impact of base-metal mining at the Tui Mine, Te Aroha, New Zealand, *Jour. Roy. Soc. N.Z.*, 30, 197–208.

Sadler, P.J. 1976. The biological chemistry of gold: a metallo-drug and heavy-atom label with variable valency, *Structure Bonding*, 29, 171–215.

Samecka-Cymerman, A. and A.J. Kempers. 1998. Bioindication of gold by aquatic bryophytes, *Acta Hydrochim. Hydrobiol.*, 26, 90–94.

Shakeshaft, J., A.K. Clarke, M.J. Evans, and S.C. Lillicrap. 1993. X-ray fluorescence determination of gold *in vivo*, in *Human Body Composition*, J.D. Eastman and K.J. Ellis, (Eds.), Plenum, New York, 307–310.

Suarez, I., M. Ginarte, V. Fernandez-Redondo, and J. Toribio. 2000. Occupational contact dermatitis due to gold, *Contact Dermatitis*, 43, 367–368.

Terashima, S., H. Katayama, and S. Itoh. 1991. Geochemical behavior of gold in coastal marine sediments from the southeastern margin of Japan Sea. *Mar. Mining*, 10, 247–257.

Terashima, S., S. Nakao, N. Mita, Y. Inouchi, and A. Nishimura. 1995. Geochemical behavior of Au in terrigenous and pelagic marine sediments, *Appl. Geochem.*, 10, 35–44.

Tsuruta, K., K. Matsunaga, K. Suzuki, R. Suzuki, H. Akita, Y. Washimi, A. Tomitaka, and H. Ueda. 2001. Female predominance of gold allergy, *Contact Dermatitis*, 44, 55–56.

Vamnes, J.S., T. Morken, S. Helland, and N.R. Gjerdet. 2000. Dental gold alloys and contact hypersensitivity, *Contact Dermatitis*, 42, 128–133.

Van de Velde, K., C. Barbante, G. Cozzi, I. Moret, T. Bellomi, C. Ferrari, and C. Boutron. 2000. Changes in the occurrence of silver, gold, platinum, palladium and rhodium in Mont Blanc ice and snow since the 18th century, *Atmospher. Environ.*, 34, 3117–3127.

Weber, V.A., G. Lehrberger, and G. Morteani. 1997. Gold und Arsen in Pilzen, Moosen und Baumnadeln — biogeochemische Aspekte einer "geogenen Atlast„ im Moldanubikum des Oberpfalzer Waldes bei Oberviechtach, *Geol. Bavarica*, 102, 229–250 (in German, English summary).

Zhuk, L.I., I.N. Mikholskaya, E.A. Danilova, and A.E. Kist. 1994. Mapping using human blood composition data, *Biol. Trace Elem. Res.*, 43-45, 371–381.

CHAPTER 6

The Effects of Gold on Plants and Animals

Lethal and sublethal effects of Au^0, Au^+, and Au^{+3} are summarized for aquatic organisms and laboratory mammals. Gold accumulations from solution are documented for microorganisms and other living resources under various physicochemical conditions.

6.1 AQUATIC ORGANISMS

This section summarizes lethal and sublethal effects of Au^+ and Au^{+3} on aquatic microorganisms, plants, fishes, and amphibians.

6.1.1 Monovalent Gold

Monovalent gold is toxic to aquatic biota at comparatively elevated concentrations of 7.9 mg Au/L and higher (Nomiya et al. 2000). Toxicity of gold to microorganisms is affected by concentration and oxidation state of gold, presence of competing metal ions in solution, pH, and composition of the growth medium (Savvaidis et al. 1998). Exposure to gold may induce cell adaptation and cell resistance, as has been demonstrated for monovalent gold chloride, sodium aurothiomalate, and auranofin. Cellular adaptation is a potential mechanism for gold resistance (Savvaidis et al. 1998). Antimicrobial activities of two isomeric Au^+-triphenylphosphine compounds were documented for two species of Gram-positive bacteria (*Bacillus subtilis*, *Staphylococcus aureus*) and one species of yeast (*Candida albicans*) at concentrations as low as 7.9 mg Au^+/L for bacteria and 250.0 mg/L for yeast (Nomiya et al. 2000). Growth inhibition of *Tetrahymena pyriformis*, a ciliate protozoan, is reported after 24 hours in 99 to 296 mg Au^+/L (as gold sodium aurothiomalate), and prolonged cell generation time at 390 to 2960 mg/L in 24 hours (Nilsson 1993). At 1576 mg Au^+/L, no cells died in 24 hours; although endocytosis and cell proliferation were inhibited; after 2 days however, the cell density of the culture was sufficiently high to permit recovery (Nilsson 1993). Exposure of *Tetrahymena* to 3050 mg Au^+/L (as gold sodium aurothiomalate) for 24 hours, equivalent to eight normal cell generations, resulted in a growth reduction of 50% and visible amounts of gold accumulated (Nilsson 1997). Gold remained detectable for at least 24 hours. After dilution to a low cell density, gold turnover was slow except in rapidly proliferating cells. The protozoan recovers fully after heavy accumulation of Au^+, but only in low-density cultures. Proliferating *Tetrahymena* have a high metabolic rate associated with high lysosomal enzyme activity, which are presumed to be the prerequisite for a rapid turnover of accumulated gold (Nilsson 1997). The electrochemiluminescence response induced from body fluids and homogenized tissues of American oysters (*Crassostrea virginica*) and several species of tunicates (*Molgula occidentalis, Styela plicata, Diplosoma macdonaldi*) was severely inhibited in a dose-dependent fashion by monovalent gold ions and other strongly oxidizing metal ions — especially Ag^+, Cu^{+2}, and Hg^{+2} — at concentrations of 100 mg/L and higher (Bruno et al. 1996).

Intact single fibers of skeletal muscle of bullfrogs (*Rana catesbeiana*) were subjected to varying concentrations of Au^+ as gold sodium thiomalate. At 500 μM (98.5 mg/kg), Au^+ decreased tension amplitude by 27% after 30 minutes, and resting membrane potential by 5.3% after 22 minutes (Oba et al. 1999). Results suggest that Au^+, as gold sodium thiomalate, could be used as an antirheumatic drug without severe side effects on skeletal muscle and that coexistent thiomalate probably contributes to the protection of muscle function from the side effects of Au^+ (Oba et al. 1999).

6.1.2   Trivalent Gold

Trivalent gold is significantly more toxic to aquatic biota than monovalent gold. $Gold^{+3}$, as tetrachloroaurate ($AuCl_4^-$), depressed chlorophyll concentrations, photosynthetic rates, and thiol levels at concentrations greater than 98.5 μg Au^{+3}/L over a 21-day period in *Amphora coffeaeformis*, a marine diatom (Robinson et al. 1997). Cells were able to recover at concentrations less than 985 μg Au^{+3}/L due to cellular and photoreduction of the $AuCl_4^-$. Adverse effects were exacerbated by Cu^{+2}. Uptake of Au^{+3} by *Amphora* is apparently not an energy-dependent process. At 394 to 985 μg Au^{+3}/L, only 30% of the total gold uptake after 24 h was internal, although increased uptake by heat-killed cells and uptake by illuminated cells suggest otherwise. It was concluded that algal cells, alive or dead, rapidly accumulated Au^{+3} and begin to reduce it to Au^0 and Au^+ within 2 days (Robinson et al. 1997).

Growth inhibition of yeast (*Saccharomyces cerevisiae*) was observed in 40 hours at the lowest concentration tested of 20 mg Au^{+3}/L, with no growth observed at 50 mg/L. Both calcium and magnesium enhanced the inhibitory effect of gold on the yeast cells (Karamushka and Gadd 1999).

Results of acute toxicity bioassays of 96 hours' duration with adults of *Fundulus heteroclitus*, an estuarine cyprinodontiform killifish, and salts of various metals and metalloids showed that gold, as auric chloride (Au^{+3}), was comparatively lethal, with 50% dead in 96 hours at <0.8 mg/L. The relative order of lethality, with silver (Ag), most toxic and lithium (Li) least toxic was: Ag^+, Hg^{+2}, Au^{+3}, Cd, followed by As^{+3}, Be, Al, Cu, Zn, Y, Tl, Fe, La, Cr^{+6}, Ni, Co, Sb, and Li. Salts of 13 additional elements tested to *Fundulus* were less toxic than were salts of Li, including Rb, Si, Mo, Re, Ba, Mn, Ca, Sr, K, and Na, in that order (Eisler 1986, unpublished).

When bullfrog skeletal muscle fibers previously pretreated with 98.5 mg/kg Au^+ (as gold sodium thiomalate) were subjected to 2.0 mg Au^{+3}/kg (as $NaAuCl_4$), the fibers lost their ability to contract upon electrical stimulation, as was the case for 2.0 mg Au^{+3}/kg alone (Oba et al. 1999). However, in the presence of thiomalic acid, Au^{+3} did not completely block tetanus tension, even at 10 mg Au^{+3}/kg. Thiomalic acid also inhibited Au^{+3}-induced membrane depolarization (Oba et al. 1999). In bullfrogs, skeletal muscle fibers spontaneously produced phasic and tonic contractures upon addition of 5 to 20 μM Ag^+ or more than 50 μM Au^{+3} (9 mg Au^{+3}/L; Nihonyanagi and Oba 1996). Simultaneous application of 5 μM Ag^+ and 20 μM Au^{+3} inhibited contractures induced by Ag^+. Trivalent gold applied immediately after development of Ag^+-induced contractures shortened the duration of the phasic contracture and markedly decreased the tonic contracture through modification of the Ca^{+2} release channel. It was concluded that extracellular Au^{+3} at comparatively low concentrations inhibits the silver (Ag^+)-induced contractions in skeletal muscle and that intracellular Au^{+3} activates the sarcoplasmic reticulum Ca^{+2} release channel to partially contribute to the tonic contractions (Nihonyanagi and Oba 1996).

6.2 ACCUMULATION

Extraction of gold from solutions is under active investigation using a variety of physical, chemical, and biological processes. Recovery of ionic gold from dilute solutions usually involves either precipitation by zinc dust, carbon adsorption, solvent extraction, or ion exchange resins. All of these are of low selectivity and comparatively expensive (Suyama et al. 1996; Niu and Volesky 1999). Chemical methods for the recovery of gold from ores include cyanidation and thiourea leaching, which present environmental and health risks (Eisler et al. 1999; Gardea-Torresdey et al. 2000; Fields 2001). Biorecovery of dissolved gold from solution presents fewer environmental risks than chemical methods, and is documented for microorganisms (Puddephatt 1978; Kai et al. 1992; Lindstrom et al. 1992; Claassen 1993; Maturana et al. 1993; Agate 1996; Tsezos et al. 1996; Xie et al. 1996; Pethkar and Paknikar 1998; Rawlings 1998; Savvaidis 1998; Savvaidis et al. 1998; Gonzalez et al. 1999; Karamushka and Gadd 1999; Niu and Volesky 1999, 2000; Gardner and Rawlings 2000; Kashefi et al. 2001), algae (Ting et al. 1995; Savvaidis et al. 1998), water ferns (Antunes et al. 2001), peat (Wagener and Andrade 1997), alfalfa (Gardea-Torresdey et al. 2000), seaweeds (Kuyucak and Volesky 1989; Zhao et al. 1994; Niu and Volesky 1999, 2000), fungi (Gomes and Linardi 1996; Gomes et al. 1998, 1999a; Pethkar and Paknikar 1998; Niu and Volesky 1999, 2000; Ting and Mittal 1999); yeasts (Savvaidis 1998), crab exoskeletons (Niu and Volesky 2001), and chicken feathers and other animal fibrous proteins (Suyama et al. 1996; Ishikawa and Suyama 1998). This section briefly reviews the potential of living and dead plants and animals to accumulate gold from solution, and some of the processes involved — including biooxidation, dissolution, bioreduction, bacterial leaching, and biosorption.

6.2.1 Microorganisms, Fungi, and Higher Plants

Biomining processes are used successfully on a commercial scale for the recovery of gold and other metals, and are based on the activity of obligate chemoauto-lithotrophic bacteria that use iron or sulfur as their energy source and grow in highly acidic media (Rawlings 1998). Biooxidation of difficult to treat gold-bearing arse-nopyrite ores occurs in aerated, stirred tanks and rapidly-growing, arsenic-resistant bacterial strains of *Thiobacillus ferrooxidans*, *Leptospirillium ferrooxidans*, and *Thiobacillus thiooxidans*. These bacterial species obtain their energy through the oxidation of ferrous to ferric iron (*T. ferrooxidans*, *L. ferrooxidans*) or through the reduction of inorganic sulfur compounds to sulfate (*Thiobacillus* spp.). Monetary costs of biooxidation are reported to be about 50% lower than roasting or pressure oxidation (Agate 1996; Rawlings 1998). Adding *Thiobacillus ferrooxidans* into the thiourea leaching solution produces a 20% increase in the extraction of gold. The reaction describing gold dissolution in an acidic solution of thiourea in the presence of ferric ion is described by Kai et al. (1992) as:

$$Au^0 + Fe^{+3} + 2CS(NH_2)_2 \rightarrow Au[CS(NH_2)_2]_2^+ + Fe^{+2}$$

The use of bacteria in pretreatment processes to degrade recalcitrant gold-bearing arsenopyrite ores and concentrates is well established (Lindstrom et al. 1992; Agate 1996; Gardner and Rawlings 2000). Recalcitrant ores are those in which the gold is enclosed in a matrix of pyrite and arsenopyrite, and cannot be solubilized by direct cyanidation. Bacterial decomposition of arsenopyrite assists in opening the molec-ular mineral structure, permitting access of the gold to cyanide. However, greater quantities of cyanide are required to solubilize gold after bacterial treatment when ores contain high quantities of gold. A possible cause of this excessive cyanide is the presence of the enzyme rhodanese, produced by *Thiobacillus caldus*, a common species of bacterium encountered in biooxidation facilities (Gardner and Rawlings 2000). Optimum microbiological leaching by *Thiobacillus* spp. and *Sulfolobus* spp. of refractory sulfide ores for recovery of gold in tanks is possible under controlled conditions of pH, dissolved oxygen, carbon dioxide, sulfur balance, redox potential, toxic metal concentrations, and rate of leaching (Lindstrom et al. 1992).

In one case, refractory gold-bearing sulfides scavenged from cyanidation tailings of an Ontario, Canada, gold mine produced a pyrite–arsenopyrite concentrate at a rate of 15 tons daily, containing about 30 g Au/t (Chapman et al. 1993). The high arsenic sulfide concentrate (7.9% As, 28.9% S) was amenable to biooxidation treat-ment to enhance gold extraction, with gold extraction enhanced from about 5% for the pretreated flotation concentrate to >90% for the final bioleached product (Chap-man et al. 1993).

Several species of Fe^{+3}-reducing bacteria (*Bacteria* spp., *Archaea* spp.) can precipitate gold by reducing Au^{+3} to Au^0 with hydrogen as the electron donor (Kashefi et al. 2001). Rate of bacterial oxidation by *Thiobacillus ferrooxidans* and *Leptospir-illium ferrooxidans* of three South African refractory gold ores of varying gold–arsenopyrite composition was dependent mainly on crystal structure (Claassen 1993). These gold ores were classified as refractory due to the presence of gold inclusions in arsenopyrite and pyrite, and submicroscopic gold mainly in arsenopyrite. Refractory gold occurs at sites that are preferentially leached by the bacteria. The rate of gold liberation from sulfides is enhanced during the early stages of bacterial oxidation. Defects in crystal structure influence the rate of biooxidation and are directly related to the crystal structure of the sulfide mineral, the crystallographic orientation of the exposed surfaces, and differences in chemical composition and mechanical deviations in the crystals (Claassen 1993). Pretreatment of refractory gold concentrates with the bacterium *Thiobacillus ferrooxidans* ultimately results in sulfur and sulfide oxidation by ferric ions from bacterial oxidation of ferrous ions. The maximum concentration of attached *Thiobacillus* increases with increasing concentration of Fe^{+2} and decreases with increasing size of the refractory gold concentrate particles (Gonzalez et al. 1999). In Chile, which produced 30,000 kg of gold in 1990, *Thiobacillus ferrooxidans* was used to recover gold from a complex ore under laboratory conditions (Maturana et al. 1993). The ore contained 8.2% Fe, 0.78% Cu, 0.88% As, and 3.5 g Au/t, with pyrite, hematite, arsenopyrite, and chalcopyrite as the main metal-bearing minerals. Initial gold recovery by conventional cyanidation on a crushed ore sample was 54%; concentration by flotation improved recovery to 56%. Concentrated samples (17.0 g Au/t) were leached in reactors at pH 1.8. In the presence of bacteria, all dissolved iron was present as ferric ion; gold recovery by cyanidation increased from 13% for the initial concentrate to 97% after 10 days of bacterial leaching. To further increase gold recovery, flotation tailings were submitted to cyanidation (Maturana et al. 1993).

Some microorganisms isolated from gold-bearing deposits are capable of dissolving gold; dissolution was aided by the presence of aspartic acid, histidine, serine, alanine, glycine, and metal oxidants (Puddephatt 1978). Bacteriform gold is well known, with uptake of Au^{+3} from chloride solutions documented for at least seven genera of freshwater cyanobacteria (Dyer et al. 1994). Some bacteriform gold is biogenic — the result of precipitation by bacteria — and may be a useful indicator of gold deposits and of processes of gold accumulation. *Plectonema terebrans*, a species of filamentous marine cyanobacteria, accumulates gold in its sheath from an aqueous solution of $AuCl_3$. Sheaths are among the few structures likely to be preserved in some form in microfossils of ancient bacteria. In marine media, it is expected that $AuCl_3$ (2.0 g Au/L) will form $AuCl_4^-$, AuO_2^-, and $AuCl_2^-$ (Dyer et al. 1994). Biosorption of Au^{+3}, as $AuCl_4^-$, by dried *Pseudomonas* strains of bacteria was inhibited by palladium, as Pd^{+2}, and possibly other metal ions (Tsezos et al. 1996).

Gold adsorption from cyanide solutions by dead biomass of bacteria (*Bacillus subtilis*), fungus (*Penicillium chrysogenum*), or seaweed (*Sargassum fluitans*) at pH 2 were 1.8 g Au/kg DW for bacteria, 1.4 g/kg DW for fungus, and 0.6 g Au/kg DW for seaweed. Anionic $AuCN_2^-$ adsorption was the major mechanism in gold biosorption from cyanide solutions, being most efficient at lower pH values (Niu and Volesky 1999). L-cysteine increased gold–cyanide biosorption of *Bacillus*, *Penicillium*, and *Sargassum* (Niu and Volesky 2000). At pH 2, the maximum gold uptakes were 4.0 g Au/kg DW for bacteria, 2.8 g/kg for fungus, and 0.9 g/kg for seaweed, or 150 to 250% greater than in the absence of cysteine. The anionic gold cyanide species were adsorbed by ionizable functional groups on cysteine-loaded biomass; deposited gold could be eluted from gold-loaded biomass at pH 5.0 (Niu and Volesky 2000).

Gold-resistant strains of bacteria that also accumulate gold are documented, although the fundamental mechanism of resistance to gold in microorganisms is neither known nor understood (Savvaidis et al. 1998). One strain of *Burkholderia* (*Pseudomonas*) *cepacia* contained millimolar concentrations of Au^+ thiolates. *Burkholderia* cells were large, accumulated polyhydroxybutyrate and gold, and excreted thiorin, a low-molecular-weight protein, into the culture medium. This effect was not observed with the Au^{+3} complexes tested, which were reduced to metallic gold in the medium. Gold-resistant strains of fungi and heterotrophic bacteria are also known (Savvaidis et al. 1998).

Rapid recovery of gold from gold–thiourea solutions was documented for waste biomass of yeasts (*Saccharomyces cerevisiae*), cyanobacteria (*Spirulina platensis*), and bacteria (*Streptomyces erythralus*; Savvaidis 1998). The process is pH-dependent for yeast and bacteria, and pH-independent for *Spirulina*. Of all strains of microorganisms examined, *Spirulina platensis* has the highest affinity and capacity for gold, even at low pH values. Gold uptake by *Spirulina* was 7.0 g Au/kg biomass DW in 1 to 2 hours at pH 2.0, and about 3.0 g Au/kg DW in 15 minutes at pH 2 through 7 (Savvaidis 1998).

Metabolically active fungal cells of *Aspergillus fumigatus* and *A. niger* removed gold from cyanide leach liquor of a Brazilian gold extraction plant more efficiently than did dried fungal biomass or other species of *Aspergillus* tested. These two species of fungi removed 35 to 37% of gold from solutions containing 2.8 mg Au/L in 84 hours (Gomes and Linardi 1996). Gold removal from cyanide-containing solutions is documented for a strain of *Aspergillus niger*, a fungus isolated from the gold extraction plant at Nova Linda, Brazil (Gomes et al. 1996, 1998, 1999a). The leach liquor contained, in mg/L, 181.0 cyanide, 1.3 gold, 0.4 silver, 7.1 copper, 5.2 iron, and 4.5 zinc. After 60 to 72 hours of incubation, *A. niger* removed from solution, probably by adsorption, 64% of the gold, 100% of the silver, 59% of the copper, 80% of the iron, and 74% of the zinc; all gold was removed after 120 hours. Use of this fungus to develop a bioprocess to reduce metal and cyanide levels, as well as recovery of valuable metals, shows promise (Gomes et al. 1998, 1999a, 1999b). Uptake patterns of gold from Au^{+3} solutions by dead fungal biomass followed mathematical uptake models of Langmuir and Freundlich; biomass was prepared from the fruiting body of a mushroom collected from the forests of Kerala, India (Ting and Mittal 1999). Dried fungus, *Cladosporium cladosporoides*, mixed with keratinous material of natural origin to form a bead, proved effective in absorbing gold from solution (Pethkar and Paknikar 1998). The biosorbent beads adsorbed 100.0 g Au/kg beads from a solution containing 100.0 mg Au/L. Maximum biosorption of 80% occurred at acid pH (1 to 5) in less than 20 minutes. The biosorbent beads degraded in soil in about 140 days. The beads also removed 55% of the gold from electroplating solutions containing 46.0 mg Au/L, with observed gold loading capacity of 36.0 g/kg beads (Pethkar and Paknikar 1998). Dried biosorbents encapsulated in polysulfone were prepared from microorganisms isolated from pristine or acid mine drainage environments (Xie et al. 1996). Biosorbent material rich in exopolysaccharides from the acid mine drainage site bound Au^{+3} three times more effectively than did other materials, and removed 100% of the Au^{+3} from solutions containing 1.0 mg Au/L within 16 hours at 23°C and pH 3.0.

Algal cells, alive or dead, rapidly accumulate Au^{+3} and begin to reduce it to Au^0 and Au^+ within 2 days (Robinson et al. 1997). Uptake of Au^{+3} by *Chlorella vulgaris*, a unicellular green alga, from solutions containing 10.0 or 20.0 mg Au^{+3}/L is documented (Ting et al. 1995). *Chlorella* accumulated up to 16.5 g Au/kg DW. Inactivating the algal cells by various treatments resulted in some enhancement in uptake capacity over the pristine cells. Inactivation by heat treatment yielded up to 18.8 g/kg DW; for alkali treatment, this was 20.2 g/kg DW; for formaldehyde treatment, 25.5 g/kg DW; and for acid treatment, 25.4 g/kg DW. Elemental gold (Au^0) was measured by x-ray photoelectron spectroscopy on the cell surface, indicating that a reduction had occurred (Ting et al. 1995). Studies with living *Chlorella vulgaris* suggest that accumulated Au^{+3} is rapidly reduced to Au^+, followed by a slow reduction to Au^0 (Savvaidis et al. 1998). With dead algae, Au^0 initiates a seeding process that results in the formation of elemental gold.

Sequestering metal ions using living or dead plants is a proposed economical means of removing gold and other metals via intracellular accumulation or surface adsorption. However, in the case of live plants, this is frequently a relatively slow and time-consuming process. Nonliving plant material for surface adsorption offers several advantages over live plants, including reduced cost, greater availability, easier regeneration, and higher metal specificity (Gardea-Torresdey et al. 2000). In South African mining effluents, gold usually ranges between 1.0 and 10.0 mg/L. In studies of 180-minute duration, dried red water ferns, *Azolla filiculoides*, removed 86 to 100% of Au^{+3} from solutions containing 2.0 to 10.0 mg Au^{+3}/L; removal increased with increasing initial concentration of Au^{+3} (Antunes et al. 2001). The biomass gave > 95% removal efficiency at all biomass concentrations measured. Optimum (99.9%) removal of gold occurred within 20 minutes at pH 2, 42% removal at pH 3 and 4, 63% at pH 5, and 73% removal at pH 6; removal efficiency seemed independent of temperature (Antunes et al. 2001). Similar results were observed by Zhao et al. (1994) with four species of ground dried seaweeds (*Sargassum* sp., *Gracilaria* sp., *Eisenia* sp., and *Ulva* sp.). Treated seaweeds removed 75 to 90% of the gold within 60 minutes at pH 2 from solutions containing 5.0 mg Au^{+3}/L. Gold (Au^{+3}) can be sequestered from acid solutions by dead biomass of a brown alga, *Sargassum natans*, and deposited in its elemental form, Au^0 (Kuyucak and Volesky 1989). The cell wall of *Sargassum* was the major locale for gold deposition, with carbonyl groups ($C = O$) playing a major role in binding, and N-containing groups a lesser role. Like activated carbon, the biomass of *Sargassum natans* is extremely porous, reportedly more than most biomaterials, and accounts, in part, for its ability to accumulate gold (Kuyucak and Volesky 1989). Dried ground shoots of alfalfa, *Medicago sativa*, were effective in removing gold from solution (Gardea-Torresdey et al. 2000). The accumulation process involved the reduction of Au^{+3} to colloidal Au^0, and was most efficient at elevated temperatures and acid pH. In solutions containing 60.0 mg Au^{+3}/L, about 90% of the Au^{+3} was bound to dried alfalfa shoots in about 2 hours at pH 2 and 55°C. The mechanisms to account for this phenomenon are unknown but may involve reduction of Au^{+3} to Au^+, the latter being unstable in water to form Au^0 and Au^{+3} (Gardea-Torresdey et al. 2000). Dried peat from a Brazilian bog accumulated up to 84.0 g Au/kg DW within 60 minutes from solutions containing 30.0 mg Au^{+3}/L (Wagener and Andrade 1997).

6.2.2   Aquatic Macrofauna

Except for crab exoskeletons, gold recovery from the medium by various species of living molluscs, crustaceans, and fishes is negligible (Eisler 2003). ─

Certain chitinous materials, such as exoskeletons of the swamp ghost crab, *Ucides cordatus*, can remove and concentrate gold from anionic gold cyanide solutions over a wide range of pH values (Niu and Volesky 2001). The maximum $AuCN_2^-$ uptake occurred at pH 3.7, corresponding to a final value of 4.9 g Au/kg DW; exoskeletons burned in a non-oxidizing atmosphere removed 90% of the gold at pH 10. Phenolic groups created during the heat treatment seemed to be the main functional group responsible for $AuCN_2^-$ binding by burned, acid-washed crab shells (Niu and Volesky 2001).

Bioconcentration factors (BCFs) were recorded for carrier-free [198]Au+ (physical half-life of 2.7 days) in freshwater organisms after immersion for 21 days in a medium containing 25,000 pCi/L = 675,700 Bq/L (Harrison 1973). In goldfish, *Carassius auratus*, the highest BCFs measured were <1 in muscle (i.e., less than 675,700 Bq/kg FW muscle), 10 in viscera, and 9 in whole fish. In the freshwater winged floater clam, *Anodonta nuttalliana*, the maximum BCF was 7 in soft parts; for crayfish (*Astacus* sp.), BCFs were <1 in muscle and 14 in viscera. For marine organisms immersed for 26 days in synthetic seawater containing 33,000 pCi/L = 891,900 Bq/L, maximum BCFs measured were 4 in muscle and 16 in viscera of the red crab, *Cancer productus*, 11 in soft parts of the butter clam, *Saxidomus giganteus*, 12 in soft parts of the common mussel, *Mytilus edulis*, and <1 in muscle and 1 in a whole gobiid fish, the longjaw mudsucker, *Gillichthys mirabilis* (Harrison 1973). Maximum stable gold concentrations recorded in soft tissues of marine molluscs and crustaceans ranged from 0.3 to 38.0 µg Au/kg DW; for fish muscle, the mean concentrations were 0.1 µg/kg DW and 2.6 µg/kg ash weight (Eisler 1981). In studies with the American oyster, *Crassostrea virginica*, the blue crab, *Callinectes sapidus*, and the mummichog *Fundulus heteroclitus*, an estuarine cyprinodontiform fish, all species were exposed in cages under field conditions to sediment-sorbed, carrier-free [198]Au+ (Duke et al. 1966). The maximum level of radiogold in the caged organisms was detected in oysters 17 hours after contact with [198]Au-spiked sediments. Indigenous organisms collected 41 hours after contact with the [198]Au-labeled sediments contained no detectable radioactivity (Duke et al. 1966). In a 25-day study with blue crab, northern quahog clam *Mercenaria mercenaria*, and the sheepshead minnow *Cyprinodon variegatus*, all species were maintained in a 1000-L aquarium containing bentonite clay and seawater spiked with carrier-free [199]Au (physical half-life of 3.2 days) as $AuCl_3$; crabs accumulated the most radioactivity, followed by clams, clay, and fish, in that order (Duke et al. 1966).

Bioconcentration factors (BCFs) for metals and aquatic organisms derived from carrier-free radiotracers in the medium are probably artificially high and should be interpreted with caution (Eisler 1981, 2000). For metals, it is a general observation that high BCFs are associated with low concentrations in the medium, and that BCFs are especially high when they are derived from carrier-free radioisotopes. Typically, BCFs for metals ─ and other chemicals studied ─ reach a plateau before declining with increasing concentrations in solution (Eisler 1981, 2000). The maximum concentration of stable gold measured in tissues of living marine organisms was 38.0 μg/kg FW (Eisler 1981).

6.2.3 Animal Fibrous Proteins

Gold recovery is proposed using animal fibrous proteins such as egg shell membrane, chicken feathers, wool, silk, elastin, and other stable water-soluble fibers with high surface area (Suyama et al. 1996; Ishikawa and Suyama 1998). All animal fibrous proteins tested accumulated gold–cyanide ion from aqueous solution. Adsorption was highest at pH 2; accumulations were up to 9.8% of the dry weight for wool, 8.6% for egg shell membrane, 7.1% for chicken feathers, and <3.9% for other materials. In the case of egg shell membrane, adsorbed gold was desorbed with 0.1 M NaOH and the material can be used repeatedly. Egg shell membrane could remove gold–cyanide ion at concentrations near 1 μg/L.

6.3 LABORATORY MAMMALS

No satisfactory animal model studies exist that show the same responses to gold complexes as those of human rheumatoid arthritis patients (Brown and Smith 1980). The models generally used included rats with adjuvant arthritis and resistance to penicillamine, rats with kaolin paw edema, and guinea pigs with erythema. In animal gold studies — as in human gold studies — gold was widely distributed in tissues, with major gold accumulations in kidney, liver, spleen, skin, lymph, and bone marrow. Significant gold accumulations were found in most other tissues examined, including brain. In rat liver cells, gold uptake from sodium gold thiomalate was via membrane binding to lysosomes, possibly to thiols; however, in blood plasma, it was complexed to albumin. And in guinea pigs, different gold distributions occurred depending on oral or parenteral route of administration (Brown and Smith 1980).

6.3.1 Metallic Gold

Submicroscopic gold particles (0.05 to 0.10 microns in diameter) in colloidal suspension when injected intravenously (i.v.) into rabbits (*Oryctolagus* sp.) at 2 mg/kg BW (total dose of 6 to 8 mg Au) produced significant elevation of rectal temperatures over a 7-hour postinjection observation period (Eisler et al. 1955). Similar observations were recorded with colloidal suspensions of glass, iron oxide, quartz, and thorium dioxide. The fine state of particle division, shown by all materials tested, was the factor which rendered them thermogenic (Eisler et al. 1955). The distribution of colloidal gold coupled with albumin within lymph nodes of rats up to 10 hours following intrapleural injection was studied (Glazyrin et al. 1995) using x-ray fluorescence analysis-synchrotron radiation beams (XFA-SR). Potentially, XFA-SR can detect very low concentrations of gold and other elements, and microscopical SR analysis can demonstrate differences in elemental concentrations within single cells.

Gold appeared in the lysosomes of the follicular reticular cells 4 hours postinjection; colloidal gold concentrations in the node periphery were maximal after 6 to 8 hours (Glazyrin et al. 1995).

Dose enhancement in tumor therapy is reported at interfaces between high- and low-atomic-number materials, which is significantly intense for low-energy photon beams. Gold microspheres suspended in cell culture or distributed in tumorous tissues exposed to kilovoltage beams produced an increased biologically effective dose, with increasing tumor cell death related to increasing concentration of microspheres 1.5 to 3.0 μm in diameter; the mean effective dose increase in solutions that contained 1% gold particles was 42 to 43% for 200 kv x-rays (Herold et al. 2000).

Tissue injury in mice from intraperitoneal (ip) insertion of gold implants initiated an inflammatory response, involving the activation of the humoral and cellular defense systems, that terminated in healing or rejection (Nygren et al. 1999). The early inflammatory reaction *in vivo* to gold was measured by the adherence and activation of inflammatory cells during ip implantation. After 1 hour, gold implants inserted ip into mice had 18% of the surface covered with white blood cells. It was concluded that peritoneal leukocytes adhering to foreign materials produced a respiratory burst response via a phospholipase D-dependent and protein kinase C-independent pathway (Nygren et al. 1999). Subcutaneous implantation of gold (1.000 fine) and gold alloys in rats caused only a mild tissue reaction when compared with other dental restorative materials, inducing relatively few inflammatory cells (Scott et al. 1995).

6.3.2 Monovalent Gold: Obese Mouse Model

Gold thioglucose ($C_6H_{11}O_5SAu$) was initially developed and marketed as a therapeutic agent for the treatment of arthritis and rheumatism. However, a single subcutaneous or intraperitoneal injection of gold thioglucose (GTG), equivalent to 0.5 to 0.6 mg Au^+/kg body weight (BW), in juveniles of certain strains of mice produced irreversible hyperphagia and obesity 10 to 12 weeks later, with many of the characteristics of human obesity. In contrast to genetically obese mice, GTG-injected mice were relatively tolerant of gold (Heydrick et al. 1995; Bergen et al. 1996; Blair et al. 1996a, 1996b; Marks et al. 1996; Bryson et al. 1999a, 1999b; Challet et al. 1999). The effect of GTG on the brain of mice is specific (Blair et al. 1996a). Other gold thiol compounds tested — including gold thiogalactose, gold thiosorbitol, gold thiomalate, gold thiocaproate, gold thioglycoanilide, and gold thiosulfate — do not induce the brain damage that results from the administration of GTG, and neither obesity nor increased appetite occur, although all were toxic (Blair et al. 1996b).

Injected mice developed a hypothalamic lesion within 24 hours of GTG administration (Marks et al. 1996). Gold thioglucose induced bilateral necrosis of the ventromedial hypothalamus region of the brain and caused damage to the supraoptic nuclei, ventromedial nuclei, arcuate nuclei, and median eminence (Bergen et al. 1996; Blair et al. 1996b). These GTG-induced lesions in the hypothalamus impaired regulation of food intake and body weight. The degree of obesity induced by mice is dependent on the dose of GTG administered and the strain of mouse. Administration of a range of doses of GTG induces variable weight gain and death in the C58, RIII,

DBA, and BALB/C strains of mice when compared with the CBA strain. CBA mice show a uniform response with respect to obesity and survival. In rats, GTG produced a hypothalamic lesion that was similar to that observed in mice; however, GTG doses that induced obesity in rats were usually fatal (Blair et al. 1996b).

In addition to hyperphagia, the GTG brain lesion also induces hyperglycemia, hyperinsulinemia, insulin resistance, triglyceride accumulation, and a range of tissue-specific changes in regulatory enzyme activities of glucose and lipid metabolic pathways (Blair et al. 1996b; Marks et al. 1996). Hypersecretion of insulin is evident from an early stage in the development of GTG-induced obesity. Hyperinsulinemia is an early abnormality in many animal models of obesity and non-insulin dependent diabetes mellitus, including the GTG-injected obese mouse (Blair et al. 1996a). The hyperinsulinemia of GTG mice was accompanied by a developing insulin resistance in fat and skeletal muscle, and was evident in both young (age 8 weeks) and old (age 24 weeks) GTG-obese mice (Heydrick et al. 1995; Blair et al. 1996b). Insulin resistance at the level of phosphatidylinositol 3-kinase (PIK) occurs very early both in muscle and adipose tissue at a time when alterations in glucose transport were moderate or absent (Heydrick et al. 1995). Removal of glucocorticoid hormones alters insulin release and glucose metabolism in both lean control and GTG-obese mice (Blair et al. 1996a). GTG mice that were adrenalectomized and examined 1 week later for glucose tolerance and insulin secretion showed reductions in body weight, liver glycogen content, and plasma glucose. But adrenalectomy normalized plasma insulin concentrations (Blair et al. 1996a). Injection of GTG into C57BL/6J mice also damages glucose receptive neurons in the ventromedial hypothalamus, preventing metabolic regulation of circadian responses to light during shortage of glucose availability (Challet et al. 1999).

The role of neuropeptides and leptins in GTG-induced obesity have been considered. Neuropeptide Y (NPY) is a 36-amino-acid neuropeptide that is widely distributed in the mammalian brain, especially in the hypothalamus, and elevated levels are thought to be involved in the etiology of genetic models of obesity. In GTG-injected mice, however, NPY levels were reduced and, therefore, unlikely to be a key factor causing obesity in this model, It is probable that other ventromedial hypothalamic factors altered by the GTG lesion were the major contributors to obesity (Marks et al. 1996). Leptin affects glucose and lipid metabolism in GTG-obese mice and lean controls within 2 hours of leptin administration. Plasma leptin levels were strongly related to the degree of adiposity, with hyperleptinemia being associated with hyperinsulinemia (Bryson et al. 1999a, 1999b). GTG-injected mice showed excess insulin production resulting in abnormally low blood sugar prior to a rise in plasma leptin levels; this is consistent with the role of leptin as an indicator of energy supplies.

6.3.3 Monovalent Gold: Other

In addition to the obese mouse model, selected studies show that monovalent organogold compounds affect survival, carcinogenicity, teratogenicity, histopathology, metabolism, immune function, disease resistance, and gold accumulation dynamics.

High survival of the BALB/C mouse strain — a nonresistant organomonovalent gold strain — was reported after administration of comparatively high doses of organomonovalent gold compounds used in chrysotherapy (Koide et al. 1998). No deaths were reported after 20 days in 4-week-old mice given either a single subcutaneous injection of 10 mg Au$^+$/kg BW, three subcutaneous injections of 2 mg Au$^+$/kg BW at 48-hour intervals, or 30 mg Au$^+$/kg BW daily per os for 10 days. By comparison, monovalent organogold salts used to treat humans for rheumatoid arthritis were usually administered at the rate of 0.36 mg Au$^+$/kg BW every 7 to 14 days (Koide et al. 1998).

Survival was reduced in adult mice infected with various viruses — including Semliki Forest virus, various strains of yellow fever virus, and West Nile virus — after intraperitoneal injection of gold sodium thiomalate (Gibson et al. 1990). Adverse effects of therapeutic monovalent gold compounds may be linked to their ability to induce membrane proliferation (Mehta et al. 1990). For example, the virulent strain of Semliki Forest virus in adult mice is characterized by the development of numerous membrane vesicles in brain with mature virus budding from these structures. In contrast, infection of the avirulent strain of Semliki Forest virus results in the formation of very few membrane vesicles and no mature virus particles in adult mouse brain. Proliferation of smooth membrane vesicles from whole mouse brain was induced in mice treated with gold sodium thiomalate. In certain virus infections, smooth membranes are a prerequisite for virus RNA synthesis and maturation. The ability of a virus to stimulate the smooth membranes may be the limiting factor in determining the extent of viral RNA synthesis and maturation. This mechanism could also be responsible for the striking variation in strains of Semliki Forest virus in adult mice. The concept of membrane proliferation might also be relevant to the increased virulence of encephalitogenic viruses in childhood when the brain is still developing and cellular membrane proliferation is more abundant (Mehta et al. 1990).

Carcinogenicity and teratogenicity of myochrisine ($C_4H_3AuNa_2O_4S$), the disodium salt of gold sodium thiomalate ($C_4H_4AuNaO_4S$), is reported at high doses in rats and rabbits (Sifton 1998). Renal adenomas are documented in rats at 2 mg Au/kg BW weekly for 45 weeks followed by 6 mg/kg BW daily for 47 weeks; this dosage is equivalent to twice that administered to humans at the low dose and 42 times at the high dose. Adenomas produced were similar to those produced in rats by chronic administration of other gold compounds, lead, and other heavy metals. There is no report of myochrisine-associated renal adenomas in humans. Teratogenic effects were observed in rats and rabbits when given myochrisine during the organogenetic period at doses of 140 times higher than that of humans for rats and 175 times higher for rabbits. Hydrocephaly and microphthalmia malformations were observed in rats given subcutaneous injection dose levels of 25 mg Au$^+$/kg BW daily from days 6 through 15 of gestation. In rabbits, limb malformations and gastroschisis were observed when subcutaneous injection doses were 20 to 45 mg Au/kg BW from gestation days 6 through 18 (Sifton 1998).

Antitumor activity of gold sodium thiomalate was documented in mice when given subcutaneously or per os (Kamei et al. 1998). Mice inoculated with Meth A tumor cells were given gold sodium thiomalate either in sc injections every other day for three injections, or in drinking water daily for 2 weeks. Survival was improved 50% by 30 mg/kg BW via injection, or 75 mg/kg BW daily per os. No significant toxicity of gold was observed in afflicted mice at doses up to 125 mg Au$^+$ kg/BW daily via drinking water; however, adverse effects were observed at 6 mg/kg BW daily and lethality at 125 mg/kg BW daily via injection (Kamei et al. 1998). Malignant tumor cells from humans and from mice *in vitro* showed growth inhibition after 4 days when subjected to 2 mg Au$^+$/L as gold sodium thiomalate, with 50% inhibition recorded between 10 and 50 mg Au$^+$/L (Koide et al. 1998). Antitumor activity of the metal-bound aromatic cation Au$^+$ (1,2-bis(diphenylphosphino)ethane) was reported in mice (McKeage et al. 2000). However preclinical development was abandoned after the identification of severe hepatotoxicity in dogs after administration of this cation, and was attributed to alterations in mitochondrial function. Antitumor activity of other gold phosphino complexes is a function of the drug's lipophilicity. Alteration of lipophilicity of aromatic cationic antitumor Au$^+$ drugs greatly affects cellular uptake and binding to plasma proteins. Changes in lipophilicity also affect host toxicity, and optimal lipophilicity may be a critical factor in the design of gold analogues with high antitumor activity (McKeage et al. 2000).

Renal injury was documented in male Wistar rats weighing 130 grams and receiving a single ip injection of 25 mg of gold sodium thiomalate. Nephrotoxic effects were observed mainly in the renal tubular region. Changes in the activity of urinary enzymes suggested that the renal tubules were selectively injured by gold salts (Ogura et al. 1996). Similar results were observed in adjuvant arthritic rats (produced by a single intradermal injection of *Mycobacterium butyricum*) receiving 11 intramuscular injections of gold sodium thiomalate every other day for 21 days (Takahashi et al. 1995). Gold concentrations in these rats after the final injection were 316 mg/kg FW in kidney and 44 mg/kg FW in liver. Selected chelating agents reduced gold concentrations in kidney and liver by about 51% (Takahashi et al. 1995). Effects of various chelating agents on distribution, excretion, and renal toxicity of gold sodium thiomalate in rats was measured immediately after intravenous injection of 0.026 mmol gold sodium thiomalate/kg BW (Takahashi et al. 1994). Of all agents tested, tiopronin and captopril successfully ameliorated gold sodium thiomalate-induced renal toxicity. DMPS (2,3-dimercaptopropanesulfonate) was the most effective in removing gold from the kidney and in protecting against renal toxicity after gold sodium thiomalate injection of 0.2 mmol/kg BW (Takahashi et al. 1994).

Human patients with rheumatoid arthritis sometimes developed skin reactions, glomerulonephritis, and increased serum IgE concentrations when treated with Au$^+$ salts (Kermarrec et al. 1995, 1996). [Note: elevated levels of immunoglobulin E indicate an increased probability of an IgE-mediated hypersensitivity, responsible for allergic reactions.] Brown Norway rats (*Rattus norvegicus*) injected with allochrysine (gold sodium thiopropanosulfonate) showed an increase in serum IgE concentration, produced anti-laminin antibodies, and developed glomerular liver immunoglobulin deposits; Lewis strain rats were resistant. Genetic studies of gold salt-induced immune disorders in brown Norway rats showed that susceptibility to allochrysine was linked mainly to histocompatibility complex genes. Like humans, brown Norway rats injected with allochrysine had an autoimmune glomerulonephritis and increased serum IgE concentration, with few complications. Homologous genes encoding for cytokines, genes involved in sulfoxidation or in the control of glutathione synthetase, and genes encoding for nuclear factors regulating the IgE response could be implicated in allochrysine manifestation in rats and in the regulation of IgE levels in humans (Kermarrec et al. 1995, 1996). Brown Norway rats injected with allochrysine developed an autoimmune syndrome (Saoudi et al. 1995). Major histocompatibility complex (MHC) T-cell lines were derived from gold-treated rats. On transfer into normal brown Norway rats, the T-cells produced an autoimmune syndrome similar to, or more severe than, that observed in the active gold model, including an increase in serum IgE concentration, and production of anti-DNA and anti-laminin antibodies. The T-helper cell lines may induce an autoantibody-mediated disease and may be responsible for cell-mediated immunity (Saoudi et al. 1995).

Gold[+] salts injected subcutaneously over a 10-day period induced an autoimmune syndrome similar to that of $HgCl_2$ in brown Norway rats (Qasim et al. 1997). The autoimmune syndrome in both cases was characterized by a marked increase in IgE production, lymphoproliferation, T-cell-dependent polyclonal B-cell activation, hypergammaglobulinemia, and tissue injury with necrotizing leucocytoclastic vasculitis in the gut at day 15 postinjection. Gold also induced granulomata and neutrophil infiltrates in the lung at day 15 postinjection (Qasim et al. 1997).

Gold toxicity as a consequence of chrysotherapy has been treated by various chelating agents to remove accumulated gold from the body (Kojima et al. 1992). Gold in urine and bile of rats after gold sodium thiomalate administration was mainly bound to high-molecular-weight compounds. After administration of chelating agents, gold in urine was bound to the sequestering agents. In bile, the gold was excreted into the feces primarily as a gold-chelating agent compound and secondarily as gold-L-cysteine and high-molecular-weight compounds (Kojima et al. 1992). Incidentally, gold accumulation rates in the rat kidney differed markedly between administration routes (Ueda 1998). Renal concentrations of gold in rats administered auranofin orally were 33 times lower than those given equivalent gold dosages from sodium gold thiomalate parenterally. These differences persisted for at least 1 year (Ueda 1998) and spotlight administration route as a factor in gold accumulation dynamics.

Distribution of radiogold-198 in rats 2 hours postadministration was mainly in urine (40.1% of injected dose), small intestine (39.6%), liver (8.2%), and carcass (8.0%) (Berning et al. 1998; Table 6.1). Histological examination of thyroid, adrenal, suprarenal, and testicular glands of rats given a total of 540 mg of allochrysine via 9 subcutaneous injections over a 3-week period showed gold–sulfur compounds in aurosomes of all glands (Manoubi et al. 1994). In gold-containing tissues, gold was localized intracellularly and selectively concentrated in lysosomes as nonsoluble crystalline precipitates of a constant S/Au ratio (Manoubi et al. 1994).

Monovalent gold salts impacted metabolism of selenium, copper, and zinc. Intravenous Au[+] may adversely affect the availability of selenium for synthesis of selenoenzymes. Rats given gold sodium thiomalate i.v. at 25 or 50 µmol/kg BW had significantly altered selenium deposition, as measured by radioselenium-75 (Gregus et al. 2000). These effects included the almost complete cessation of ^{75}Se exhalation as dimethyl sulfide and the accumulation of ^{75}Se in blood plasma. Direct

Table 6.1   Distribution of [198]Au+ in Anesthetized Sprague-Dawley Rats*

Organ	Percent
Brain	0.01
Blood	1.7[a]
Heart	0.03
Lung	0.2
Liver	8.2
Spleen	0.03
Large intestine	0.3
Small intestine	39.6
Kidneys	2.2
Urine	40.1
Muscle	0.01
Pancreas	0.2
Carcass	8.0[b]
Total	100.3

*Values are in percent of injected dose two hours post-injection

[a] Other [198]Au-phosphino compounds tested showed significant blood activity (11–40%) at 2 h post-injection, suggesting instability of these complexes *in vivo* due to formation of colloidal gold or trans-methylation to plasma proteins.

[b] Other [198]Au-phosphino compounds had a high (40–43%) degree of retention in carcass, indicating that [198]Au+ is present as an immobilized form for reasons given above.

Source: Modified from Berning et al. 1998.

chemical reaction with nucleophilic selenium metabolites in the body may underlie these alterations (Gregus et al. 2000). Auranofin inhibited selenium–glutathione peroxidase in bovine erythrocytes; this enzyme protects the cell from initiation and propagation of free radical reactions (Roberts and Shaw 1998). High doses of gold alter copper and zinc metabolism in liver and kidney of Sprague-Dawley rats (Hinck-Kneip and Alsen-Hinrichs 1996). These, in turn, were controlled by metallothioneins, low-molecular-weight proteins that serve as oxygen scavengers and as metal sequestering agents. At low doses of 0.5 μg Au+/kg BW, given via intramuscular injection, the main changes occurred in the kidney where an increase of gold was found 0.5 hours postinjection, followed by an increase in copper and metallothionein concentrations after 6 hours. Zinc homeostasis did not change. Authors concluded that gold induces an increase of metallothionein-like peptides in the kidney cytosol, accompanied by an increase in copper bound to these peptides (Hinck-Kneip and Alsen-Hinrichs 1996).

Gold compounds were among the few classes of antiarthritic drugs that retarded rheumatoid arthritis and are now widely used to treat this disease and other chronic immunologically mediated inflammatory conditions in humans and dogs (Bloom et al. 1988). Healthy dogs given either 0.6 to 3.6 mg auranofin orally every 24 hours, or gold sodium thiomalate via intramuscular injection of 0.5 to 2.0 mg/kg BW every 3 days, were examined for changes in 13 immune functions. None of the changes in these aspects of immune function, previously attributed to treatment with auranofin or gold sodium thiomalate, could be demonstrated in normal dogs after treatment with either drug after treatment for 6 to 7 years. It seems that long-term administration of auranofin or gold sodium thiomalate to normal dogs at doses up to 30 times therapeutic doses in humans does not result in many of the alterations in lymphocyte, monocyte, and neutrophil functions observed in previous studies using other animal models or in human patients with rheumatoid arthritis. Accordingly, the changes previously reported in these variables may not be due to the effects of monovalent gold compounds (Bloom et al. 1988).

6.3.4   Trivalent Gold

Organometallic Au^{+3} complexes hold promise as possible antitumor agents, as judged by cytotoxic activity *in vitro* against cultured tumor cell lines and appreciable tumor inhibiting properties *in vivo* toward human tumors grown in xenografted rats (Messori et al. 2001). DNA is the probable target for newly developed cytotoxic Au^{+3} complexes. *In vitro* tests with five representative organogold complexes and calf thymus DNA showed that all complexes interacted with DNA and modified its behavior in solution (Messori et al. 2001). Another potential antitumor agent is chlorobischolylglycinatogold^{+3} [$(C_{52}H_{84}N_2)_{12}AuCl$], a synthetic bile acid (Carrasco et al. 2001). This compound is soluble in water, methanol, ethanol, and dimethylsulfoxide and can inhibit the growth of a variety of cell lines. The cytostatic effect was mild against human hepatoma, mouse hepatoma, rat hepatoma, and human colon adenocarcinoma cell lines, but stronger against mouse sarcoma S180-II and mouse leukemia L-1210 cells. The appearance of colloidal gold during the process of hydrolysis under physiological conditions may account for the limited cytostatic activity (Carrasco et al. 2001).

Studies with sodium tetrachloroaurate^{+3} showed that gold accumulates mainly in kidneys and in the reticuloendothelial cells of tissues, especially in the renal cortex. Male rats given daily intraperitoneal injections of 1 mg Au^{+3}/kg BW as sodium tetrachloroaurate dihydrate [$Na(AuCl_4) \cdot 2H_2O$] and killed one day after the last injection had elevated concentrations of gold in kidneys (55.1 mg/kg FW) and liver (4.0), as was the case for copper and zinc (Saito et al. 1997). Rats given a single ip injection of sodium tetrachloroaurate dihydrate at 5, 10, or 20 mg Au^{+3}/kg BW and killed 14 hours postinjection showed dose-dependent increases of gold and metallothioneins in kidneys (Saito 1996; Saito and Kojima 1996). The increasing amount of gold was attributed to high-molecular-weight proteins and the metallothionein fractions. About 14% of the increased gold in renal cytosols of gold-injected rats was bound to metallothioneins and 79% to high-molecular-weight fractions (Saito and Kojima 1996). Trivalent gold had stronger binding affinity to metallothioneins than did zinc or cadmium (Saito and Kurasaki 1996). Zinc concentrations and metallothionein content in rat livers showed a dose-dependent increase in response to Au^{+3} injections of 5, 10, or 20 mg/kg/BW, but copper was unaffected (Saito and Yoshida 1998). Increased zinc was due to high-molecular-weight proteins and the comparatively low-molecular-weight metallothioneins. About 68% of the increased zinc in the hepatic cytosol of Au^{+3}-injected rats was bound to metallothioneins, suggesting that the role of metallothioneins in zinc accumulation in the liver was similar for both Au^{+3}- and Zn^{+2}-injected rats (Saito and Yoshida 1998).

Gold sensitivity in mice may be genetically determined (Merchant 1998). Chronic treatment of inbred mice strains with monovalent gold sodium thiomalate produced immune responses to Au^{+3} in 100% of A.SW mice, 70% of C57BL/6 mice, and zero in DBA/2 mice. The dose-dependent immune responses were directed toward the trivalent gold compounds $AuCl_3$ and $HAuCl_4$ and were T-cell dependent and specific. Thus, in order to induce T-cell sensitization, gold has to exist in the Au^{+3} state; however, mechanisms governing gold immunogenicity are imperfectly understood (Merchant 1998). Popliteal lymph node (PLN) assay reactions to Au^{+3} in mice were dose- and T-cell dependent (Schumann et al. 1990). When Au^{+3} was reduced to Au^+ by addition of disodium thiomalate or methionine before testing in the PNL assay, its sensitizing capacity was significantly decreased, suggesting that Au^+ of disodium thiomalate was oxidized to Au^{+3} before T-cells were sensitized and adverse immunological reactions developed (Schumann et al. 1990).

6.4 SUMMARY

Monovalent gold compounds have negligible impact on aquatic organisms at medium concentrations less than 98 mg Au^+/L; however, trivalent Au^{+3} compounds are significantly more toxic than Au^+ compounds and inhibit growth of diatoms at 98 μg Au^{+3}/L, kill marine teleosts at less than 800 μg Au^{+3}/L, inhibit contraction ability of frog muscle at 2 mg Au^{+3}/L, and reduce growth of yeasts at 20 mg Au^{+3}/L. Accumulation of ionic and metallic gold from a wide variety of solutions by selected species of bacteria, yeasts, fungi, algae, and higher plants is documented. Gold accumulations are up to 7.0 g/kg dry weight (DW) in various species of bacteria, 25.0 g/kg DW in freshwater algae, 84.0 g/kg DW in peat, and 100.0 g/kg DW in dried fungus mixed with keratinous material. Crab exoskeletons accumulate up to 4.9 g Au/kg DW; however, gold accumulations in various tissues of living teleosts, decapod crustaceans, and bivalve molluscs are negligible. Uptake patterns are significantly modified by the physicochemical milieu. The mechanisms of accumulation — which include oxidation, reduction, dissolution, leaching, and sorption — are not known with certainty and merit additional research.

In tests with small laboratory mammals, injected doses of colloidal gold are associated with increased body temperatures, gold accumulations in reticular cells, and dose enhancement in tumor therapy. In certain strains of gold-resistant mice, a single injection of sodium gold thioglucose at 0.5 to 0.6 mg Au^+/kg body weight destroyed the appetite center in the hypothalamus portion of the brain, resulting in obese mice with many of the characteristics of human obesity. Other strains of mice and other species of rodents tested either died or produced variable responses to gold thioglucose. High doses of monovalent organogold compounds administered parenterally are usually tolerated by mice, but survival is reduced in adult mice infected with various viruses. Carcinogenicity, teratogenicity, and other effects are documented in small laboratory mammals given extremely high injected doses of various monovalent organogold compounds; effects include renal adenomas, limb malformations, nephrotoxicity, elevated gold accumulations in kidney (316 mg/kg FW) and liver (44 mg/kg FW), increased serum immunoglobulin E, and altered metabolism of selenium, copper, and zinc. At comparatively low doses, monovalent organogold compounds may have antitumor properties. Exposure of rodents to trivalent inorganic gold compounds results in high accumulations in the renal cortex and other tissues, increased metallothionein concentrations, and variable immune responses dependent on rodent strain tested.

LITERATURE CITED

Agate, A.D. 1996. Recent advances in microbial mining, *World Jour. Microbiol. Biotechnol.*, 12, 487–495.

Antunes, A.P.M., G.M. Watkins, and J.R. Duncan. 2001. Batch studies on the removal of gold (III) from aqueous solution by *Azolla filiculoides*, *Biotechnol. Lett.*, 23, 249–251.

Bergen, H.T., N. Monkman, and C.V. Mobbs. 1996. Injection with gold thioglucose impairs sensitivity to glucose: evidence that glucose-responsive neurons are important for long-term regulation of body weight, *Brain Res.*, 734, 332–336.

Berning, D.E., K.V. Katti, W.A. Volkert, C.J. Higginbotham, and A.R. Ketring. 1998. [198]Au-labeled hydroxymethyl phosphines as models for potential therapeutic pharmaceuticals, *Nucl. Med. Biol.*, 25, 577–583.

Blair, S.C., I.D. Caterson, and G.J. Cooney. 1996a. Glucose tolerance and insulin secretion after adrenalectomy in mice made obese with gold thioglucose, *Jour. Endocrinol.*, 148, 391–398.

Blair, S.C., I.D. Caterson, and G.J. Cooney. 1996b. Glucose and lipid metabolism in the gold thioglucose injected mouse model of diabesity, in *Lessons from Animal Diabetes VI*, E. Shafrir, (Ed.), Birkhauser, Boston, 237–265.

Bloom, J.C., P.A. Thiem, L.K. Halper, L.Z. Saunders, and D.G. Morgan. 1988. The effect of long-term treatment with auranofin and gold sodium thiomalate on immune function in the dog, *Jour. Rheum.*, 15, 409–417.

Brown, D.H. and W.E. Smith. 1980. The chemistry of the gold drugs used in the treatment of rheumatoid arthritis, *Chem. Soc. Rev.*, 9, 217–240.

Bryson, J.M., J.L. Phuyal, D.R. Proctor, S.C. Blair, I.D. Caterson, and G.J. Cooney. 1999a. Plasma insulin rise precedes rise in *ob* mRNA expression and plasma leptin in gold thioglucose-obese mice, *Amer. Jour. Physiol.*, 276, E358–E364.

Bryson, J.M., J.L. Phuyal, V. Swan, and I.D. Caterson. 1999b. Leptin has acute effects on glucose and lipid metabolism in both lean and gold-thioglucose-obese mice, *Amer. Jour. Physiol.*, 277, E417–E422.

Bruno, J.G., S.B. Collard, D.J. Kuch, and J.C. Cornette. 1996. Electrochemiluminescence from tunicate, tunichrome-metal complexes and other biological samples, *Jour. Biolumin. Chemilumin.*, 11, 193–296.

Carrasco, J., J.J. Criado, R.I.R. Macias, J.L Manzano, J.J.G. Marin, M. Medarde, and E. Rodriguez. 2001. Structural characterization and cytostatic activity of chloro-bischolylglycinatogold(III), *Jour. Inorg. Biochem.*, 84, 287–292.

Challet, E., D.J. Bernard, and F.W. Turek. 1999. Gold-thioglucose-induced hypothalamic lesions inhibit metabolic modulation of light-induced circadian phase shifts in mice, *Brain Res.*, 824, 18–27.

Chapman, J.T., P.B. Marchant, R.W. Lawrence, and R. Knopp. 1993. Bio-oxidation of a refractory gold bearing high arsenic sulphide concentrate: a pilot study, *Feder. Eur. Microbiol. Soc. Microbiol. Rev.*, 11, 243–252.

Claassen, R. 1993. Mineralogical controls on the bacterial oxidation of refractory Barberton gold ores, *Feder. Eur. Microbiol. Soc. Microbiol. Rev.*, 11, 197–206.

Duke, T.W., J.P. Baptist, and D.E. Hoss. 1966. Bioaccumulation of radioactive gold used as a sediment tracer in the estuarine environment, *U.S. Fish. Bull.*, 65, 427–436.

Dyer, B.D., W.E. Krumbein, and D.J. Mossman. 1994. Accumulation of gold in the sheath of *Plectonema terebrans* (filamentous marine cyanobacteria), *Geomicrobiol. Jour.*, 12, 91–98.

Eisler, R. 1981. *Trace Metal Concentrations in Marine Organisms*. Pergamon, New York, 687 pp.

Eisler, R. 1986. Use of *Fundulus heteroclitus* in pollution studies, *Amer. Zool.*, 26, 283–288.

Eisler, R. 2000. *Handbook of Chemical Risk Assessment: Health Hazards to Humans, Plants, and Animals. Volumes 1–3.* Lewis Publishers, Boca Raton, FL, 1903 pp.

Eisler, R. 2003. Biorecovery of gold, *Indian Jour. Exper. Biol.*, 41, 967–971.

Eisler, R., D.R. Clark Jr., S.N. Wiemeyer, and C.J. Henny. 1999. Sodium cyanide hazards to fish and other wildlife from gold mining operations, in *Environmental Impacts of Mining Activities: Emphasis on Mitigation and Remedial Measures*, J.M. Azcue, (Ed.), Springer-Verlag, Berlin, 55–67.

Eisler, R., H.C. Moeller, and M.I. Grossman. 1955. Febrile response of rabbits to intravenous injection of colloidal substances, U.S. Army Medical Nutrition Lab. Rep. 167, 14 pp.

Fields, S. 2001. Tarnishing the earth: gold mining's dirty secret, *Environ. Health Perspect.*, 109, A474–A482.

Gardea-Torresdey, J.L., K.J. Tiemann, G. Gamez, K. Dokken, I. Cano-Aguilera, L.R. Furenlid, and M.W. Renner. 2000. Reduction and accumulation of gold(III) by *Medicago sativa* alfalfa biomass: X-ray absorption, spectroscopy, pH, and temperature dependence, *Environ. Sci. Technol.*, 34, 4392–4396.

Gardner, M.N., and D.E. Rawlings. 2000. Production of rhodanese by bacteria present in bio-oxidation plants used to recover gold from arsenopyrite concentrates, *Jour. Appl. Microbiol.*, 89, 185–190.

Gibson, C.A., M.R. Wills, E.A. Gould, P.G. Saunders, and A.D.T. Barrett. 1990. Effect of administration of sodium aurothiomalate on the virulence of yellow fever viruses in adult mice, *Vaccine*, 8, 590–594.

Glazyrin, A.L., S.I. Kolesnikov, G.N. Dragun, E.L. Zelentsov, K.V. Zolotarev, Y.A. Sorin, G.N. Kulipanov, V.N. Gorchakov, and I.P. Dolbnya. 1995. Distribution of colloidal gold tracer within rat parasternal lymph nodes after intrapleural injection, *Anat. Rec.*, 241, 175–180.

Gomes, N.C.M., E.R.S. Camargos, J.C.T. Dias, and V.R. Linardi. 1998. Gold and silver accumulation by *Aspergillus niger* from cyanide-containing solution obtained from the gold mining industry, *World Jour. Microbiol. Biotechnol.*, 14, 149.

Gomes, N.C.M., M.M. Figueira, E.R.S. Camargos, L.C.S. Mendonca-Hagler, J.C.T. Dias, and V.R. Linardi. 1999a. Cyano-metal complexes uptake by *Aspergillus niger*, *Biotechnol. Lett.*, 21, 487–490.

Gomes, N.C.M. and V.R. Linardi. 1996. Removal of gold, silver and copper by living and nonliving fungi from leach liquor obtained from the gold mining industry, *Revista Microbiol.*, 27, 218–222.

Gomes, N.C.M., C.A. Rosa, P.F. Pimental, V.R. Linardi, and L.C.S. Mendonca-Hagler. 1999b. Uptake of free and complexed silver ions by yeast from a gold mining industry in Brazil, *Jour. Gen. Appl. Microbiol.*, 45, 121–124.

Gonzalez, R., J.C. Gentina, and F. Acevedo. 1999. Attachment behaviour of *Thiobacillus ferrooxidans* cells to refractory gold concentrate particles, *Biotechnol. Lett.*, 21, 715–718.

Gregus, Z., A. Gyurasics, and I. Csanaky. 2000. Effects of arsenic-, platinum-, and gold-containing drugs on the disposition of exogenous selenium in rats, *Toxicol. Sci.*, 57, 22–31.

Harrison, F.L. 1973. Availability to aquatic animals of short-lived radionuclides from a Plowshare cratering event, *Health Physics*, 24, 331–343.

Herold, D.M., I.J. Das, C.C. Stobbe, R.V. Iyer, and J.D. Chapman. 2000. Gold microspheres: a selective technique for producing biologically effective dose enhancement, *Int. Jour. Radiat. Biol.*, 76, 1357–1364.

Heydrick, S.J., N. Gautier, C. Olichon-Berthe, E. Van Obberghen, and Y.L. Marchand-Brustel. 1995. Early alteration of insulin stimulation of PI 3-kinase in muscle and adipocyte from gold thioglucose obese mice, *Amer. Jour. Physiol.*, 268, E604-E612.

Hinck-Kneip, C., and C. Alsen-Hinrichs. 1996. Influences of gold on zinc, copper and metallotheionein kinetics in liver and kidney of the rat, *Human Exper. Toxicol.*, 15, 518–522.

Ishikawa, S-I. and K. Suyama. 1998. Recovery and refining of Au by gold-cyanide ion biosorption using animal fibrous proteins, *Appl. Biochem. Biotechnol.*, 70–72, 719–728.

Kai, T., K. Yamasaki, and T. Takahashi. 1992. Application of iron oxidizing bacteria in thiourea leaching of gold bearing ores, *Biorecovery*, 2, 83–93.

Kamei, H., T. Koide, T. Kojima, Y. Hashimoto, and M. Hasegawa. 1998. Effect of gold on survival of tumor-bearing mice, *Cancer Biother. Radiopharmaceut.*, 13, 403–406.

Karamushka, V.I. and G.M. Gadd. 1999. Interaction of *Saccharomyces cerevisiae* with gold: toxicity and accumulation, *Biometals*, 12, 289–294.

Kashefi, K., J.M. Tor, K.P. Nevin, and D.R. Lovley. 2001. Reductive precipitation of gold by dissimilatory Fe (III)-reducing *Bacteria* and *Archaea*, *Appl. Environ. Microbiol.*, 67, 3275–3279.

Kermarrec, N., C. Blanpied, L. Pelletier, N. Feingold, C. Mandet, P. Druet, and F. Hirsch. 1995. Genetic study of gold-salt-induced immune disorders in the rat, *Nephrol. Dial. Transplant.*, 10, 2187–2191.

Kermarrec, N., C. Dubay, B. de Gouyon, C. Blanpied, D. Gauguier, K. Gillespie, P.W. Mathieson, P. Druet, M. Lathrop, and F. Hirsch. 1996. Serum IgE concentration and other immune manifestations of treatment with gold salts are linked to the MHC and IL4 regions in the rat, *Genomics*, 31, 111–116.

Koide, T., T. Kojima, and H. Kamei. 1998. Antitumor effect of gold as revealed by growth suppression of cultured cancer cells, *Cancer Biother. Radiopharmaceut.*, 13, 189–192.

Kojima, S., Y. Takahashi, M. Kiyozumi, T. Funakoshi, and H. Shimada. 1992. Characterization of gold in urine and bile following administration of gold sodium thiomalate with chelating agents to rats, *Toxicology*, 74, 1–8.

Kuyucak, N. and B. Volesky. 1989. The mechanism of gold biosorption, *Biorecovery*, 1, 219–235.

Lindstrom, E.B., E. Gunneriusson, and O.H. Tuovinen. 1992. Bacterial oxidation of refractory sulfide ores for gold recovery, *Brit. Rev. Biotechnol.*, 12, 133–155.

Manoubi, L., M.H. Jaafoura, A. Skhiri-Zhioura, A. El Hil, J.P. Berry, and P. Galle. 1994. Subcellular localization of gold in suprarenal testicle and thyroid glands after injection of allochrysine in rats, *Cell. Molec. Biol.*, 40, 483–487.

Marks, J.L., K. Waite, D. Cameron-Smith, S.C. Blair, and G.J. Cooney. 1996. Effects of gold thioglucose on neuropeptide Y messenger RNA levels in the mouse hypothalamus, *Amer. Jour. Physiol.*, 270, R1208–R1214.

Maturana, H., U. Lagos, V. Flores, M. Gaete, L. Cornejo, and J.V. Wiertz. 1993. Integrated biological process for the treatment of a Chilean complex gold ore, *Fed. Eur. Microbiol. Soc. Microbiol. Rev.*, 11, 215–220.

McKeage, M.J., S.J Berners-Price, P. Galettis, R.J. Bowen, W. Brouwer, L. Ding, L. Zhuang, and B.C. Baguley. 2000. Role of lipophilicity in determining cellular uptake and antitumour activity of gold phosphine complexes, *Cancer Chemother. Pharmacol.*, 46, 343–350.

Mehta, S., S. Pathak, and H.E. Webb. 1990. Induction of membrane proliferation in mouse CNS by gold sodium thiomalate with reference to increased virulence of the avirulent Semliki virus, *Biosci. Rep.*, 10, 271–180.

Merchant, B. 1998. Gold, the noble metal and the paradoxes of its toxicology, *Biologicals*, 26, 49–59.

Messori, L., P. Orioli, C. Tempi, and G. Marcon. 2001. Interactions of selected gold (III) complexes with calf thymus DNA, *Biochem. Biophys. Res. Comm.*, 281, 352–360.

Nihonyanagi, K. and T. Oba. 1996. Gold ion inhibits silver ion induced contracture and activates ryanodine receptors in skeletal muscle, *Eur. Jour. Pharmacol.*, 311, 271–276.

Nilsson, J.R. 1993. Effects of a gold salt and its intracellular distribution in *Tetrahymena*, *Acta Protozool.*, 32, 141–150.

Nilsson, J.R. 1997. *Tetrahymena* recovering from a heavy accumulation of a gold salt, *Acta Protozool.*, 36,111–119.

Niu, H. and B. Volesky. 1999. Characteristics of gold biosorption from cyanide solution, *Jour. Chem. Technol. Biotechnol.*, 74, 778–784.

Niu, H. and B. Volesky. 2000. Gold-cyanide biosorption with L-cysteine, *Jour. Chem. Technol. Biotechnol.*, 75, 436–442.

Niu, H. and B. Volesky. 2001. Gold adsorption from cyanide solution by chitinous materials, *Jour. Chem. Technol. Biotechnol.*, 76, 291–297.

Nygren, H., S. Kanagara, M. Braide, C. Eriksson, and I. Lundstrom. 1999. Characterization of cellular response to thiol-modified gold surfaces implanted in mouse peritoneal cavity, *Jour. Biomed. Mater. Res.*, 45, 117–124.

Oba, T., T. Ishikawa, and M. Yamaguchi. 1999. Different effects of two gold compounds on muscle contraction, membrane potential and ryanodine receptor, *Europ. Jour. Pharmacol.*, 374, 477–487.

Ogura, T., M. Takaoka, T. Yamauchi, T. Oishi, Y. Mimura, M. Hashimoto, N. Asano, M. Yamamura, F. Otsuka, H. Makino, Z. Ota, and K. Takahashi. 1996. Changes in urinary enzyme activity and histochemical findings in experimental tubular injury induced by gold sodium thiosulfate, *Jour. Med.*, 27, 41–55.

Pethkar, A.V. and K.M. Paknikar. 1998. Recovery of gold from solutions using *Cladosporium cladosporoides* biomass beads, *Jour. Biotechnol.*, 63, 121–136.

Puddephatt, R.J. 1978. *The Chemistry of Gold.* Elsevier, Amsterdam, 274 pp.

Qasim, F.J., S. Thiru, and K. Gillespie. 1997. Gold and D-penicillamine induce vasculitis and up-regulate mRNA for IL-4 in the brown Norway rat: support for a role for Th2 cell activity, *Clin. Exp. Immunol.*, 108, 438–445.

Rawlings, D.E. 1998. Industrial practice and the biology of leaching of metals from ores, *Jour. Indus. Microbiol. Biotechnol.*, 20, 268–274.

Roberts, J.R. and C.F. Shaw III. 1998. Inhibition of erythrocyte selenium-glutathione peroxidase by auranofin analogues and metabolites, *Biochem. Pharmacol.*, 55, 1291–1299.

Robinson, M.G., L.N. Brown, and B.D. Hall. 1997. Effect of gold (III) on the fouling diatom *Amphora coffeaeformis*: uptake, toxicity and interactions with copper, *Biofouling*, 11, 59–79.

Saito, S. 1996. The effect of gold on copper and zinc in kidney and in metallothionein, *Res. Comm. Molecul. Pathol. Pharmacol.*, 93, 171–176.

Saito, S. and Y. Kojima. 1996. Relative gold-binding capacity of metallothionein: studies in renal cytosols of gold-injected rats, *Res. Comm. Molecul. Pathol. Pharmacol.*, 92, 119–126.

Saito, S. and M. Kurasaki. 1996. Gold replacement of cadmium, zinc-binding metallothionein, *Res. Comm. Molecul. Pathol. Pharmacol.*, 93, 101–117.

Saito, S., M. Okabe, and M. Kurasaki. 1997. Localization of renal Cu-binding metallothionein induced by Au injection into rats, *Biochim. Biophys. Acta*, 1335, 335–338.

Saito, S. and K. Yoshida. 1998. The effect of gold on zinc in liver and in metallothionein, *Res. Comm. Molecul. Pathol. Pharmacol.*, 100, 83–91.

Saoudi, A., M. Castedo, D. Nochy, C. Mandet, R. Pasquier, P. Druet, and L. Pelletier. 1995. Self-reactive anti-class II T helper type 2 cell lines derived from gold salt-injected rats trigger B cell polyclonal activation and transfer autoimmunity in CD8-depleted syngeneic recipients, *Eur. Jour. Immunol.*, 25, 1972–1979.

Savvaidis, Y. 1998. Recovery of gold from thiourea solutions using microorganisms, *Biometals*, 11, 145–151.

Savvaidis, Y., V.I. Karamushka, H. Lee, and J.T. Trevors. 1998. Micro-organism-gold interactions, *Biometals*, 11, 69 –78.

Schumann, D., M. Kubicka-Muranyi, J. Mirtschewa, J. Gunther, P. Kind, and E. Gleichmann. 1990. Adverse immune reactions to gold. I. Chronic treatment with an Au(I) drug sensitizes mouse cells not to Au(I), but to Au(III) and induces autoantibody formation, *Jour. Immunol.*, 145, 2132–2139.

Scott, F.R., A.P. Dhillon, J.F. Lewin, W. Flavell, and I.M.Laws. 1995. Gold granuloma after accidental implantation, *Jour. Clin. Pathol.*, 48, 1070–1071.

Sifton, D.W. (ed.). 1998. *Physicians' Desk Reference*, (52nd ed.), Medical Econ. Co., Montvale, NJ, 3223 pp.

Suyama, K., Y. Fukazawa, and H. Suzumura. 1996. Biosorption of precious metal ions by chicken feather, *Appl. Biochem. Biotechnol.*, 57/58, 67–74.

Takahashi, Y., T. Funakoshi, H. Shimada, and S. Kojima. 1994. Comparative effects of chelating agents on distribution, excretion, and renal toxicity of gold sodium thiomalate in rats, *Toxicology*, 90, 39–51.

Takahashi, Y., T. Funakoshi, H. Shimada, and S. Kojima. 1995. The utility of chelating agents as antidotes for nephrotoxicity of gold sodium thiomalate in adjuvant-arthritic rats, *Toxicology*, 97, 151–157.

Ting, Y.P. and A.K. Mittal. 1999. An evaluation of equilibrium and kinetic models for gold biosorption, *Resource Environ. Biotechnol.*, 2, 311–326.

Ting, Y.P., W.K. Teo, and C.Y. Soh. 1995. Gold uptake by *Chlorella vulgaris*, *Jour. Appl. Phycol.*, 7, 97–100.

Tsezos, M., E. Remoudaki, and V. Angelatou. 1996. A study of the effects of competing ions on the biosorption of metals, *Inter. Biodeter. Biodegrad.*, 38, 19–29.

Ueda, S. 1998. Nephrotoxicity of gold salts, D-penicillamine, and allopurinol, in *Clinical Nephrotoxins: Renal Injury from Drugs and Chemicals*, M.E. De Broe, G.A. Porter, W.M. Bennett, and G.A. Verpooten, (Eds.), Kluwer, Dordrecht, 223–238.

Wagener, A.D.L.R. and W.D.O. Andrade. 1997. On the binding capacity of peat to several metal ions of environmental and economic concern, *Cien. Cult. Jour. Brazil. Assoc. Adv. Sci.*, 49, 48–53.

Xie, J.Z., H.L. Chang, and J.J. Kilbane II. 1996. Removal and recovery of metal ions from wastewater using biosorbents and chemically modified biosorbents, *Bioresour. Technol.*, 57, 127–136.

Zhao, Y., Y. Hao, and G.J. Ramelow. 1994. Evaluation of treatment techniques for increasing the uptake of metal ions from solution by nonliving seaweed algal biomass, *Environ. Monitor. Assess.*, 33, 61–70.

PART 2

Human Health Impacts

CHAPTER 7

Health Risks of Gold Miners

Health problems are documented for gold miners who worked mainly underground with little exposure to elemental mercury in Australia, North America, South America, Europe, and Africa. Major problems examined included life expectancy, cancer frequency, and pleural diseases. Health problems of miners who worked mainly on the surface and with extensive exposure to elemental mercury owing to its use in amalgamating and extracting gold, are reported in Australia, the Philippines, Brazil, and Venezuela; emphasis is on mercury residues in tissues, air, and diet and their significance when compared with existing mercury criteria for human health protection (Eisler 2000b, 2003). Health risks to gold miners from the use of cyanide in heap leaching and vat leaching gold recovery techniques were comparatively low, unlike effects on wildlife and the landscape (Eisler et al. 1999; Eisler 2000a), which are discussed in Chapter 11.

7.1 HISTORICAL BACKGROUND

Since before recorded time, gold has been mined, collected from alluvial deposits, or separated from the ores of silver, copper, and other metals (Merchant 1998). Gold is the first metal mentioned in the Old Testament in Genesis 2:11. One gold mine in Saudi Arabia has been mined for more than 3000 years (Kirkemo et al. 2001). Artisans of ancient civilizations used gold lavishly in decorating tombs and temples, and gold objects made more than 5000 years ago have been found in Egypt (Kirkemo et al. 2001). Among the most productive gold fields in ancient times were those in Egypt, where in the deep mines the slave laborers were maltreated, and in Asia Minor near the River Pactolus, the source of Croesus' wealth. The Romans obtained much of their gold from Transylvania (Rose 1948). Slaves were used to mine gold in Brazil from 1690 to 1850 (Lacerda 1997b). From 1850 to 1860, gold production in the United States and Australia was at its peak. In the 1890s, the placers of the Canadian Klondike and Alaska were prominent gold producers. By 1927, the Transvaal (Republic of South Africa) had been the richest gold field in the world for many years, although there were important gold fields in every con-tinent and in most countries (Rose 1948). Major population shifts as a result of gold discoveries are documented for Chile in 1545; in Brazil between 1696 and the 1970s; in Siberia between 1744 and 1866; in the United States in 1799 (North Carolina), 1847 (California), 1858 (Colorado), 1859 (Nevada), 1862 (Idaho), 1864 (Montana), and 1884 (Alaska); in Canada between 1857 and 1896; in Australia between 1850 and 1893; in New Zealand from 1862 to 1865; and in South Africa between 1873 and 1886 (Nriagu and Wong 1997).

The use of mercury in the mining industry to amalgamate and concentrate precious metals dates from about 2700 BCE when the Phoenicians and Carthaginians used it in Spain. Amalgamation became widespread by the Romans in 50 CE and is similar to the process employed today (Lacerda 1997a). In South America, for example, mercury was used extensively by the Spanish colonizers to extract gold, releasing nearly 200,000 metric tons of mercury to the environment between 1550 and 1880 as a direct result of this process (Malm 1998). At the height of the Brazilian gold rush in the 1880s, more than 6 million people were prospecting for gold in the Amazon region alone (Frery et al. 2001). In modern Brazil, where there has been a gold rush since 1980, at least 2000 tons of mercury have been released, with subsequent mercury contamination of sediments, soils, air, fish, and human tissues; a similar situation exists in Colombia, Venezuela, Peru, and Bolivia (Malm 1998). Recent estimates of global anthropogenic total mercury emissions range from 2000 to 4000 metric tons per year of which 460 tons are from small-scale gold mining (Porcella et al. 1995, 1997). Major contributors of mercury to the environment as a result of gold mining activities include Brazil (3000 tons since 1979), China (596 tons since 1938), Venezuela (360 tons since 1989), Bolivia (300 tons since 1979), the Philippines (260 tons since 1986), Columbia (248 tons since 1987), the United States (150 tons since 1969), and Indonesia (120 tons since 1988; Lacerda 1997a).

Adverse health effects from occupational and environmental acute inhalation exposure to mercury include cough, dyspnea, chest pain, bronchitis, pneumonitis, and pulmonary edema (Rojas et al. 2001). Chronic exposure produces gastrointestinal, neurological, and renal effects; and in the mouth, stomatitis, gingivitis, discolored gums, and loose teeth. Neurological symptoms observed include tremors — typically in fingers, arms, legs, and eyelids — fatigue, weakness, depression, headache, insom-nia, drowsiness, inability to concentrate, and loss of memory. Personality changes are common and take the form of shyness, moodiness, excitability, and timidity. Some individuals have developed a sensitivity to mercury, resulting in dermatitis (Rojas et al. 2001).

The total number of gold miners in the world using mercury amalgamation to produce gold ranges from 3 to 5 million, including 650,000 from Brazil, 250,000 from Tanzania, 250,000 from Indonesia, and 150,000 from Vietnam (Jernelov and Ramel 1994). To provide a living — marginal at best — for this large number of miners, gold production and mercury use would come to thousands of tons annually; however, official figures account for only 10% of the production level (Jernelov and Ramel 1994). At least 90% of the gold extracted by individual miners in Brazil is not registered with authorities for a variety of reasons, some financial. Accordingly, official gold production figures reported in Brazil and probably most other areas of the world are grossly underreported (Porvari 1995). Cases of human mercury contamination have been reported from various sites around the world ever since mercury was introduced as the major mining technique to produce gold and other precious metals in South America hundreds of years ago (de Lacerda and Salomons 1998). Contamination is reflected by elevated mercury concentrations in air, water, and diet, and in hair, urine, blood, and other tissues. However, only a few studies actually detected symptoms or clinical evidence of mercury poisoning.

Indigenous peoples of the Amazon living near gold mining activities have elevated levels of mercury in hair and blood. Other indigenous groups are also at risk from mercury contamination, as well as from malaria and tuberculosis (Greer 1993). The miners, mostly former farmers, are also victims of hard times and limited opportunities. Small-scale gold mining is appealing to them as it offers an income, and an opportunity for upward mobility (Greer 1993). Throughout the Brazilian Amazon, about 650,000 small-scale miners are responsible for about 90% of Brazil's gold production and for the discharge of 90 to 120 tons of mercury to the environment every year. About 33% of the miners had elevated concentrations in tissues over the tolerable limit set by the World Health Organization [WHO] (Greer 1993). In Brazil, it is alleged that health authorities are unable to detect conclusive evidence of mercury intoxication due to difficult logistics and the poor health conditions of the mining population, which may mask evidence of mercury poisoning. There is a strong belief that a silent outbreak of mercury poisoning has the potential for regional disaster (de Lacerda and Salomons 1998).

7.2 HEALTH RISKS: UNDERGROUND MINERS

Health problems of gold miners from selected locations in Australia, North America, South America, Europe, and Africa are briefly documented.

7.2.1 Australia

Australian gold miners are vulnerable to dengue fever (a mosquito-borne acute infectious viral disease characterized by headache, severe joint pain, and rash), silicosis (massive fibrosis of the lungs marked by shortness of breath and caused by inhalation of silica dusts, usually SiO_2), and phthisis (a historical term used to describe a wasting condition, possibly pulmonary tuberculosis).

Gold miners were the first recorded victims of dengue fever in 1885 in tropical northeastern Queensland (Russell et al. 1996). In the dengue epidemic of 1993, 2% of the population was infected despite source reduction of surface mosquito breeding grounds. In 1994, larvae and pupae of the dengue vector mosquito *Aedes aegypti* were found in flooded unused shafts of gold mines more than 45 meters below ground. Copepods, *Mesocyclops aspericornis*, were also found in some flooded shafts and were found to be effective predators of mosquito larvae in the laboratory. Copepods (N = 50) were added to about half the mosquito-infested wells and the rest were untreated controls. After 9 months, all copepod-inoculated shafts were free of mosquitos and all untreated wells contained *A. aegypti* larvae. The use of *M. aspericornis* has been recommended as an effective control agent of *Aedes aegypti*, especially in comparatively inaccessible breeding sites, such as flooded gold mine shafts (Russell et al. 1996).

Gold miners from Bendigo suffered — in epidemic proportions for 100 years, from the 1860s to the 1960s — a wasting disease, possibly silicosis or pulmonary tuberculosis. Eventually, it was treated as an occupational sickness, with social, economic, and political implications that resulted in marked improvements in working conditions, better medical treatment, and improved productivity (Kippen 1995). In Western Australia, three retired gold miners were diagnosed with asbestos-related pleural disease after working in gold mines for 5 to 17 years (Lee et al. 1999). They had no other significant known asbestos exposure except for possible asbestos contamination of gold mine dust. Although air from these mines contained measurable concentrations of asbestos fibers, this is the first report of asbestos-related diseases among gold miners. In view of the large number of potentially exposed workers, additional assessment is recommended on the relation between dust exposure from gold mining and asbestos-related lung disease (Lee et al. 1999).

In Western Australia, 2297 gold miners were examined in 1961, 1974, 1985, and 1993 for lung cancer and silicosis (de Klerk and Musk 1998). The incidence of silicosis was clearly related to exposure to silica, and the onset of silicosis conferred a significant increase in risk for subsequent lung cancer. But there was no evidence that exposure to silica caused lung cancer in the absence of silicosis. Silica has recently been reclassified as carcinogenic to humans based largely on the observed increase in rates of lung cancer in patients with silicosis. The International Agency for Research on Cancer has reclassified crystalline silica inhaled in the form of quartz or cristobalite from occupational sources as carcinogenic to humans (Class 1). Previously, silica was in Class 2A, that is, carcinogenic to animals and probably carcinogenic to humans (de Klerk and Musk 1998).

7.2.2   North America

Canadian gold miners had an increased risk of cancer of the trachea, bronchus, lung, and stomach. In the United States, gold miners had significantly higher rates of lung cancer, silicosis, and tuberculosis when compared with the general population, and elevated risks for several debilitating diseases including diseases of the blood, skin, and musculoskeletal system.

A significant excess of mortality from carcinoma of the stomach was demonstrated in gold miners from Ontario, Canada, when compared with other miners (Kusiak et al. 1993). The increased frequency of stomach cancer appeared 5 to 19 years after they began gold mining in Ontario. Twenty or more years after the gold miners started work, stomach cancer cases were significantly greater in miners born outside North America when compared with a reference population, but not in those native born. This late increase is similar to the excess of gastric carcinoma evident in residents of Ontario born in Europe. Possible explanations for the excess of stomach cancer in Canadian gold miners include exposures to arsenic, chromium, mineral fiber, diesel emissions, and aluminum powder. Diesel emissions and aluminum powder were rejected because gold miners and uranium miners were exposed to both agents but excess stomach cancer was noted only in gold miners. Exposure to dust was significant and the time-weighted duration of exposure to dust in gold mines was found in miners under age 60. A statistically significant time-weighted correlation for chromium — but not arsenic or mineral fiber — occurred, especially among gold miners under age 60. Exposure to chromium is associated with the development of the intestinal, rather than the diffuse, type of gastric cancer (Kusiak et al. 1993). Gold miners in Ontario with 5 or more years of gold mining experience before 1945 had a significantly increased risk of primary cancer of the trachea, bronchus, or lung (Kabir and Bilgi 1993). A minimum of 15 years' latency was recorded between first employment in a dusty gold mining occupation and diagnosis of primary lung cancer. For purposes of occupational exposure assessment in establishing work-relatedness, authors concluded that primary lung cancer in Ontario gold miners was related to exposure to silica, arsenic, and radon decay products and was consistent with miners' age at first exposure, length of exposure to dust, and latency (Kabir and Bilgi 1993).

In the United States, the health of 3328 miners who worked underground in a South Dakota gold mine for at least 1 year (average time spent was 9 years) between 1940 and 1965 was analyzed through 1990, with emphasis on exposures to silica and nonasbestiform minerals, by death certificates and radiographic surveys (Steenland and Brown 1995a, 1995b). Miners had been exposed to a median silicon level of 0.05 mg/m^3 after 1930 and 0.15 mg/m^3 for those hired before 1930. The risk of silicosis was less than 1% with a cumulative exposure under 0.5 mg/m^3-yr, increasing to 68 to 84% for the highest cumulative exposure category of more than 4 mg/m^3-yr. Cumulative exposure was the best predictor of silicosis, followed by duration of exposure and average exposure. After adjustment for competing causes of death, a 45-year exposure under the current U.S. Occupational Safety and Health Administration (OSHA) standard of 0.09 mg Si/m^3 would lead to a lifetime risk of silicosis of 35 to 47%, suggesting that the current OSHA silicon exposure level is unacceptably high (Steenland and Brown 1995b). The lung cancer rate of these miners was 13% higher than that of the general U.S. population, 25% higher when the county was the referral group, and 27% higher 30 years postexposure. Miners had significantly higher frequencies of tuberculosis and silicosis with clear exposure–response trends. Renal disease associated with silica exposure was elevated for those hired as young men, and also showed a positive correlation with length of exposure. This group also had significant excesses of arthritis, musculoskeletal diseases, skin diseases, diseases of autoimmune origin, and diseases of the blood and hematopoietic organs (Steenland and Brown 1995a).

7.2.3 South America

Death from mining accidents in Colombia, increased prevalence of malaria in Brazil, and increased frequency of attacks by rabid vampire bats (*Desmodus rotundus*) in Venezuela are documented.

In Colombia, at least 28 gold miners were killed by landslides and dozens reported missing while digging at a condemned strip mine. The victims were said to be poor people who had ignored government warnings that erosion had made the mine unsafe (Toro 2001). This incident was documented in a newspaper, and also, perhaps, in official mining records that were difficult to obtain; however, it is reasonable to conclude that gold mining fatalities are probably grossly under-reported. The prevalence of malaria in Brazil has increased dramatically since the 1980s, particularly in Amazonian gold mining areas where increased colonization and deforestation are recorded (de Andrade et al. 1995). About 600,000 cases of malaria are reported annually in Brazil. The Amazon River Basin accounts for 99% of the cases in Brazil and for about 50% of all cases in the Americas. Infections by *Plasmodium vivax* protozoans represent about 58% of the cases, followed by *Plasmodium falciparum* (41%) and *Plasmodium malariae* (1.0%). Many of the infected miners have no obvious symptoms of malaria and often do not take prescribed antimalarial agents. Malarial control programs rely on early detection and treatment; however, special problems are associated with limited access to gold mining areas, the high mobility of the mining population, and the steady increase in drug-resistant *Plasmodium* species. These alluvial gold mining sites are important reservoirs of drug-resistant *P. falciparum* and other parasites, and nonminers (Indians, farmers, loggers) who live there are at increased risk of malaria (de Andrade et al. 1995). An outbreak of attacks by rabid vampire bats (154 cases in 4 months in a population of about 1500) was documented for the gold mining village of Payapal in south-eastern Venezuela (Caraballo, 1996). Cattle and horses were bitten by vampire bats in the 2-month period preceding the human attacks. The outbreak may be due to loss of normal prey habitat of bats from mining, deforestation, and housing con-struction, with human blood providing an alternative food source.

7.2.4   Europe

A high incidence of neoplasms of the respiratory system among gold extraction and refinery workers in Solsigne, France, was first reported in 1977, and again in 1985, and appears related to occupational exposure (Simonato et al. 1994). Mine and smelter workers at this location were twice as likely to die from lung cancer than the general population. Soluble and insoluble forms of arsenic in combination with other risk factors, such as radon and silica in the mine, are likely determinants of the lung cancer excess (Simonato et al. 1994).

7.2.5   Africa

Gold miners in Africa show increased prevalence of various bacterial and viral diseases (Gabon), noise-induced hearing loss (Ghana), lung cancer (Zimbabwe), carbon monoxide poisoning (Kenya), and in the Republic of South Africa — the largest producer of gold in the world — almost the entire spectrum of mining-related health problems, especially lung diseases and cancer.

Residents of five gold-panning villages in northeastern Gabon were analyzed for seroprevalence of leptospirosis and Ebola virus, both of which can cause lethal hemorrhagic fever (Bertherat et al. 1999). The villages surveyed were remote, isolated communities and their economy was entirely dependent on gold. The seroprevalence was 15.7% for leptospirosis (14.7% of gold miners, 0% of fishermen) and 10.2% for Ebola virus (11.3% of miners, 25.0% of fishermen), demonstrating the persistence of this infection among the endemic population and the need to consider it a potential cause of hemorrhagic fever in Gabon. In another survey, residents from these same villages had elevated (up to 8.5%) blood serum titers for spotted fever and typhus group *Rickettsia* bacteria (Bertherat et al. 1998). The influence of *Rickettsia* on public health in Africa remains unknown, but victims sometimes die as a result of infection by louse and flea vectors (Bertherat et al. 1998).

Noise pollution laws are usually not enforced in developing countries. This was the case at a large gold mining company in central Ghana where 20% of all workers experienced significant noise-induced hearing loss, with frequency rates of 34% for miners, 20% for machine operators, and zero percent for office workers (Amedofu et al. 1996). In general, hearing loss increased with increasing age and noise exposure. The authors concluded that mining companies need to implement hearing conservation programs to protect workers exposed to hazardous noise levels. Lung cancers were reported in gold miners from Zimbabwe, with silica dust and arsenic considered relevant exposures (Boffetta et al. 1994). In Kenya, carbon monoxide is responsible for many deaths underground; in 1980, for example, seven miners died underground due to suffocation from carbon monoxide released from a faulty water pump (Ogola et al. 2002).

The gold mining industry in the Republic of South Africa (RSA) began around 1886 when gold was discovered on the Witwatersrand (Butchart 1996). By 1920, about 200,000 migrant African laborers were employed in the RSA gold mines; in 1961, this number was 427,000, and in 1988 just over 500,000. Most worked underground at depths up to 3500 meters. Until the mid-1970s, when recruiting patterns began to shift toward domestic sources of migrant labor, most workers were recruited from Mozambique and Malawi, with smaller numbers coming from Angola, Botswana, Zambia, and Zimbabwe. In the 1970s, critical studies appeared on the conditions of extreme social and physical deprivation governed by monetary interests and racist policies. These conditions reportedly rendered the labor force excessively prone to tuberculosis and pneumonia, parasitic infections, and traumatic injury or death as a result of poor safety procedures in the mines. The culture of violence from housing in ethnically segregated, single-sex hostels also contributed to the difficulties the miners faced. During this period, mining medicine improved to sustain productivity, although it was widely perceived by black miners as yet another means to repress the African persona (Butchart 1996).

Black miners in RSA comprise approximately 85% of all gold miners in that country (Murray et al. 1996). Between 1975 and 1991, and based on 16,454 case histories, the prevalence of tuberculosis (TB) increased from 0.9% in 1975 to 3.9% in 1991; for silicosis, these values were 9.3% in 1975 and 12.8% in 1991. The frequency of both diseases increased with age and duration of service. Silicosis was the most significant predictor of TB. Lowering of dust levels in the mines was recommended to prevent an increased disease burden (Murray et al. 1996). In a 7-year study, it was shown that miners with chronic simple silicosis had a nearly threefold greater risk of developing TB than did their fellow workers of similar age who did not show radiographic evidence of silicosis at the start of the study; about

25% of the miners with silicosis will have developed TB by age 60 years (Cowie 1994). Death rates of black RSA gold miners from pulmonary TB and silicosis were higher than those from their white counterparts, possibly because of greater severity of silicosis and a high rate of HIV infection (Hnizdo and Murray 1998). By 1996, the death rate from tuberculosis among black migrant miners had risen to 2476 per 100,000, accounting for the largest single cause of death among this group, apart from trauma in the workplace (Churchyard et al. 1999). Concomitantly, HIV prevalence in RSA miners with TB increased from 15% in 1993 to 45% in 1996; HIV is known to interfere with the accuracy of radiological TB screening programs. TB is likely to remain the most important health hazard in RSA mines during the new millennium, necessitating greater commitment to TB control and reduction of risk factors, such as silicosis and HIV infection (Churchyard et al. 1999). The role of HIV, a retrovirus that infects human T-cells and causes acquired immune deficiency syndrome (AIDS) — a condition of deficiency of certain leukocytes resulting in infections and cancer — is discussed later.

During 1980 to 1989, cancer deaths of black male gold miners were due primarily to liver cancer followed by esophageal and lung cancers (Boffetta et al. 1994; McGlashan and Harington 2000). Primary liver cancer during this period was the fourth leading cause of death in the RSA, but first among black gold miners who worked underground (McGlashan and Harington 2000). From 1990 to 1994, esophageal cancer had overtaken liver cancer in numbers of deaths. New cases of esophageal cancer had doubled. New cases of respiratory cancer had also doubled. The reasons for these trends are uncertain but may be associated with repatriation of transient workers to their homelands outside RSA where health care was not as extensive (McGlashan and Harington 2000). In another study, pulmonary dysfunction was measured in black South African gold miners with reactive airways (Cowie 1989). Reactive airways were found in 12% of 1197 older miners, and were not related to extent of exposure to the underground environment. However, those so afflicted were more susceptible to bronchial tree problems after correction for age, tobacco smoking, and presence of silicosis.

White South African miners who had spent at least 85% of their working life in gold mines and had worked underground for at least 15% of their shifts had a 30% chance of dying sooner than the general population due to higher frequencies of lung cancer (140%), heart disease (124%), pulmonary disease (189%), and cirrhosis of the liver (155%). However, very little of this increase could be attributed to gold mining and was instead associated with their unhealthy lifestyle when compared with other South African white males, particularly in smoking and excessive alcohol consumption (Reid and Sluis-Cremer 1996). There is, however, growing evidence that white RSA gold miners — like their black counterparts — were also vulnerable to silicosis (Hnizdo and Sluis-Cremer, 1993), emphysema (Hnizdo et al. 2000), lung cancer (Hnizdo et al. 1997), asthma (Cowie and Mabena 1996), and pulmonary tuberculosis (Hnizdo and Murray 1998).

RSA gold miners have among the highest rates of TB in the world. This is attributed, in part, to the high endemic rate of TB in rural regions from which miners are recruited, crowding, silica dust exposure, increasing age of the work force, and HIV infection. Rates are rising, despite cure rates that meet WHO targets in patients

with new TB (Godfrey-Faussett et al. 2000). The incidence of pulmonary tuberculosis in RSA gold miners increased from 686 per 100,000 workers in 1989 to more
than 1800 per 100,000 in 1995 (Sonnenberg et al. 2000). Changes were associated
with longer service and a rise in the average age of the work force. Miners with
pulmonary mycobacterial disease were more likely to have nontuberculosis mycobacteria (NTM) than *Mycobacterium tuberculosis* (TB) if they had worked longer
underground, had silicosis, or had been treated previously for TB. Attempts to reduce
the incidence of all pulmonary mycobacterial disease among gold miners should
include early diagnosis and treatment (Sonnenberg et al. 2000). Despite a control
program that cures 86% of new cases, most TB in this mining community is due to
ongoing transmission from persistently infectious individuals who have previously
failed treatment and may be responsible for as many as one third of TB cases
(Godfrey-Faussett et al. 2000). There is a low incidence of NTM isolates and diseases
in developed countries; however, this incidence is 27% in RSA miners (Corbett et
al. 1999a) and is largely attributable to chronic chest disease from silicon dust
inhalation and prior tuberculosis (Corbett et al. 1999c). Previous studies have shown
that isolates of the most common NTM species, *M. kansasii* and *M. scrofulaceum*,
occur with high incidence and are more often associated with NTM risk factors such
as silicosis and lung diseases than with patients with TB or control patients (Corbett
et al. 1999a). During the study period, NTM were isolated from 118 patients of
whom 40 (34%) were HIV positive (Corbett et al. 1999a). HIV infection has recently
become an additional risk factor for mycobacterial disease in miners and is likely to
become increasingly important as the HIV epidemic progresses (Corbett et al. 1999c).

The association between silicosis and pulmonary tuberculosis (PTB) is well
established (Hnizdo and Murray 1998). Epidemiological and case studies show that
workers exposed to silica dust have increased morbidity and mortality from PTB.

Silicosis reflects a failure in adequate control of occupational dust exposure
(Cowie 1998). The risk of silicosis in a cohort of 2235 white RSA gold miners with
an average of 24 years of mining experience between 1940 and 1970 was followed
up to 1991 for radiological signs of onset of silicosis (Hnizdo and Sluis-Cremer
1993). About 14% of the miners developed silicosis at an average age of 56 years,
with radiological signs appearing, on average, 7.4 years after mining exposure
ceased. The risk of silicosis was strongly dose-dependent, although the latency period
was variable. Silicosis risk increased exponentially with the cumulative dust dose,
the accelerated increase occurring after 7 mg/m^3-yr. At the highest exposure level
of 15 mg/m^3-yr — equivalent to about 37 years of gold mining exposed to an average
respirable dust concentration of 0.4 mg/m^3 — the cumulative risk for silicosis
reached 77% (Hnizdo and Sluis-Cremer 1993). There is also a positive association
between exposure to silica dust and risk of lung cancer (Boffetta et al. 1994); risks
were higher among those exposed to higher dust exposures and also diagnosed with
silicosis (Hnizdo et al. 1997). Miners who had withdrawn from dusty occupations
showed declines in lung function similar to those who continued to work underground for 5 years (Cowie 1998). RSA gold miners with chronic obstructive airway
disease from working in a dusty atmosphere in designated mines or works were
entitled to workmen's compensation, as judged by lung function tests for airflow
obstruction (Hnizdo et al. 2000).

The association between silicosis and pulmonary tuberculosis (PTB) is well
established (Hnizdo and Murray 1998). Epidemiological and case studies show that
workers exposed to silica dust have increased morbidity and mortality from PTB.

In one study, a cohort of 2255 white RSA gold miners was evaluated for increased risk of PTB from 1968 to 1971, when they were 45 to 55 years of age, to December 1995. During the followup, 1592 (71%) of this cohort died. Of these, 1296 (81%) were necropsied to determine the presence of silicosis and PTB. It was concluded that exposure to silica dust is a risk factor for the development of PTB in the absence of silicosis, even after exposure to silica dust ends, and that the risk of PTB increases with the presence of silicosis. In miners without silicosis, but with increasing exposure to dust, the severity of silicosis was associated with increasing risk of PTB (Hnizdo and Murray 1998). In addition to silicosis, TB, and obstructive airway disease, RSA gold miners showed a high prevalence of previously undiagnosed and untreated pneumoconiosis, a lung disease caused by habitual inhalation of irritant mineral or metallic particles (Trapido et al. 1998).

South Africa currently harbors one of the fastest-growing HIV epidemics in the world (Bredell et al. 1998). The prevalence of HIV-1 in pregnant women has increased from 0.76% in 1990 to 14.1% in 1996, with more than 2.5 million South Africans infected. Migrant workers employed as RSA gold miners were found infected with HIV-1 (Bredell et al. 1998). HIV infection and silicosis are powerful risk factors for TB and are associated with an increased risk of death among RSA gold miners. The incidence of TB was almost five times greater in HIV-positive than HIV-negative miners (Corbett et al. 2000). Among RSA gold miners with TB, the prevalence of HIV infection increased rapidly to about 50% of all cases between 1993 and 1997 (Churchyard et al. 2000). NTM disease incidence, morbidity, and mortality are likely to increase further among miners as the HIV epidemic progresses (Corbett et al. 1999b).

RSA gold miners have a high prevalence of HIV infection (Campbell 1997). Most are migrants from rural areas within South Africa, and others are from surrounding countries such as Lesotho, Botswana, and Mozambique. The vast majority of these workers are housed in single-sex hostels close to their workplace. Despite extensive education from mine operators on the consequences of unprotected sex, this group perceives condom use as a diminishment of their masculinity and continues to practice risky behaviors with sex workers, and the incidence of sexually transmitted diseases in these men is extremely high (Corbett et al. 1999c). Many workers commented that the risk of HIV/AIDS appears minimal compared to the risks of death or injury underground and that this was the reason why many mine workers did not bother with condoms. It remains unclear how best to communicate risks of HIV and prevent transmission by altering risky behaviors in African populations (Campbell 1997).

7.3 HEALTH RISKS: SURFACE MINERS WHO USE MERCURY

In general, mercury concentrations in drinking water, soils, sediments, tailings, or edible fish near gold mining operations exceeded national and international limits in the Philippines (Appleton et al. 1999; Akagi et al. 2000), Kenya (Ogola et al. 2002) and Brazil (Malm et al. 1990, 1995b), but not in fish diets of farmers and miners in Tanzania (van Straaten 2000; Campbell et al. 2003) and Colombia (Olivero et al. 1997). Selected case histories of mercury intoxication associated with gold mining activities are shown for Australia, the Philippines, Brazil, and Venezuela. Measured mercury concentrations in hair, urine, blood, and other human tissues in the vicinity of gold amalgamation extraction and refining activities are shown together with similar data for ambient air and fish diets.

7.3.1 Case Histories

Most countries recognize mercury hazards to health, although occupational health problems and direct inorganic mercury toxicosis rarely appear (Porcella et al. 1995, 1997). The following case histories provide anecdotal evidence for mercury poisoning associated with the amalgamation process, although large-scale epidemiological evidence for the association does not exist.

A 19-year-old male in Queensland, Australia, developed hand tremors and fatigue after starting work at a placer gold mine where he was exposed to Au–Hg amalgam (Donoghue 1998). His mercury urine level of 143 µg/L exceeded the recommended no-adverse-effect-level of 50 µg/L. Seven weeks after removal from the work environment, his mercury urine level had fallen to 32 µg/L and the tremors had almost disappeared. Contaminated air was the source of the exposure. Smelting of retorted gold with previously unrecognized mercury resulted in peak air levels of 0.53 mg Hg/m^3 (vs. a recommended threshold level of 0.05 mg Hg/m^3). Engineering and procedural controls were instituted to prevent further occurrences (Donoghue 1998).

In 1987, 11 Filipinos became ill and one died after 8 hours spent blowtorching about 2 kg of a Au–Hg amalgam indoors (Greer 1993; de Lacerda and Salomons 1998), strongly indicating the need to burn amalgam and melt gold in closed vessels (Drasch et al. 2001). It was alleged that Filipinos who lived within 500 meters of a similar mercury source for 30 months had a 75% probability of exhibiting symptoms of clinical mercury poisoning (Greer 1993). Latent effects of mercury exposure on health in Philippine gold miners were attributed to occupational exposure during the 1960s, 30 years earlier (Akagi et al. 2000). The United Nations Environment Program recommends the banning of elemental mercury in amalgamation processes (Greer 1993), but in the Philippines and elsewhere, little appears to have been accomplished by attempts to educate miners about mercury's hazards or by regulations concerning its use as long as the price of gold makes extraction and processing economically viable (Greer 1993).

In another case from the Philippines, school children from an area of intense mercury amalgamation activities were frequently under height, under weight, and presented with gum discolorations and skin abnormalities (Akagi et al. 2000). Blood mercury concentrations were elevated (up to 57 µg total Hg/L; 47 µg methylmercury/L), as were mercury concentrations in hair (up to 20 mg total Hg/kg dry weight [DW]; up to 18 mg methylmercury/kg DW). Symptoms were attributed to ingestion of mercury-contaminated fish containing up to 0.44 mg total Hg/kg fresh weight (FW) muscle, or 0.38 mg methylmercury/kg FW (Akagi et al. 2000).

In Brazil, armed force by the military has proven ineffective in stopping the illegal use of mercury by more than an estimated one million gold miners (Veiga et al. 1995). It is postulated that mercury pollution is sufficiently severe to cause adverse effects on brain development, as evidenced by elevated (>10 mg/kg) mercury concentrations in hair of Brazilian children ages 7 to 12 (Grandjean et al. 1999). About 80% of the children examined had >10 mg Hg/kg hair. Decreasing test scores of children subjected to neuropsychological tests of motor function, attention span, and visuospatial performance were associated with increasing hair mercury levels (Grandjean et al. 1999). Mercury discharges from informal gold mining activities in the Amazon region are attributed to a lack of concern of the miners for the environment as well as poor knowledge about efficient gold extraction techniques. Moreover, individuals in contact with these miners, including priests, physicians, hygienists, social workers, nurses, mining inspectors, union personnel, and others, also lack information about mercury transformations in the environment. Education on the toxicological properties of mercury and its transformation products is strongly recommended as a viable solution to reducing emissions and for creating opportunities for illegal miners to form their own legal companies (Veiga et al. 1995; Veiga and Meech 1995). Current approaches to protect human health in Brazil against mercury intoxication include prohibition of elemental mercury in gold mining activities and the temporary suspension of gold mining operations (Malm et al. 1995a). But these solutions are not realistic for economic and political reasons. Minimizing mercury emissions and limiting consumption of larger carnivorous fish seem reasonable recommendations (Malm et al. 1995a).

Venezuelan gold miners ($N = 40$) exposed to elevated air concentrations of mercury exceeded hair and urine mercury occupational exposure guidelines (Rojas et al. 2001). Overall, mercury guidelines for air were exceeded in 18% of the measurements, in urine 48% of the cases, and in hair 24% of the individuals. Despite substantial occupational exposure to mercury for some individuals, few adverse health effects were observed that were clearly related to mercury exposure (Rojas et al. 2001).

7.3.2 Mercury in Tissues

Data from gold miners and gold mining communities show conclusively that mercury concentrations in urine, blood, hair, and breast milk exceed by a wide margin the most conservative criteria of mercury contamination proposed by various national and international regulatory agencies. However, the large variability in concentrations of mercury in blood, urine, and hair among individuals exposed to mercury vapor, inorganic mercury, and methylmercury, does not seem to adequately monitor mercury burdens of brain and other target tissues (Drasch et al. 2001).

In urine, mercury concentrations from different gold mining sites and groups occupationally exposed to mercury emissions frequently exceeded the WHO levels of 50 µg/L, the maximum acceptable concentration, and 100 µg/L, the minimum concentration before developing symptoms of mercury poisoning (de Kom et al. 1998; de Lacerda and Salomons 1998; Drake et al. 2001; Drasch et al. 2001). Workers in gold trader shops at three Brazilian locations had maximum levels of 79, 160, and 1168 µg Hg/L urine (Malm et al. 1995b; de Lacerda and Salomons 1998). About 22% of all Brazilian gold shop workers exposed to mercury vapors had urinary levels in excess of 50 µg Hg/L (Santa Rosa et al. 2000). In Tanzania, where it is estimated that 250,000 people are involved in small-scale gold mining using mercury amalgamation, mercury levels in urine of miners reached 241 μg/L vs. 3 μg/L in a control group (Ikingura and Akagi, 1996); about 36% of Tanzanian gold miners working with amalgam exceeded the WHO guideline for mercury in urine (van Straaten 2000; Campbell et al. 2003). Elevated mercury levels in urine were positively correlated with increasing fish consumption, alcohol consumption on a yearly basis, number of hours worked daily, and number of dental amalgam fillings (Santa Rosa et al. 2000). In China, mean mercury concentrations in urine of villagers from a gold mining community ranged between 38 and 87 μg/L, reaching maximum concentrations of 540 μg/L in boys under 16,418 μg/L in adult women, 290 μg/L in adult men, and 195 μg/L in girls under 16; for the entire population, 91% exceeded 50 μg Hg/L urine (Lin et al. 1997).

In hair, mercury concentrations, as methylmercury, should not exceed 4 to 7 mg/kg; higher concentrations of 10 to 20 mg/kg are associated with abnormal infant development, and 50 to 100 mg/kg with paraesthesia (de Lacerda and Salomons 1998). In Brazilian gold mining populations that consume more than 100 grams of fish daily, levels of methylmercury in hair of women of child-bearing age should be monitored because 83% exceeded 10 mg/kg DW and 8.3% exceeded 30 mg methylmercury/kg hair (Kehrig et al. 1997). Elevated concentrations of total mercury (25 to 37 mg/kg) were observed in hair of inhabitants from fishing villages near gold mining areas, but not in gold miners (4.1 mg/kg); more than 90% of the total mercury in hair from both groups was in the form of methylmercury (Akagi et al. 1995a). To reach 50 mg total Hg/kg hair — the recommended maximum acceptable limit — it is necessary to consume 330 g of fish muscle daily containing 1 mg Hg/kg FW (Aula et al. 1994). Total mercury levels in hair, up to 71 mg/kg, are recorded in Brazilian gold mining areas, and up to 176 mg/kg in fishing villages from consumption of mercury-contaminated fish (Akagi et al. 1995b; Malm et al. 1995a, 1997; Lacher and Goldstein 1997; Kehrig et al. 1997). In one case, a hair mercury level of 240 mg/kg was measured in a fisherman who consumed 14 fish meals weekly from fish captured in an artificial reservoir that received gold mining wastes vs. 8.5 mg/kg in a nearby population that consumed, on average, only two fish meals weekly (Leino and Lodenius 1995). Elevated (>10 mg/kg) hair mercury concentrations attributed to gold mining activities, especially to consumption of mercury-contaminated fish, are reported in Bolivia (Maurice-Bourgoin et al. 2000), Tanzania (Harada et al. 1999), the Philippines (Appleton et al. 1999; Akagi et al. 2000; Drasch et al. 2001), French Guiana (Frery et al. 2001), and Columbia (Olivero et al. 1998). In French Guiana, for example, 57% of the native Amerindians had hair mercury concentrations that exceeded the WHO recommended limit of 10 mg/kg (Frery et al. 2001). Some Tanzanian subjects who showed a high total mercury level in hair made habitual use of toilet soap containing about 2% mercuric iodide (Harada et al. 1998; van Straaten 2000; Campbell et al. 2003), and this needs to be considered in evaluating mercury risk assessment in gold mining areas. To reduce mercury loadings in the hair of gold miners, fisherman, and their families, a change in diet is recommended from predatory to herbivorous fishes, and a reduction in mercury emissions from current gold mining practices (Leino and Lodenius 1995).

Blood mercury concentrations in gold miners are primarily related to exposure to metallic mercury vapor in the air and to consumption of mercury-contaminated fish diets (Barbosa 1997). Blood mercury levels greater than 10 µg/L (the recommended maximum level) were exceeded by 33% of gold prospectors in the Brazilian Amazon (de Lacerda and Salomons 1998). Gold miners and refiners contained 25 (8 to 159) µg Hg/L blood when exposure to mercury was within the past two days; blood mercury levels were 7.6 (2.2 to 19.4) µg/L when this group had been exposed 2 to 60 days previously; and 5.6 (3 to 14) µg/L when exposed to mercury vapors more than 60 days earlier (Aks et al. 1995). Blood mercury concentrations up to 108 µg/L were measured in ball mill workers from gold mining areas of Mindanao, Philippines (Drasch et al. 2001), and up to 57 µg/L (47 µg/L as methylmercury) in school children age 5 to 17 years studying near eleven Philippine gold mills and processing plants (Akagi et al. 2000).

Ingestion of mercury-contaminated diets as a result of gold mining activities in Amazonia is alleged to be the cause of elevated mercury concentrations in breast milk of nursing mothers (Barbosa and Dorea, 1998). Concentration of total mercury in breast milk from nursing mothers in the Amazon Basin, Brazil, ranged from 0.0 to 24.8 µg/kg with a mean of 5.8 µg/kg (vs. 0.9 µg/kg in the U.S.). Of the infants measured, 53% fed milk from these mothers may receive more than 0.5 µg Hg/kg body weight (the tolerable daily mercury intake recommended for adults by the WHO). The mercury concentration in breast milk was not significantly correlated with mercury content in hair of the mother or infant (Barbosa and Dorea 1998).

7.3.3   Mercury in Air and in Fish Diet

Air and fish diet — the main routes of mercury entry into gold miners — are typically grossly contaminated with mercury. In Brazil, for example, more than 130 tons of mercury are discharged into the biosphere each year as a result of gold mining activities; 55% is discharged into the atmosphere and 45% into aquatic ecosystems (Frery et al. 2001). Within the biogeochemical cycle of the metal, Hg^0 can be oxidized to inorganic Hg^{2+} and then methylated by biotic (bacteria) and abiotic (humic acids as methyl group donors) processes (Eisler 2000b; Frery et al. 2001).

Inhalation of metallic mercury vapor is the main route of human occupational exposure, and gold shop workers are at the greatest risk (Malm et al. 1995b; Hacon et al. 1997). Concentrations of mercury in ambient air of gold dealer shops and workplaces in gold mining areas exceeded the generally recommended exposure level of 50 µg Hg/m³ in the Philippines (1664 µg/m³), Brazil (292 µg/m³), and elsewhere (de Lacerda and Salomons 1998). In Venezuela, where ambient air concentrations of occupational gold workers reached a maximum of 6315 µg/m³ and averaged 183 µg/m³, about 20% of the exposures exceeded 50 µg/m³ — the recommended exposure limit of the U.S. National Institute of Occupational Safety and Health — and 26% exceeded 25 µg/m³ — the threshold limit value recommended by the American Conference of Governmental Industrial Hygienists (Drake et al. 2001). About 72% of Philippine workers who refined amalgam through burning were classified as mercury-intoxicated (Drasch et al. 2001). In Dexing County, Jiangxi Province, China, about 200 small-scale gold mines were in operation between 1990 and 1995 using mercury amalgamation to extract gold (Lin et al. 1997). Gold firing was usually conducted in private residences with excessive mercury contamination of the air in workrooms (up to 2600 $\mu g/m^3$) and workshops (up to 1000 $\mu g/m^3$). Since September 1996, most small-scale gold mining activities have been prohibited through China's national environmental legislation (Lin et al. 1997).

In the Mato Grosso's Pantanal, the world's largest wetland, fish muscle contained mercury concentrations up to 24 times higher than the level considered safe (<0.5 mg total Hg/kg FW muscle) for human consumption by the WHO (Greer 1993). Mercury-contaminated fish were detected up to 590 km downstream from some gold mining areas (Greer 1993). In the Amazonian region of Bolivia, regular consumption of fish contaminated by mercury from gold mining activities is considered to be a major threat to public health (Bidone et al. 1997; Maurice-Bourgoin et al. 2000). Mercury concentrations in muscle of edible fish collected near various mercury-amalgamation gold mining sites were routinely elevated in Colombia (Olivero et al. 1998), French Guiana (Frery et al. 2001), the Philippines (Appleton et al. 1999), and Brazil (Malm et al. 1990, 1995a, 1997; Akagi et al. 1995a; Porvari 1995; Bidone et al. 1997). Mercury concentrations in muscle of Brazilian fishes sampled from 1997 to 1998 were highest in carnivorous species, lowest in herbivores, and inter-mediate in omnivores; concentrations were lower with increasing distance from gold mining areas (Aula et al. 1994; Lima et al. 2000). In some areas of the Amazonian region of Brazil, the estimated intake of mercury from fish consumption is 114 μg daily, or about three to four times higher than the daily tolerable intake recommended by the WHO; the estimated exposure level of 1.6 μg Hg/kg body weight for adult consumers was more than five times higher than the recommended consumption limit of 0.3 μg Hg/kg BW daily (Bidone et al. 1997). Variations in mercury content of muscle from some Amazon fish species could not be related to nearby gold mining activities (Jernelov and Ramel 1994).

7.4 SUMMARY

Health problems of gold miners who worked underground include decreased life expectancy; increased frequency of cancer of the trachea, bronchus, lung, stomach, and liver; increased frequency of pulmonary tuberculosis, silicosis, and pleural diseases; increased frequency of insect-borne diseases, such as malaria and dengue fever; noise-induced hearing loss; increased prevalence of certain bacterial and viral diseases; and diseases of the blood, skin, and musculoskeletal system. These prob-lems are briefly documented in gold miners from Australia, North America, South America, Europe, and Africa. In general, HIV infection or excessive alcohol and tobacco consumption tended to exacerbate existing health problems. Miners who used elemental mercury to amalgamate and extract gold were heavily contaminated with mercury. Among individuals exposed occupationally, concentrations of mercury in their air, fish diet, hair, urine, blood, and other tissues significantly exceeded all criteria proposed by various national and international regulatory agencies for pro-tection of human health. However, large-scale epidemiological evidence of severe mercury-associated health problems in this cohort was not demonstrable.

To protect the health of underground workers, authorities recommend continued intensive monitoring of atmospheric dust levels in order to conform to recognized safe occupational levels, implementation of more frequent medical examinations with emphasis on early detection and treatment of disease states, and continuation of educational programs on hazards of risky behaviors outside of the mine environment; these recommendations can be implemented satisfactorily through mine management. Miners who use elemental mercury to extract gold need to control mercury emissions in confined environments and to be made cognizant of mercury hazards through educational programs administered by informed personnel, as well as to limit their consumption of larger carnivorous fishes. Intensive monitoring by physicians and toxicologists of populations at high risk from mercury poisoning is strongly recommended in order to provide evidence of adherence to existing criteria, as is re-examination of current mercury criteria to protect human health.

LITERATURE CITED

Akagi, H., E.S. Castillo, N. Cortes-Maramba, A.T. Francisco-Rivera, and T.D. Timbang. 2000. Health assessment for mercury exposure among schoolchildren residing near a gold processing and refining plant in Apokon, Tagum, Davao del Norte, Philippines, *Sci. Total Environ.*, 259, 31–43.

Akagi, H., O. Malm, F.J.P. Branches, Y. Kinjo, Y. Kashima, J.R.D. Guimaraes, R.B. Oliveira, K. Haraguchi, W.C. Pfeiffer, Y. Takizawa, and H. Kato. 1995a. Human exposure to mercury due to goldmining in the Tapajos River Basin, Amazon, Brazil: speciation of mercury in human hair, blood and urine, *Water Air Soil Pollut.*, 80, 85–94.

Akagi, H., O. Malm, Y. Kinjo, M. Harada, F.J.P. Branches, W.C. Pfeiffer, and H. Kato. 1995b. Methylmercury pollution in the Amazon, Brazil, *Sci. Total Environ.*, 175, 85–95.

Aks, S.E., T. Erickson, F.J.P. Branches, C. Naleway, H.N. Chou, P. Levy, and D. Hryhorczuk. 1995. Fractional mercury levels in Brazilian gold refiners and miners, *Clin. Toxicol.*, 33, 1–10.

Amedofu, G.K., G.W. Brobby, and G.A. Ocansey. 1996. Occupational hearing loss among workers at a large gold mining company in Ghana (west Africa), *Austral. Jour. Audiol.*, 18, 47–48.

Appleton, J.D., T.M. Williams, N. Breward, A. Apostol, J. Miguel, and C. Miranda. 1999. Mercury contamination associated with artisanal gold mining on the island of Mindanao, the Philippines, *Sci. Total Environ.*, 228, 95–109.

Aula, I., H. Braunschweiler, T. Leino, I. Malin, P. Porvari, T. Hatanaka, M. Lodenius, and A. Juras. 1994. Levels of mercury in the Tucurui reservoir and its surrounding area in Para, Brazil, in *Mercury Pollution: Integration and Synthesis*, C.J. Watras and J.W. Huckabee, (Eds.), Lewis Publishers, CRC Press, Boca Raton, FL, 21–40.

Barbosa, A.C. 1997. Mercury in Brazil: present or future risks? *Cien. Cult. Jour. Brazil. Assoc. Adv. Sci.*, 49, 111–116.

Barbosa, A.C. and J.G. Dorea. 1998. Indices of mercury contamination during breast feeding in the Amazon Basin, *Environ. Toxicol. Pharmcacol.*, 6, 71 –79.

Bertherat, E., R. Nabias, A.J. Georges, and A. Renaut. 1998. Seroprevalence of *Rickettsia* in a gold-panning population in north-eastern Gabon, *Trans. Roy. Soc. Trop. Med. Hyg.*, 92, 393–394.

Bertherat, E., A. Renault, R. Nabias, G. Dubreuil, and M.C. Georges-Courbot. 1999. Leptospirosis and ebola virus infection in five gold-panning villages in northeastern Gabon, *Amer. Jour. Trop. Med. Hyg.*, 60, 610–615.

Bidone, E.D., Z.C. Castilhos, T.J.S. Santos, T.M.C. Souza, and L.D. Lacerda. 1997. Fish contamination and human exposure to mercury in Tartarugalzinho River, Amapa State, Northern Amazon, Brazil. A screening approach, *Water Air Soil Pollut.*, 97, 9–15.

Boffetta, P., M. Kogevinas, N. Pearce, and E. Matos. 1994. Cancer, in *Occupational Cancer in Developing Countries*, Int. Agen. Res. Cancer, IARC Sci. Publ. 129, Oxford University Press, New York, 111–126.

Bredell, H., C. Williamson, P. Sonnenberg, D.J. Martin, and L. Morris. 1998. Genetic characterization of HIV Type 1 from migrant workers in three South African gold mines, *AIDS Res. Human Retrovir.*, 14, 677–684.

Butchart, A. 1996. The industrial panopticon: mining and the medical construction of migrant African labour in South Africa, 1900–1950, *Social Sci. Med.*, 42, 185–197.

Campbell, C. 1997. Migrancy, masculine identities and AIDS: the psychosocial context of HIV transmission on the South African gold mines, *Social Sci. Med.*, 45, 273–281.

Campbell, L.M., D.G. Dixon, and R.E. Hecky. 2003. A review of mercury in Lake Victoria, East Africa: implications for human and ecosystem health, *Jour. Toxicol. Environ. Health*, 6B, 325–356.

Caraballo, H.A.J. 1996. Outbreak of vampire bat biting in a Venezuelan village, *Rev. Saude Publica*, 30, 483–484.

Churchyard, G.J., I. Kleinschmidt, E.L. Corbett, J. Murray, J. Smit, and K.M. De Cock. 2000. Factors associated with an increased case-fatality rate in HIV-infected and non-infected South African gold miners with pulmonary tuberculosis, *Int. Jour. Tubercul. Lung Dis.*, 4, 705–712.

Churchyard, G.J., I. Kleinschmidt, E.L. Corbett, D. Mulder, and K.M. De Cock. 1999. Mycobacterial disease in South African gold miners in the era of HIV infection, *Int. Jour. Tubercul. Lung Dis.*, 3, 791–798.

Corbett, E.L., L. Blumberg, G.J. Churchyard, N. Moloi, K. Mallory, T. Clayton, B.G. Williams, R.E. Chaisson, R.J. Hayes, and K.M. De Cock. 1999a. Nontuberculosis mycobacteria: defining disease in a prospective cohort of South African miners, *Amer. Jour. Respir. Crit. Care Med.*, 160, 15–21.

Corbett, E.L., G.J. Churchyard, M. Hay, P. Herselman, T. Clayton, B. Williams, R. Hayes, D. Mulder, and K.M. De Cock. 1999b. The impact of HIV infection on *Mycobacterium kansasii* disease in South African gold miners, *Amer. Jour. Respir. Crit. Care Med.*, 160, 10–14.

Corbett, E.L., G.J. Churchyard, T. Clayton, P. Herselman, B. Williams, R. Hayes, D. Mulder, and K.M. De Cock. 1999c. Risk factors for pulmonary mycobacterial disease in South African gold miners, *Amer. Jour. Respir. Crit. Care Med.*, 159, 94–99.

Corbett, E.L., G.J. Churchyard, T.C. Clayton, B. G. Williams, D. Mulder, R.J. Hayes, and K.M. De Cock. 2000. HIV infection and silicosis: the impact of two potent risk factors on the incidence of mycobacterial disease in South African miners, *AIDS*, 14, 2757–2768.

Cowie, R.L. 1989. Pulmonary dysfunction in gold miners with reactive airways, *Brit. Jour. Indus. Med.*, 46, 873–876.

Cowie, R.L. 1994. The epidemiology of tuberculosis in gold miners with silicosis, *Amer. Jour. Respir. Crit. Care Med.*, 150, 1460–1462.

Cowie, R.L. 1998. The influence of silicosis on deteriorating lung function in gold miners, *Chest*, 113, 340–343.

Cowie, R.L. and S.K. Mabena. 1996. Asthma in goldminers, *South Afr. Med. Jour.*, 86, 804–807.

de Andrade, A.L.S.S., C.M.T. Martelli, R.M. Oliveira, J.R. Arias, F. Zicker, and L. Pang. 1995. High prevalence of asymptomatic malaria in gold mining areas in Brazil, *Clin. Infect. Dis.*, 20, 475.

de Klerk, N.H. and A.W. Musk. 1998. Silica, compensated silicosis, and lung cancer in Western Australian goldminers, *Occup. Environ. Medic.*, 55, 243–248.

de Kom, J.F.M., G.B. van der Voet, and F.A. de Wolff. 1998. Mercury exposure of maroon workers in the small scale gold mining in Suriname, *Environ. Res.*, 77A, 91–97.

de Lacerda, L.D. and W. Salomons. 1998. *Mercury from Gold and Silver Mining: A Chemical Time Bomb?* Springer, Berlin, 146 pp.

Donoghue, A.M. 1998. Mercury toxicity due to the smelting of placer gold recovered by mercury amalgam, *Occup. Med.*, 48, 413–415.

Drake, P.L., M. Rojas, C.M. Reh, C.A. Mueller, and F.M. Jenkins. 2001. Occupational exposure to airborne mercury during gold mining operations near El Callao, Venezuela, *Int. Arch. Occup. Environ. Health,* 74, 206–212.

Drasch, G., S. Bose-O'Reilly, C. Beinhoff, G. Roider, and S. Maydl. 2001. The Mt. Diwata study on the Philippines 1999 — assessing mercury intoxication of the population by small scale gold mining, *Sci. Total Environ.*, 267, 151–168.

Eisler, R. 2000a. *Handbook of Chemical Risk Assessment: Health Hazards to Humans, Plants, and Animals. Volume 2. Organics.* Lewis Publishers, Boca Raton, FL, 903–959.

Eisler, R. 2000b. *Handbook of Chemical Risk Assessment: Health Hazards to Humans, Plants, and Animals. Volume 1. Metals.* Lewis Publishers, Boca Raton, FL, 313–409.

Eisler, R. 2003. Health risks to gold miners: a synoptic review, *Environ. Geochem. Health*, 25, 325–345.

Eisler, R., D.R. Clark Jr., S.N. Wiemeyer, and C.J. Henny. 1999. Sodium cyanide hazards to fish and other wildlife from gold mining operations, in *Environmental Impacts of Mining Activities: Emphasis on Mitigation and Remedial Measures*, J.M. Azcue, (Ed.), Springer-Verlag, Berlin, 55–67.

Frery, N., R. Maury-Brachet, E. Maillot, M. Deheeger, B. de Merona, and A. Boudou. 2001. Gold-mining activities and mercury contamination of native Amerindian communities in French Guiana: key role of fish in dietary uptake, *Environ. Health Perspec.*, 109, 449–456.

Godfrey-Faussett, P., P. Sonnenberg, S.C. Shearer, M.C. Bruce,C. Mee, L. Morris, and J. Murray. 2000. Tuberculosis control and molecular epidemiology in a South African gold-mining community, *Lancet*, 356, 1066–1071.

Grandjean, P., R.F. White, A. Nielsen, D. Cleary, and E.C. de Oliveira Santos. 1999. Methylmercury neurotoxicity in Amazonian children downstream from gold mining, *Environ. Health Perspec.*, 107, 587–591.

Greer, J. 1993. The price of gold: environmental costs of the new gold rush, *The Ecologist*, 23 (3), 91–96.

Hacon, S., E.R. Rochedo, R. Campos, G. Rosales, and L.D. Lacerda. 1997. Risk assessment of mercury in Alta Floresta, Amazon Basin — Brazil, *Water Air Soil Pollut.*, 97, 91–105.

Harada, M., S. Nakachi, T. Cheu, H. Hamada, Y. Ono, T. Tsuda, K. Yanagida, T. Kizaki, and H. Ohno. 1999. Monitoring of mercury pollution in Tanzania: relation between head hair mercury and health, *Sci. Total Environ.*, 227, 249–256.

Hnizdo, E. and J. Murray. 1998. Risk of pulmonary tuberculosis relative to silicosis and exposure to silica dust in South African gold miners, *Occup. Environ. Med.*, 55, 496–502.

Hnizdo, E., J. Murray, and A. Davison. 2000. Correlation between autopsy findings for chronic obstructive airways disease and in-life disability in South African gold miners, *Int. Arch. Occup. Environ. Health*, 73, 235–244.

Hnizdo, E., J. Murray, and S. Klempman. 1997. Lung cancer in relation to exposure to silica dust, silicosis and uranium production in South African gold miners, *Thorax*, 52, 271–275.

Hnizdo, E. and G.K. Sluis-Cremer. 1993. Risk of silicosis in a cohort of white South African gold miners, *Amer. Jour. Indus. Med.*, 24, 447–457.

Ikingura, J.R. and H. Akagi. 1996. Monitoring of fish and human exposure to mercury due to gold mining in the Lake Victoria goldfields, Tanzania, *Sci. Total Environ.*, 191, 59–68.

Jernelov, A. and C. Ramel. 1994. Mercury in the environment, *Ambio*, 23, 166.

Kabir, H. and C. Bilgi. 1993. Ontario gold miners with lung cancer, *Jour. Occup. Med.*, 35, 1203–1207.

Kehrig, H.A., O. Malm, and H. Akagi. 1997. Methylmercury in hair samples from different riverine groups, Amazon, Brazil, *Water Air Soil Pollut.*, 97, 17–29.

Kippen, S. 1995. The social and political meaning of the silent epidemic of miners' phthisis, Bendigo 1860–1960, *Social Sci. Med.*, 41, 491–499.

Kirkemo, H., W.L. Newman, and R.P. Ashley. 2001. *Gold*. U.S. Geological Survey, Denver, 23 pp.

Kusiak, R.A., A.C. Ritchie, J. Springer, and J. Muller. 1993. Mortality from stomach cancer in Ontario miners, *Brit. Jour. Indus. Med.*, 50, 117–126.

Lacerda, L.D. 1997a. Global mercury emissions from gold and silver mining, *Water Air Soil Pollut.*, 97, 209–221.

Lacerda, L.D. 1997b. Evolution of mercury contamination in Brazil, *Water Air Soil Pollut.*, 97, 247–255.

Lacher, T.E., Jr. and M.I. Goldstein. 1997. Tropical ecotoxicology: status and needs, *Environ. Toxicol. Chem.*, 16, 100–111.

Lee, Y.C.G., N.H. De Klerk, and A.W. Musk. 1999. Asbestos-related pleural disease in Western Australia gold-miners, *Med. Jour. Austral.*, 170, 263–265.

Leino, T. and M. Lodenius. 1995. Human hair mercury levels in Tucurui area, state of Para, Brazil, *Sci. Total Environ.*,175, 119–125.

Lima, A.P.S., R.C.S. Muller, J.E.S. Sarkis, C.N. Alves, M.H.S. Bentes, E. Brabo, and E.O. Santos. 2000. Mercury contamination in fish from Santarem, Para, Brazil, *Environ. Res.*, 83A, 117–122.

Lin, Y., M. Guo, and W. Gan. 1997. Mercury pollution from small gold mines in China, *Water Air Soil Pollut.*, 97, 233–239.

Malm, O. 1998. Gold mining as a source of mercury exposure in the Brazilian Amazon, *Environ. Res.*, 77A, 73–78.

Malm, O., F.J.P. Branches, H. Akagi, M.B. Castro, W.C. Pfeiffer, M. Harada, W.R. Bastos, and H. Kato. 1995a. Mercury and methylmercury in fish and human hair from the Tapajos river basin, Brazil, *Sci. Total Environ.*, 175, 141–150.

Malm, O., M.B. Castro, W.R. Bastos, F.J.P. Branches, J.R.D. Guimaraes, C.E. Zuffo, and W.C. Pfeiffer. 1995b. An assessment of Hg pollution in different goldmining areas, Amazon Brazil, *Sci. Total Environ.*, 175, 127–140.

Malm, O., J.R.D. Guimaraes, M.B. Castro, W.R. Bastos, J.P. Viana, F.J.P. Branches, E.G., Silveira, and W.C. Pfeiffer. 1997. Follow-up of mercury levels in fish, human hair and urine in the Madeira and Tapajos basin, Amazon, Brazil, *Water Air Soil Pollut.*, 97, 45–51.

Malm, O., W.C. Pfeiffer, C.M.M. Souza, and R. Reuther. 1990. Mercury pollution due to gold mining in the Madeira River Basin, Brazil, *Ambio*, 19, 11–15.

Maurice-Bourgoin, L., I. Quiroga, J. Chincheros, and P. Courau. 2000. Mercury distribution in waters and fishes of the upper Madeira rivers and mercury exposure in riparian Amazonian populations, *Sci. Total Environ.*, 260, 73–86.

McGlashan, N.D. and J.S Harington. 2000. Cancer in black gold miners, 1980–89 and 1990–94: the Chamber of Mines of South Africa's records of cancer, *South African Jour. Sci.*, 96, 249–251.

Merchant, B. 1998. Gold, the noble metal and the paradoxes of its toxicology, *Biologicals*, 26, 49–59.

Murray, J., D. Kielkowski, and P. Reid. 1996. Occupational disease trends in black South African gold miners, *Amer. Jour. Respir. Crit. Care Med.*, 153, 706–710.

Nriagu, J., and H.K.T. Wong. 1997. Gold rushes and mercury pollution, in *Mercury and Its Effects on Environment and Biology*, A. Sigal and H. Sigal, (Eds.), Marcel Dekker, New York, 131–160.

Ogola, J.S., W.V. Mitullah, and M.A. Omulo. 2002. Impact of gold mining on the environment and human health: a case study in the Migori gold belt, Kenya, *Environ. Geochem. Health*, 24, 141–158.

Olivero, J., V. Navas, A. Perez, B. Solano, I. Acosta, E. Arguello, and R. Salas. 1997. Mercury levels in muscle of some fish species from the Dique Channel, Columbia, *Bull. Environ. Contam. Toxicol.*, 58, 865–870.

Olivero, J., B. Solano, and I. Acosta. 1998. Total mercury in muscle of fish from two marshes in goldfields, Columbia, *Bull. Environ. Contam. Toxicol.*, 61, 182–187.

Porcella, D.B., C. Ramel, and A. Jernelov. 1997. Global mercury pollution and the role of gold mining: an overview, *Water Air Soil Pollut.*, 97, 205–207.

Porvari, P. 1995. Mercury levels of fish in Tucurui hydroelectric reservoir and in River Moju in Amazonia, in the state of Para, Brazil, *Sci. Total Environ.*, 175, 109–117.

Reid, P.J., and G.K. Sluis-Cremer. 1996. Mortality of white South African gold miners, *Occup. Environ. Medic.*, 53, 11–16.

Rojas, M., P.L. Drake, and S.M. Roberts. 2001. Assessing mercury health effects in gold workers near El Callao, Venezuela, *Jour. Occup. Environ. Med.*, 43, 158–165.

Rose, T.K. 1948. Gold. *Encyclopaedia Britannica*, 10, 479–485.

Russell, B.M., L.E. Muir, P. Weinstein, and B.H. Kay. 1996. Surveillance of the mosquito *Aedes aegypti* and its biocontrol with the copepod *Mesocyclops aspericornis* in Australian wells and gold mines, *Med. Veterin. Entomol.*, 10, 155–160.

Santa Rosa, R.M.S., R.C.S. Muller, C.N. Alves, J.E.deS. Sarkis, M.H.daS. Bentes, E. Brabo, and E.S. de Oliveira. 2000. Determination of total mercury in worker's urine in gold shops of Itaituba, Para State, Brazil, *Sci. Total Environ.*, 261, 169–176.

Simonato, L., J.J. Moulin, B. Jvelaud, G. Ferro, P. Wild, R. Winkelmann, and R. Saracci. 1994. A retrospective mortality study of workers exposed to arsenic in a gold mine and refinery in France, *Amer. Jour. Indus. Med.*, 25, 625–633.

Sonnenberg, P., J. Murray, J.R. Glynn, R.G. Thomas, P. Godfrey-Faussett, and S. Shearer. 2000. Risk factors for pulmonary disease due to culture-positive *M. tuberculosis* or nontuberculous mycobacteria in South African gold miners, *Euro. Respir. Jour.*, 15, 291–296.

Steenland, K. and D. Brown. 1995a. Mortality study of gold miners exposed to silica and nonasbestiform amphibole minerals: an update with 14 more years of follow-up, *Amer. Jour. Indus. Med.*, 27, 217–229.

Steenland, K. and D. Brown. 1995b. Silicosis among gold miners: exposure-response analyses and risk assessment, *Amer. Jour. Public Health*, 85, 1372–1377.

Toro, J.T. 2001. 28 gold miners killed by mudslide in Columbia. *Washington Post*, November 23, 2001:A37.

Trapido, A.S., N.P. Mqoqi, B.G. Williams, N.W. White, A. Solomon, R.H. Goode, C.M. Macheke, A.J. Davies, and C. Panter. 1998. Prevalence of occupational lung disease in a random sample of former mineworkers, Libode District, eastern Cape Province, South Africa, *Amer. Jour. Indus. Med.*, 34, 305–313.

van Straaten, P. 2000. Human exposure to mercury due to small scale gold mining in northern Tanzania, *Sci. Total Environ.*, 259, 45–53.

Veiga, M.M. and J.A. Meech. 1995. Gold mining activities in the Amazon: clean-up techniques and remedial procedures for mercury pollution, *Ambio*, 24, 371–375.

Veiga, M.M., J.A. Meech, and R. Hypolito. 1995. Educational measures to address mercury pollution from gold-mining activities in the Amazon, *Ambio*, 24, 216–220.

CHAPTER 8

Human Sensitivity to Gold

Effects of various gold compounds on human health are documented in this chapter, except effects associated with the use of gold drugs to treat rheumatoid arthritis, which are covered in detail in Chapter 9. This chapter specifically reviews (1) the history of gold drugs in medicine; (2) adverse reactions to gold treatments, including possible lethal, carcinogenic, and teratogenic effects; (3) case histories documenting hypersensitivity, Goldschlager syndrome, and other effects; and (4) dental aspects of gold, including allergic and sensitization reactions documented by selected case histories.

8.1   HISTORY

Monovalent organogold compounds have been used extensively to treat a variety of human diseases (other than rheumatoid arthritis), including psoriatic arthritis (Schwartzman et al. 1995; Quarenghi et al. 1998; Lacaille et al. 2000); pemphigus (Pandya and Dyke 1998); tumors (Kamei et al. 1999); HIV (Shapiro and Masci 1996); bronchial asthma (Suzuki et al. 1995); and inflammatory polyarthritis (Eardley et al. 2001) with varying degrees of success.

Psoriatic arthritis can be a chronic progressive disease responsible for damage to more than five joints in up to 40% of affected individuals and severe functional limitation in 11% (Lacaille et al. 2000). Intramuscular gold therapy for this condition was first reported in 1946, accompanied by a high frequency of side effects, especially rash. Intramuscular gold injections are now safer and more tolerated in the treatment of psoriatic arthritis, but are still considered inferior to other compounds tested in achieving a clinical response and in permitting long-term treatment. Nevertheless, injectable organogold[+] salts allegedly achieved a long lasting satisfactory response in 35% of patients, making them a reasonable alternative for the treatment of psoriatic arthritis in patients who experienced adverse effects with other compounds (Lacaille et al. 2000). Membranous glomerulonephritis can complicate gold salt therapy in psoriatic arthritis patients (Quarenghi et al. 1998). In one case, however, glomerulonephritis was a consequence of oral gold therapy in a patient treated for psoriatic arthritis. The nephrotic syndrome disappeared after discontinuation of oral gold preparations (Quarenghi et al. 1998). In another case, a 41-year-old male with psoriatic arthritis developed progressive shortness of breath and airflow obstruction after 4 months of gold therapy (Schwartzman et al. 1995). Open lung biopsy revealed bronchiolitis obliterans of the constrictive type, an inflammatory disease of the airways characterized pathologically by fibrosis of the bronchiolar lumina and physiologically by progressive airflow obstructions. Psoriatic arthritis had not previously been associated with this pulmonary condition. Because this disease is usually irreversible, clinicians need to pursue respiratory complaints in patients receiving gold therapy (Schwartzman et al. 1995).

Patients afflicted with disabling psoriatic arthritis, as well as human immunodeficiency virus (HIV), have limited gold treatment options because of the risk of exacerbating the immune suppression associated with HIV infection (Shapiro and Masci 1996). In one case, a 42-year-old female with psoriatic arthritis tested positive for HIV during the first trimester of pregnancy. The reported risk factor was sexual contact with her spouse, who was HIV positive. Oral gold treatment (auranofin) was initiated 9 months later at 3 mg per os. Skin lesions and arthritis resolved after treatment and she remained free of opportunistic infections during a 24-month followup (Shapiro and Masci 1996).

Intramuscular gold injections over a 12-month period (sodium gold thiomalate at 50 mg Au^+ weekly) were effective in 16 of 26 patients as a primary treatment for pemphigus (large blisters on skin and mucous membranes, usually with itching or burning), although 42% of the patients had some adverse side effects (Pandya and Dyke 1998). Treatment was discontinued if significant toxic effects were observed (protein in urine, pruritus) or if a total dose of 1000 mg was reached without beneficial effect (Pandya and Dyke 1998).

Sodium gold thiomalate (Au^+) was used to treat two patients with a history of cancer (Kamei et al. 1999). One patient, who had had a tongue carcinoma removed 8 years before and showed consistent high levels of tumor-associated antigens — suggesting recurrence of cancer — received weekly intramuscular injections of 25 mg for 10 weeks. Another patient, who had been treated with radiation therapy for pulmonary carcinoma 5 years earlier, but who had consistent elevated levels of tumor-associated antigens, received 25 mg of sodium gold thiomalate every other week for 30 injections. Levels of tumor-associated antigens declined in both patients to normal levels, with no adverse side effects observed on blood chemistry or kidney function (Kamei et al. 1999).

Sodium gold thiomalate may be capable of controlling eosinophil function regulated by interleukin-5 (IL-5) in patients with bronchial asthma (Suzuki et al. 1995). Eosinophils are considered to be the main effector cells in the pathogenesis of bronchial asthma, destroying bronchial epithelium. Various functions of eosinophils are regulated by cytokines, such as IL-3, IL-5, interferon, and granulocyte–macrophage colony stimulating factor. IL-5 affects eosinophil differentiation, adhesion, effector function, and survival, and is considered the most important cytokine in eosinophil regulation. High concentrations of sodium gold thiomalate inhibited IL-5-mediated eosinophil survival in blood from patients with bronchial asthma *in vitro* (Suzuki et al. 1995).

A 25-year-old female with inflammatory polyarthritis was treated with sodium gold thiomalate after unsuccessful treatment with methotrexate, prednisolone, and diclofenac (Eardley et al. 2001). The patient received 10 mg of gold[+] the first week and 50 mg the second week. Two days later, the patient developed septicemia and intravascular coagulation, which was relieved by antibiotics. The patient may have been afflicted with the rare Adult Onset Still's Syndrome (AOSD), with features similar to those of juvenile idiopathic arthritis. Gold may acutely precipitate multi-organ failure and nephrotic syndrome in AOSD victims (Eardley et al. 2001).

8.2   ADVERSE REACTIONS

Adverse side effects of various gold treatments, as well as generalized reactions to gold and gold compounds are listed next.

8.2.1   Suicide Attempt

A suicide attempt by a 27-year-old male was made by ingesting about 4 mL of a gold potassium cyanide solution (Wu et al. 2000). He developed vomiting and abdominal pain within 3 hours and was sent to a nearby hospital. Vital signs and respiration were stable and the blood cyanide test was negative. Blood amylase was elevated and a liver biopsy showed centrilobular cholestasis. After 24 hours, gold levels were measured and found to be grossly elevated in whole blood (4.36 mg Au/L), serum (6.01 mg Au/L), and in urine (0.429 mg excreted daily). Authors concluded that ingestion of gold potassium cyanide solution results in significant systemic toxicity of gold; the mechanism of action was not known (Wu et al. 2000).

8.2.2   Teratogenicity and Carcinogenicity

Although there are no adequate studies of teratogenicity for gold sodium thiomalate in pregnant humans, a potential risk to the fetus exists because gold was found in the serum and red blood cells of a nursing infant (Sifton 1998).

Trivalent gold complexes were potentially attractive as anticancer agents because of their cytotoxic effects on established human tumor cell lines (Calamai et al. 1998). All tested Au^{+3} complexes substantially retained their antitumor potency against platinum-resistant tumor cell lines for leukemia and ovarian cancer. Cytotoxicity of these compounds *in vitro* is attributed to binding with DNA and modification and subsequent impairment of replication and transcription processes. The paucity of data on Au^{+3} complexes probably derives from their high redox potential and relatively poor stability, which makes their use problematical under physiological conditions (Calamai et al. 1998).

8.2.3   Hypersensitivity

Proverbially stable and generally considered inert, gold was long overlooked as an allergen, and overt hypersensitivity to the metal was observed so rarely as to be virtually unknown (Hostynek 1997). Gold is now gaining recognition as a major factor in the etiology of cellular and humoral immunity owing to increasing systemic exposure for therapeutic purposes and to new patterns of intimate cutaneous contact. Characteristic immunological responses to gold hypersensitivity include late reactions to challenge, extraordinary persistence of clinical effects, formation of intracutaneous nodules and immunogenic granulomas unresponsive to conventional steroid therapy, the occurrence of eczema at sites distant from the contact site, and flareups of eczema upon systemic provocation with allergen characteristic of drug-induced therapy (Hostynek 1997). Gold salts take one of the top positions among drugs causing cutaneous side effects, and gold dermatitis may have many presentations, including eczematous, lichenoidal, toxicodermal, and pityriasis rosea-like eruptions (Moller et al. 1996a).

In 2001, gold was selected as the contact allergen of the year by the American Contact Dermatitis Society (Fowler 2001). In the United States, Europe, and Japan, gold is now ranked among the ten most frequent allergens; the greatest majority of those sensitized were women (Hostynek 1997). The prevalence of gold allergy worldwide, as determined by patch tests with various gold salts, might be as high as 13%, with 9.5% the most recent estimate in North America. Positive reactions to gold salts may appear in 7 to 10 days, or longer, after testing. Most patients with positive gold patch tests have dental gold (Fowler 2001). In Sweden, gold is now considered the second most common metal allergen after nickel (Hostynek 1997), as based on sensitivity to gold sodium thiomalate in patch tests (Bruze et al. 1994). In Sweden, hypersensitivity to gold sodium thiomalate was more frequent in patients with oral restorative materials containing gold and was associated with distal eczema (Hostynek 1997). Since the 1980s, there have been increasing reports of gold causing dermatitis at sites of jewelry contact and eyelid dermatitis from gold allergy (Guin 1999; Fowler 2001).

The clinical picture of allergic contact dermatitis to gold usually consists of a toxicoderma-like rash at the site of contact and transient fever (Moller et al. 1999). Cell-mediated allergic responses to gold were accompanied by positive lymphocyte transformation and proliferation tests; gold was selectively accumulated in Langerhans cells of the epidermis (Hostynek 1997). Intramuscular injections of gold sodium thiosulfate into patients allergic to gold are accompanied by immunological tissue reactions and release in blood of cytokines and acute phase reactants, including plasma tumor necrosis factor-alpha, soluble tumor necrosis factor receptor 1, interleukin-1 receptor antagonist, and neutrophil gelatinase associated lipocalin (Moller et al. 1999). Results of patch tests with gold sodium thiosulfate among Swedish dermatitis patients should take longer than 3 days — the usual postobservation period — in order to fully evaluate the findings (Bruze et al. 1995a). Only 46% of the positive patch test reactions appeared within 3 days; the rest appeared within 10 days. Reactions were still readable after 2 months in about a third of the tests. Authors recommend a supplemental reading of patch test results at 3 weeks postexposure (Bruze et al. 1995a).

The most common outcome of female patients who had a positive allergic response to gold sodium thiosulfate, was eczema of the head and neck (62% frequency), limbs (46%), and anus and vulva (15%). The mean duration of eczema in this group was 15.8 months. Most (54%) of the patients allergic to gold were also allergic to nickel (McKenna et al. 1995). Contact allergy to gold sodium thiosulfate in humans (unlike certain strains of mice) is hypothesized to be either lifelong or at least to last for years, although evidence is incomplete (Lee and Maibach 2001). Experimental studies with gold sodium thiosulfate in humans indicates a 100% response that lasts for at least 2 months (Lee and Maibach 2001).

Gold dermatitis from occupational exposure is rare. Gold salts are usually the cause, rarely gold objects (Estlander et al. 1998). The main exposure sources of gold contact dermatitis are personal jewelry and dental alloys (Bruze et al. 1994; McKenna et al. 1995; Suarez et al. 2000). The occupations most frequently causative of contact dermatitis due to gold are photography, chinaware or glass decorating, jewelery making, and dental alloy manufacture. Occupational allergic contact dermatitis due to gold is infrequent in automated industrial processes (Suarez et al. 2000).

Aside from medical therapeutic purposes, the use of gold in jewelry brings the greatest risk of sensitization. The risk is greatest when the gold-containing alloys are introduced and left in permanent contact with live tissues, as occurs in piercing of ears and other body parts (Hostynek 1997). Cases of contact dermatitis due to gold, especially in pierced earlobes, are increasing worldwide (Suzuki 1998). Small fragments of gold may remain in the skin lesions of pierced earlobes for at least 4 months after the 24-carat gold studs have been removed, causing prolonged irritation and various cutaneous reactions (Suzuki 1998). Insertion of gold earrings immediately following piercing may result — through gold solubilization and cellular response — in the formation of intracutaneous bodies in the earlobes at the site of piercing, with ultimate surgical removal of the nodules. The nodules were characterized by large macrophages, lymphoid cell infiltration and eosinophils, confirming the immunological nature of such nodules (Hostynek 1997). However, metallic gold (Au^0) used both in jewelry and in prostheses is ordinarily alloyed with other metals that may contribute to acute contact dermatitis (Merchant 1998). High-carat yellow gold contains minute quantities of copper and silver; low-carat yellow gold contains these metals plus zinc and small amounts of nickel. White gold usually contains palladium and nickel. The nickel in white gold alloys is a strong sensitizer, and contact dermatitis to nickel often coexists with rare instances of acute contact dermatitis following exposure to Au^0. Even the most highly purified forms of gold contain minute quantities of contaminating materials, mainly iron and sodium, which in total may represent about 0.1% or 1000 mg/kg (Merchant 1998). Defects in the gold coating on stems of some commercial ear-piercing studs, normally in contact with the pierced ears, allowed body fluids to contact the stem's substrate; the substrate contained nickel, cobalt, zinc, and copper, with cytotoxicity in at least one case attributed to copper (Rogero et al. 2000).

In contact allergy to gold, a low rate of responsiveness and mild symptoms were typical, although some people developed strong and persistent reactions (Rasanen et al. 1996). Sensitivity to gold was based on responsiveness to patches applied to the skin containing either metallic gold (Au^0), gold chloride (Au^{+3}), or various organomonovalent gold compounds (Au^+). Gold sodium thiomalate (Au^+) was the best marker of gold contact allergy because Au^0 often yielded false negative results due to the inadequate release of soluble gold, and Au^{+3} caused persistent allergic reactions more frequently than did other gold compounds (Rasanen et al. 1996). Patch tests in recent years using gold sodium thiomalate have indicated positive patch test frequencies as high as 8.6% in Asia, 10% in Europe, and 13% in North America (Ehrlich and Belsito 2000). In patch tests, some studies suggested that gold sodium thiomalate produced few positive reactions in patients hypersensitive to gold sodium thiosulfate (Bruze et al. 1995b). But in tests of intracutaneous administration of equimolar concentrations, allergic reaction rates were similar for gold sodium thiomalate and gold sodium thiosulfate, suggesting that contact allergy rates were probably similar (Bruze et al. 1995b). The efficacy of gold salt patch tests needs to be critically reexamined.

Hypersensitivity to gold is variable. Among 373 patients tested against gold sodium thiosulfate in western Scotland by routine patch testing, only 2.1% tested positive; however, these tests were based on an observation period of 4 days, which is considered an insufficient period to fully assess sensitivity to gold (Fleming et al. 1997b). Rheumatoid arthritis patients who discontinued intramuscular chrysotherapy because of adverse side effects, especially mucocutaneous reactions, were patch tested for contact sensitivity to gold sodium thiosulfate in order to determine if side effects were due to a previously unrecognized gold allergy (Fleming et al. 1998b). All patients tested negative, indicating that this procedure does not detect hypersensitivity to previous or current gold exposure (Fleming et al. 1998b). In a study of 823 patients with suspected acute contact dermatitis, 8.6% gave positive patch tests to gold sodium thiosulfate and none reacted positively to metallic gold (Merchant 1998). A positive skin test to sodium thiosulfate, in the absence of sensitivity to metallic gold, may represent a unique form of gold allergy that is clinically irrelevant (Merchant 1998).

It is suggested that Au^0 toxicity may be associated, in part, with the formation of the more reactive Au^+ and Au^{+3} species (Eisler 2004); however, this has not yet been verified. Additional research is warranted at the molecular level of the unusual mechanisms of action induced by gold dermotoxicity (Hostynek 1997).

8.3   CASE HISTORIES

Selected case histories documenting various hypersensitive reactions to gold or gold compounds are presented below.

8.3.1   Hypersensitivity

In one case history, a 22-year-old male working in the electrolytic gold-plating section at a cutlery factory had — for the past 2 years while employed there — dermatitis over the backs of his hands and fingers (Suarez et al. 2000). At work, he handled, without gloves, a solution containing 5% gold trichloride and 0.006% cobalt and nickel. He had no dental restorations and no previous history of metal sensitivity. The patient tested mildly positive to cobalt and strongly positive to gold sodium thiosulfate. He was removed from that section and all symptoms disappeared within

4 months. Gold trichloride appeared to be the cause of dermatitis because the level of cobalt in the electroplating solution was low and variable (Suarez et al. 2000).

One study concluded that there were no significant differences in prevalence of hypersensitivity to gold sodium thiosulfate, as judged by patch tests, attributable to age, sex, or exposure to gold in jewelry, dental restoration, or occupation (Fleming et al. 1998a). In that study, 1203 patients from three hospitals and 105 volunteers were screened by routine patch testing for sensitivity to 0.5 and 0.05% gold sodium thiosulfate. A total of 38 patients (3.2%) and five volunteers (4.8%) tested positive (Fleming et al. 1998a). Most studies showed that females were usually more sensitive to gold than were males. In Portugal, 2583 patients were routinely patch-tested for contact allergy to gold sodium thiosulfate in 1995 (Silva et al. 1997). Only 22 (0.7%) tested positive (all females). All reactors had had their ears pierced and had been exposed to gold jewelry, mainly earrings; most of the 22 patients also tested positive to nickel (Silva et al. 1997). Of 54 Japanese patients who tested positive to gold sodium thiosulfate, 17.3% were female and 3.3% were male; similar results were reported in Sweden and the U.K. (Tsuruta et al. 2001). Gold dental alloys, gold earrings, and other gold jewelry were the presumptive sources of gold sensitization. Exposure to gold jewelry is clinically relevant in persons hypersensitive to gold (Ahnlide et al. 2000). Effects of exposure to metallic gold were evaluated in 60 female patients with pierced earlobes who tested positive to gold sodium thiosulfate. Half the patients received earrings with a surface layer of 24K gold and the other 30 received earrings with a surface layer of titanium nitride. After 8 weeks, 17 of the 60 had skin reactions, 12 of these had received gold earrings and 5 titanium. Earlobe reaction was observed in 11 patients: 7 from the gold group and 4 from the titanium group (Ahnlide et al. 2000). Studies have shown frequencies of 4.6 to 10% of contact dermatitis to gold sodium thiomalate (Sabroe et al. 1996). Of 100 patients routinely attending a contact dermatitis clinic in Bristol, England, 13 tested positive in patch tests to gold sodium thiomalate. Of these, 11 were female and 12 had pierced ears. Only 7 of the 13 had symptoms. There was a high incidence of nickel sensitivity (33%) in the 100 patients, but eczema on the ring fingers and neck was significantly more common in the group positive to gold sodium thiomalate (Sabroe et al. 1996).

A 27-year-old woman presented persistent painless nodules at multiple sites of ear piercing with gold earrings done ten years previously (Armstrong et al. 1997). At that time, when 17 years of age, she noted tenderness and swelling of these sites within 6 weeks. Despite removing her earrings and avoiding further gold contact, she developed discrete nodules at each pierced site which remained unchanged. The woman tested strongly positive to a gold sodium thiosulfate patch test. To account for the continued swelling, it was postulated that the ear contained gold inclusions, as had been documented in other recent cases (Armstrong et al. 1997). Painless nodules of the earlobes in a 20-year-old woman was attributed to her wearing 14K gold earrings 4 months earlier (Park et al. 1999). At that time she noticed pruritus, tenderness, and swelling at these sites a few days after wearing them. Despite removing her earrings and avoiding further contact, dome-shaped subcutaneous nodules developed on the earlobes and continued to enlarge. The earlobes were treated successfully. A patch test indicated sensitivity to gold sodium thiosulfate. Authors concluded that allergic contact dermatitis from gold earrings appears clinically as discrete nodules at the sites of piercing in gold-sensitive individuals, and usually remains despite avoidance of further gold contact (Park et al. 1999).

Lymphomatoid allergic contact dermatitis from gold is rare and characterized by nodules at sites of piercing with gold jewelry (Fleming et al. 1997a). In one case, a 24-year-old woman with ears pierced at age 13, complained of mild dermatitis after wearing gold earrings. In a standard patch test, she tested positive to gold sodium thiosulfate, but not to four other gold compounds including gold leaf. Contact allergy to gold sodium thiosulfate is variable, ranging from no reaction in resistant individuals to lymphatomoid responses in those with persistent dermal gold exposure or abnormal gold immunoreactivity. Intermediate responses include positive patch tests to gold regardless of history of contact dermatitis (Fleming et al. 1997a). Of 345 patients in Singapore subjected to a standard patch test series over a 6-month period, 22 were highly sensitive to gold sodium thiosulfate 0.5% in petrolatum; however, only 3 of the 22 who patch tested positive had chemically relevant reactions that could be traced to gold jewelry (Leow and Goh 1999).

Gold is a relatively common allergen that appears to induce dermatitis about the face and eyelids, as well as at sites of direct skin contact. Gold-sensitive individuals ($N = 15$), as determined by patch testing, were reevaluated 2 months after contact with gold jewelry was discontinued (Ehrlich and Belsito 2000). Dermatitis cleared in 7 of the 15, and another 4 needed to discontinue contact with other allergens for improvement. None of the patients required the removal of dental gold (Ehrlich and Belsito 2000). Occupational allergic contact dermatitis of the skin and eyelids was recorded for a male, age 26 years, working in the electroplating department of a metal factory (Estlander et al. 1998). For the previous 3 months he had been exposed to both gold-plating solutions and metallic gold. Symptoms were alleviated during weekends and disappeared in a week away from work. He was not sensitive to nickel-, silver-, or tin-plating solutions. Tests showed that he was sensitive to gold sodium thiosulfate, but not to other metals tested. It was necessary for him to get a new job elsewhere with no exposure to gold salts. On follow-up, 3 months later, he was symptomless (Estlander et al. 1998).

Due to suspicion of gold contact allergy caused by jewelry or dental restorations, nine female patients with no previous history of gold treatment were given gold sodium thiomalate patch tests, and were also tested intradermally to gold sodium thiomalate (Kalimo et al. 1996). Only six tested positive in patch tests, but all tested positive via intradermal injection route. However, five of eight patients injected intradermally developed skin papules at the injection site. The papules persisted for up to 20 months. Histological examination of the surgically excised lesions showed pseudolymphoma of cells containing follicular structures. By electron microscopy, the macrophages were found to contain gold-bearing endosomes. Authors concluded that gold sodium thiomalate binds persistently in the skin after intradermal injection, accumulating in the macrophages of susceptible individuals and inducing pseudolymphoma formation (Kalimo et al. 1996).

In a study conducted in Israel, 34 of 406 patients (8.4%) tested positive in gold sodium thiomalate patch tests (Trattner and David 2000). None of the patients who tested positive had suspected gold allergy before testing. Most (23 of 34) of the patients who tested positive to gold sodium thiomalate also tested positive to nickel (47%), chromate (26%), cobalt (15%), or various organic substances (53%). Of the 34 who tested positive, 73% had direct skin contact with gold objects (vs. 50% in those who tested negative), and 79% (vs. 48%) had pierced ears (Trattner and David 2000).

Auranofin ointment is a significant contact sensitizer with gold as its allergic component (Marks et al. 1995). Auranofin — an organogold complex composed of Au^+, thiosugar, and triethylphosphine — has been used successfully in the treatment of rheumatoid arthritis. More recently, a crude 0.18% auranofin ointment was used to treat psoriasis, resulting in clearing of lesions in some patients. However, contact dermatitis developed at the treatment site in 17 of 76 (22%) patients treated with auranofin ointment (Marks et al. 1995).

Contact allergy to gold is frequent (10.4%) among patients with rheumatoid arthritis before gold therapy (Moller et al. 1997). Rheumatoid arthritis patients (N = 20) with a contact allergy to gold sodium thiosulfate were challenged with an intramuscular injection of either gold sodium thiomalate or a placebo (Moller et al. 1996a). Patients given gold sodium thiosulfate showed epidermal and dermal flare-up of healed patch test reactions to the gold salt, and a high (104.0°F, 40°C), but transient, rise in body temperature; no effect was seen in patients receiving a placebo. Skin tests, both patch and intradermal, with gold sodium thiosulfate, gold sodium thiomalate, and auranofin (oral gold triethylphosphine) are recommended prior to gold therapy in order to avoid early hypersensitivity reactions (Moller et al. 1997). A rheumatoid arthritis patient intended for gold therapy showed contact allergy to both gold sodium thiosulfate and gold sodium thiomalate (Moller et al. 1996b). An intramuscular test dose of gold sodium thiomalate induced a flare-up of previously positive epicutaneous and intradermal test reactions compatible with that of an allergic contact dermatitis. The patient had no dental gold and had been using gold jewelry without significant problems; however, a gold necklace would occasionally give rise to slight irritation and red patches on the neck, appearing hours or days after she started to wear it, and disappearing rapidly after removal. Authors recommend that patients intended for chrysotherapy should be examined prior to treatment with appropriate skin tests (Moller et al. 1996b). A positive skin test to gold may not necessarily contraindicate further treatment with gold preparations if carefully selected low dosages are used (Moller et al. 1997).

8.3.2 Goldschlager Syndrome

The ingestion of gold-containing liquor beverages can result in allergic-type reactions similar to those seen after gold-allergic individuals are exposed to gold through medications or jewelry. In all cases, the rashes disappeared after discontinuation of the product; time to rash resolution ranged from days to several months and was directly proportional to the duration of gold ingestion. In one case, a 31-year-old female previously sensitized to gold jewelry developed a rash after ingesting 90 to 120 mL of Goldschlager (one of several brands of a cinnamon-flavored schnapps containing 53% ethanol and 10 to 23 mg of flake gold/L) the previous evening. Her serum gold level at the time of admission was negative. Treatment was with antihistamines and was resolved in 2 weeks (Guenther et al. 1998, 1999).

Several brands of gold-containing cinnamon schnapps are available in the United States. Analysis of five 750-mL bottles showed 8 to 17 mg of gold flakes per bottle (75% gold by weight) and about 2.8 mg Au/L dissolved in the liquid portion. The gold flakes were allegedly added to enhance the appearance of the product (Russell et al. 1996, 1997). A survey of bartenders and liquor distributors in Nashville, Tennessee, showed that gold-containing liquors are popular with college students and young to middle-aged adults. Gold has been approved for use in alcoholic beverages since at least 1982 and the gold-containing cinnamon schnapps consumed by all patients in the three case histories that follow has been available in the United States since 1993.

In the first case, a 24-year-old male bartender presented with skin eruptions on the forearms, shins, ankles, and buccal mucosa consistent with lichen planus. Lichen planus is a papulosquamous eruption that typically occurs in middle-aged persons, although drug-induced lichen planus has been reported after the administration of numerous medications, including gold-containing compounds. The patient had regularly consumed gold-containing schnapps for about a year at 200 to 300 mL weekly. The initial serum gold level, measured 3 months after he had last consumed the gold-containing beverage, was 0.4 mg Au/L (normal = 0.0 to 0.1 mg/L); the urinary gold excretion level was 86 μg/daily (normal = 0.0 to 1.0 μg/daily). Three months after the first measurement (6 months since last Goldschlager consumption), the pruritic eruptions gradually cleared and serum and urine measurements were within the normal range (Russell et al. 1996, 1997). The second case was a 47-year-old female with papular eruptions on her lower legs that began 8 weeks after she first consumed gold-containing cinnamon schnapps. She consumed about 150 mL of the beverage weekly for about 7 months with no other gold intake. Serum and urine gold levels measured 6 weeks after her last ingestion of Goldschlager were normal. Patch testing to gold sodium thiomalate was negative. There was a gradual clearing of her pruritus and dermatitis 3 months after she stopped ingestion of the gold-containing liquor (Russell et al. 1997). In the last case, a 58-year-old female developed an itchy papular eruption on the lower legs 14 to 16 weeks after first consuming gold-containing liquor, with total consumption of about 400 mL before the eruption started. The patient had several gold crowns and amalgam fillings, and these were surrounded by prominent reticulated white plaques. Four months after the last intake of the gold-containing liquor, serum and urine gold levels were normal, and the reticulated plaques on her buccal mucosa receded to the area opposite the gold crowns (Russell et al. 1997).

8.3.3 Prostheses

Gold (0.999 fine) has been used successfully in synthetic middle ear prostheses (Gjuric and Schagerl 1998). Implant rejection was rarely encountered and gold implants showed high biocompatability. However, in one study conducted between

November 1993 and February 1996 wherein 59 patients underwent tympanoplasty with ossicular chain reconstruction using gold implants, prostheses extrusion occurred in 11 cases (19%) 7 to 21 months after tympanoplasty combined with a significant retraction of the tympanic membrane. Authors are developing new types of gold implants (Gjuric and Schagerl 1998).

Gold implants have been inserted into the upper eyelids to compensate for lagophthalmos with a success rate of about 90%. Although rare, instances of allergic responses following direct exposure to implanted gold were documented (Merchant 1998).

8.3.4   Protective Effect of Gold Rings

Protective effects of gold rings against rheumatoid arthritis are documented, but unresolved (Belt and Kaarela 1998). A gold ring worn on the left ring finger of patients with rheumatoid arthritis seems to protect against articular erosion at the left-hand ring joint and adjacent joints when compared with patients with rheumatoid arthritis (RA) who had never worn a ring on this finger (Mulherin et al. 1997). The authors aver — without evidence — that gold could pass from a gold ring through skin and local lymphatics to nearby metacarpophalangeal joints in sufficient quantities to delay articular erosion. Bolosiu (1998) does not exclude the possibility of a local chemical action, but suggests that lack of joint deformity in RA patients wearing a gold ring is due to a physical explanation, namely the weight of the ring (Bolosiu 1998) and the protective effect of other fingers (Belt and Kaarela 1998).

8.4   DENTAL ASPECTS

Allergic and sensitization effects of dental gold together with selected case histories are discussed below.

8.4.1   Allergic Reactions and Sensitization

Gold allergy was overrepresented in those having dental gold (Bruze et al. 1994), and sensitization to gold seems to be more common than previously anticipated (Rasanen et al. 1996). Dental patients (N = 52) had a 12.4% positive patch test reaction to gold sodium thiomalate; of those who tested positive, 73% responded to gold compounds *in vitro* in the lymphocyte proliferation test (Rasanen et al. 1996). When sensitization does occur, it is usually from exposure to the salts of mercury or other metals in dental alloys and may manifest itself as oral lichenoid lesions; replacement of the gold filling with non-metallic restorations frequently leads to resolution of the inflammatory lesion (Merchant 1998). Many metals that are used today in dentistry may be hazardous to certain genetically predisposed individuals, and as such could limit their use (Stejskal et al. 1994). In one case, oral mucosal problems suspected to be associated with release of metal ions from dental restorations, coupled with chronic fatigue, was reported in a patient occupationally exposed to metals while working in a dental practice. Lymphocyte assays indicated that the patient was sensitive to gold, mercury, and palladium (Stejskal et al. 1994). Patients (N = 397) claiming various subjective symptoms related to dental restoration materials were tested for sensitivity to 19 metals by patch test; sensitivity was 23% to gold sodium thiomalate, 22% to nickel sulfate, and <8% for all other metals (Marcusson 1996).

Dental restorations made of dissimilar metals may undergo a series of electrogalvanic reactions of corrosion when brought together, causing short-lived, but severe pain in a few patients (Williamson 1996). In opposing teeth, when one with amalgam and another with gold restoration are in contact, the galvanic current generated by the gold restoration is always smaller than that in the tooth restored with amalgam. If pain persists, treatment may consist of replacing the amalgam restoration with a composite restoration to break the interproximal dissimilar metal contact (Williamson 1996). Oral fluids slowly dissolve elemental gold used in dental restorative materials (Hostynek 1997). Because gold is used in alloys with copper, silver, zinc, platinum, and palladium, solubilization can be accelerated by galvanic reactions with other adjoining restorative metals. The salts thus formed may provoke allergic response of the delayed type as they are absorbed through the mucous membrane. Oral lesions are seen as a consequence including erythema, mucosal erosions, lichen planus, and stomatitis (Hostynek 1997).

The safety of amalgam, containing about 50% mercury, as restorative material in dentistry is controversial, and its use has been restricted in several countries (Begerow et al. 1999). Alternatives to dental amalgam include alloys in which gold is partly replaced by palladium (gold-reduced alloys) or the more expensive high-gold alloys. Insertion of high-gold dental alloys containing platinum and palladium did not contribute to increased gold or palladium in urine over a 3-month period in three non-occupationally exposed volunteers. Platinum content of urine, however, was significantly elevated when compared to pre-insertion levels. *In vitro* release studies of gold from four different types of artificial alloys containing 4, 51, 70, or 74% gold into either artificial saliva or 1% lactic acid solutions showed that gold, as well as Pt and Pd are released from noble metal-containing dental alloys by corrosion. The possibility exists that the release of noble metals from dental alloys may cause local or systemic effects (Begerow et al. 1999).

Interleukin production can be inhibited by Au^{+3} (Rausch-Fan et al. 2000). Interleukin-1 (IL-1) is an immunoregulatory polypeptide cytokine produced mainly by mononuclear phagocytes. IL-1 acts as a key mediator in the host response to microbial invasion, inflammation, immunological variations, and mesenchymal tissue remodeling. The predominant form of IL-1 released upon stimulation of macrophages is IL-1β. When IL-1β gains access to the circulation, it induces systemic changes in neurologic, hematologic, metabolic, and endocrinologic systems. Dental amalgams and various cations affect IL-1β expression by peripheral blood mononuclear cells from healthy human donors. After 72 hours' incubation, there was a decrease in IL-1β production by freshly prepared amalgam, but not by amalgam aged for 6 weeks; high inhibition by 7 mg Hg^{+2}/kg; and a dose-dependent inhibition by Au^{+3}, as $AuCl_3$, between 0 and 330 nmol/L (Rausch-Fan et al. 2000).

Contact allergy to gold sodium thiomalate was reported in 4.6% of females with suspected contact dermatitis (Kilpikari 1997). The most frequent sites of eczema were the head and neck, with 62% frequency. Up to 10% of patients tested positive to gold sodium thiomalate in patch-tested eczema patients and seemed to reflect true contact allergy. In both studies, many tested patients were also allergic to nickel. Allergy to gold sodium thiomalate was overrepresented in those having dental gold. Gingivitis caused by gold in teeth, without eczema, is also reported (Kilpikari 1997). Contact allergy to dental gold can also lead to glossitis, oral lichen planus, and chrysiasis (Estlander et al. 1998).

8.4.2 Case Histories

There is an overrepresentation of gold allergies among those with dental restorations containing gold. Of 172 patients referred to in the Norwegian National Adverse Reaction Group, 33 (19%) showed a positive reaction to gold sodium thiosulfate (Vamnes et al. 2000). There was a significant correlation ($p = 0.002$) between the presence of dental gold and a positive patch test to gold, although there were no clinical correlates to positive patch tests to gold (Vamnes et al. 2000).

Lichen planus is a well-defined disease of the skin and mucous membranes and is related to various drugs and chemicals, including gold salts, color film developers, and salts of mercury, copper, and palladium. Oral lichenoid lesions in dental patients with gold amalgam fillings were associated with sensitivity to gold sodium thiomalate; replacement of amalgam fillings with composite resin or gold fillings resulted in an improved or total clearing of the condition 1 to 9 months postreplacement (Koch and Bahmer 1995, 1999).

Hypersensitivity to gold may also be combined with hypersensitivity to other metals, such as mercury, nickel, and palladium (Wiesner and Pambor 1998). In one case, allergic contact dermatitis from gold jewelry — earrings, wedding ring, necklace, bracelet — was reported for a 34-year-old female who tested positive for gold sodium thiomalate and mercury. This woman was referred by her dentist with suspected mercury allergy one month after all her dental fillings of mercury amalgam had been redone. Removal of new fillings and replacement with silver–palladium alloys alleviated her condition of redness of tongue and erosions of the oral mucosa (Wiesner and Pambor 1998).

Patients with local and general symptoms attributed to their gold restorations are rare (Vamnes et al. 2000). One 34-year-old female with dental gold restorations who tested positive to gold sodium thiomalate complained of itching in the mouth, loss of taste, burning sensation of the oral mucosa, and facial dermatitis. She became symptom free after replacement of all gold restorations with titanium/ceramics (Vamnes et al. 2000). In another case, also positive for gold sodium thiomalate, a previously healthy 50-year-old male had itchy lichenoid dermatitis on his trunk and thighs. The dermatitis lasted for about a year. There was a mucosal lesion adjacent to his only gold-containing crown installed 5 years earlier. He also complained of loss of taste, mucosal itching, and a burning sensation of the oral mucosa. All symptoms disappeared within 5 months after the gold crown was replaced by a titanium crown and he remained asymptomatic for at least 3 years (Vamnes et al. 2000).

A florid granulomatous reaction to gold dental alloy accidentally implanted in the oral mucosa of the lip 20 years earlier was documented in a 66-year-old male (Scott et al. 1995). These painless oral swellings had been evident for 18 months. Granulomatous inflammation is a distinctive reaction by tissue to irritant nondegradable material and a florid reaction to gold or gold alloy was unusual. Gold deposition is documented in the dermis following chrysotherapy and in the liver after treatment for rheumatoid arthritis with injectable intramuscular gold compounds. However, gold is an uncommon finding in oral lesions (Scott et al. 1995).

Chronic severe pharyngeal and laryngeal disorders in a 51-year-old diesel locomotive engineer were diagnosed after several years as severe gold allergy (Kilpikari 1997). Symptoms disappeared after removal of gold from his teeth, and he remained asymptomatic for at least 2 years after gold removal. More research is needed to evaluate the effect of replacing gold restorations with titanium or other materials in patients with positive patch tests to gold and otherwise unexplained symptoms (Vamnes et al. 2000).

8.5 SUMMARY

In humans — especially among females wearing body-piercing gold objects — there is increasing documentation of allergic contact dermatitis and other effects to gold from jewelry, dental restorations, and occupational exposure, as judged by patch tests with monovalent organogold salts; one estimate of the prevalence of gold allergy worldwide is 13%. Eczema of the head and neck was the most common response of individuals hypersensitive to gold, and sensitivity may last for at least several years. Ingestion of beverages containing flake gold can result in allergic-type reactions similar to those seen in gold-allergic individuals exposed to gold through medication or jewelry. The toxic action of Au^0 may be attributed, in part, to the formation of the more reactive Au^+ and Au^{+3} species, although this has not been verified. Gold salts may also be lethal or teratogenic. In one case, deliberate ingestion of gold potassium cyanide resulted in systemic toxicity; in another case, a potential risk to the fetus existed because gold was found in the blood of an infant nursing from a mother receiving gold drug therapy. In both cases, the mechanisms of action were not known.

LITERATURE CITED

Ahnlide, I., B. Bjorkner, M. Bruze, and H. Moller. 2000. Exposure to metallic gold in patients with contact allergy to gold sodium thiosulfate, *Contact Dermatitis*, 43, 344–350.

Armstrong, D.K.B., M.Y. Walsh, and J.F. Dawson. 1997. Granulomatous contact dermatitis due to gold earrings, *Brit. Jour. Dermatol.*, 136, 776–778.

Begerow, J., J. Neuendorf, M. Tarfeld, W. Raab, and L. Dunemann. 1999. Long-term urinary platinum, palladium, and gold excretion of patients after insertion of noble-metal dental alloys, *Biomarkers*, 4, 27–36.

Belt, E.A. and K. Kaarela. 1998. Gold and ring finger, *Ann. Rheum. Dis.*, 57, 323.

Bolosiu, H.D. 1998. Protective effect of gold rings and rheumatoid arthritis, *Ann. Rheum. Dis.*, 57, 323.

Bruze, M., B. Bjorkner, and H. Moller. 1995a. Skin testing with gold sodium thiomalate and gold sodium thiosulfate, *Contact Dermatitis*, 32, 5–8.

Bruze, M., H. Hedman, B. Bjorkner, and H. Moller. 1995b. The development and course of test reactions to gold sodium thiomalate, *Contact Dermatitis*, 33, 386–391.

Bruze, M., B. Edman, B. Bjorkner, and H. Moller. 1994. Clinical relevance of contact allergy to gold sodium thiosulfate, *Jour. Amer. Acad. Dermatol.*, 31, 579–583.

Calamai, P., S. Carotti, A. Guerri, T. Mazzei, L. Messori, E. Mini, P. Orioli, and G.P. Speroni. 1998. Cytotoxic effects of gold (III) complexes on established human tumor lines sensitive and resistant to cisplatin, *Anti-Cancer Drug Design*, 13, 67–80.

Eardley, K.S., K. Raza, D. Adu, and R.D. Situnayake. 2001. Gold treatment, nephrotic syndrome, and multi-organ failure in a patient with adult onset Still's disease, *Ann. Rheum. Dis.*, 60, 4–5.

Ehrlich, A. and D.V. Belsito. 2000. Allergic contact dermatitis to gold, *Cutis*, 65, 323–326.

Eisler, R. 2004. Mammalian sensitivity to elemental gold (Au^0), *Biol. Trace Element Res.* (in press).

Estlander, T., O. Kari, R. Jolanki, and L. Kanerva. 1998. Occupational allergic contact dermatitis and blepharoconjunctivitis caused by gold, *Contact Dermatitis*, 38, 40–41.

Fleming, C., D. Burden, M. Fallowfield, and R. Lever. 1997a. Lymphomatoid contact reaction to gold earrings, *Contact Dermatitis*, 37, 298–299.

Fleming, C., A. Forsyth, and R. MacKie. 1997b. Prevalence of gold contact hypersensitivity in the west of Scotland, *Contact Dermatitis*, 36, 302–304.

Fleming, C., T. Lucke, A. Forsyth, S. Rees, R. Lever, D. Wray, R. Aldridge, and R. MacKie. 1998a. A controlled study of gold contact hypersensitivity, *Contact Dermatitis*, 38, 137–139.

Fleming, C., D. Porter, and R. MacKie. 1998b. Absence of gold sodium thiosulfate contact hypersensitivity in rheumatoid arthritis, *Contact Dermatitis*, 38, 55–56.

Fowler, J.F., Jr. 2001. Gold, *Amer. Jour. Contact Dermatitis*, 12, 1–2.

Gjuric, M. and S. Schagerl. 1998. Gold prostheses for ossiculoplasty, *Amer. Jour. Otol.*, 19, 273–276.

Guenther, T., C. Stork, and R. Cantor. 1998. Goldschlager allergy in a gold allergic patient, *Jour. Toxicol.*, 36, 499.

Guenther, T., C.M. Stork, and R.M. Cantor. 1999. Goldschlager allergy in a gold allergic patient, *Veterin. Human Toxicol.*, 41, 246.

Guin, J.D. 1999. Black dermographism and gold dermatitis, *Contact Dermatitis*, 41, 114–115.

Hostynek, J.J. 1997. Gold: an allergen of growing significance, *Food Chem. Toxicol.*, 35, 839–844.

Kalimo, K., L. Rasanen, H. Aho, J. Maki, U.P. Mustikkamaki, and I. Rantala. 1996. Persistent cutaneous pseudolymphoma after intra dermal gold injection, *Jour. Cutan. Pathol.*, 23, 328–334.

Kamei, H. T. Koide, T. Koijima, Y. Hashimoto, and M. Hasegawa. 1999. Effect of gold on tumor-associated antigens, *Cancer Biother. Radiopharmaceut.*, 14, 403–406.

Kilpikari, I. 1997. Contact allergy to gold with pharyngeal and laryngeal disorders, *Contact Dermatitis*, 37, 130–131.

Koch, P. and F.A. Bahmer. 1995. Oral lichenoid lesions, mercury hypersensitivity and combined hypersensitivity to mercury and other metals: histologically-proven reproduction of the reaction by patch testing with metal salts, *Contact Dermatitis*, 33, 323–328.

Koch, P. and F.A. Bahmer. 1999. Oral lesions and symptoms related to metals used in dental restorations: a clinical, allergological, and histologic study, *Jour. Amer. Acad. Dermatol.*, 41, 422–430.

Lacaille, D., H.B. Stein, J. Raboud, and A.V. Klinkhoff. 2000. Longterm therapy of psoriatic arthritis: intramuscular gold or methotrexate? *Jour. Rheum.*, 27, 1922–1927.

Lee, E.E. and H.L. Maibach. 2001. Is contact allergy in man lifelong? An overview of patch test follow-ups, *Contact Dermatitis*, 44, 137–139.

Leow, Y-H. and C-L. Goh. 1999. Contact allergy in Singapore, *Asian Pac. Jour. Allergy Immunol.*, 17, 207–217.

Marcusson, J.A. 1996. Contact allergies to nickel sulfate, gold sodium thiosulfate and palladium chloride in patients claiming side effects from dental alloy components, *Contact Dermatitis*, 34, 320–323.

Marks, J.G., Jr., K.F. Helm, G.G. Krueger, C.E.M. Griffiths, C.A. Guzzo, and J.J. Leyden. 1995. Contact dermatitis from topical auranofin, *Jour. Amer. Acad. Dermatol.*, 32, 813–814.

McKenna, K.E., O. Dolan, M.Y. Walsh, and D. Burrows. 1995. Contact allergy to gold sodium thiosulfate, *Contact Dermatitis*, 32, 143–146.

Merchant, B. 1998. Gold, the noble metal and the paradoxes of its toxicology, *Biologicals*, 26, 49–59.

Moller, H., B. Bjorkner, and M. Bruze. 1996a. Clinical reactions to systematic provocation with gold sodium thiomalate in patients with contact allergy to gold, *Brit. Jour. Dermatol.*, 135, 423–427.

Moller, H., A. Larsson, B. Bjorkner, M. Bruze, and A. Hagstam. 1996b. Flare-up at contact allergy sites in a gold-treated rheumatic patient, *Acta Derm. Venereol.*, 76, 55–58.

Moller, H., K. Ohlsson, C. Linder, B. Bjorkner, and M. Bruze. 1999. The flare-up reactions after systemic provocation in contact allergy to nickel and gold, *Contact Dermatitis*, 40, 200–204.

Moller, H., A. Svensson, B. Bjorkner, M. Bruze, Y. Lindroth, R. Manthorpe, and J. Theander. 1997. Contact allergy to gold and gold therapy in patients with rheumatoid arthritis, *Acta Derm. Venereol.*, 77, 370–373.

Mulherin, D.M., G.R. Struthers, and R.D. Situnayake. 1997. Do gold rings protect against articular erosion in rheumatoid arthritis? *Ann. Rheum. Dis.*, 56, 497–499.

Pandya, A.G. and C. Dyke. 1998. Treatment of pemphigus with gold, *Arch. Dermatol.*, 134, 1104–1107.

Park, Y.M., H. Kang, H.O. Kim, and B.K. Cho. 1999. Lymphomatoid eosinophilic reaction to gold earrings, *Contact Dermatitis*, 40, 216–217.

Quarenghi. M.I., L.D. Vecchio, D. Casartelli, P. Manunta, and R. Rossi. 1998. MPO antibody-positive vasculitis in a patient with psoriatic arthritis and gold-induced membranous glomerulonephritis, *Nephrol. Dial. Transplant.*, 13, 2104–2106.

Rasanen, L., K. Kalimo, J. Laine, O. Vainio, J. Kotiranta, and I. Pesola. 1996. Contact allergy to gold in dental patients, *Brit. Jour. Dermatol.*, 134, 673–677.

Rausch-Fan, X., A. Schedle, A. Franz, A. Spittler, A. Gornikiewicz, E. Jensen-Jarolim, W. Sperr, and G. Boltz-Nitulescu. 2000. Influence of dental amalgam and heavy metal cations on *in vitro* interleukin-1 production by human peripheral blood mononuclear cells, *Jour. Biomed. Mater. Res.*, 51, 88–95.

Rogero, S.O., O.Z. Higa, M. Saiki, O.V. Correa, and I. Costa. 2000. Cytotoxicity due to corrosion of ear piercing studs, *Toxicology in Vitro*, 14, 497–504.

Russell, M.A., L.E. King, Jr., and A.S. Boyd. 1996. Lichen planus after consumption of a gold-containing liquor, *New England Jour. Med.*, 334, 603.

Russell, M.A., M. Langley, A.P. Truett III, L.E King, Jr., and A.S. Boyd. 1997. Lichenoid dermatitis after consumption of gold-containing liquor, *Jour. Amer. Acad. Dermatol.*, 36, 841–844.

Sabroe, R.A., L.A. Sharp, and R.D.G. Peachey. 1996. Contact allergy to gold sodium thio-sulfate, *Contact Dermatitis*, 34, 345–348.

Schwartzman, K.J., D.M. Bowie, C. Yeadon, R. Fraser, E.D. Sutton, and R.D. Levy. 1995. Constrictive bronchiolitis obliterans following gold therapy for psoriatic arthritis, *Eur. Respir. Jour.*, 8, 2191–2193.

Scott, F.R., A.P. Dhillon, J.F. Lewin, W. Flavell, and I.M. Laws. 1995. Gold granuloma after accidental implantation, *Jour. Clin. Pathol.*, 48, 1070–1071.

Shapiro, D.L. and J.R. Masci. 1996. Treatment of HIV associated psoriatic arthritis with oral gold, *Jour. Rheumatol.*, 23, 1818–1820.

Sifton, D.W. (Ed.). 1998. *Physicians' Desk Reference,* (52nd ed.). Medical Econ. Co., Montvale, NJ, 3223 pp.

Silva, R., F. Pereira, O. Bordalo, E. Silva, A. Barros, M. Goncalo, T. Correia, G. Pessoa, A. Baptista, and M. Pecegueiro. 1997. Contact allergy to gold sodium thiosulfate. A comparative study, *Contact Dermatitis*, 37, 78–81.

Stejskal, V.D.M., K. Cederbrant, A. Lindvall, and M. Forsbeck. 1994. MELISA — an *in vitro* tool for the study of metal allergy, *Toxicol. in Vitro*, 8, 991–1000.

Suarez, I., M. Ginarte, V. Fernandez-Redondo, and J. Toribio. 2000. Occupational contact dermatitis due to gold, *Contact Dermatitis*, 43, 367–368.

Suzuki, S. 1998. Nickel and gold in skin lesions of pierced earlobes with contact dermatitis. A study using scanning electron microscopy and X-ray microanalysis, *Arch. Dermatol. Res.*, 290, 523–527.

Suzuki, S., M. Okubo, S. Kaise, M. Ohara, and R. Kasukawa. 1995. Gold sodium thiomalate selectivity inhibits interleukin-5-mediated eosinophil survival, *Jour. Allergy Clin. Immunol.*, 96, 251–256.

Trattner, A., and M. David. 2000. Gold sensitivity in Israel — consecutive patch test results, *Contact Dermatitis*, 42, 301–302.

Tsuruta, K., K. Matsunaga, K. Suzuki, R. Suzuki, H. Akita, Y. Washimi, A. Tomitaka, and H. Ueda. 2001. Female predominance of gold allergy, *Contact Dermatitis*, 44, 55–56.

Vamnes, J.S., T. Morken, S. Helland, and N.R. Gjerdet. 2000. Dental gold alloys and contact hypersensitivity, *Contact Dermatitis*, 42, 128–133.

Wiesner, M. and M. Pambor. 1998. Allergic contact dermatitis from gold, *Contact Dermatitis*, 38, 52.

Williamson, R. 1996. Clinical management of galvanic current between gold and amalgam, *Gen. Dentist.*, 44, 70–73.

Wu, M.L., W.J. Tsai, J. Ger, and J.F. Deng. 2000. Hepatitis and hyperamylasemia caused by gold potassium cyanide, *Jour. Toxicol.*, 38, 552.

CHAPTER 9

Chrysotherapy

About 100 million individuals are afflicted with rheumatoid arthritis (RA), including 8 million Americans. This painful disease is associated with high morbidity and mortality. The causes of RA are unknown and there is no known cure. Chrysotherapy, or the treatment of RA by monovalent gold thiol drugs, has been practiced since 1929, although the mechanisms of action are not known with certainty. In general, the gold drugs were most efficacious — and toxic — during the first 2 years of treatment. This chapter synthesizes information on (1) the history of gold drugs used in the treatment of RA; (2) proposed modes of action of gold drugs; and (3) chrysotherapy treatment regimes, adverse effects, and case histories, as modified from Eisler (2003).

9.1   HISTORY

Rheumatoid arthritis (RA) is an inflammatory condition that leads to progressive erosion of the articular cartilage lining the interfaces of bones in joints (Shaw 1999a). Unchecked, the bones will eventually fuse after cartilage loss is complete. Inflammation occurs first in the synovial membrane which surrounds the joints and then moves into the synovial cavity between the bones. Damage to the tissues is effected by lysosomal enzymes including collagenase and other proteases that are released because of the inflammatory condition. The resulting tissue destruction releases cell and tissue fragments, which stimulates further inflammation and the migration of immune cells, including macrophages, into the inflamed area. Thus, a cycle is established of degradation and further release of destructive enzymes (Shaw 1999a). About 1 to 2% of the world's population (about 60 to 120 million individuals) are affected by RA (Gromer et al. 1998), including an estimated 8 million Americans (Roberts and Shaw 1998). Rheumatoid arthritis is a disease associated with high morbidity and mortality from infection. Death from infection is nine times more common in patients with RA than the general population; the majority of these deaths are due to pneumonia (Snowden et al. 1996).

Active RA is defined as an unremitting disease for at least 6 months with at least three of the following characteristics: (1) six or more tender joints; (2) three or more swollen joints; (3) early morning joint stiffness lasting for at least 1 hour; and (4) altered erythrocyte sedimentation rate or c-reactive protein concentrations greater than 20 mg/L (Zeidler et al. 1998). For most patients with RA, irreversible joint damage frequently occurs within 2 years if left untreated (Zeidler et al. 1998). The pathogenesis of RA is predicated on T-cell and monocyte and macrophage activation with the release of monokines and tumor necrosis factor, which secondarily leads to the release of cytokines from other cells, such as synoviocytes (Yanni et al. 1994). Collectively, the release of these and other inflammatory mediators, as well as the activation of synoviocytes, monocytes, macrophages, chondrocytes, and osteoclasts brings about the inflammatory and destructive changes which are characteristic of chronic immune-mediated inflammatory diseases, such as RA (Yanni et al. 1994)

Because the cause of RA is unknown, therapy has largely been directed at suppressing the inflammatory process, with the aim of diminishing symptoms and preventing joint damage (Hashimoto et al. 1992). There is no known cure for RA; however, a small number of disease-modifying antirheumatic drugs have demonstrated ability to slow or arrest the progress of the disease (Roberts and Shaw 1998). Injectable gold compounds have been used to treat RA since the 1920s (Jones and Brooks 1996; Canumalla et al. 2001). In general, gold was equivalent to other widely used second-line agents in terms of efficacy and was most toxic and efficacious in the first 2 years of treatment; there was a dose–response relation for both efficacy and toxicity (Jones and Brooks 1996). Complex Au^+ salts, especially those of gold thiomalate, gold thioglucose, and gold thiosulfate were used frequently in the treatment of RA to produce durable remissions of this chronically progressive disease (Merchant 1998). After absorption, either from tissues or from the gastrointestinal tract, gold salts were bound to albumin and globulin and carried by the plasma to practically every tissue in the body. Gold salts remained in the plasma for many months and more than 75% of the injected dose was ultimately fixed in the kidneys, liver, spleen, marrow, skin, hair, and nails. Gold salts and their metabolites were commonly contained within phagolysomes called aurosomes. Aurosomes may attain elevated skin gold concentrations sufficient to cause a blue-grey discoloration (chrysiasis), which is most pronounced on the face, hands, and other body areas exposed to sunlight. Most Au^+ salts not stored in the tissues were excreted in the urine. The highest concentrations of gold salts were found in organs rich in reticuloendothelial cells, namely, lymph nodes, liver, bone, marrow, and spleen. Although used as a treatment for RA, the concentrations of gold salts in articular and para-articular structures — including synovium, bone, muscle, and cartilage — were substantially lower in treated patients than in those with normal and untreated arthritic tissues (Merchant 1998).

Only monovalent gold salts showed therapeutic activity in the treatment of rheumatoid arthritis (Asperger and Cetina-Cizmek 1999). A characteristic feature of RA treatment with gold compounds is a delayed onset of clinical effects and an ability frequently to slow or prevent the progress of the disease (Hirohata 1996). The most important antirheumatics with gold are trisodium bis (thiosulfato) aurate (I)

Table 9.1   Gold Thiolate Compounds Used Medicinally

Name	Formula	H_2O Soluble
Myocrisin, Myochrisine, Myochrisis	Sodium gold[+] thiomalate [Na_2AuST_m], a variable mixture of the mono-($C_4H_4AuNaO_4S$) and disodium ($C_4H_3AuNa_2O_4S$) salts	Yes
Solganol	Sodium gold[+] thioglucose; gold 4-aminomethylsulphinic-acid-2-mercaptobenzene-1-sulphonic acid	Yes
Sanocrysin, Crisalbine, Aurothion, Sanocrysis	Sodium gold thiosulfate [$Na_3Au(S_2O_3)_2$]	Yes
Allocrysine	Sodium gold thiopropanolsulfonate	Yes
Krysolgan	Sodium gold 4-amino-2-mercaptobenzoic acid	Yes
Auranofin	3-triethylphosphinegold[+]-2,3,4,6-tetra-O-acetyl-1-thio-β-D-glucopyranoside; 2,3,4,6-tetra O-acetyl-1-thio-B-D-pyranosato-S-(triethylphosphine)-Au[+]	No

Source: Data from Sadler 1976; Brown and Smith 1980; Windholz 1983; Jones and Brooks 1996; Shaw 1999b; Tiekink 2003.

(Sanocrysin or Sanocrysis); disodium (thiomalato-S) aurate (I) (Myocrisin, Myochrisis); (thioglucose-) gold (I) (Solganol); and 2,3,4,6,-tetrabis-O-acetyl-1-thi-B-D-glucopyranosato-S (triethyphosphine) gold (I) (Auranofin) (Table 9.1; Roberts and Shaw 1998; Asperger and Cetina-Cizmek 1999; Tiekink 2003).

Parenteral gold salts modified the course of RA when given in sufficient dose over a longer period of time (Sander et al. 1999). Gold salts, such as sodium gold thiomalate, have been used in the treatment of patients with RA since 1929 and are still considered as important disease modifying antirheumatic drugs (ten Wolde et al. 1995; Choy et al. 1997; Hamilton et al. 2001). Around 50% of patients receiving these are usually markedly improved or cured, but toxic side effects are observed in about 40% of the cases (Sadler 1976). When chrysotherapy was discontinued, it was usually due to adverse mucocutaneous side effects during the first 2 years and inefficacy after 4 years (Bendix and Bjelle 1996). The frequencies of mucocutaneous effects — including pruritus, dermatitis, and stomatitis — were 29% after 6 months, 42% after 1 year, 55% after 2 years, 74% after 5 years, and 92% after 10 years. Moderate dose parenteral gold treatment carried negligible risks of serious side effects. The major problem in long-term maintenance treatment was inefficacy. Nevertheless, many patients with RA did well during many years on this treatment (Bendix and Bjelle 1996).

Following parenteral therapy that involved relatively high doses, gold persisted for months before being carried by plasma bound to α-globulin to practically every tissue, including skin, hair, and nails (Hostynek 1997). Gold particles once deposited in the dermis remained in place permanently, and they have also been located around blood vessels and in dermal macrophages (Hostynek 1997).

Gold has a useful place in the treatment of RA, but adverse side effects in 30 to 45% of treated patients resulted in discontinuation (Nilsson 1997; van Gestel et al.

1994; Goebel et al. 1995; ten Wolde et al. 1995; Choy et al. 1997; Ueda 1998; Hamilton et al. 2001). Adverse side effects of chrysotherapy included skin rashes; protein in the urine; inflammation of the mouth; reduction in the number of circulating leukocytes; decreased number of blood platelets; aplastic anemia due to organ damage; lung abnormalities; adverse immune reactions, such as stomatitis, eosinophilia, lymphadenopathy, and hypergammaglobulinemia; severe hypotension, angina, myocardial infarction, nephrotoxicity, and nephrotic syndrome; hepatitis; colitis; and chrysiasis (pigmentation) of the cornea, lens, and skin (Pickl et al. 1993; van Gestel et al. 1994; Jones and Brooks 1996; Jain and Lipsky 1997; Tilelli and Heinrichs 1997; Merchant 1998; Ueda 1998; Kiely et al. 2000). The most common side effect of chrysotherapy was skin toxicity, accounting for up to 60% of all adverse reactions, especially lichenoid eruptions and nonspecific dermatitis (Choy et al. 1997; Raza and Phillipps 1997; Merchant 1998; Rasanen et al. 1999).

Sodium gold thiomalate, a water-soluble gold salt, is used extensively in RA therapy (Nilsson 1997). Gold binds to sulfhydryl-containing ligands, affects the release and activity of some lysosomal enzymes, affects the immune system by inhibiting monocyte function, and enhances virus infections (Nilsson 1997). Injectable gold salts, such as sodium gold thiomalate and gold thioglucose, were the initial drugs of choice in chrysotherapy; however, introduction of auranofin (triethylphosphine gold tetra-acetyl glucopyranoside), an orally administered gold compound, may produce side effects of lower frequency and severity than the injectable compounds (Ueda 1998). But auranofin was not as effective in treating RA as injectable gold compounds (Jain and Lipsky 1997). Vasomotor reactions were encountered with all therapeutic gold treatments, but most commonly with sodium gold thiomalate (Hill et al. 1995). Reactions were usually mild and transient, accompanied by nausea, vomiting, facial flushing, and hypotension. Serious consequences have been reported infrequently, including myocardial infarction and stroke-like syndromes (Hill et al. 1995).

9.2 PROPOSED MODES OF ACTION

Gold drugs have well-documented anti-inflammatory activity, which they share with organic medicines such as aspirin and ibuprofen, but they also have more profound effects on the underlying joint destruction that simple anti-inflammatory agents lack (Shaw 1999a). Accordingly, some RA patients receiving chrysotherapy obtained long-lasting remission of the disease, with sustained and less frequent treatment. Many mechanisms have been proposed for chrysotherapy, but none are considered definitive by the medical community. The proposed mechanisms range from broad and sometimes vague systemic effects, such as immunomodulatory activity, to very specific biochemical processes that act through a progressively multiplied effect on a larger system. It is not currently possible to design new drugs that combine the effectiveness of the injectable Au^+ thiolates with the safety of gold$^+$ complexes taken orally until a mechanism of action is well established (Shaw 1999a).

Although the exact mode of action of organomonovalent gold salts in chrysotherapy is not known with certainty, it does include a number of associations in reducing inflammation. These are arbitrarily grouped into four sometimes overlapping associations:

1. Gold metabolites, including $Au(CN)_2^-$ (Sadler 1976; Whitehouse and Graham 1996; Graham and Kettle 1998; Shaw 1999a; Canumalla et al. 2001), Au^{+3} (Goebel et al. 1995; Whitehouse and Graham 1996; Choy et al. 1997; Merchant 1998; Shaw 1999a; Zou et al. 1999), Au^0 (Abraham and Himmel 1997; Zou et al. 1999)
2. Immunomodulatory activity, including alterations of peptides, macrophage phagocytosis, histamine release, monocyte migration, T-cells, B-cells, and immunoglobulins (Hashimoto et al. 1992; Yanni et al. 1994; Goebel et al. 1995; Griem and Gleichmann 1996; Hirohata 1996; Snowden et al. 1996; Hirohata et al. 1997; Jain and Lipsky 1997; Wang et al. 1997; Ueda 1998; Hirohata et al. 1999; Kiely et al. 2000; Sanders 2000)
3. Tumor necrosis factor (Yanni et al. 1994; Yadav et al. 1997; Bratt et al. 2000; Mangalam et al. 2001)
4. All others

This last category includes bone resorption (Hall et al. 1996); leukocyte infiltration (Newman et al. 1994) and leukocyte antigens (Pickl et al. 1993; ten Wolde et al. 1995); lysosomal enzymes (Sadler 1976; Nilsson 1997); protein-gold complexes (Sadler 1976; Sato et al. 1995; Shaw 1999a); polymorphonuclear neutrophils (Whitehouse and Graham 1996; Heimburger et al. 1998); sulphydryl binding sites (Sadler 1976; Handel 1997); superoxide ion radical (Asperger and Cetina-Cizmek 1999); and thioredoxin reductase, a selenoenzyme (Gromer et al. 1998).

9.2.1 Au^+ and Au^+ Metabolites

Antirheumatic gold complexes may be activated by their conversion to aurocyanide, $Au(CN)_2^-$ (Graham and Kettle 1998). Aurocyanide is produced by two processes involving the formation of hypothiocyanite and hypochlorous acid. Aurocyanide is an effective inhibitor of the respiratory burst of neutrophils and monocytes, and the proliferation of lymphocytes. Therefore, sodium gold thiomalate may attenuate inflammation by acting as a pro-drug which is reliant on neutrophils and monocytes to produce hypothiocyanite. When the hypocyanite decays to hydrogen cyanide, the pro-drug is converted to aurocyanide, which then suppresses further oxidant production by these inflammatory cells. The neutrophil enzyme myeloperoxidase converts gold thiomalate to aurocyanide through the oxidation of thiocyanate. Thiocyanate is a major substrate of myeloperoxidase, and this appears to be the dominant route by which aurocyanide is formed from the gold complexes *in vivo* (Graham and Kettle 1998).

Traces of $Au(CN)_2^-$ have been identified as a common metabolite of all Au^+ gold drugs used clinically in the United States to treat RA (Shaw 1999a). It occurs in treated RA patients at concentrations of 1 to 5 µg/L in blood and at higher concentrations in urine. Research on the effect of gold compounds on the immune system showed that thiocyanate $(SCN)^-$ can be converted into cyanide by hypochlorite generated during the oxidative burst of immune cells. The formation of $Au(CN)_2^-$ may be a critical metabolite at the inflammatory sites in the arthritic joints of patients.

The $Au(CN)_2^-$ is taken up by the cells and limits the extent of the oxidative burst. The $Au(CN)_2^-$ transport appears to be via the sulfhydryl shuttle mechanism. Intact $Au(CN)_2^-$ ions bind to serum albumin at one strong and three weak binding sites. The newly found role of $Au(CN)_2^-$ may end the long uncertainty of the action of gold drug metabolites on inflamed joints and immune systems of patients (Shaw 1999a).

Recent advances in biochemical and cellular mechanisms of action for gold complexes during chrysotherapy highlight the roles of $Au(CN)_2^-$, Au^{+3}, and protein–gold complexes in the metabolism of gold compounds (Shaw 1999a). These medicinal agents are actually pro-drugs, and this is consistent with the hypothesis that the bioinorganic chemistry of gold complexes requires the formation of common metabolites from the parenteral drugs and auranofin. More research is needed on the gold metabolites to determine whether they are either the active species or generate it at the presently unidentified sites of action (Shaw 1999a).

Antirheumatoid organoAu$^+$ complexes may be converted either to Au^{+3} complexes or to aurocyanide, $Au(CN)_2^-$ by myeloperoxidase in polymorphonuclear leukocytes (Whitehouse and Graham 1996). Inflammatory cells have the potential to transform organoAu$^+$ compounds to yield active species, the degree of transformation being governed by the intensity of the inflammation. In one scenario, the stable complex aurocyanide is produced through the simultaneous availability of gold thiolate and thiocyanate $[(SCN)^-]$, followed by oxidation of the gold complex. The thiocyanate anion is a normal component of plasma, being derived from dietary items that contain $(SCN)^-$ or cyanide derivatives that are metabolized to thiocyanate in the liver. Thiocyanate is a substrate for several mammalian peroxidases, including myeloperoxidase acting on hydrogen peroxide and $(SCN)^-$ in polymorphonuclear leukocytes. Cyanide is made available by the oxidation of $(SCN)^-$ by myeloperoxidase, displacing thiol ligands from the gold in polymeric gold complexes. Cyanide would be expected to displace gold from its binding site on albumin, thus generating $Au(CN)_2^-$. At least part of the aurocyanide is released from polymorphic leukocytes (PMN), inhibiting the proliferation of lymphocytes produced by PMN incubated with sodium gold thiomalate. Thus, aurocyanide does not normally act through the release of cyanide, and is considered an active metabolite of gold thiolate complexes. Aurocyanide-plasma albumin interactions and thiocyanate-PMN interactions are confounding factors (Whitehouse and Graham 1996).

The oxidation of $Au^+(CN)_2^-$, a gold metabolite, and further cyanidation of Au^{+3} products to $Au^{+3}(CN)_4^-$ is postulated as a possible mechanism of an immunologically generated Au^+/Au^{+3} redox cycle *in vivo* (Canumalla et al. 2001). In that study, hypochlorite ion, an oxidant released during the oxidative burst of immune cells, in combination with Au^{+3} cyanates, generates mixed dicyanoaurate^{+3} complexes, *trans*-$(Au^+(CN)_2 X_2)^-$, where X^- represents equilibrating hydroxide and chloride ligands, and establishes the chemical feasibility of dicyanoaurate$^+$ oxidation by OCl$^-$ to Au^{+3} species, suggesting a new procedure for synthesis of $H(Au(CN)_2Cl_2)$. Reaction of *trans*-$(Au^{+3}(CN)_2X_2)^-$ — where $X = Cl^-$ or Br^- — or $Au^{+3}Cl_4^-$ with HCN in aqueous solution at pH 4 leads directly to $Au^{+3}(CN)_4^-$ without detection of the anticipated $Au(CN)_xX_{4-x}$ intermediates and is attributed to the *cis* and *trans* accelerating effects of the cyanides. The reduction of $Au^{+3}(CN)_4^-$ by glutathione and other thiols is a complex, pH-dependent process that proceeds through two intermediates and ultimately generates $Au^+(CN)_2^-$ (Canumalla et al. 2001).

Gold^{+3} can cause the oxidation of methionine residues to sulfoxides, leading to denaturation of protein and production of Au^+ (Merchant 1998). It is proposed that the oxidation of proteins eventually leads to a complete and continued conversion of offending auto-antigens, but this hypothesis remains untested. According to Merchant (1998), once Au^+ salts are taken up by lysosomes, myeloperoxidases and other lysosomal enzymes collaborate in oxidizing the gold in Au^+ to auric chloride, $AuCl_3$ (Au^{+3}). Since Au^{+3} is a better oxidant than Au^+, it dominates both the anti-inflammatory and toxic effects of gold salts. *In vitro* studies with phagolysomes suggest that a redox system might be operative, and that $AuCl_3$ might be formed from Au^+ following an oxidative burst in phagocytic immune cells. The reduction of $AuCl_3$ and its derivatives by serum albumin and various thiols and thiol-ethers suggests that reduction occurs over a longer period than does the hypochlorite oxidation of Au^+. When oxidation occurs rapidly, the oxidant diffuses away from its site of generation and reacts with any protein reductants encountered, which subsequently triggers the immunological consequences associated with clinical benefits and side effects of chrysotherapy (Merchant 1998).

Injectable Au^+ complexes are oxidized by hypochlorite (generated by myeloperoxidase) to Au^{+3} species in the phagolysomes of activated macrophages and granulocytes (Whitehouse and Graham 1996; Choy et al. 1997). Trivalent gold, but not Au^+, can induce proliferation of lymphocytes in up to 73% of patients with Au^+-induced cutaneous eruptions. The oxidized species may contribute to the toxicity of Au^+ complexes by sensitization of T-cell lymphocytes.

Griem and Gleichmann (1996) conclude that Au^+ and Au^{+3} each exert specific effects on several distinct functions of macrophages and the activation of T-cells. These effects may explain the anti-inflammatory and the adverse effects of antirheumatic gold drugs. Griem and Gleichmann (1996) present three scenarios to account for the action of disodium gold thiomalate. The first involves selective inhibition of T-cell receptor-mediated antigen recognition by rodent CD4$^+$ T-cell hybridomas specific for antigenic peptides containing at least two cysteine residues. In this scenario, Au^+ acts as a chelating agent forming linear complexes (cysteine–Au^+–cysteine) which prevents correct antigen processing and peptide recognition by the T-cell receptor. In the second scenario, Au^+ is oxidized to Au^{+3} in mononuclear phagocytes, such as macrophages. Because Au^{+3} rapidly oxidizes protein and is then reduced to Au^+, this introduces an Au^+/Au^{+3} redox system into phagocytes which scavenge reactive oxygen species, such as hypochlorous acid (HOCl), and inactivates lysosomal enzymes. Finally, authors showed that a model protein antigen–bovine ribonuclease A (RNaseA) system subjected to Au^{+3} induced novel antigenic determinants recognized by CD4$^+$ T-lymphocytes. Analysis of the "Au^{+3}-specific" T-cells shows that they reacted to RNaseA peptides, but that recognition of these cryptic peptides did not require the presence of gold (Griem and Gleichmann 1996). The CD4$^+$ cells are important in the pathogenesis of RA because sodium gold thiomalate can inhibit mitogen-induced DNA synthesis of CD4$^+$ T-cells in a dose-dependent manner (Hashimoto et al. 1992).

In both humans and mice, gold is predominantly accumulated in mononuclear phagocytes (Goebel et al. 1995). Inside the cells, gold is stored mainly in lysosomes (aurosomes). Monocytes and certain stages of macrophages contain peroxidase-positive granules, indicating myeloperoxidase. In the presence of H_2O_2 and Cl^- the isolated enzyme myeloperoxidase could generate Au^{+3} from the Au^+ of disodium gold thiomalate, probably through the formation of hypochlorous acid, a powerful oxidant. Au^{+3} is very short-lived in the presence of reducing agents, such as thiols and proteins, because it is reduced to Au^+. This process leaves behind specifically altered self proteins which may sensitize T-cells. Goebel et al. (1995) envisage three anti-inflammatory mechanisms of Au^{+3}:

1. Generation in phagocytic cells of Au^{+3} from Au^+ scavenges reactive oxygen species, such as hypochlorous acid.
2. Au^{+3} irreversibly denatures proteins nearest the site of oxidation, possibly lysosomal enzymes which nonspecifically enhance inflammation when released from cells.
3. Interference of Au^{+3} with reduced production of arthritogenic peptides.

All of these anti-inflammatory effects of Au^+/Au^{+3} might be extended by the fact that in the presence of protein, Au^{+3} is re-reduced to Au^+ within minutes. If the Au^+/Au^{+3} redox system in phagocytic systems is correct, the anti-inflammatory actions of Au^+/Au^{+3} outlined above could be effective over a prolonged period (Goebel et al. 1995).

Trivalent gold (Au^{+3}) can be generated from Au^+ *in vivo* and may be responsible for Au^+ toxicity in the treatment of RA (Zou et al. 1999). Ionic $[AuCl_4]^-$ interacts with glycine at acidic and neutral pH; Au^{+3} cleaves glycine to produce $(NH_4)^-$ and glyoxylic acid. The glyoxylic acid is oxidized by Au^{+3} to yield formic acid and CO_2. Metallic gold formation was observed in all reactions (Zou et al. 1999).

The oxidizing capacity of phagocytic cells is responsible for the generation of immunogenic drug metabolites, especially those that cause extrahepatic immune pathological lesions (Goebel et al. 1995). Adverse immune reactions to Au^+ drugs are elicited by T-cell sensitization to Au^{+3}, which apparently is a reactive intermediate metabolite formed *in vivo* through oxidation of Au^+. In the case of disodium gold thiomalate, oxidation of Au^+ to Au^{+3} seems to be responsible for the adverse immune reactions that may develop during gold therapy. The reactive metabolite Au^{+3} may be generated by mononuclear phagocytes exposed to Au^+. This mechanism was analyzed in mice using T-lymphocytes previously sensitized to Au^{+3}. Further, mononuclear phagocytes exposed to Au^+ *in vitro* also proved capable of eliciting a specific secondary response of Au^{+3}-primed T-cells. Goebel et al. (1995) conclude that the mononuclear phagocytes exposed to Au^+ generate the reactive intermediate Au^{+3} which, through oxidation of proteins, sensitizes T-cells. Because mononuclear phagocytes are found in many organs and communicate with T-cells, their capacity to generate Au^{+3} may account for the various extrahepatic adverse immune reactions induced by Au^+ salts (Goebel et al. 1995).

Abraham and Himmel (1997) concluded that colloidal gold (Au^0) could become an effective and safer alternative to the aurothiolates now used almost exclusively in the management of RA patients. Because aurothiolates have only limited success in the treatment of RA and were associated with a high incidence of side effects, and because Au^0 was reportedly generated *in vivo* from Au during oxidation of Au^{+3}, authors postulated that the active ingredient in chrysotherapy was Au^0 and that the side effects were caused by Au^{+3}. To test this postulate, 10 RA patients with long-standing erosive bone disease not responding to previous treatment were recruited from a private practice. Patients were given 30 mg of colloidal Au^0 daily for 24 weeks and afterward weekly for 4 weeks and monthly for an additional 5 months; there was no clinical or laboratory evidence of toxicity in any of the patients. Tenderness and swelling of joints were rapid and dramatic, with a significant decrease in both parameters after the first week which persisted for the study period. By week 24, patients scored about 10 times lower on tenderness and swelling than pretreatment levels. Evaluated individually, 9 of the 10 patients improved markedly after 24 weeks of colloidal gold at 30 mg daily. The cytokines interleukin-6 (IL-6) and tumor necrosis factor alpha (TFN-α), the immune complexes IgG and IgM, and rheumatoid factor were significantly suppressed by the colloidal gold (Abraham and Himmel 1997).

9.2.2 Immunomodulatory Activity

In the treatment of RA, efforts have focused on the effects of organoAu$^+$ compounds on T-lymphocytes and macrophages, the cells that seem to be the most important in the pathogenesis of RA (Wang et al. 1997). Observed effects of gold preparations include inhibition of interferon production by monocytes, partial inhibition of neutrophil protein kinase C activity, and inhibition of lymphocyte membrane activity. The precise action of gold compounds commonly used to treat RA remains uncertain. Monovalent gold preferentially forms complexes with thiolate and thioether ligands, binding to a large number of proteins, particularly those having nucleophilic sulfhydryl groups in accessible sites. Wang et al. (1997) list a number of alternate mechanisms to account for Au$^+$ efficacy. These include (1) inhibition of peptide antigen presentation to T-cells. T-cell activation by peptides containing cysteine was inhibited owing to a direct interaction between Au$^+$ and thiol groups within the stimulating peptides; (2) inhibition of activity of certain transcription factors via interactions between gold and specific cysteine thiol groups regulated by intracellular redox potential. This involves inhibition of progesterone and glucocorticoid receptors; and (3) inhibition of activity of a protein–tyrosine phosphatase (CD4$^+$) considered essential for antigen receptor lymphocyte signaling. *In vitro* inhibition by disodium gold thiomalate was 50% effective at 236 μg Au/kg using a papain substrate. Other protein–tyrosine phosphatases were inhibited at 709 μg Au/kg (Wang et al. 1997).

At pharmacologically relevant concentrations, injectable Au$^+$ compounds are reported to (1) inhibit immunoglobulin levels and rheumatoid factor titers; (2) reduce the number of circulating lymphocytes; (3) inhibit antigen- and mitogen-induced proliferation of human lymphocytes indirectly by inhibiting the accessory function of monocytes; (4) inhibit monocyte capacity to produce superoxide anions and components after activation; and (5) inhibit the differentiation of monocytes into effector cells (Hashimoto et al. 1992). The biochemical basis for various cellular effects of gold compounds may be associated with protein kinase C activation (which is involved in T-cell activation), inositol phospholipid breakdown, and mobilization of internal Ca^{+2} stores. Hashimoto et al. (1992) maintain that the most significant immunomodulatory effects of chrysotherapy result from its capacity to inhibit protein kinase C activity in a dose-dependent manner. Sodium gold thiomalate inhibited protein kinase C *in vitro* and *in vivo* in T-cells, probably by interacting with thiol groups. Interactions of sodium gold thiomalate and protein kinase C and inhibition of catalytic activity of protein kinase C in T-cells may be responsible, in part, for the therapeutic antirheumatic action of sodium gold thiomalate. It is unknown whether sodium gold thiomalate inhibits protein kinase C activity in other inflammatory cells (Hashimoto et al. 1992).

Moderate suppression of serum immunoglobulin levels occurred in about 50% of patients treated for RA with parenteral gold salts (Snowden et al. 1996; Kiely et al. 2000). Pathological immunoglobulin deficiency is generally considered to be a rare consequence of gold therapy, although cases have been documented. Antibody deficiency, as judged by subnormal (three standard deviations below normal) serum immunoglobulin levels, were reported in 22 RA patients undergoing chrysotherapy with sodium gold thiomalate. Duration of treatment ranged between 0.5 and 13.0 years (mean of 4 years), and total dose of gold received ranged between 0.5 and 10 g (mean 2.5 g; Snowden et al. 1996). There was no apparent relation between duration of administration or dose and antibody deficiency. Patterns of antibody deficiency were either mild or severe. In the mild reaction, only one immunoglobulin isotype was affected and specific antibody production was normal; gold treatment was continued. In the severe reaction, two or three immunoglobulin isotypes were affected and specific antibody production was defective, resulting in infections; gold treatment was discontinued with normal antibody production recovered in all but one patient (Snowden et al. 1996).

One of the characteristic features of RA is chronic stimulation of B-cells, evidenced clinically by the production of rheumatoid factors (RF) which play an important role in the pathogenesis of the disease (Hirohata 1996; Hirohata et al. 1999). Serum immunoglobulin (Ig) levels and rheumatoid factor (RF) titers often decrease in RA patients treated with gold compounds. Inhibition of Ig production by sodium gold thiomalate is not due to the thiomalate component but most likely due to its gold component because thiomalate alone did not inhibit Ig production. Experimental studies indicate that sodium gold thiomalate preferentially inhibits the function of B-cells at concentrations much lower than those that inhibit the function of T-cells through interfering with the initial activation of B-cells. The direct inhibitory effects of sodium gold thiomalate on human B-cell activation contributes, in part, to its therapeutic effect in treating RA (Hirohata 1996; Hirohata et al. 1999). *In vivo*, Au^+ dissociates from its carrier molecule, such as thiomalate, and binds to the thiol groups of proteins. Sodium gold thiomalate also likely interacts with thiol groups of enzymes that participate in the activation of human B-cells. It is postulated that B-cells take up gold components bound to cysteine residues of exogenous antigens more rapidly than T-cells. In patients with RA, it has been suggested that the production of RF appears to be T-cell dependent. Hirohata (1996) and Hirohata et al. (1999) conclude that low concentrations of gold compounds directly suppress B-cell function by interfering with the initial activation of B-cells, as evidenced by the decrease in serum Ig levels and RF titers in RA patients treated with gold compounds. Synergistic inhibitory effects of sodium gold thiomalate and auranofin with thiols on *in vitro* B-cell activation suggest the therapeutic efficacy of combinations of these compounds in the treatment of RA (Hirohata et al. 1999).

Since RA is a chronic inflammatory disease characterized by hyperplasia of synovial lining cells, Hirohata et al. (1997) suggest that synovial hyperplasia in RA might be a result of continuous recruitment of bone marrow-derived monocytes into the synovium. An adequate supply of peripheral blood monocytes, granulocytes, and platelets is necessary for an optimal inflammatory process. Generation of monocyte-lineage cells from the bone marrow is accelerated in patients with RA. The generation of monocyte cells from RA bone marrow was significantly suppressed *in vitro* by sodium gold thiomalate, and is consistent with the hypothesis that sodium gold thiomalate interferes with monocyte differentiation in the bone marrow. Further, sodium gold thiomalate is a potent inhibitor of monopoeisis in RA patients (Hirohata et al. 1997).

Jain and Lipsky (1997) aver that gold salts act by inhibiting mononuclear phagocyte function within the inflamed synovium. Endothelial cell proliferation may also be inhibited. Each of these effects may relate to gold-mediated inhibition of the enzyme protein kinase C, which plays a central role in cellular activation (Pickl et al. 1993; Jain and Lipsky 1997).

9.2.3 Tumor Necrosis Factor

The adhesion of circulating leukocytes to vascular endothelium is pivotal for the inflammatory response in various diseases, including RA (Bratt et al. 2000). The pro-inflammatory cytokine tumor necrosis factor-alpha (TFN-α) is known to activate endothelial cells to express adhesive and activation molecules for leukocytes. This can lead to an activation of the cytotoxic capacity of polymorphic neutrophils, which may result in injury to the endothelial cells that are dependent on adhesion molecules. Both auranofin (orally administered) and sodium gold thiomalate (injectable) gold salts are known to reduce human umbilical vein endothelial cell (HUVEC) adhesion molecule expression and neutrophil adherence. Both compounds reduced TNF-α-mediated neutrophil-dependent toxicity of human epithelial cells in a dose-dependent manner. The clinical importance of auranofin and sodium gold thiomalate as anti-rheumatic agents may result from their ability to inhibit the expression of adhesion molecules involved in leukocyte recruitment (Bratt et al. 2000).

Increased levels of TNF-α were demonstrated in synovial fluids of patients with RA, along with increased amounts of RNA in cells infiltrating the synovium (Yadav et al. 1997). Sodium gold thiomalate can modulate TNF-α production by peripheral blood mononuclear cells (PBMC) through pathways that are also activated by lipopolysaccharides (LPSs). *In vitro* studies show that sodium gold thiomalate inhibits

TNF-α production in patients with RA, as well as in healthy individuals when the PBMC are pre-activated by polysaccharides (Yadav et al. 1997). TNF-α has emerged in the last decade as the major pro-inflammatory cytokine in the pathogenesis of RA (Mangalam et al. 2001). Sodium gold thiomalate stimulated spontaneous TNF-α production and inhibited LPS-stimulated TNF-α production in certain cases. The action of gold complexes on TNF-α production may be through inhibition of transcription factors. The effect of sodium gold thiomalate on spontaneous and LPS-stimulated TNF-α production on human PBMC is to inhibit LPS-stimulated TNF-α production through inhibition of transcription factors that regulated sphingomyelin pathways or downregulation of inflammatory cytokines. Further studies are recommended on the intracellular pathways modulated by Au^+ and leading to the design of more effective therapeutic molecules for the treatment of RA (Mangalam et al. 2001).

9.2.4   Bone Resorption

Gold salts inhibit osteoclastic bone resorption (Hall et al. 1996). Loss of bone mass is commonly associated with RA and may be due to increased activity of bone resorbing osteoclasts. *In vitro* tests with Au^+ compounds showed a decreased survival of osteoclasts on bovine bone slices at therapeutic concentrations. Inhibition of osteoclastic bone resorption by gold salts may, in part, account for their beneficial effects in treating RA (Hall et al. 1996).

9.2.5   Leukocyte Infiltration

Leukocyte infiltration is a key component of chronic infiltration (Newman et al. 1994). Adhesion molecules on endothelial cells play an important role in targeting and facilitating leukocyte migration out of the blood vessels in response to an inflammatory stimulus. Sodium gold thiomalate has a direct effect on endothelial adhesion molecule expression, namely, inhibition of cytokine-stimulated expression of vascular cell adhesion molecules on endothelial cells, and this may contribute to its anti-inflammatory activity (Newman et al. 1994). The genetic basis of RA may involve human leukocyte antigens (HLA; Pickl et al. 1993). HLA-DR1 seems to be a marker for the susceptibility of gold adverse reactions. Patients with RA had higher frequencies of DR1 and DR4 positives (38 to 43%) than healthy controls (20 to 26%). An association of thrombocytopenia and proteinuria with HLA-DR3 indicates a possible role of HLA molecules in these adverse reactions (Pickl et al. 1993). However, the association between HLA, especially HLA-type DR3, and gold toxicity is inconsistent, and HLA typing is not considered helpful in predicting the therapeutic response to parenteral gold therapy (ten Wolde et al. 1995).

9.2.6   Lysosomal Enzymes

Joint damage in RA could be caused by the release of hydrolytic lysosomal enzymes within the inflamed joint; inhibition of these enzymes by Au^+-thiols is suggested by Sadler (1976). Nilsson (1997) found that the majority of cytosolic gold was concentrated in lysosomes where it bound to several species of macromolecules and inhibited release of lysosomal enzymes, and this may account, in part, for the therapeutic efficacy of Au+-thiol drugs.

9.2.7 Macrophages

Macrophages are assumed to be the target of gold drugs. Mammalian cells respond to a wide variety of adverse conditions by inducing the synthesis of stress proteins (Sato et al. 1995). Stress proteins induced in peritoneal macrophages by sodium gold thiomalate were identified as heme oxygenase and MSP23, both anti-oxidants inhibit macrophage maturation. Increased synthesis of these proteins may have a role in mediating the pharmacologic effect of the Au+ agents (Sato et al. 1995).

9.2.8 Polymorphic Neutrophils

Gold+-containing drugs modulated the adhesiveness of endothelial cells and polymorphonuclear neutrophils (PMN) by different mechanisms involving surface adhesion molecules, this being the first step for PMN migration to inflammatory lesions (Heimburger et al. 1998). Sodium gold thiomalate *in vitro* impaired the ability of interleukin endothelial cells to bind PMN. Auranofin, however, hampered interleu-kin-induced hyper-adhesiveness with a net effect of reduction of cytokine adhesiveness (Heimburger et al. 1998). The inhibitory effects of Au+ compounds on pro-inflammatory cytokine synthesis, especially interleukin 1, offer a plausible mechanism for their inhibitory effects on bone erosion and joint destruction in RA (Sanders 2000).

9.2.9 Sulfhydryl Binding Sites

Sodium gold thiomalate's anti-inflammatory mode of action probably involves sulfhydryl binding sites (Sadler 1976). The Au+ enters many cells but localizes within the lysosomes of the phagocytic cells called macrophages where it inhibits enzymes important in inflammation. Monovalent gold may discriminate between SH groups with the pool based on steric formation and state of protonation. Although $Au(CN)_2^-$ is too toxic for clinical use, it is one of the most stable Au+ ions in solution. The simple Au+ cation does not exist in water and most Au+ compounds are insoluble in water and unstable in its presence. However, mercaptides stabilize Au+ in water and make possible the widespread use of sodium gold thiomalate in treating RA (Sadler 1976). Monovalent gold drugs are reactive with thiols (Handel 1997). Thiol groups in pro-inflammatory transcription factors AP-1 and NF-κB are targets for at least some of the therapeutic effects of disease-modifying antirheumatic drugs. Developments in understanding the transcriptional effects of glucocorticoid and retinoid receptors sug-gest that they also act, in part, via inhibition of AP-1 and NF-κB. There probably exists a common therapeutic mode of action between steroids, retinoids, and disease-modi-fying antirheumatic drugs. But these drugs have different adverse effects, and probably cannot be mediated by a common mode of action, although a pure inhibitor of NF-κB and AP-1 should be effective with minimal drawbacks (Handel 1997).

9.2.10 Superoxide Ion

Gold compounds used in RA inhibit the action of superoxide ion O_2^-, which if left unchecked can lead to the degradation of proteins (Asperger and Cetina-Cizmek 1999). Different oxidizing agents can easily transform the superoxide ion radical into very reactive singlet oxygen; however, Au^+ deactivates the singlet oxygen.

9.2.11 Thioredoxin Reductase

Gold compounds used in the treatment of RA exert at least some of their pharmacologic effects by strongly inhibiting the activity of thioredoxin reductase, a selenoenzyme (Gromer et al. 1998). Because RA is considered to be an autoimmune condition initiated by various agents — the Epstein-Barr virus (EBV) being a prime candidate — and because lymphocytes infected with EBV or other viruses secrete thioredoxin, then inhibition of the thioredoxin system may control certain immune processes. The activity of thioredoxin as a cytokine, however, depends on its dithiol state (Gromer et al. 1998), which needs clarification.

9.3 TREATMENT REGIMES, CASE HISTORIES, AND ADVERSE EFFECTS

Monovalent organogold salts suppress the rheumatic process in 50 to 70% of RA patients and induce remission in about 20% (ten Wolde et al. 1995; Shaw 1999a). Others did not benefit from gold or suffered side effects that required cessation of treatment (Shaw 1999a). Factors used to evaluate the success of various treatments on RA patients included the number of swollen joints, grip strength, patient assessment of pain and mobility, erythrocyte sedimentation rate, c-reactive protein, and hemoglobin concentration. Treatment failure was usually indicated by more than six swollen joints, more than nine tender points, and elevated sedimentation rate (Rau et al. 1998a). Rheumatoid arthritis is usually treated with a combination of different antirheumatic drugs including nonsteroidal anti-inflammatory drugs, corticosteroids, and slow-acting antirheumatic drugs — including oral and injectable Au^+ compounds at 10 to 50 mg weekly (Jain and Lipsky 1997; Rau et al. 1998a). Patients were typically monitored weekly or monthly for blood gold levels, which should be less than 3.0 mg/L in order to minimize the accumulation of gold in tissues and the resulting side effects (Shaw 1999a).

Radiographs are primary in the assessment of disease progression in RA because of their objective, quantitative depiction of bone and joint damage (Rau et al. 1998b). The American College of Rheumatology now includes the assessment of radiographs in the preliminary case set of disease activity measures for RA clinical trials (Rau et al. 1998b). Radiographs were important in the evaluation of antimalarial drugs and Au^+ salts in RA treatment, showing that the former had minimal effect on halting erosion and joint destruction, and the latter showed promise in slowing bone erosion and joint destruction (Sanders 2000). Based on radiograph evaluation, sodium gold thiomalate was more effective in controlling bone erosion than placebos, about equal to methotrexate, but probably not as effective as cyclophosphamide or azathioprine (Sanders 2000).

9.3.1 Treatment Regimes

Patients receiving intramuscular injections of sodium gold thiomalate for the treatment of RA typically received 50 mg of the drug in 0.5 mL of water every week. About 85% of the gold in circulation is bound to serum albumin, a protein of molecular weight 65,000. At high serum gold levels of 3 to 7 mg Au/L, there is a significant binding to other serum proteins such as immunoglobulins (Sadler 1976). Sodium gold thiomalate was effective in reducing clinical and biochemical disease activity in patients with longstanding destructive RA when administered at 50 mg weekly until a cumulative total of 2000 mg was reached, then 50 mg every 2 weeks (Rau et al. 1998a). About 30% of each dose is excreted within 1 week (Sadler 1976). Improvement is usually noted after 3 to 4 months. Serum gold levels are kept high at about 3.0 mg Au/L serum. Some patients have been maintained on gold treatment for as long as 20 years, but this is unusual as adverse side effects result in early discontinuation in 40% of the patients. Eventually, gold enters almost every cell in the body and is removed slowly. In one patient, a serum gold level of 220 µg/L was found 9 years after treatment had ceased (Sadler 1976).

Both sodium gold thiomalate and methotrexate show equivalent efficacy in treating a population with active RA from a deprived area (Hamilton et al. 2001). Although toxic side effects were more common with the gold treatment, it was considered a useful alternative to patients in whom methotrexate was contraindicated. Methotrexate has the advantage of sustained treatment of more than 2 years in at least 50% of the patients, whereas fewer than 20% of gold-treated patients were maintained beyond 5 years. One common parenteral treatment regimen for sodium gold thiomalate consisted of 10 mg the first week, 50 mg weekly for the next 20 weeks or until a clinical response was attained, and 50 mg injections thereafter at 2-, 3-, or 4-week intervals for the duration of a 48-week treatment cycle. In one study, toxic side effects induced by sodium gold thiomalate treatment — including skin rash, mouth ulcers, low white cell or platelet counts, and proteinuria — were observed in 71% of the patients during a 48-week treatment cycle, with 43% of the patients experiencing sufficiently severe side effects to discontinue treatment before the end of the cycle (Hamilton et al. 2001).

Parenteral gold therapy may protect against bacterial ulcer disease in RA patients (Janssen et al. 1992; Paimela et al. 1995). The presence of *Helicobacter pylori* in the gastric mucosa is responsible for most cases of gastritis and may be a major factor in the pathogenesis of peptic ulcer disease. Sodium gold thiomalate is reported to have *in vitro* activity against this bacterium, and patients with RA receiving intramuscular (im) injections of gold had a decreased level of antibodies against *H. pylori*, suggesting that treatment with Au⁺ compounds decreases *H. pylori* colonization (Janssen et al. 1992). However, another study of 12-months' duration with RA patients with high seroprevalence of *H. pylori* demonstrated no significant effect of chrysotherapy on bactericidal activity (Paimela et al. 1995).

9.3.2 Case Histories

Selected case histories follow, involving as few as 10 RA patients to as many as 1019. During the period 1933 to 1979, RA patients (N = 1019) treated with monovalent organic gold salts for up to 7 years were evaluated for toxicity (Lockie and Smith 1985). Patient status at the start of therapy was 25% mild, 46% moderate, and 29% severe. At the conclusion of the study, or at the discontinuation of therapy, there were 4% resolved (discontinued therapy, in remission), 49% mild, 30% moderate, and 18% severe, representing a 95% increase in the milder/resolved group and a 35% decrease in the moderate/severe group. Episodes of toxicity, usually skin reactions, appeared in as few as 1 to 6 months (60 to 710 mg Au received) to 10 months to 10 years (890 to 2715 mg Au). There was no increase in toxicity in relation to age, sex, rheumatoid factor, or gold dosage. Authors concluded that gold compounds were efficacious when used as part of the general program in the management of RA (Lockie and Smith 1985).

A 5-year study with 440 RA patients treated with sodium gold thiomalate concluded that only patients diagnosed with RA during the first 2 years of onset have a long-lasting improvement of their functional ability after starting intramuscular gold treatment (Munro et al. 1998). Evaluation was by subjective patient response to pain, joint articulation, duration of morning sickness, and laboratory parameters such as erythrocyte sedimentation rate and c-reactive protein. About a third of the patients (160) completed 5 years of treatment and received 4.24 g of gold (range 3.2 to 5.4) during that interval at about 50 mg weekly. The actual dose rate was 10 mg of sodium gold thiomalate the first week, then 50 mg weekly until a positive response was reached, then 50 mg every fourth week (Munro et al. 1998).

In a study with 187 patients afflicted with severe RA for less than 3 years, patients were given weekly im injections of sodium gold thiomalate of 10 to 50 mg over an 18-month period until a cumulative dose of 1.0 g was reached (Zeidler et al. 1998). Afterwards, patients received 50 mg every 2 to 4 weeks. There was a high (54%) withdrawal rate from the program by the 15th month, mostly from adverse skin and gastrointestinal disorders. The remaining cases of early active severe RA showed retarded progression of joint damage in response to chrysotherapy (Zeidler et al. 1998).

In another study, 174 patients with active early erosive RA received weekly intramuscular injections of either 15 mg methotrexate or 50 mg of sodium gold thiomalate for 12 months (Rau et al. 1998b). Radiographs of joints, hands, wrists. and toes were evaluated after 6 and 12 months with no statistically significant differences between treatments. A similar pattern was observed for the number of joints with erosions. Both treatments reduced the slope of radiographic progression in patients with active erosive RA, and this confirmed the importance of radiographic outcomes in RA, particularly when evaluating disease-modifying drugs during clinical trials (Rau et al. 1998b).

Patients (N = 93) who had been in chrysotherapy for 3 to 13 years were observed yearly by the same physician (Graudal et al. 1994). After the first year, the median number of swollen joints had decreased from six to two, and the median number of tender joints from five to two; this improvement was maintained throughout the period. Reductions in the erythrocyte sedimentation rate were evident for 10 years; hemoglobin was reduced for 7 years. And the number of joints with limited motion increased from 4 to 20. All patients received either gold sodium thiosulfate (37% Au) or sodium gold thiomalate (50% Au) at dose rates of 10 to 25 mg Au weekly for about 6 months, then once every 4 weeks. At year 4 — after receiving a cumulative dose of 1027 mg of Au — adverse side effects were noted in 43% of the patients and 15% had dropped out. At year 10 — after receiving a total dose of 2368 mg of Au — 70% had dropped out and side effects were noted in 68% of the remaining patients. Authors concluded that early symptomatic improvement of RA during gold treatment remained stable for several years, although joint relief seemed to continue over the entire treatment period (Graudal et al. 1994).

Intramuscular injections of sodium gold thiomalate were used to treat 87 patients with active early erosive RA over a 3-year period at 50 mg weekly the first year and 25 mg weekly thereafter (Menninger et al. 1998). Adverse effects were noted in 89% of the patients; 53% of the patients withdrew from the study due to side effects after a mean time of 6.1 months. Of the remaining patients, clinical remission was evident in 38% and marked improvement (more than 50% reduction in number of swollen or tender joints) in 87% (Menninger et al. 1998).

To compare the efficacy of sodium gold thiomalate and methotrexate, two groups of 72 patients — each with early, active, and erosive RA — received either 50 mg im weekly of sodium gold thiomalate or 15 mg im of methotrexate weekly for a period of 6 years (Sander et al. 1999). If one drug had to be withdrawn because of adverse effects, treatment was continued with the other in still active disease state patients. Within the first 36 months, 38 patients withdrew from the gold treatment, and 23 from methotrexate. At the time of withdrawal, 40 to 70% of the patients showed improvement in all parameters measured (erythrocyte sedimentation rate, C reactive protein, swollen and tender joints, radiological progression) when compared with baseline. Over the next 3 years the same rate of improvement was seen in patients who withdrew from gold treatment, but those withdrawing from methotrexate experienced a deterioration of their disease. Authors concluded that patients with early RA who stopped gold treatment because of side effects (and substituted methotrexate) showed almost the same sustained improvement as patients continuing gold or methotrexate, but patients withdrawn from methotrexate experienced a reactivation of their disease (Sander et al. 1999).

Sodium gold thiomalate was successful in controlling RA in 12 patients, age 18 to 65 years, as judged by clinical and laboratory parameters over a 3-year period (Biasi et al. 2000). During recent years, aggressive therapeutic approaches have been proposed in order to control the disease activity promptly and to avoid joint destruction. The 12 who completed the 3-year study were diagnosed with RA within 12 months of the onset of the study; at that time, all had six or more swollen joints, morning discomfort of >45 minutes' duration, and no previous treatment for RA. Sodium gold thiomalate was administered parenterally at 50 mg weekly for the first year; during years 2 and 3, the interval between injections was progressively increased to 4 weeks (Biasi et al. 2000).

Intramuscular injection of sodium gold thiomalate may suppress RA disease activity by diminishing monocyte and macrophage number, and consequently monokine production, in the synovial membrane (Yanni et al. 1994). This conclusion was reached based on 10 patients of average age 64.3 years and RA disease duration of 7.4 years, given a dose of 10 mg followed by 50 mg weekly until a total dose of 1.0 g was reached, then 50 mg monthly. After 12 weeks, there was significant improvement as evidenced by a decrease in pain and an increase in grip strength (Yanni et al. 1994).

9.3.3 Adverse Effects

Early detection of adverse reactions in RA patients who have recently started chrysotherapy is a high-priority research goal (Duro and Andreu 1995). A partial list of major side effects of chrysotherapy, arranged in alphabetical order, follows.

Accumulations

After injection of sodium gold thiomalate, gold was rapidly absorbed with maximal levels in plasma at 2 hours postinjection (Jones and Brooks 1996). In general, gold concentrations in blood and blood components were unsatisfactory in predicting impending toxicity or in evaluating appropriate regimens for continued courses of therapy (Merchant 1998). In chrysotherapy, about 70% of the administered gold is retained in the body and may be detectable in patients for up to 23 years posttreatment, suggesting a capture by proteins before storage in the intracellular compartments from which turnover may be slow (Jones and Brooks 1996; Nilsson 1997). Gold was usually excreted in the feces and urine (Zuazua et al. 1996).

Serum gold levels in RA patients treated with sodium gold thiomalate were in the range of 2 to 5 mg/L, equivalent to the gold in 4 to 10 mg of sodium gold thiomalate/L; synovial tissues of RA patients treated with injectable gold contained 21 to 25 mg Au/kg FW tissue, equivalent to 42 to 50 mg sodium gold thiomalate/L (Hashimoto et al. 1992). Gold sodium thioglucose injected over a 5-year period into an RA patient resulted in gold accumulations in Kupfer cells, renal proximal tubules, alveolar macrophages, and bone marrow cells; in all cases, gold accumulated in lysosomes (aurosomes) in the form of high-density microneedles (Manoubi et al. 1994). After intramuscular injections of sodium gold thiomalate (48% gold by weight), the gold is quickly absorbed and concentrated in the reticuloendothelial cells of the lymph nodes, adrenal gland, liver, kidneys, bone marrow, and spleen; during lengthy courses of treatment, gold deposits have also been described in the eyes, breasts, and parotid glands (Zuazua et al. 1996). Gold concentrations in hair, nails, and skin of normal untreated adults ranged between 0.1 and 1.1 mg/kg DW, with a mean of 0.35 mg/kg DW (Merchant 1998). After 12 months of chrysotherapy for RA, increases of two- to fivefold were recorded for these values in treated patients (Merchant 1998).

Aplastic Anemia

Aplastic anemia is a rare, but often fatal, complication of gold salt therapy (Doney et al. 1988). Neither the total dose of gold nor other known side effects of gold therapy have been predictive for the development of aplasia. Of 12 patients treated for gold-induced aplastic anemia with immunosuppressive therapy for 4 to 10 days, five recovered fully, one had a partial response, one spontaneously recovered a year later, and five died 6 to 67 days after beginning treatment. Survivors lived at least 4.3 years posttreatment (Doney et al. 1988).

Blood Chemistry

Patients with RA receiving intramuscular injections of gold drugs for 12 months had lipid profiles characterized by an overall increase in total cholesterol and tri-glycerides, and lower HDL levels (Munro et al. 1997). Authors advise that gold therapy may be contraindicated in RA patients who have significant cardiovascular risk factors.

Cancer

The cancer risk from chrysotherapy in 305 Swedish patients was negligible (Bendix et al. 1995). Patients receiving sodium gold thiomalate for a mean period of 19 months and observed over a 7.5-year posttreatment period had no increased risk of total malignancies when compared with the regional cancer register. There was, however, an increased risk of lymphoma and leukemia — not correlated to dosage or duration of therapy — and this is consistent with increased risk of hematopoietic malignancies in RA patients regardless of treatment (Bendix et al. 1995).

Chrysiasis

Chrysiasis and chrysoderma are terms used to describe the permanent slate-gray pigmentation of skin areas exposed to sunlight after parenteral administration of gold salts. The first observed case of chrysiasis was in 1928 in a patient treated with gold salts for pulmonary tuberculosis (Fleming et al. 1996; Miller et al. 1997; Smith and Cawley 1997). Chrysiasis is not serious pathologically, but it is underreported because it is uncommon and the symptoms largely unrecognized (Smith et al. 1995; Smith and Cawley 1997). In one study, chrysiasis was found in 31 of 40 Caucasian patients with RA who had received intramuscular injections of sodium gold thiomalate (Smith et al. 1995). The majority of affected patients were Caucasian females, with severity of pigmentation associated with dose-dependence over a cumulative threshold of 20 mg Au/kg body weight (BW), equivalent to 50 mg sodium gold thiomalate/kg BW (Smith et al. 1995; Smith and Cawley 1997). The most important factors in chrysiasis were total dose and exposure to light, that is, exposure to sunlight in RA patients receiving a total dose of more than 2.5 g of gold in parenteral therapy (Fleming et al. 1996). In one case, chrysiasis — as confirmed by light microscopy, transmission electron microscopy, and radiographic microanalysis — developed in an RA patient who received only 1.05 g of gold, and this may have been due to the patient's exposure to the intense UV light in Australia (Fleming et al. 1996).

With the advent of antibiotics, the use of gold salts and the reports of chrysiasis in the literature decreased. Chrysiasis is now considered a rare complication of chrysotherapy (Miller et al. 1997). A case was reported in 1997 in a 70-year-old

Caucasian female receiving 50 mg of sodium gold thiomalate every other week for the previous 7 years as treatment for RA, for a total cumulative dose of >9 g (>3.6 g of gold). Biopsied tissues showed a characteristic orange-red biorefringence of gold when viewed under polarized light. There is no known safe treatment to remove the pigment, and the patient was counseled to avoid sunlight and apply a covering makeup. The patient elected to discontinue gold therapy and try other medications to treat her RA (Miller et al. 1997).

Death

Fatal effects from all causes of chrysotherapy are estimated at 0.016 deaths per million prescriptions (Doney et al. 1988). Injected gold preparations are of different potency, and confusion in administration may have fatal consequences. These gold preparations may differ in their suspensory vehicles and bioavailability and should not be considered generic equivalents (Tilelli and Heinrichs 1997). In one case, a 71-year-old female undergoing chrysotherapy died following a hypotensive crisis and cerebral ischemia. The patient had been receiving sodium gold thioglucose, but the fatal incident occurred when the injected compound was sodium gold thiomalate. Authors recommend that elderly patients with demonstrated sensitivity to sodium gold thiomalate, as was the case here, would be better served by sodium gold thioglucose or by an oral gold preparation (Tilelli and Heinrichs 1997).

Dermatitis

Diverse forms of dermatitis are the most frequently recorded reactions to chrysotherapy, with up to 66% of patients displaying some form of dermatitis at the usual recommended dosages (Merchant 1998). The most common toxicities to Au^+ salts include hypersensitivity reactions of skin and mucous membranes, including pruritus, rash, eosinophilia, chronic papular eruptions, contact sensitivity, erythema nodosum, allergic contact purpura and pityriasis rosea, and lichenoid and exfoliative dermatitis. Adverse dermal reactions do not appear to be related to gold concentrations in the skin and rarely occur in patients who received less than 250 mg of Au^+ salts; moreover, many patients given more than 1000 mg of Au^+ salts did not experience adverse effects, and some received more than 5000 mg without such reactions (Merchant 1998).

Following chrysotherapy, skin toxicities are often the first adverse effects to appear (Merchant 1998). Lichen planus-like eruptions are the most common histologically classifiable form of dermatitis; however, they generally include atypical histological features, including parakeratosis and perivascular infiltrates with eosinophils and plasma cells in the deep and superficial corium. Cessation of chrysotherapy usually will result in recovery within 3 months, depending on their extent and severity. These dermal reactions are usually dose-related. This form of dermatitis is viewed as non-allergenic because after the original eruption has cleared, patients who experienced it can frequently be returned to chrysotherapy without the redevelopment of dermatitis (Merchant 1998).

Of 74 patients with RA of recent onset and treated with gold thioglucose, 39 (53%) experienced gold dermatitis; treatment can often be continued with dose reduction and local steroids (van Gestel et al. 1994). Gold concentrations in skin following chrysotherapy were frequently lower in rash-affected than in unaffected areas from the same individual (Merchant 1998). A roughly 90-fold range of values between rash-affected areas in different patients indicated that the absolute concentrations of gold were not directly related to rash initiation and that the mechanisms precipitating the development of skin rashes were probably not strictly dose-dependent (Merchant 1998). Based on studies with RA patents without side effects from gold sodium thiomalate, gold dermatosis patients, and healthy controls, authors concluded that gold dermatosis was mediated, in part, by allergic mechanisms; the lymphocyte proliferation test was recommended in the diagnosis of gold dermatosis (Rasanen et al. 1999).

Skin of the external ear canal and tympanic membrane may be involved in a generalized gold reaction (Raza and Phillips 1997). A female patient treated for RA with intramuscular gold developed a widespread exfoliative dermatitis shortly after her third weekly injection of sodium gold thiomalate. She also had tympanic membrane perforations in both ears with significant hearing loss; authors attribute these effects to gold toxicity (Raza and Phillips 1997).

Certain tumor necrosis factor microsatellite markers identified by markers TNFa5b5 and TNFa6b5 may be indicators of gold-induced cutaneous reactions, and are recommended for prediction of gold-related adverse effects (Evans et al. 1999).

The first report on nail shedding after gold dermatitis was documented in the Netherlands (ter Borg et al. 2000). A 73-year-old female RA patient developed severe generalized dermatitis after receiving weekly intramuscular injections of 50 mg of sodium gold thioglucose for 5 weeks, at which time gold therapy was discontinued. After 12 weeks, nail shedding of the fingers and toes was observed, and these were replaced by new, normal nails. Nail shedding is documented in many skin diseases after a severe systemic upset, after cyclical chemotherapy, and during treatment with various drugs (ter Borg et al. 2000).

Nickel is a significant contaminant of the gold preparations used in chrysotherapy, accounting for up to a total of 650 ng after 6 months of treatment, suggesting that a significant percentage of nonspecific dermatitis during chrysotherapy may be due to nickel contamination of the gold preparation (Choy et al. 1997).

Enterocolitis

Enterocolitis is a rare and potentially fatal complication of parenteral gold treatment for RA after patients received a total dose of 150 mg of gold (Stillman and Dubey 1988; Dorta et al. 1993; Duro and Andreu 1995). Histological examination of stomal cells in the colon of a patient with gold-induced enterocolitis revealed granular yellow-to-black inclusions, which were identified as metallic gold (Mohr and Gorz 2000). Enterocolitis-associated diarrhea may be due to increased fluid secretion caused by inhibition of the intestinal sodium pump. Afflicted patients were treated successfully with the antisecretory drug octreotide (Dorta et al. 1993).

Four Jewish female RA patients who developed gold-induced enterocolitis were examined for possible genetic predisposition (Evron et al. 1995). Results indicated that the DRB1*0404 allele — present in three of the four patients — may be associated with risk for development of gold-induced enterocolitis. Because the prevalence of this allele among the Ashkenazi Jewish populations of RA patients without colitis was 9.2%, authors concluded that HLA DNA typing should be considered in Jews who are contemplating chrysotherapy (Evron et al. 1995).

Immune Function

Immunoglobulin (Ig) deficiency is a widely described complication of gold treatment (Kiely et al. 2000). Deficiencies of IgG subclasses were found in 30% of gold-treated RA patients vs. 8.5% of RA controls (Kiely et al. 2000). Numerous effects of Au^+ drugs on the immune system are reported, most of which involve impaired function of macrophages, T-, and B-cells (Schumann et al. 1990). These immuno-suppressive effects contrast with the unusually high frequency of adverse immune reactions to Au^+ drugs, most of which involve a stimulation of the immune system. Adverse immune reactions required discontinuation of gold therapy in up to one third of the patients after several months of treatment (Schumann et al. 1990).

Organ Damage

Acute pancreatitis, severe hepatitis, and neuropathy developed in a 63-year-old male with RA receiving 150 mg of sodium gold thioglucose (Ben-Ami et al. 1999). The patient tested positive in a lymphocyte transformation gold test, suggesting a cell-mediated hypersensitivity to the drug. Recovery was complete after cessation of gold therapy. Comparatively rare complications induced by gold treatment included impairment of bile, prolonged cholesterosis, fatal hepatic necrosis, and neurological problems (Ben-Ami et al. 1999). Gold hepatotoxicity is uncommon (Koryem et al. 1998). Egyptian RA patients (n = 40) treated with sodium gold thiomalate for at least 40 weeks over a 4-year period showed no evidence of hepatotoxicity during the course of chrysotherapy. However, hepatitis was evident shortly after the last dose of gold, at a cumulative range of 35 to 2900 mg. Potential risk of hepatotoxicity is best evaluated through genetic typing (Koryem et al. 1998).

Parotid Glands

A patient with swollen parotid glands had been treated with parenteral sodium gold thiomalate (48% gold) for RA over a 10-year period, receiving a cumulative total of 7.52 g (Zuazua et al. 1996). The cause of inflammation was attributed to mechanical, non-tumoral blockage of the excretory ducts of the parotids by gold salt compound. The symptoms disappeared within a few days after gold treatment was stopped.

Proteinuria

RA patients who discontinued chrysotherapy (50 mg Au weekly) after 4 to 10 months of treatment because of protein in the urine could resume gold treatment at lower dosages of 25 mg Au every one or two weeks. Reinstitution of gold therapy at the lower dose was effective in most patients after proteinuria had resolved after 4 to 12 months (Klinkhoff and Teufel 1997). The peak incidence of proteinuria occurred between 4 to 6 months of chrysotherapy treatment but may develop anytime between 1 week and 39 months after the start of treatment (Ueda 1998).

There is an association between classic or definite RA and the HLA-DRw4 antigen (Wooley et al. 1980). Those who were positive for HLA-DRw3 or DRw2 had an increased risk of toxic manifestations during treatment with sodium gold thiomalate. Of 91 patients with RA given sodium gold thiomalate, 71 had adverse reactions. Of 15 patients who had protein in their urine, 14 tested positive for HLA-B8 and HLA-Drw3 antigens, suggesting that toxicity during sodium gold thiomalate treatment for RA may be under genetic control (Wooley et al. 1980).

Pulmonary Complications

Pulmonary complications caused by gold compounds were first reported in 1948 (Tomioka and King 1997). A total of 140 cases of gold-induced pulmonary disease were identified from 110 reports. In 81% of the patients, gold was being used to treat RA patients, 6% to treat bronchial asthma, 5% for pemphigus, and 9% for all other ailments. Side effects were common and included skin rash (38%), liver dysfunction (15%), and proteinuria (22%). Gold-induced pulmonary disease most often followed improvement in RA, presumably induced by gold therapy. Gold-induced lung disease was distinguishable from rheumatoid lung disease and usually improved with cessation of therapy or by treatment with corticosteroids (Tomioka and King 1997).

Psychiatric/Neurological

Reports of psychiatric and neurological complications of chrysotherapy appear in the German and French literature, and cases of neuropathy and encephalopathy have been reviewed by others (Dubowitz et al. 1991). In one case, a 56-year-old male treated with sodium gold thiomalate at 50 mg weekly became anorectic and lost 20 kg in weight after 2 months. He had esophagal and gastric ulceration that was successfully treated and stopped losing weight, but he remained anorectic. After 5 months of chrysotherapy (total dose of 1200 mg of gold), the patient developed muscular twitching and pain in his hands and feet. At this time, the patient appeared confused and disoriented, and exhibited bizarre and inappropriate behavior and significant intellectual impairment. Gold therapy was stopped. Gold-induced neuro-encephalopathy was diagnosed. Chelation therapy resulted in rapid improvement, and he regained his appetite and normal behavior patterns. After 10 days he was discharged alert and oriented. A 4-month followup indicated improved nerve conduction but still slightly abnormal EEG (Dubowitz et al. 1991). In another study, neurological side effects appeared in RA patients after treatment with sodium gold thiomalate, but effects disappeared when gold thioglucose was substituted (Hill et al. 1995).

Renal

Gold may damage tubular epithelial cells. Fatal acute tubular necrosis is a rare type of gold-induced kidney damage in patients allergic to gold (Ueda 1998). Nephrotic syndrome and glomerulonephritis were documented in some RA patients receiving oral administration of an organic gold compound (Hostynek 1997). Auranofin, a compound in which gold is complexed to trialkylphosphines, produced circulatory immune complexes. As an electrophile reacting with native protein, gold formed a complete antigen which induced specific antibodies associated with glomerular injury (Hostynek 1997). However, no adverse renal effects were observed in 32 RA patients receiving auranofin (Ueda 1998). Renal elimination of auranofin was <15% vs. >70% for parenterally administered sodium gold thiomalate (Ueda 1998). Chronic administration of monovalent gold compounds to patients with RA increased urinary enzyme secretion and renal tubular cell rates; however, there was no evidence of kidney disease from increased renal tubular cell turnover and a low level of renal tubular injury (Ganley et al. 1989).

9.4 SUMMARY

Numerous mechanisms of action have been proposed to account for the anti-inflammatory action of gold[+] thiol drugs in the treatment of rheumatoid arthritis. These mechanisms include the following four arbitrary, and sometimes overlapping, groupings: (1) action of Au^+ on inhibition of T-lymphocytes, macrophages, immunoglobulin levels, rheumatoid factor titers, tumor necrosis factor, monocyte capacity to produce superoxide anions, bone resorption, vascular cell adhesion molecules on endothelial cells, release of lysosomal enzymes, release of stress proteins from macrophages, surface adhesion of polymorphonuclear neutrophils, activity of thioredoxin reductase, and pro-inflammatory cytokine synthesis, especially interleukins; the chelating action of Au^+ is believed to interfere with correct antigen processing and peptide recognition of T-cell receptors; (2) conversion to aurocyanide $[Au(CN)_2]^-$, which suppresses oxidant production of inflammatory cells, including neutrophils, monocytes, and lymphocytes; (3) formation of auric chloride ($Au^{+3}Cl_3$), which is a better oxidant than Au^+ and dominates both the anti-inflammatory and toxic effects of gold salts; Au^{+3} rapidly oxidizes proteins and is then reduced to Au^+, forming an Au^+/Au^{+3} redox system; and (4) formation of colloidal gold (Au^0) during oxidation of Au^+ to Au^{+3}, which — like Au^+ — suppresses activity of inflammatory-associated cytokines, tumor necrosis factor, immune complexes, and rheumatoid factors.

Chrysotherapy dose and dosing interval are variable. One chrysotherapy treatment regimen using, for example, sodium gold thiomalate is 50 mg (25 mg Au) in 0.5 mL of water given weekly by intramuscular injection until a total of 2000 mg

(1000 mg Au) is reached, then 50 mg every other week. Treatment efficacy was evaluated using radiographs; measurements of erythrocyte sedimentation rate, c-reactive proteins, and serum gold levels; and patient assessment of pain and mobility, as well as the number of tender and swollen joints. Selected case histories have been presented. Adverse reactions to chrysotherapy occurred in up to 45% of patients, necessitating discontinuation of treatments. These reactions included altered blood chemistry and immune function, high accumulations of gold, aplastic anemia, chrysiasis, dermatitis, enterocolitis, psychiatric and neurological complications, and damage to the lungs, kidneys, liver, and parotid glands.

LITERATURE CITED

Abraham, G.E. and P.B. Himmel. 1997. Management of rheumatoid arthritis: rationale for the use of colloidal metallic gold, *Jour. Nutrit. Environ. Med.*, 7, 295–305.

Asperger, S. and B. Cetina-Cizmek. 1999. Metal complexes in tumour therapy, *Acta Pharmaceut.*, 49, 225–236.

Ben-Ami, H., S. Pollack, P. Nagachandran, I. Lashevsky, D. Yarnitsky, and Y. Edoute. 1999. Reversible pancreatitis, hepatitis, and peripheral polyneuropathy associated with parenteral gold therapy, *Jour. Rheum.*, 26, 2049–2050.

Bendix, G. and A. Bjelle. 1996. A 10-year follow up of parenteral gold therapy in patients with rheumatoid arthritis, *Ann. Rheum. Dis.*, 55, 169–176.

Bendix, G., A. Bjelle, and E. Holmberg. 1995. Cancer morbidity in rheumatoid arthritis patients treated with Proresid or parenteral gold, *Scandin. Jour. Rheumatol.*, 24, 79–84.

Biasi, D., P. Caramaschi, A. Carletto, M.L. Pacor, and L.M. Bambara. 2000. Combination therapy with hydroxychloroquine, gold sodium thiomalate and methotrexate in early rheumatoid arthritis. An open 3-year study, *Clin. Rheum.*, 19, 505–507.

Bratt, J., J. Belcher, G.M. Vercellotti, and J. Palmblad. 2000. Effects of anti-rheumatic gold salts on NF-kB mobilization and tumor necrosis factor-α(TNF-α)-induced neutrophil-dependent cytotoxicity for human endothelial cells, *Clin. Exper. Immunol.*, 120, 79–84.

Brown, D.H. and W.E. Smith. 1980. The chemistry of the gold drugs used in the treatment of rheumatoid arthritis, *Chem. Soc. Rev.*, 9, 217–240.

Canumalla, A.J., N. Al-Zamil, M. Phillips, A.A. Isab, and C.F. Shaw III. 2001. Redox and ligand exchange reactions of potential gold (I) and gold (III)-cyanide metabolites under biomimetic conditions, *Jour. Inorg. Biochem.*, 85, 67–76.

Choy, E.H.S., L. Gambling, S.L. Best, R.E. Jenkins, E. Kondeatis, R. Vaughan, M.M. Black, P.J. Sadler, and G.S. Panayi. 1997. Nickel contamination of gold salts: link with gold-induced skin rash, *Brit. Jour. Rheum.*, 36, 1054–1058.

Doney, K., R. Storb, C.D. Buckner, and E.D.Thomas. 1988. Treatment of gold-induced aplastic anaemia with immunosuppressive therapy, *Brit. Jour. Haematol.*, 68, 469–472.

Dorta, G., J.F. Schnegg, E. Saraga, and P.A. Schmeid. 1993. Treatment of gold-induced enteritis with octreotide, *Lancet*, 342, 179.

Dubowitz, M.N., P.J. Hughes, R.J.M. Lane, and J.P.H. Wade. 1991. Gold-induced neuro-encephalopathy responding to dimercaprol, *Lancet*, 337, 850–851.

Duro, J.C. and M. Andreu. 1995. Gold induced colitis, *Jour. Rheumatol.*, 22, 572–573.

Eisler, R. 2003. Chrysotherapy: a synoptic review, *Inflammation Res.*, 52, 487–501.

Evans, T.I., R.E. Small, T.W. Redford, J. Han, and G. Moxley. 1999. Tumor necrosis factor microsatellite markers TNFa5b5 and TNFa6b5 influence adverse reactions to parenteral gold in caucasians, *Jour. Rheum.*, 26, 2302–2309.

Evron, E., C. Brautbar, S. Becker, G. Fenakel, Y. Abend, Z. Sthoeger, P. Cohen, and D. Geltner. 1995. Correlation between gold-induced enterocolitis and the presence of the HLA-DRB1*0404 allele, *Arthritis Rheum.*, 38, 755–759.

Fleming, C.J., E.L.C. Salisbury, P. Kirwan, D.M. Painter, and F.S. Barnetson. 1996. Chrysiasis after low-dose gold and UV light exposure, *Jour. Amer. Acad. Dermatol.*, 34, 349–351.

Ganley, C.J., S.A. Paget, and M.M Reidenberg. 1989. Increased renal tubular cell excretion by patients receiving chronic therapy with gold and with nonsteroidal anti-inflammatory drugs, *Clin. Pharmacol. Therapeut.*, 46, 51–55.

Goebel, C., M. Kubicka-Muranyi, T. Tonn, J. Gonzalez, and E. Gleichmann. 1995. Phagocytes render chemicals immunogenic: oxidation of gold (I) to the T cell-sensitizing gold (III) metabolite generated by mononuclear phagocytes, *Arch. Toxicol.*, 69, 450–459.

Graham, G.G. and A.J. Kettle. 1998. The activation of gold complexes by cyanide produced by polymorphonuclear leukocytes. III. The formation of aurocyanide by myeloperoxidase, *Biochem. Pharmacol.*, 56, 307–312.

Graudal, I.K., N. Graudal, and A.G. Jurik. 1994. On the course of seropositive rheumatoid arthritis during and after long-term gold therapy, *Scandin. Jour. Rheumatol.*, 23, 223–230.

Griem, P. and E. Gleichmann. 1996. Das Antirheumatikum Gold: Erwunschte und unerwuschte Wirkungen von Au(I) und Au(II) auf das Immunsystem, *Z. Rheumatol.*, 55, 348–358.

Gromer, S., L.D. Arscott, C.H. Williams Jr., R.H. Schirmer, and K. Becker. 1998. Human placenta thioredoxin reductase, isolation of the selenoenzyme, steady state kinetics, and inhibition by therapeutic gold compounds, *Jour. Biol. Chem.*, 273, 20096–20101.

Hall, T.J., H. Jeker, H. Nyugen, and M. Schaeublin. 1996. Gold salts inhibit osteoclastic bone resorption *in vitro*, *Inflamm. Res.*, 45, 230–233.

Hamilton, J., I.B. McInnes, E.A. Thomason, D. Porter, J.A. Hunter, R. Madhok, and H.A. Capell. 2001. Comparative study of intramuscular gold and methotrexate in a rheumatoid arthritis population from a socially deprived area, *Ann. Rheum. Dis.*, 60, 566–572.

Handel, M.L. 1997. Transcription factors AP-1 and NF-B: where steroids meet the gold standard of anti-rheumatic drugs, *Inflamm. Res.*, 46, 282–286.

Hashimoto, K., C.E. Whitehurst, T. Matsubara, K. Hirohata, and P.E. Lipsky. 1992. Immunomodulatory effects of therapeutic gold compounds. Gold sodium thiomalate inhibits the activity of T cell protein kinase C, *Jour. Clin. Invest.*, 89, 1839–1848.

Heimburger, M., R. Lerner, and J. Palmblad. 1998. Effects of antirheumatic drugs on adhesiveness of endothelial cells and neutrophils, *Biochem. Pharmacol.*, 56, 1661–1669.

Hill, C., K. Pile, D. Henderson, and B. Kirkham. 1995. Neurological side effects in two patients receiving gold injections for rheumatoid arthritis, *Brit. Jour. Rheum.*, 34, 989–990.

Hirohata, S. 1996. Inhibition of human B cell activation by gold compounds, *Clin. Immunol. Immunopathol.*, 81, 175–181.

Hirohata, S., K. Nakanishi, T. Yanagida, M. Kawai, H. Kikuchi, and K. Isshi. 1999. Synergistic inhibition of human B cell activation by gold sodium thiomalate and auranofin, *Clin. Immunol.*, 91, 226–233.

Hirohata, S., T. Yanagida, H. Hashimoto, T. Tomita, T. Ochi, H. Nakamura, and S. Yoshino. 1997. Differential influences of gold sodium thiomalate and bucillamine on the generation of CD14+ monocyte-lineage cells from bone marrow of rheumatoid arthritis patients, *Clin. Immunol. Immunopathol.*, 84, 290–295.

Hostynek, J.J. 1997. Gold: an allergen of growing significance, *Food Chem. Toxicol.*, 35, 839–844.

Jain, R. and P.E. Lipsky. 1997. Treatment of rheumatoid arthritis, *Adv. Rheumatol.*, 81, 57–84.

Janssen, M., B.A.C. Dijkmans, J.P. Vandenbroucke, W. van Duijn, A.S. Pena, and C.B.H.W. Lamers. 1992. Decreased levels of antibodies against *Helicobacter pylori* in patients with rheumatoid arthritis receiving intramuscular gold, *Ann. Rheum. Dis.*, 51, 1036–1038.

Jones, G. and P.M. Brooks. 1996. Injectable gold compounds: an overview, *Brit. Jour. Rheum.*, 35, 1154–1158.

Kiely, P.D.W., M.R. Herlbert, J. Miles, and D.B.G. Oliveira. 2000. Immunosuppressant effect of gold on IgG subclasses and IgE; evidence for sparing of the Th2 responses, *Clin. Exper. Immunol.*, 120, 369–374.

Klinkhoff, A.V. and A. Teufel. 1997. Reinstitution of gold after gold induced proteinuria, *Jour. Rheumatol.*, 24, 1277–1279.

Koryem, H.K., K.M. Taha, I.K. Ibrahim, and L.K. Younes. 1998. Liver toxicity profile in gold-treated Egyptian rheumatoid arthritis patients, *Int. Jour. Clin. Pharm. Res.*, 18, 31–37.

Lockie, L.M. and D.M. Smith. 1985. Forty-seven years experience with gold therapy in 1,019 rheumatoid arthritis patients, *Semin. Arthritis Rheum.*, 13, 238–246.

Mangalam, A.K., A. Aggrawal, and S. Naik. 2001. Mechanism of action of disease modifying anti-rheumatic agent, gold sodium thiomalate (GTM), *Int. Immunopharmacol.*, 1, 1165–1172.

Manoubi, L., M.H. Jaafoura, A. Skhiri-Zhioura, A. El Hil, J.P. Berry, and P. Galle. 1994. Subcellular localization of gold in suprarenal testicle and thyroid glands after injection of allochrysine in rats, *Cell. Molec. Biol.*, 40, 483–487.

Menninger, H., G. Herborn, O. Sander, J. Blechschmidt, and R. Rau. 1998. A 36-month comparative trial of methotrexate and gold sodium thiomalate in the treatment of early active and erosive rheumatoid arthritis, *Brit. Jour. Rheum.*, 37, 1060–1068.

Merchant, B. 1998. Gold, the noble metal and the paradoxes of its toxicology, *Biologicals*, 26, 49–59.

Miller, M.L., R.R. Harford, J.K. Yeager, and F. Johnson. 1997. A case of chrysiasis, *Cutis*, 59, 256–258.

Mohr, W. and E. Gorz. 2000. Morphology of gold deposits in stromal cells of the colonic mucosa after chrysotherapy, *Akt. Rheumatol.*, 25, 103–105 (in German, English abstract).

Munro, R., R. Hampson, A. McEntegart, E.A. Thomson, R. Madhok, and H. Capell. 1998. Improved functional outcome in patients with early rheumatoid arthritis treated with intramuscular gold: results of a five year prospective study, *Ann. Rheum. Dis.*, 57, 88–93.

Munro, R., E. Morrison, A.G. McDonald, J.A. Hunter, R. Madhok, and H.A. Capell. 1997. Effect of disease modifying agents on the lipid profiles of patients with rheumatoid arthritis, *Ann. Rheum. Dis.*, 56, 374–377.

Newman, P.M., S.S.T. To, B.G. Robinson, V.J. Hyland, and L. Schreiber. 1994. Effect of gold sodium thiomalate and its thiomalate component on the *in vitro* expression of endothelial cell adhesion molecules, *Jour. Clin. Invest.*, 94, 1864–1871.

Nilsson, J.R. 1997. *Tetrahymena* recovering from a heavy accumulation of a gold salt, *Acta Protozool.*, 36, 111–119.

Paimela, L., M. Leirisalo-Repo, and T.U. Kosunen. 1995. Effect of long term intramuscular gold therapy on the seroprevalence of *Helicobacter pylori* in patients with early rheumatoid arthritis, *Ann. Rheum. Dis.*, 54, 437.

Pandya, A.G. and C. Dyke. 1998. Treatment of pemphigus with gold, *Arch. Dermatol.*, 134, 1104–1107.

Pickl, W.F., G.F. Fischer, I. Fae, G. Kolarz, and O. Scherak. 1993. HLA-DR1-positive patients suffering from rheumatoid arthritis are at high risk for developing mucocutaneous side effects upon gold therapy, *Human Immunol.*, 38, 127–131.

Rasanen, L., O. Kaipiainen-Seppanen, R. Myllykangas-Luosujarvi, T. Kasanen, P. Pollari, P. Saloranta, and M. Horsmanheimo. 1999. Hypersensitivity to gold in gold sodium thiomalate-induced dermatosis, *Brit. Jour. Dermatol.*, 141, 683–688.

Rau, R., G. Herborn, H. Menninger, and O. Sangha. 1998a. Progression in early erosive rheumatoid arthritis: 12 month results from a randomized controlled trial comparing methotrexate and gold sodium thiomalate, *Brit. Jour. Rheum.*, 37, 1220–1226.

Rau, R., B. Schleusser, G. Herborn, and T. Karger. 1998b. Longterm combination therapy of refractory and destructive rheumatoid arthritis with methotrexate (MTX) and intra-muscular gold or other disease modifying antirheumatic drugs compared to MTX monotherapy, *Jour. Rheum.*, 25, 1485–1492.

Raza, S.A. and J.J. Phillips. 1997. Bilateral tympanic membrane perforation — a result of gold toxicity? *Brit. Jour. Dermatol.*, 136, 479–480.

Roberts, J.R. and C.F. Shaw III. 1998. Inhibition of erythrocyte selenium-glutathione perox-idase by auranofin analogues and metabolites, *Biochem. Pharmacol.*, 55, 1291–1299.

Sadler, P.J. 1976. The biological chemistry of gold: a metallo-drug and heavy-atom label with variable valency, *Structure Bonding*, 29, 171–215.

Sander, O., G. Herborn, E. Bock, and R. Rau. 1999. Prospective six year follow up of patients withdrawn from a randomised study comparing parenteral gold salt and methotrexate, *Ann. Rheum. Dis.*, 58, 281–287.

Sanders, M. 2000. A review of controlled clinical trials examining the effects of antimalarial compounds and gold compounds on radiographic progression in rheumatoid arthritis, *Jour. Rheum.*, 27, 523–529.

Sato, H., M. Yamaguchi, T. Shibasaki, T. Ishii, and S. Bannai. 1995. Induction of stress proteins in mouse peritoneal macrophages by the anti-rheumatic agents gold sodium thiomalate and auranofin, *Biochem. Pharmacol.*, 49, 1453–1457.

Schumann, D., M. Kubicka-Muranyi, J. Mirtschewa, J. Gunther, P. Kind, and E. Gleichmann. 1990. Adverse immune reactions to gold. I. Chronic treatment with an Au (I) drug sensitizes mouse cells not to Au (I), but to Au (III) and induces autoantibody forma-tion, *Jour. Immunol.*, 145, 2132–2139.

Shaw, C.F., III. 1999a. Gold complexes with anti-arthritic, anti-tumour and anti-HIV activity, in *Uses of Inorganic Chemistry in Medicine*, N.C. Farrell, (Ed.), Royal Society of Chemistry, Cambridge, UK, 26–57.

Shaw, C.F., III. 1999b. The biochemistry of gold, in *Gold: Progress in Chemistry, Biochemistry and Technology*, H. Schmidbaur, (Ed.), John Wiley & Sons, New York, 260–308.

Smith, R.W., and M.I.D. Cawley. 1997. Chrysiasis, *Brit. Jour. Rheum.*, 36, 3–5.

Smith, R.W., B. Leppard, N.L. Barnett, G.H. Millard-Sadler, F. McCrae, and M.I.D. Cawley. 1995. Chrysiasis revisited: a clinical and pathological study, *Brit. Jour. Dermatol.*, 133, 671–678.

Snowden, N., D.M. Dietch, L.S. Teh, R.C. Hilton, and M.R. Haeney. 1996. Antibody defi-ciency associated with gold treatment: natural history and management in 22 patients, *Ann. Rheum. Dis.*, 55, 616–621.

Stillman, A.E. and D.P. Dubey. 1988. Immune functions and gold-induced enterocolitis, *Digest. Dis. Sci.* 33, 1046–1047.

Tange, R.A., A.J.G. de Bruijn, and W. Grolman. 1998. Experience with a new pure gold piston in stapedotomy for cases of otosclerosis, *Auris Nasus Larynx*, 25, 249–253.

ten Wolde, S., B.A.C. Dijkmans, J.J. van Rood, F.H.J. Claas, R.R.P. de Vries, J.M.W. Hazes, P.L.C.M. van Riel, A. van Gestel, and F.C. Breedveld. 1995. Human leucocyte antigen phenotypes and gold-induced remissions in patients with rheumatoid arthritis, *Brit. Jour. Rheum.*, 34, 343–346.

ter Borg, E.J., C.G. Ramselaar, and E.H.H. Wiltink. 2000. Nail shedding (Beau's lines) after severe gold dermatitis, *Arthritis Rheum.*, 43, 1420.

Tiekink, E.R.T. 2003. Phosphinegold(I) thiolates — pharmacological use and potential, *Bioinorg. Chem. Applica.*, 1, 53–67.

Tilelli, J.A. and M.M. Heinrichs. 1997. Adverse reactions to parenteral gold salts, *Lancet*, 349, 853.

Tomioka, H. and T.E. King, Jr. 1997. Gold-induced pulmonary disease: clinical features, outcome, and differentiation from rheumatoid arthritis, *Amer. Jour. Respir. Crit. Care Med.*, 155, 1011–1020.

Ueda, S. 1998. Nephrotoxicity of gold salts, D-penicillamine, and allopurinol, in *Clinical Nephrotoxins: Renal Injury from Drugs and Chemicals*, M.E. De Broe, G.A. Porter, W.M. Bennett, and G.A. Verpooten, (Eds.), Kluwer, Dordrecht, 223–238.

van Gestel, A., R. Koopman, M. Wijnands, L. van de Putte, and P. van Riel. 1994. Mucocutaneous reactions to gold: a prospective study of 74 patients with rheumatoid arthritis, *Jour. Rheumatol.* 21, 1814–1819.

Wang, Q., N. Janzen, C. Ramachandran, and F. Jirik. 1997. Mechanism of inhibition of protein-tyrosine phosphatases by disodium aurothiomalate, *Biochem. Pharmacol.*, 54, 703–711.

Whitehouse, M.W. and G.G. Graham. 1996. Is local biotransformation the key to understanding the pharmacological activity of salicylates and gold drugs? *Inflamm. Res.*, 45, 579–592.

Windholz, M. (Ed.). 1983. *The Merck Index*, 10th ed. Merck & Co., Rahway, NJ.

Wooley, P.H., J. Griffin, G.S. Panyi, J.R. Batchelor, K.I. Welsh, and T.J. Gibson. 1980. HLA-DR antigens and toxic reaction to sodium aurothiomalate and *D*-penicillamine in patients with rheumatoid arthritis, *New England Jour. Med.*, 303, 300–302.

Yadav, R., R. Misra, and S. Naik. 1997. *In vitro* effect of gold sodium thiomalate and methotrexate on tumor necrosis factor production in normal healthy individuals and patients with rheumatoid arthritis, *Inter. Jour. Immunopharmac.*, 19, 111–114.

Yanni, G., M. Nabil, M.R. Farahat, R.N. Poston, and G.S. Panayi. 1994. Intramuscular gold decreases cytokine expression and macrophage numbers in the rheumatoid synovial membrane, *Ann. Rheum. Dis.*, 53, 315–322.

Zeidler, H.K., T.K. Kvien, P. Hannonen, F.A. Wollheim, O. Forre, H. Geidel, I. Hafstrom, J.P. Kaltwasser, M. Leirisalo-Repo, B. Manger, L. Laasonen, E. R. Markert, H. Prestele, and P. Kurk. 1998. Progression of joint damage in early active severe rheumatoid arthritis during 18 months of treatment: comparison of low-dose cyclosporin and parenteral gold, *Brit. Jour. Rheum.*, 37, 874–882.

Zou, J., Z. Guo, J.A. Parkinson, Y.Chen, and P.J. Sadler. 1999. Gold (III)-induced oxidation of glycine: relevance to the toxic side-effects of gold drugs, *Jour. Inorg. Biochem.*, 74, 352.

Zuazua, J.S., A.M.G. Fuente, J.C.M. Rodriguez, G.B. Garcia, and A.P. Rodriguez. 1996. Obstructive sialadenitis caused by intraparotid deposits of gold salts, *Oral Surg. Oral Med. Oral Pathol. Oral Radiol. Endod.*, 81, 649–651.

PART 3

Effects of Gold
Extraction on Ecosystems

CHAPTER 10

Gold Mine Wastes: History, Acid Mine Drainage, and Tailings Disposal

Of the major metal mining industries, gold mining is the most waste intensive (Da Rosa and Lyon 1997). Refined gold consists of but 0.00015% of all raw materials used in the gold-mining process. It is estimated that it takes 2.8 tons of gold ore to produce the gold in a single wedding band, the rest being waste (Da Rosa and Lyon 1997). After waste rock is removed and the ore extracted, the ore is processed to separate the gold from the valueless portion of remaining rock which is known as tailings. Mine tailings and waste rock contain heavy metals and acid-forming minerals. Tailings can also contain chemicals used in ore processing. Amounts of toxicants in tailings — including arsenic, lead, cyanide, and sulfuric acid — are deleterious to fish and other wildlife. Tailings are usually stored in piles on land or in containment ponds, but sometimes are pumped back into the underground space from which the ore was mined. Dumping of mine tailings directly into rivers or other water bodies is no longer allowed in the United States, but occurs with some frequency elsewhere, especially in developing countries (Da Rosa and Lyon 1997).

This chapter presents an overview of gold mining and gold mining wastes, with emphasis on acid mine drainage effects and mitigation, and tailings disposal into various ecosystems. Later chapters deal with gold mining wastes of arsenic (Chapter 11), cyanide (Chapter 12), and mercury (Chapter 13).

10.1 OVERVIEW

The mining process consists of exploration, mine development, mining or extraction, mineral processing or beneficiation, and reclamation for closure (USNAS 1999). Modern exploration involves various types of sophisticated geochemical sampling, geophysical techniques, satellite remote sensing, and other methodologies for identifying deeply buried mineral deposits. After mining rights are acquired, exploration continues with testing, usually drilling, which disturbs surface and sub-surface environments, although effects are usually minor. The area required for a large mine and its facilities, including waste dumps and tailings ponds, sometimes exceeds 1000 ha, and in the United States often involves a combination of federal and private lands for a single mine. When an economic deposit has been identified from the exploration and the required permits are obtained, the deposit is prepared for extraction. This involves installation of power, roads, water, and physical support facilities including offices, fuel bays, and materials handling systems. Surface locations are marked and prepared for storage of overburden materials, tailings, and other wastes (USNAS 1999). In the United States, any citizen can locate and file a mining claim on public land — usually administered by the U.S. Bureau of Land Management — entitling the prospector to mineral rights of a certain tract, usually 20 acres (9.1 ha). One part of the claim stipulates that mining operations must not interfere with fish migration and spawning seasons (Petralia 1996). Proper design of a tailings disposal system is essential to the economic success of the operation as well as to the preservation of wilderness, hunting, fishing, trapping, and agriculture (Ripley et al. 1996).

Near-surface deposits in open-pit mines are prepared for production by removing the overlying waste material (USNAS 1999). Deeper deposits involve construction of shafts and tunnels. Mine development has the potential for significant environmental damage. Most mines use the same basic operations in extracting ores: drilling, blasting, loading, and hauling. After blasting, the fragmented rock is transported to a mineral processing facility. Continued mining activities result in growing waste dumps. Mineral processing or beneficiation usually involves crushing and grinding the ore, separating the valuable minerals by physical and chemical methods, and transporting the concentrate to a smelter or refinery. The waste or unwanted minerals (tailings) are stored in tailings ponds near the mine site. Tailings usually contain small amounts of gold not completely recovered during beneficiation, undesirable toxic minerals, waste rock minerals, and residual chemicals. Environmental damage may be substantial if stored wastes from tailings dams, ponds, leached rock, or leach solutions are discharged or otherwise released (Ripley et al. 1996; USNAS 1999; Fields 2001). Reclamation returns the mining and processing site to beneficial use after mining. In some cases, however, complete reclamation may not be possible and long-term monitoring will be necessary. Current reclamation practices include reducing slope angles on the edges of waste rock dumps and heaps to minimize erosion; capping these piles and tailings with soil; planting grasses or other vegetation that will benefit wildlife or grazing stock and help prevent erosion; directing water flows to minimize contact with potential acid-generating sulfides in the dumps, heaps, and piles; and removing buildings and roads (USNAS 1999).

Adverse effects of gold extraction include land disturbance, erosion, and the disruption of riverine ecosystems (Ripley et al. 1996). Discharges of water containing suspended solids and runoff from disturbed land affects local streams through increased turbidity and reduced light penetration, channel alteration, and altered stream flow rates and course. Heavily mined streams had a reduction in algal species diversity and avoidance by predatory fish. Sediment deposition adversely affected fish behavior, inhibited reproduction, and lowered dissolved oxygen levels. Physiological effects of suspended solids on Arctic grayling (*Thymallus arcticus*) are extensive and include abnormal gill development, reduced feeding activity, and altered pigmentation patterns. If left untended, sedimented streams in the Yukon area of Canada may take as long as 20 years for recovery of water quality and 30 to 70 years for habitat restoration (Ripley et al. 1996).

10.1.1 Lode Mining

Where the gold is still held in the host rock, it is known as lode gold and its extraction is called lode or hardrock mining. Commercial operations tunnel into the mountain or dig a tunnel or shaft to extract the ore, perhaps blasting out the surrounding material. The ore-bearing rock is then crushed to free the gold (Petralia 1996). The average tenor of gold ore is 0.2 to 0.3 troy ounces per metric ton (Stone 1975). Profits depend upon the amount of ore, current price, and the costs associated with mining, treating, transporting, and marketing. Access is probably the most important economic factor, and excessive costs of road building can make a fairly rich ore deposit uneconomical. Permissible lode mining claims — as filed with a county clerk — are usually limited to 1500 ft (457 meters) along the vein and not more than 300 ft (91 meters) on each side of the vein (Stone 1975).

Lode mining accounts for about 97% of the ore tonnage extracted by hardrock mining in the United States (Da Rosa and Lyon 1997). Lode mining may take the form of strip mining, open-pit mining, and underground mining. Strip mining is the stripping away of layers of soil and waste rock over a mineral deposit. Open-pit mining involves excavating the surface in a concentrated location to access the underlying mineral ore body, including gold. To reach these deposits, the pit is dug in a progressive series of stages. The walls are usually terraced, 13 to 20 meters high, and the steps are 5 meters wide. Open pit mines can exceed 1.6 km across and 1000 meters in depth. Open-pit mines create large quantities of waste rock, usually stored on the surface in piles exceeding 100 meters in height. These wastes are usually not returned to the pit when the mine closes. Underground mine operators dig shafts for access and ventilation and horizontal tunnels (adits) for access and drainage to reach the ore. The extracted ore is carried to the surface through the shafts and adits by truck, rail car, and other conveyances. The development of new technologies for moving vast amounts of earth and for extracting gold from low-grade ores has created large quantities of new and potentially toxic mining wastes (Da Rosa and Lyon 1997).

The main environmental effects of lode gold mining are related to the discharge of liquid effluents that adversely impact aquatic life (Ripley et al. 1996). In Canada, in 1986, for example, 35 million m^3 of water used in auriferous-quartz mining were ultimately discharged to water courses together with about 16 million m^3 of mine water. Discharges were generally alkaline with pH 7.5 to 8.0, but sometimes they were acidic with pH range 1.7 to 4.9 (Ripley et al. 1996). Gold mine tailings frequently exceeded maximum allowable concentrations set by various regulatory agencies for cyanide (Eisler 1991) and metals (Eisler 2000). At Yellowknife, Canada, gold mine tailings effluents contained, in mg/L, 84.0 for total cyanide (vs. 2.0 for maximum allowable concentration); for other components in the waste stream these values were 4.7 for arsenic (1.0), 5.0 for copper (0.6), 0.4 for nickel (0.2), and 20.0 for zinc (1.0; Ripley et al. 1996).

In 1992, about 75% of the lode gold mines in Canada operated underground (Ripley et al. 1996). The gold in these auriferous-quartz deposits is usually recovered using crush and grind, cyanide leach, zinc precipitation, or carbon in pulp extraction processes followed by refining. Some operations roast the ore prior to cyanidation in order to free gold particles enclosed in arsenopyrite for leaching, with subsequent release of arsenic (Ripley et al. 1996). Arsenic wastes and wastes from the cyanidation process are discussed in more detail in Chapters 11 and 12, respectively.

10.1.2 Placer Mining

A placer deposit is the formation caused by the natural erosion of lode ore from its original location, with transport most likely by water or glacier. The word *placer* is thought to be derived from the American Spanish *placer* (sandbank), the Catalonian *plassa*, or the Latin *platea* (a place; Krause 1996). Placer gold, with purity of 70 to 90%, ranges in size from flour grains to nuggets and is usually alloyed with other metals (Petralia 1996). Two types of placer mining are common on federal lands (USNAS 1999). The first involves use of mechanized earth-moving equipment, typically involving removal of a 650-meter stretch of stream, removal of the vegetative mat or soil, gold removal from gravels with sluices that separate dense from light minerals, and reclamation by replacement of gravel and the vegetative mat or soil (USNAS 1999). The second uses suction dredging in streams whereby stream materials are removed, passed over a sluice box to sort out the gold, and discarded as tailings over another area of bed (Harvey and Lisle 1998). Placer mining in active streams may adversely affect habitat for benthic macrobiota and spawning habitat of aquatic animals (USNAS 1999).

Placer gold mining in the United States began in the eastern states during the late 1700s and in the southern Appalachian region in the early 1800s (West 1971). After the richer deposits were exhausted, interest turned to New Mexico where gold placer mining was documented in 1828. In early 1848, a major strike was made on the American River, California, and triggered the first of the great domestic gold rushes. In Alaska, gold mining was reported as early as 1848. In Canada, gold was found in the Yukon Region in 1878. Rich finds were reported in the Canadian Klondike region of the Yukon in 1897 to 1898. Gold was mined in Nome, Alaska, in 1898, and in Fairbanks in 1962 (West 1971).

The occurrence of valuable substances (including tungsten, rare earths, garnets, precious stones, gemstones) in gold placers is well known (Buryak 1993). Although economically feasible to extract these materials together with gold, with an overall reduction in mining costs, the practice is not common.

Panning and Sluicing

Panning and sluicing are simple forms of placer mining that depend on low-cost labor (Krause 1996; Da Rosa and Lyon 1997). Many of the early gold prospectors mined by panning, which involves swirling streambed gravels and sands in a shallow metal pan to trap the denser gold particles. Another placer mining technique is to pour the stream gravel into a long trough or sluice that contains a series of riffles along the bottom. The denser gold particles are trapped in the riffles while the less dense sediments are washed away.

Hydraulicking

As the amount of gold which could be recovered easily by stream panning dwindled, a new form of capital-intensive placer mining was practiced. Commonly called hydraulicking and first used in California in 1853, this technique involved spraying gravel banks of rivers with pressurized water and capturing the runoff in long sluices to recover the gold particles (Nriagu and Wong 1997; Da Rosa and Lyon 1997; USNAS 1999). The high-pressure nozzles used in hydraulic operations consumed water at the rate of about 20,000 m^3 per hour, washing out large portions of the river banks (Da Rosa and Lyon 1997). To obtain the large quantities of water needed, mining companies constructed dams and more than 8200 km of water delivery systems to transport the water from reservoirs to mining sites. The large amounts of sediment mobilized by hydraulicking choked natural streambeds with mud and sand, causing flooding that impacted agricultural crops, fisheries, and drinking water for livestock and humans. In 1882, agricultural interests in Marysville, California — after a series of hydraulicking-induced floods — initiated legal action against mining companies. In 1884, Judge Alonzo Sawyer ruled in favor of farming interests by enjoining the mining companies from discharging debris into the flooded waterways and its tributaries. The Sawyer decision started the decline of the hydraulic mining period in California. In 1893, the U.S. Congress passed the Caminetti Act which provided for restricted hydraulic mining in California under the control of the California Debris Commission and required hydraulicking operations to impound all debris (Da Rosa and Lyon 1997). Today hydraulicking is practiced in only a few places in the United States, and these operations need to comply with state and federal water quality discharge requirements (USNAS 1999).

Dredging

Placer dredging consists of digging underwater deposits by a rotating cutterhead and suction line or by rotating a cutting bucket line (Nriagu and Wong 1997). The dredged material is delivered onto a floating platform into a revolving screen or shaking table, and disaggregated using a jet of water. The fluid mixture falls through perforations in the screen or table onto a series of sluices equipped with gold-saving riffles, mats, and mercury. Primitive forms of dredging were used in West Africa in the 1700s and the first steam engine for dredge service was constructed in England in 1795 (Nriagu and Wong 1997). The first successful bucket line dredge in the United States was operated in 1896 in southwestern Montana (West 1971). Placer mining in most areas of western North America benefitted from the introduction of the dredge in 1898, making possible consolidation of many small claims into large leases (Nriagu and Wong 1997). In Alaska, gold dredging began in 1903; by 1914, 42 dredges were in operation, with a peak of 49 reached in 1910 (West 1971). California had 63 operating dredges in 1910. Dredging was interrupted by World War II in 1941; after the war, in 1945, dredging costs were prohibitively high and only a few of the deactivated dredges were returned to service (West 1971).

In 1986, however, the world's largest bucket line offshore dredge began operations on 85 km² (21,000 acres) of the State of Alaska through leases brokered by the U.S. Minerals Management Service (Barker et al. 1990). The operation produced about 1.1 tons (36,000 ounces) of gold in 1987 worth US $34.5 million; a similar result occurred in 1988. Fine gold is also mined along the coast and sea floor off Nome, Port Clarence, Tuksuk Channel, Cook Inlet, Yakutat, and other locations (Barker et al. 1990).

Suction dredging and associated activities have various effects on stream ecosystems, and most are not well understood (Harvey and Lisle 1998). Suction dredging is common during the summer in many river systems in western North America and reportedly adversely affects aquatic and riparian organisms, channel stability, and use of river ecosystems for other human activities. Suction dredging is subject to federal and state regulations, but additional regulations seem needed to protect threatened or endangered aquatic species in dredged areas, incubation of embryos in gravel substrates, or spawning runs followed by high flows (Harvey and Lisle 1998). Suction dredge gold mining in a northern California stream in 1983 did not significantly affect mean numbers of benthic invertebrates or diversity indices; however, some taxa were adversely affected at selected sites (Somer and Hassler 1992). Dredging dislodged aquatic insects that were eaten by young coho salmon (*Oncorhynchus kisutch*) and steelhead trout (*Oncorhynchus mykiss*). Sedimentation rates and organic fractions were elevated downstream from the dredging. In 1984, coho salmon and steelheads were observed spawning in areas that had been dredged in 1983 (Somer and Hassler 1992).

10.2   ACID MINE DRAINAGE

Gold mines in the United States and Canada — some more than 100 years old, some recently closed, and some still active — are leaking metal-rich acidic water into the environment, resulting in hundreds of millions of dollars in remediation costs annually (Da Rosa and Lyon 1997; USNAS 1999; Fields 2001). This acidic drainage, often referred to as acid mine drainage or AMD, is derived from sulfide-containing rock excavated from an underground mine or open pit. The sulfur reacts with water and oxygen to form sulfuric acid (H_2SO_4). Iron pyrite (FeS_2) is the most common rock type that reacts to form AMD, but marcasites and pyrrhotites also contribute significantly. On exposure to air and water, the acid will continue to leach from the source rock until the sulfides are leached out — a process that can last for centuries. The sulfur is released by weathering, oxidation, and erosion, with concurrent production of sulfuric acid. The rate of acid production from inorganic oxidation of iron sulfides is enhanced by various species of acidophilic bacteria, especially *Thiobacillus ferrooxidans*. The acidity of the water and its proximity to metal in the ore may generate waters of low pH that are high in copper, cadmium, iron, zinc, aluminum, arsenic, selenium, manganese, chromium, mercury, lead, and other elements released from the ores with increasing acidity. The resulting solution is sufficiently acidic to dissolve iron tools in underground mines and kill migratory waterfowl that shelter overnight in pit lakes. AMD seeps out of tailings, overburden, and rock piles being processed for gold removal. If left unchecked, it can contaminate groundwater. AMD is often transported from the mining site by rainwater or surface drainage into nearby watercourses where it severely degrades water quality, killing aquatic life and making water virtually unusable (Da Rosa and Lyon 1997; USNAS 1998; Fields 2001).

Anthropogenic AMD dates back to at least the Middle Ages, but new techniques in gold mining have produced a virtual flood of acid water throughout the American West, Canada (Fields 2001), and elsewhere (Cidu et al. 1997; Ogola et al. 2002). Naturally occurring acid rock drainage can produce a trickle of acidic waste that stains rock faces red from iron. Mining, however, accelerates the process by exposing very reactive components — potentially unstable thermodynamically with respect to oxygen — to surface atmospheres (Fields 2001). Underground gold mines puncture ore bodies with adits, mine tunnels, and shafts that allow air and water to enter and react with sulfide materials that are exposed inside the mine (Da Rosa and Lyon 1997). AMD can leach from underground mine openings into streams and aquifers. In open-pit mines, sulfide minerals on the exposed sides of the pit excavation are moistened by precipitation or by groundwater seeps, generating intense AMD flows (Da Rosa and Lyon 1997).

10.2.1 Effects

Aquatic ecosystems are considered the most sensitive to the effects of AMD waters, toxic heavy metals, and sediments from mining. Collectively, these contaminants cause disrupted reproduction, altered feeding, inhibited growth, habitat loss, decreased respiration, death, and chronic degradation of the aquatic environment (Da Rosa and Lyon 1997). Massive fish kills are reported after a major spill or sudden storm which adds additional pollutants to streams. In many AMD-impacted streams, there is no life for several kilometers downstream of a mine except for the most acid-resistant species. Land animals, such as mink (*Mustela vison*) and otters (*Lutra* spp.), dependent on aquatic systems for food and habitat are also affected by AMD, with population declines reported near affected streams (Da Rosa and Lyon 1997).

AMD is usually first recognized when streams or pools appear orange (Da Rosa and Lyon 1997). Acid waters dissolve and mobilize many metals, including iron, copper, aluminum, cadmium, and lead. These, especially the iron, precipitate with decreasing acidity and coat stream bottoms with an orange-, red-, or brown-colored slime or cement (Da Rosa and Lyon 1997). The cement physically embeds gravels, impairing streambed habitat for fishes and macroinvertebrates (USNAS 1999). When the spaces between gravels are embedded with fine-grained sediments or floc, egg survival of trout, salmon, and other benthic spawners is threatened by lack of oxygen (USNAS 1999).

Below a pH level of 4.0, most aquatic organisms die (Da Rosa and Lyon 1997). Many streams receiving AMD are 10 to 100 times more acidic (pH 2 to 3) than the concentration lethal to most species of aquatic plants and animals, except select species of acid-tolerant bacteria. Heavy rains can flush large amounts of acidified mine wastes into streams, causing massive fish kills. The main physiological mechanisms for fish death in acid water are osmoregulatory failure and impaired oxygen uptake. At pH 3.5 to 4.0, only about half the frog and salamander embryos tested had survived. Most freshwater fish species were unable to survive when water pH was less than 4.2. At pH levels less than 4.5, most benthic species of animals died. At sublethal pH levels less than 5.0, most aquatic plants were impaired and acid-tolerant plants tended to dominate. Heavy metals and sediments associated with AMD exacerbated the toxic effects of low acidity (Da Rosa and Lyon 1997).

One gold mine in California discharged AMD into the Sacramento River for about 100 years until mining was halted in 1963. Fish kills of hundreds of thousands of salmon and trout have been documented at this site since the 1920s. Unless remediation is implemented, low AMD pollution may persist for hundreds of years (Da Rosa and Lyon 1997). A gold mine that opened in 1988 in the Black Hills of South Dakota began generating AMD in 1992. In 1994, and again in 1995, AMD flooded offsite into a nearby creek, creating a low pH (2.1) environment lethal to fish and invertebrates (Da Rosa and Lyon 1997). At Spirit Mountain, Montana, AMD contaminated the drinking water supply of about 1000 nearby residents with lead, arsenic, and cadmium (Fields 2001). When consumed in high doses, sulfates mobilized during AMD can cause diarrhea and other gastrointestinal problems, especially in children (Da Rosa and Lyon 1997).

10.2.2 Mitigation

Acid will continue to be generated until the iron sulfides are leached from the mine waste material, or until steps are taken to completely seal off the sulfide rock source from oxygen and water (Da Rosa and Lyon 1997).

Methods for prevention of acid drainage include those that prevent acid generation from starting and those that treat the acid generation at the source so that no drainage occurs (Da Rosa and Lyon 1997; USNAS 1999; Fields 2001). Prevention of acid generation usually includes capping and sealing acid-generating rock to prevent air and water from reaching the rock and initiating the generation of acid. In dry climates, a less effective seal and a good vegetative cover may allow evapotranspiration of most of the water infiltrating into the pile. Another variation on capping — and one practiced widely outside the United States — is to bury acid-generating materials in water to prevent contact with air. This is accomplished by placing the waste in a closed body of water or by covering the top of a tailings pond with water once tailings deposition is completed. Subaqueous tailings disposal of acidic mine wastes is used at several Canadian mines wherein wastes are discharged under water into a prepared impoundment or a natural body of water, such as a lake or the ocean floor — although discharge into natural waters is prohibited in the United States. Some mine operators, both domestic and foreign, place potential acid-generating materials into pits that are expected to fill with water. Once the pit lake is formed, the material is no longer exposed to air. However, covering the rocks and tailings may not prevent oxygen from reacting with sulfides in the rocks because substantial amounts of oxygen can be trapped in waste rocks and tailings and oxygenated water can infiltrate the area from other sources. To reduce the availability of sulfides to both water and air, new techniques are under investigation, including autoclaving and encapsulating the rock in materials such as silica.

The use of chemical additives to prevent acid generation when applied to waste rock or spent ore piles is economically feasible (USNAS 1999; Fields 2001). The most common method for treating in place to prevent acid drainage is to add lime or other neutralizing materials to acid-generators. The neutralizing materials need to be in sufficient concentration to counteract all the acid-generating potential. The long-term effectiveness of this type of mixing is unknown, and the relative rates of acid generation and neutralization are not well documented. Other cost-effective processes to prevent acid drainage include separation of acid-generating portions of the ore from other components, and these portions can be treated more efficiently than the larger volume of spent ore material (USNAS 1999; Fields 2001).

Acid drainage that contains metals is a potential long-term water quality issue at some mine sites. The factors that create acid drainage and that minimize its impacts are well understood; however, few long-term monitoring data are available to predict the extent of damage at a specific mine site. Further, it is difficult to predict when acid drainage will start, the degree of acidity, and the total amounts of metals involved (USNAS 1999). One procedure used to predict AMD is acid–base accounting, which is based on estimations of acid-generating and acid-neutralizing materials in the waste rocks (Da Rosa and Lyon 1997). Minerals containing sulfur, especially pyrites, have the potential to generate acidity when exposed to water and oxygen. Buffering or neutralizing-acid minerals include carbonates, especially $CaCO_3$. The acid-generating and acid-neutralizing potentials are expressed as numerical values and are compared to predict the potential for generation of AMD. However, acid–base accounting does not include the potential role of bacteria and other variables in producing AMD. In one case, a gold mine near Elko, Nevada, has been combatting a serious AMD problem since 1990, when surface water drainage from the mine's waste rock piles began generating acid (plus mercury and arsenic), contaminating 3.2 km of a nearby stream. Acid–base accounting tests conducted by the mine owners on rock samples indicated that no potential acid problems were expected (Da Rosa and Lyon 1997). Accordingly, kinetic testing is often used to supplement acid–base accounting and is based on acid generation from materials in a controlled chamber environment of air, water, and bacteria. In contrast to acid–base accounting, kinetic tests on mine wastes use a larger sample volume, and tests are run for extended periods of time, often months (Da Rosa and Lyon 1997).

In some mines, remediation efforts can be concentrated on specific areas within the mine. Using these techniques, problem areas can be identified and contaminated flows isolated or diverted (Hazen et al. 2002). For example, in one multiple-level underground mine in Colorado that was in gold production between 1870 and 1951, hydrometer measurements using water isotopes of hydrogen and oxygen were used to identify problem areas. Measurements showed that discharges from a central level portal increased by a factor of 10 during snowmelt runoff, but zinc concentration increased by a factor of 9.0. Less than 7% of the peak discharge of zinc was from snowmelt; the majority was from a single internal stream with high zinc (270 mg Zn/L) and low pH (3.4). New water contributed up to 79% of the flow in this high zinc source during the melt season. Diversion of this high zinc source within the mine decreased zinc flow by 91% to 2.5 mg/L (Hazen et al. 2002).

An alternative to chemical treatment of AMD is bioremediation, a set of passive treatment techniques which use bacteria or other organic agents (Da Rosa and Lyon 1997). For example, bactericides to inhibit iron-oxidizing bacteria, such as *Thiobacillus ferrooxidans*, have been used successfully to reduce the costs of treating acidic runoff from reactive waste rock piles. Another successful technique is the construction of wetlands to route mine effluents through areas stocked with metal-absorbing aquatic plants, such as cattails (*Typha latifolia*). These plants can also serve as a growing base for bacteria that function as metal collectors (Da Rosa and Lyon 1997).

10.3 TAILINGS

Most of the early metal mines in the United States dumped tailings directly into streams (Da Rosa and Lyon 1997). Damage from this practice was extensive and persisted unchecked until the 1930s. At that time, tailings were impounded, although some mines continued to flush tailings into public waterways until enforcement of the Clean Water Act in the 1970s forced them to stop. Tailings are now stored in a pond or impoundment behind an earth-fill dam. However, impoundment embankments can fail if improperly designed or located on an unstable foundation (Da Rosa and Lyon 1997). In the Philippines, for example, a tailings impoundment failed in March 1996 discharging 4 million tons of tailings — containing copper, lead, mercury, cadmium, and other contaminants — into a nearby river, blocking 26 km (Fields 2001). Mines in some countries continue to discharge tailings into nearby waterways. In Indonesia, a single gold mine has discharged 120,000 tons of tailings daily since 1972 (43.8 million tons annually) into a nearby river system, flooding more than 30 km^2 of rainforest and agricultural lands (Da Rosa and Lyon 1997). In general, metal wastes from gold mining — specifically, cadmium, zinc, and copper — exceed all current guidelines promulgated by regulatory agencies for freshwater and marine life protection via the medium, and health of humans, other mammals, and birds from ingestion of contaminated diets (Eisler 2000).

Selected examples follow on results of field studies and laboratory investigations of tailings waste disposal into freshwater, marine, and terrestrial ecosystems.

10.3.1 Freshwater Disposal

Adverse effects of gold mine tailings accidentally or deliberately introduced into freshwater environments include elevated sediment and stream water concentrations of cadmium, copper, zinc, lead, arsenic, and other elements; photosynthesis-inhibiting turbidity loadings; accumulations of lead and copper in sediments that were toxic to incubating fish eggs and benthic biota; reduced growth and population abundance of fishes; lead accumulations in bodies of fish and their diets; and elevated accumulations of arsenic, copper, and zinc in soft parts of bivalve molluscs.

Field Investigations

Gold mining in the Black Hills of South Dakota has remained an active industry since gold was first discovered there in 1874, with most of the mining associated with gold veins and placers in the northern Black Hills (May et al. 2001). At least five large gold mines involving 800 ha are still operating. Gold recovery from lode mines was originally accomplished through mercury amalgamation, with an estimated 15 kg of mercury lost daily to Whitewood Creek in the northern Black Hills. The use of mercury was discontinued in 1971 and replaced with processes relying on cyanidation. In addition to mercury, daily averages of 140 kg of cyanide, 100 kg of zinc, and 10,000 kg of arsenopyrites were also released into Whitewood Creek every day. In the 103 years between 1876 and 1978, a total of about 100 million tons of finely ground gold mill tailings were discharged into Whitewood Creek. Recent analysis of water samples from the impacted areas indicates levels of concern in water for arsenic (>50 µg/L) and selenium (>5 µg/L). Sediment concentrations, in mg/kg DW, for arsenic (1951), cadmium (3), copper (159), mercury (0.6), nickel (64), lead (176), and zinc (250), were considered sufficiently high for potential adverse ecological effects, including metals accumulation into the benthic food chain from sediment-released metals. May et al. (2001) recommend more research on the dynamics of metals transport from sediments and accumulation in food webs, and experimental studies of effects of metals-contaminated invertebrate diets on salmonids and other fishes typical of the study area.

In Montana, restoration of Whites Creek — home of the West Slope cutthroat trout (*Oncorhynchus clarki*) in the Big Belt Mountains — began in 1995 (Skidmore 1995). This trout stream had been devastated by historic gold mining activities over a 60-year period that ended in the 1940s. The surrounding area had been stripped of its gold by dredges and ground sluicing. A combination of tailings, piled high within the narrow valley, and hydraulic mining produced an eroding and unstable stream that threatened the survival of the last remaining native population of West Slope cutthroat trout.

In Alaska, most of the gold is recovered from placer deposits, and tailings are associated with turbidity and toxic metal problems (Pain 1987; Yeend 1991; LaPerriere and Reynolds 1997). The gold frequently lies in gravel over the stream bedrock. To reach the gold, the vegetation, soil, and gravel over the deposit are removed and the gold separated from the gravel, usually by washing the deposit through a sluice from a nearby stream. The most obvious damage around a placer mine is the physical destruction to the vegetation and stream banks. In one creek, where mining had ceased 60 years earlier, only about 25% of the bank supported plants. The lack of ground cover makes the banks unstable and liable to erode into the stream during storms. The water immediately below the mine contains a high proportion of fine clay particles and sand. Some of these particles are trapped in holding pools, as required by permit. But the smaller particles frequently remain in suspension and escape into the stream. Sediments in water can be divided into components, including settleable or nonsettleable solids, total solids, total suspended solids, total dissolved solids, and fixed and volatile components. Excessive sediments in water may alter the physical and chemical properties of the receiving water body, with adverse effects on the native plants and animals. Turbidity is an approximation of the amount of suspended solids in water. Increasing turbidity, for example, restricts photosynthesis, thereby limiting the base of the food chain. Increasing sedimentation may decrease algal productivity through smothering and scouring. Current regulations mandate that discharged water must be treated to reduce settleable solids to <0.2 mL/L. Alaska has set limits for turbidity of stream water and the permissible amount of suspended sediments. Turbidity is measured in nephelometric units (NTUs). NTUs are based on the amount of light scattered by a water sample and calibrated against a standard. To protect stream life, turbidity should legally not exceed 25 NTUs above background, but these values were often 100 to 1000 times higher downstream of many active mines. The increased sediment load reduces the amount of oxygen in the water as microorganisms break down the organic material from the soil. Increased sediment loadings also increase alkalinity, sequester nutrients by binding them into chemical complexes, blanket the stream bed effectively destroying the benthos, and produce or prevent plant photosynthesis (Pain 1987; Yeend 1991; LaPerriere and Reynolds 1997). At 25 to 50 NTUs, the light reaching a depth of 10 cm is about 60% that at the surface. Between 500 and 1000 NTUs — common levels in heavily mined streams — only 0.3 to 5% of the incident radiation penetrates to 10 cm. Theoretically, an increase of 5 NTUs can reduce photosynthesis in shallow streams by as much as 13%. An increase of 25 NTUs, the accepted standard in Alaska, could reduce production by 50%. Heavy mining increases turbidity to an average of 1700 NTUs and completely inhibits primary productivity (Pain 1987; LaPerrriere and Reynolds 1997).

High concentrations of trace metals (Cd, Pb, Zn, Cu) and arsenic associated with gold also enter the waste stream; all of these are known to produce toxic effects in salmon and trout at concentrations near background levels (Pain 1987; LaPerriere and Reynolds 1997). In Fairbanks, for example, some groundwaters are so contaminated with arsenic from gold mining activities 30 years earlier that they are considered unsafe for drinking. Bacteria associated with arsenic in the water draining from lode and placer gold mines oxidize iron and sulfur and probably accelerate the rate at which trace metals leach from the sediment (Pain 1987).

Recommendations for habitat restoration in Alaska caused by placer mine activities (Pain 1987) include:

1. Mechanical replacement of gravels and soils and restoration of the normal stream channel.
2. Reduction in the amounts of sediments released.
3. Encouraging the growth of vegetation along affected banks.
4. Maintaining structural integrity of settling ponds to prevent spills.
5. Recycling the wastewater through sluices that contained the gold-bearing gravels.
6. Adding chemicals that cause the fine particles to aggregate and sink (a technique at least 4000 years old). Ancient texts from India about 2400 BCE suggest adding vegetable substances to water for clarification purposes. Since 1889, U.S. water companies have treated drinking waters with chemical clarifiers; however, these may be too expensive, too inefficient, and may not work at low temperatures typical of northern Alaska. Pain (1987) suggests that polyethylene oxide, a water soluble resin, seems promising for aggregation.

In Canada, there are an estimated 6000 abandoned mine sites that pose potential tailings hazards to aquatic ecosystems. In October 1990, 300,000 metric tons of water-saturated gold mine tailings spilled into the Montreal River in northern Ontario, Canada, via a small creek, as a result of the collapse of a tailings dike

(Draves and Fox 1998). About 50 km of the river were contaminated and 5 km heavily contaminated. Bulk concentrations of copper in sediments (120 mg/kg DW) exceeded the severe effect level (110 mg/kg) listed in the Ontario Provincial Quality Guidelines; similar cases were made for cadmium (>0.6 mg Cd/kg DW sediment), chromium (>26 mg Cr/kg), manganese (>460 mg Mn/kg), nickel (>16 mg Ni/kg), lead (>31 mg Pb/kg), and zinc (>120 mg Zn/kg) (Draves and Fox 1998). In 1992 and 1993, eggs were collected from resident walleye (*Stizostedion vitreum*), artificially fertilized, and reared in incubators placed on the substrate and in the water column in control and tailings sites (Leis and Fox 1994). Mortality in the substrate incubators averaged 64% at control sites and 81% at tailings sites, and was not related to temperature, pH, dissolved oxygen, water velocity, conductivity, alkalinity, or suspended sediments. Sediments from tailings sites, however, had significantly higher concentrations, in mg/kg DW sediment, (when compared with a control site) of lead (86 to 220 vs. 20), copper (71 to 160 vs. 11), and nickel (26 to 37 vs. 15); egg mortality was significantly correlated with concentrations of lead and copper. Leis and Fox (1994) concluded that the comparatively high mortality of walleye eggs incubated at a gold mine tailings site was attributed to lead and copper toxicity, and possibly hypoxia from the resuspension and settling of mine tailings. Juvenile yellow perch *Perca flavescens*, sampled in 1992 from the most heavily contaminated area, had significantly less food in their stomachs, when compared with samples from a control site, and elevated concentrations of lead in whole fish and their diets (Draves and Fox 1998). In June 1992, latent effects of the spill on the early life history of walleye, when compared with a reference site, included a reduction in growth rate, reduced food intake, a higher proportion of empty stomachs, a decline in abundance, and a decline in prey species (Leis and Fox 1996).

Gold mining in the Province of Nova Scotia started in the 1860s (Wong et al. 1999). By the 1940s, most of the mines were closed because the low-grade ore became too expensive to process. In Goldenville, a gold mining area in Nova Scotia, large quantities of mercury were used in the gold recovery process. About 3 million tons of tailings remained from the mining activities between 1860 and 1945. Tailings contained about 470 kg of cadmium, 37,300 kg of lead, 6800 kg of mercury, and 20,700 kg of arsenic. Over time, the tailings became distributed across the stream basin to form a tailings field of about 2 km². Despite mine closures, there is a continuing release of arsenic, mercury, lead, and other metals from the tailings field resulting in contamination of downstream ecosystems including the Gegogen Harbor of the Atlantic Ocean. Metal concentrations in stream water (Table 10.1) and sediments of Lake Gegogen located downstream from the mine were elevated and toxic to sensitive species of benthic animals (Wong et al. 1999).

In Canada, elevated lead concentrations (>10 mg Pb/kg DW) were reported in wing bones from juveniles of three species of ducks across Quebec and Ontario from 1988 to 1989. Lead concentrations in bone from mallard *Anas platyrhynchos*, black duck *Anas rubripes*, and ring-necked duck *Aythya collaris* were positively correlated to a number of variables, including proximity to non-ferrous mining sites, especially gold mining sites (Scheuhammer and Dickson 1996).

In Korea, gold mining wastes contain high concentrations of various heavy metals and can pollute streams and harm agriculture in areas influenced by mining activity

Table 10.1   Metal Concentrations (in μg/L) in Stream
              Waters at Goldenville Gold Mine, Nova Scotia

Metal	Upstream vs. at Mine Discharge
Gold (Au)	Not detectable vs. 8–9
Arsenic (As)	30–50 vs. 230–250
Cadmium (Cd)	<0.01 vs. 0.5–0.85
Copper (Cu)	<0.1 vs. 0.7–1.3
Iron (Fe)	5–17 vs. 210–360
Mercury (Hg)	<0.05 vs. <0.05
Manganese (Mn)	5–10 vs. 35–95
Nickel (Ni)	<0.5 vs. 1.0–3.8
Lead (Pb)	0.2–0.5 vs. 3.3–5.5
Vanadium (V)	0.1 vs. <0.1–0.35
Zinc (Zn)	0.1–0.5 vs. 0.9–6.6

Source: Data from Wong et al. 1999.

Table 10.2   Average Concentrations of Cadmium (Cd), Copper (Cu), Lead (Pb), and
              Zinc (Zn) in Waters, Soils, and Crops near Korean Gold Mining Activities

Compartment and Units of Measurement	Cd	Cu	Pb	Zn
Stream Waters (mg/L)				
Mining areas, maximum values	0.045	0.02	0.11	0.46
Korean standard	0.01	1.0	0.1	1.0
Stream Sediments (mg/kg DW)				
Mining areas, maximum values	50	429	2790	1080
USEPA standard	6	50	200	200
Tailings (mg/kg DW), Mining Areas	22	94	870	607
Soils (mg/kg DW)				
Maximum	40	163	1560	1060
Surface	25	50	157	178
Subsurface	5.6	56	567	189
Korean standard	1.5	50	100	300
Crop Plants (mg/kg DW), Maximum Values				
Rice, *Oryza sativa*				
Grain	0.16	4.4	0.5	23
Stalk	0.66	5.7	1.6	48
Sesame, *Sesamum indicum*	0.28	17	6.8	58

Source: Data from Kim et al. 1998.

(Kim et al. 1998). Gold mining activity between 1908 and 1998 in an area about
125 km south of Seoul produced average concentrations of cadmium, copper, lead,
and zinc in stream sediments that were significantly higher than USEPA or Korean
standards (Table 10.2). Average concentrations of cadmium and lead in farm soils
in the area near mining activity were elevated for cadmium (8.2 mg/kg DW) and lead
(192 mg/kg DW). Lead concentrations were also elevated in rice grain (0.5 mg/kg
DW) and sesame (6.8 mg/kg DW) from these soils (Kim et al. 1998).

In Zimbabwe, from 1993 to 1995, seepage from a settling pond containing gold
mine tailings resulted in elevated metals concentrations in a nearby stream (Zaranyika et al. 1997). Concentrations of iron, chromium, silver, nickel, and to a lesser

extent cadmium, manganese, lead, copper, cobalt, zinc, and sulfates, were highest in the settling pond and decreased with increasing distance downstream; however, all concentrations of metals in the stream were within acceptable limits set by the World Health Organization (Zaranyika et al. 1997).

A Malaysian tributary that received gold mine effluents from the surrounding areas for at least 10 years had elevated concentrations in sediments (in mg/kg DW) of arsenic (147), mercury (52), lead (46), and zinc (44) (Lau et al. 1998). Soft parts of three species of economically important bivalve molluscs (*Brotia costula, Melanoides tuberculata, Clithon* sp.) from a station containing 6.3 mg As/kg DW sediment, 3.4 Cu, 0.02 Hg, 0.7 Pb, and 27 mg Zn/kg DW sediments — no molluscs were found in more heavily contaminated sediments — contained maximum concentrations (mg/kg DW) of 225 As, 115 Cu, 127 Zn, and negligible concentrations of cadmium, lead, and mercury. Lau et al. (1998) concluded that concentrations of arsenic in the molluscs exceeded mandatory levels for arsenic of the Malaysian Food Act of 1983 and were near the maximum allowable limits for copper and zinc.

In Colombia, South America, trace element concentrations of nickel, chromium, lead, zinc, mercury, cadmium, and arsenic were elevated in stream sediments as a result of gold mining activities (Grosser et al. 1994). The gold ores in the area contained up to 0.0032% gold, 1.5% copper, 2% zinc, and 32% arsenic. Maximum concentrations recorded in sediments, in mg/kg DW, were 5.1 for mercury, 15 for cadmium, 43 for nickel, 83 for chromium, 354 for copper, 725 for zinc, 5300 for lead, and 6300 for arsenic. The average zinc, copper, nickel, and chromium levels appear to be of minor importance from a health risk viewpoint (Grosser et al. 1994).

Laboratory Studies

Underyearling Arctic grayling (*Thymallus arcticus*) from the Canadian Yukon River system were exposed under laboratory conditions to sediments collected from an active placer gold mining area (McLeay et al. 1987). Fish were exposed for 4 days to suspensions of fine inorganic (up to 250 g/L) or organic (up to 50 g/L) sediments. Inorganic sediments containing >10 g/L caused graylings to surface. Mortalities of 10 to 20% occurred only at 5°C with inorganic sediment concentrations >20 g/L. Exposure to organic sediments as low as 0.05 g/L for 1 to 4 days adversely affected blood chemistry. Longer exposures of 6 weeks to inorganic sediments >0.1 g/L were associated with impaired feeding activity, reduced growth rate, and decreased resistance to pentachlorophenol, a reference toxicant (McLeay et al. 1987).

A heavy load of sediments associated with discharges of gold mine tailings affects fish and aquatic invertebrates by clogging the feeding apparatus of filter-feeding invertebrates and abrading fish gills (Pain 1987; LaPerriere and Reynolds 1997). In Alaska, fish usually avoid mined streams, as do all but the most tolerant groups of invertebrates, such as the mayflies, stoneflies, and blackflies. In the absence of invertebrates, comparatively tolerant fish species (graylings, sculpins) may starve to death. Sediments from placer mining smothered eggs of Arctic graylings and caused gill abrasion and starvation in older graylings in 16-day exposures. Laboratory studies with graylings quantified the avoidance response to turbid water at

>20 NTUs; at 10 NTUs, only 10% of the grayling's food supply is visible to these sight feeding fish; authors recommend <5 NTUs above natural conditions for clear water streams receiving placer mining effluents (LaPerriere and Reynolds 1997).

Turbidity can affect prey consumption by fish but in different ways (Bonner and Wilde 2002). In some species, prey consumption is unaffected by elevated turbidity and in others prey consumption is reduced. In tests with prairie stream fishes, elevated turbidity had less effect on the prey consumption of species that were adapted to highly turbid habitats than on those adapted to less turbid habitats. The high suspended sediment loads that historically are characteristic of many prairie streams may have excluded several species from main channel habitats. Reduced turbidity in many domestic prairie rivers may contribute to the replacement of species that historically occupied highly turbid main channel habitats by visually feeding species that are comparatively superior in low turbidity waters (Bonner and Wilde 2002).

10.3.2 Marine Disposal

As will be discussed later, effects of submarine disposal of gold mine tailings include avoidance of tailings by fishes and invertebrates; population reductions of benthic biota; reductions in diversity, biomass, and dominant taxa; accumulation of selected metals in tissues of shellfish; and alterations in physical habitat. Effects were reversible over time.

Field Investigations

From 1985 to 1990, tailings from an offshore marine placer gold mining operation near Nome, Alaska, were discharged into the sea at 1.5 meters below sea level immediately behind the advancing dredge (Garnett and Ellis 1995). The dredge operated in water depths between 4.8 and 21 meters. Typical tailing discharge rates per operating hour averaged 120 m^3 of solids and 340 m^3 of slurry. The sea bottom sediments excavated by the bucket-ladder dredge consisted of cobble and sand substrates. Levels of arsenic, cadmium, chromium, copper, mercury, nickel, lead, and zinc in the bottom sediments were similar to those from other areas of Norton Sound. Using seawater exclusively (7000 m^3/hour), only a minute amount of particulate gold (775 mg/m^3) was removed. The mining permit specified waste discharge controls including quantity and composition of the effluents; protection of Alaskan king crab (*Paralithodes camtschatica*) populations; seawater turbidity; and bioaccumulation of trace metals, especially mercury, a remnant of prior beach mining. According to the permit, effluent discharges were limited to a maximum of 171 million liters daily, settleable solids to 413 m^3/h, and suspended solids to 30 g/h. Maximum average monthly concentrations of selected metals — in µg/L effluent — were 1242 for As, 774 for Cd, 52 for Cu, 2520 for Pb, 38 for Hg, 1350 for Ni, and 1710 for Zn. There were major difficulties in the chemical analysis of mercury and minor to negligible problems with other variables measured. It was concluded that the impact on crab stocks was negligible; however, an impoverished benthos remained for at least 3 years after dredging. Sandy areas were able to recolonize to a highly variable fauna within 3 to 4 years, but cobble and repeatedly dredged areas recolonized more slowly (Garnett and Ellis 1995).

Effects of offshore placer gold mining by bucket dredge on benthic invertebrates of the northeastern Bering Sea during the summers of 1986 to 1990 in waters 9 to 20 meters deep included significant reduction in total abundance, biomass, and diversity at mined stations. Many of the dominant taxa that were reduced were known food items of the economically important Alaska king crab. Recovery of the biota was under way after 4 years, but was interrupted by severe storms (Jewett et al. 1999; Jewett and Naidu 2000). The total area mined by bucket dredge was 1.5 km^2 (371 acres). Effects from mining were apparent for benthic macrofauna with virtually no effects observed for Alaskan king crabs (Jewett 1997, 1998). One year after mining ceased there were reduced numbers of polychaete annelids and echinoid sand dollars in mined areas. Mining had negligible effects on Alaskan king crabs: catches, size, sex, and prey groups in stomachs were similar between mined and reference areas. Concentrations of arsenic, cadmium, chromium, copper, lead, nickel, mercury, and zinc in muscle and hepatopancreas were also the same and were below or within the range of concentrations in Alaskan king crabs from other North Pacific locations. Moreover, concentrations of these metals were not different in surface sediments upstream and downstream of mining. Authors concluded that mining affects the sediment environment and the benthic community; that there was a reduction in macrofauna total abundance, biomass, diversity, and the abundance of dominant taxa; and that effects were minor when compared with natural disturbances (Jewett 1997, 1998; Jewett and Naidu 2000).

Juvenile tanner crabs (*Chionoecetes bairdi*) were observed on submerged mine tailings in Gastineau Channel, near Juneau, Alaska (Stone and Johnson 1997). Crabs were seen buried in the sediment and frequently ingested sediments. After decades of weathering, authors concluded that tailings deposited into Gastineau Channel were not harmful to juvenile tanner crabs based on survival, growth, and tissue burdens of metals over a period of 502 days under controlled conditions (see details later); however, it is unknown if there were leaching and increased bioavailability of metals during the first few years after tailings disposal ceased (Stone and Johnson 1997).

Tailings and wastewater effluent from a proposed gold mine near Juneau, Alaska, were studied over a 22-month period field study (Kline 1998). According to Kline, "the taxonomic composition, abundance, and biomass of invertebrates that colonized tailings were similar to that of reference sediments." Kline (1998) concluded, it is "unlikely that exposure to this gold mill effluent in the ocean could be sufficient to cause acute toxicity."

In 1888, alluvial gold was discovered on Misima Island in Papua New Guinea (Jones and Ellis 1995). Lode gold was discovered in 1904 and underground mining initiated in 1915. By the end of 2000, total production of gold was 3 million ounces and for silver it was 26 million ounces. Tailings from the Misima gold and silver mine were discharged offshore at 112 meters in depth, well below the euphotic zone, onto a steep sea floor slope that led directly to a deep ocean basin. About 18,000 tons of tailings solids were discharged daily for at least 37 years (1915 to 1942) after passing through a mix tank with seawater intake from 82 meters. Prior to discharge, each tailings part is diluted sevenfold with seawater. Geophysical surveys, ocean floor sediment sampling, and video records from a remotely operated vehicle all confirm that tailings solids are confined to the floor of a basin of 1000 to 1500 meters' water depth (Jones and Ellis 1995). Sometimes, however, the submarine tailings disposal (STD) pipe can rupture, with potential harm to sensitive ecosystems (Fields 2001). In one case, an STD pipe in Papua New Guinea ruptured at 55 meters below sea level, and turbidity plumes were carried hundreds of kilometers from the intended disposal site.

Laboratory Studies

The toxicity of effluent from the milling process of a gold mine near Juneau, Alaska, to early life stages of fishes and crustaceans was studied for ability to induce immobilization, paralysis, and death (Kline and Stekoll 2000a; Table 10.3). The sensitivity of the reference species (mysid shrimp *Mysidopsis bahia*, sheepshead minnow *Cyprinodon variegatus*) bracketed that of the indigenous species tested (Alaskan king crab *Paralithodes camtschatica*, northern shrimp *Pandalus borealis*, and Pacific herring *Clupea harengus pallasi*). The most sensitive species tested was the juvenile mysid shrimp, with seawater solutions containing as little as 21% effluent inducing immobility in 24 hours and 37% effluent causing some deaths in that same period (Table 10.3). It was concluded that the source of acute toxicity of an aged gold mill effluent to mysid shrimp was excess Ca^{+2}; a deficiency of Na^+, relative to the proportion in seawater, reduced Ca^{+2} toxicity (Kline and Stekoll 2000b).

Between 1891 and 1944, more than 80 million metric tons of tailings from three gold mines were deposited in Gastineau Channel, near Juneau, Alaska (Stone and Johnson 1997). After 50 or more years of weathering or continuous submergence in seawater, the tailings — when compared with reference sediments — contained elevated levels of arsenic, cadmium, chromium, copper, nickel, lead, and zinc (Table 10.4), but availability to biota was unknown. A laboratory study was conducted with juvenile tanner crabs (*Chionoecetes bairdi*) held for 502 days on weathered mine tailings or control sediments collected about 35 km north of Juneau. Test aquaria were 500-L flow-through containers in triplicate; crabs were fed squid. Based on metal concentrations in crab gill and muscle, authors found no significant differences in uptake between crabs held on weathered tailings and reference sediments for individual tissue metal burdens (Table 10.4; Stone and Johnson 1997).

Female tanner crabs completely or partially bury in sediment up to one year while brooding eggs. They may need to oviposit in a soft substratum, characteristic of gold mine tailings, to allow for complete cementation of the eggs to the setae (Stone and Johnson 1998). Tanner crabs may initially avoid areas affected by submarine tailings disposal but later recolonize the altered sea floor and incorporate various metals into their tissues. In a 90-day exposure study of ovigerous tanner crabs in forced contact with fresh gold mine tailings, authors showed that all crabs survived, all females extruded a full clutch of ova within 36 hours of zoeae hatch, and all larvae appeared normal. However, egg mortality was significantly higher among crabs held on tailings for 90 days when compared with reference sediments. Metal concentrations in muscle and ova of female crabs were similar for control and tailings sediments after 90 days, except for lead which was higher in both tissues

Table 10.3   Acute Toxicity of Aged Gold Mill Effluent to Marine Fishes
and Crustaceans*

Species and Life Stage	Immobility[a]	Paralysis[a]	Death[a]
Mysid shrimp, *Mysidopsis bahia*, juveniles	21	37	37
Northern shrimp, *Pandalus borealis*, larvae	21	54	>94
Alaskan king crab, *Paralithodes camtschatica*, larvae	37	54	ND[b]
Pacific herring, *Clupea harengus pallasi*, larvae	54	54	54
Sheepshead minnow, *Cyprinodon variegatus*, larvae	>94	>94	>94

* Values shown are in lowest percentage of effluent causing effect in 24 hours.
[a] Effluent adjusted to equal osmolality and pH of 3.1% seawater.
[b] No data.
Source: Modified from Kline and Stekoll 2000a.

Table 10.4   Tissue Metal Burdens of Juvenile Tanner Crabs,
*Chionoecetes bairdi**

Metal	CS[a]	WT[b]	Gill C[c]	Gill T[d]	Mus C[e]	Mus T[f]
Arsenic[g]	2.5	29.7	9.8	8.9	8.9	8.1
Cadmium	<0.2	1.2	<0.2	8.6	0.2	0.2
Chromium	60.5	93.0	28.3	22.3	6.1	1.3
Copper	13.5	32.5	220.0	209.0	37.5	48.6
Nickel	13.0	21.0	2.3	2.5	0.2	0.2
Lead	10.0	61.0	2.3	3.4	0.04	0.05
Zinc	67.0	203.0	72.8	72.5	97.8	91.6

* Held for 502 days on weathered gold mine tailings or control sediments; all values are in mg/kg dry weight
[a] Control sediments
[b] Weathered tailings
[c] Gills, control
[d] Gills, tailings
[e] Muscle, control
[f] Muscle, tailings
[g] Levels of concern in edible tissues of crustaceans by the U.S. Food
and Drug Administration, in mg/kg fresh weight, are 76 for As, 3 for
Cd, 12 for Cr, 1.5 for Pb, 70 for Ni, 1 for Hg, and no data for Cu and
Zn (Jewett and Naidu 2000). Using an arbitrary wet/dry ratio of 8,
these values, in mg/kg fresh weight, for tanner crab muscle tissues
in the above study are 1.0 for arsenic, 0.02 for cadmium, 0.16 for
chromium, 0.06 for lead, and 0.03 for nickel, or below the level of
concern in all cases.
Source: Modified from Stone and Johnson 1997.

from crabs held on tailings (135 mg Pb/kg DW tailings vs. 5 mg Pb/kg DW reference
sediments); tissue concentrations were within safe consumption guidelines (Table
10.5). In this study, crabs were seldom observed buried in control sediments and
never in tailings; however, field observations with a submersible indicated otherwise
(Stone and Johnson 1998). Authors recommend that submarine tailings deposit sites
for gold mine wastes should be located in areas of low productivity and high natural
sedimentation rates, e.g., large glacial river mouths. These sites would have the least
effect on tanner crabs and high natural sedimentation would accelerate recovery of
the sea floor (Stone and Johnson 1998).

Table 10.5   U.S. Food and Drug Administration Guidance for Arsenic,
              Cadmium, Lead, and Nickel in Shellfish

	Arsenic[a]	Cadmium[b]	Lead[c]	Nickel[d]
Provisional tolerable daily intake level for adult humans, in μg	130	55	75	1200
Maximum allowable levels in marine shellfish, in mg/kg fresh weight soft parts	30.0	2.0	0.8	2.2
90th percentile consumers of shellfish, in μg daily				
Bivalve molluscs	57	9	4	14
Lobsters, shrimp	180	3	10	7

[a] Adams et al. 1993a
[b] Adams et al. 1993b
[c] Adams et al. 1993c
[d] Adams et al. 1993d

Egg-bearing tanner crabs avoided mine tailings produced in a pilot plant associated with a proposed gold mine near Juneau, Alaska (Johnson et al. 1998b). Contaminants in sediments, especially lead, zinc, cadmium, and carboxymethylcellulose (an organic milling reagent) were elevated and may have leached from the tailings, subsequently been detected by crabs, and been responsible for avoidance behavior (Johnson et al. 1998b). These findings are ecologically relevant because tanner crabs are intimately associated with benthic sediments, and ovigerous females brood their eggs for up to one year while partially buried and often ingest sediments incidentally while feeding. Avoidance of mine tailings would probably diminish with time because tanner crabs held on the same sediments for at least 500 days showed no deleterious effects. A laboratory study with ovigerous tanner crabs held in forced contact on mine tailings for the last 90 days of the brood cycle showed no adverse effects on survival of adults, eggs, and larvae. The mine tailings, when compared with control sediments, had elevated concentrations — in mg/kg DW — of cadmium (16 vs. <1), copper (46 vs. 18), lead (164 vs. 5), and zinc (744 vs. 60). Tailings form a more compact substrate than control sediments, and crabs may prefer this substrate for brooding. It is speculated — but not proven — that reduced food availability to ovigerous females due to smothering of the sea floor could lead to reduced fecundity and poor larval survival, and that stress resulting from contaminated tailings may increase vulnerability of tanner crabs to Bitter Crab Disease (Johnson et al. 1998b).

Age zero juvenile yellowfin sole (*Pleuronectes asper*) 50 to 80 mm in total length were exposed to mine tailings produced from a gold mine near Juneau, Alaska, and subsequently evaluated for effects on survival, growth, and behavior (Johnson et al. 1998a). Juvenile sole bury in soft sediments of silt or sand to avoid predators or during overwintering (Johnson et al. 1998b). Sole avoided fresh tailings in favor of natural marine sediments (control) and weathered tailings 75 years old (Johnson et al. 1998a). The fresh tailings contained elevated concentrations (when compared with controls), in mg/kg DW, of arsenic (15 vs. 7), cadmium (11 vs. 0.1), copper (37 vs. 17), lead (84 vs. 5), and zinc (400 vs. 60). When fresh tailings were covered with 2 cm of control sediments, there was no significant avoidance of the covered fresh tailings. Growth was inhibited for sole held on fresh tailings for 30 days when compared with controls; however, growth rates were similar during days 30 to 60. Survival was similar (90 to 93% survival) for fish held on all sediments (Johnson et al. 1998a). The time needed for sea floor recovery after mine closure is unknown, but may be as short as 22 to 24 months (Johnson et al. 1998b). In another study, Johnson et al. (2000a) concluded that avoidance or short-term reductions in flatfish growth may occur from submarine disposal of tailings, and that rapid burial of tailings in areas with high natural sedimentation may accelerate recovery of the sea floor.

10.3.3 Terrestrial Storage

Storage of solid tailings underground may contaminate groundwater, and large rainfall events may cause the groundwater to discharge with the surface system (Ripley et al. 1996). Recommended storage of tailings on land should be in settling ponds with appropriate liner and eventual vegetative cover to prevent erosion and to provide a suitable substrate for bacteria (Ripley et al. 1996). Metal-tolerant plants, such as *Equisetum* spp., are suggested for phytoremediation of gold mining sites contaminated with arsenic and mercury (Wong et al. 1999).

Tailings ponds eventually dry to become tailings fields; however, dry fields can generate potentially hazardous dust containing selenium, antimony, copper, arsenic, cadmium, chromium, and a variety of lung irritants. Rainfall can leach metal–cyanide complexes from dry tailings impoundments (Fields 2001). Concentrated chemical wastes, known as slimes, are usually resmelted to recover gold. After removal of precious metals, the slime is usually held in impoundments with tailings (Da Rosa and Lyon 1997).

10.4   WASTE ROCK

Some gold mines may generate as much as 4 billion tons of rock during the mine's working life (Fields 2001). Unlike tailings impoundments, waste rock piles are simple structures (Da Rosa and Lyon 1997). These piles consist of gold-devoid materials removed from the surface or underground mine and located close to the mine to minimize haulage. Because waste rock piles can be massive and often contain acid-forming rocks and metal contaminants, they pose environmental hazards when exposed to air and water. At present, billions of tons of waste rock left on the American landscape are unprotected from the elements. Contaminants in these rocks can be transported from the mining site into the biosphere by wind, rainfall, snow-melt, and stream water drainage. The piles are not lined and may also contaminate groundwater (Da Rosa and Lyon 1997), in many cases exceeding recommended levels in soils and drinking water for arsenic, cadmium, and mercury (Table 10.6).

10.5   SUMMARY

Gold mining contaminates the biosphere through erosion and sedimentation, acidic mine drainage, and tailings wastes. Suspended solids in wastewater and runoff

Table 10.6   Drinking Water Limits and Soil Threshold Values
             for Protection of Human Health

	World Health Organization	European Union	Germany
Water, µg/L			
As	10	50	10
Cd	3	5	5
Hg	1	1	1
Soils, mg/kg dry weight			
As	500	—[a]	20
Cd	500	1–3	3
Hg	—[a]	1–1.5	2

[a] No data.

Source: Data from Matschullat et al. 2000.

from disturbed land into waterways produce increased turbidity, reduce light penetration, and alter stream flow rates. Affected streams had a reduction in algal species diversity and were avoided by predatory fish. Sedimentation inhibited reproduction of benthic fauna and resulted in piscine gill damage.

Acidic metal-rich water, or acid mine drainage (AMD), can contaminate groundwater and is often transported from the mining site by rainwater or surface drainage into nearby waterways where it frequently degrades water quality — killing all aquatic life and making water unusable. Prevention of AMD includes capping and sealing, adding acid-neutralizing chemicals, and bioremediation. Prediction of AMD damage is through short-term acid–base accounting techniques and longer-term kinetic tests.

Terrestrial storage of gold mine tailings frequently results in metals-contaminated groundwater. Gold mine tailings introduced accidentally or deliberately into freshwater ecosystems were associated with elevated sediment and streamwater concentrations of cadmium, copper, zinc, lead, arsenic, and other elements; photosynthesis-inhibiting turbidity loadings; accumulations of lead and copper in sediments that were lethal to incubating fish eggs and benthic biota; reduced growth and population abundance of fishes; bioaccumulation of lead in fish and their diets; and elevated accumulations of arsenic, copper, and zinc in soft parts of bivalve molluscs. Gold mine tailings introduced into marine environments were associated with avoidance of tailings by fishes and invertebrates; population reduction of benthic biota; reductions in diversity, biomass, and dominant taxa; and bioaccumulation of metals in shellfish tissues. Submarine disposal effects were reversible over time.

LITERATURE CITED

Adams, M.A., M. Bolger, C.D. Carrington, C.E. Coker, G.M. Cramer, M.J. DiNovi, and
    S. Dolan. 1993a. *Guidance Document for Arsenic in Shellfish.* U.S. Food Drug
    Admin., Washington, D.C., 27 pp.

Adams, M.A., M. Bolger, D.D. Carrington, C.E. Coker, G.M. Cramer, M.J. DiNovi, and
    S. Dolan. 1993b. *Guidance Document for Cadmium in Shellfish.* U.S. Food Drug
    Admin., Washington, D.C., 29 pp.

Adams, M.A., M. Bolger, C.D. Carrington, C.E. Coker, G.M. Cramer, M.J. DiNovi, and S. Dolan. 1993c. *Guidance Document for Lead in Shellfish*. U.S. Food Drug Admin., Washington, D.C., 29 pp.

Adams, M.A., M. Bolger, C.D. Carrington, C.E. Coker, G.M. Cramer, M.J. DiNovi, and S. Dolan. 1993d. *Guidance Document for Nickel in Shellfish*. U.S. Food Drug Admin., Washington, D.C., 26 pp.

Barker, J.C., M.S. Robinson, and T.K. Bundtzen. 1990. Marine placer development and opportunities in Alaska, *Mining Engin.*, 42, (1), 21–25.

Bonner, T.H. and G.R. Wilde. 2002. Effects of turbidity on prey consumption by prairie stream fishes, *Trans. Amer. Fish. Soc.*, 131, 1203–1208.

Buryak, V.A. 1993. A comprehensive study and development of gold placers: a top priority goal of modern economic geology, *Geol. Pacific Ocean*, 93, 423–435.

Cidu, R., R. Caboi, L. Fanfani, and F. Frau. 1997. Acid drainage from sulfides hosting gold mineralization (Furtei, Sardinia), *Environ. Geol.*, 30, 231–237.

Da Rosa, C.D. and J.S. Lyon, (Eds.). 1997. *Golden Dreams, Poisoned Streams*. Mineral Policy Center, Washington, D.C., 269 pp.

Draves, J.F. and M.G. Fox. 1998. Effects of a mine tailings spill on feeding and metal concentrations in yellow perch (*Perca flavescens*), *Environ. Toxicol. Chem.*, 17, 1626–1632.

Eisler, R. 1991. Cyanide hazards to fish, wildlife, and invertebrates: a synoptic review, U.S. Fish Wildl. Serv. Biol. Rep., 85(1.23), 55 pp.

Eisler, R. 2000. *Handbook of Chemical Risk Assessment: Health Hazards to Humans, Plants, and Animals. Volume 1. Metals.* Lewis Publishers, Boca Raton, FL.

Fields, S. 2001. Tarnishing the earth: gold mining's dirty secret, *Environ Health Perspec.*, 109, A474–A482.

Garnett, R.H.T. and D.V. Ellis. 1995. Tailings disposal at a marine placer mining operation by WestGold, Alaska, *Mar. Georesour. Geotechnol.*, 13, 41–57.

Grosser, J.R., V. Hagelgans, T. Hentschel, and M. Priester. 1994. Heavy metals in stream sediments: a gold mining area near Los Andes, southern Columbia S.A., *Ambio*, 23, 146–149.

Harvey, B.C. and T.E. Lisle. 1998. Effects of suction dredges on streams: a review and an evaluation strategy, *Fisheries*, 23, 8–17.

Hazen, J.M., M.W. Williams, B. Stover, and M. Wireman. 2002. Characterisation of acid mine drainage using a combination of hydrometric, chemical and isotopic analyses, Mary Murphy mine, Colorado, *Environ. Geochem. Health*, 24, 1–22.

Jewett, S.C. 1997. Assessment of the benthic environment following offshore placer gold mining in Norton Sound, northeastern Bering Sea. Ph.D. thesis, Univ. Alaska, Fairbanks, 163 pp.

Jewett, S.C. 1998. Assessment of red king crabs following offshore placer gold mining in Norton Sound, *Alaska Fish. Res. Bull.*, 6, 1–18.

Jewett, S.C., H.M. Feder, and A. Blanchard. 1999. Assessment of the benthic environment following offshore placer gold mining in the northeastern Bering Sea, *Mar. Environ. Res.*, 48, 91–122.

Jewett, S.C. and S. Naidu. 2000. Assessment of heavy metals in red king crabs following offshore placer gold mining, *Mar. Pollut. Bull.*, 40, 478–490.

Johnson, S.W., S.D. Rice, and D.A. Moles. 1998a. Effects of submarine mine tailings disposal on juvenile yellowfin sole (*Pleuronectes asper*): a laboratory study, *Mar. Pollut. Bull.*, 36, 278–287.

Johnson, S.W., R. P. Stone, and D.C. Love. 1998b. Avoidance behavior of ovigerous tanner crabs *Chionoecetes bairdi* exposed to mine tailings: a laboratory study, *Alaska Fish. Res. Bull.*, 5, 39–45.

Jones, S.G. and D.V. Ellis. 1995. Deep water STD at the Misima gold and silver mine, Papua, New Guinea, *Mar. Georesour. Geotechnol.*, 13, 183–200.

Kim, K.W., H.K. Lee, and B.C. Yoo. 1998. The environmental impact of gold mines in the Yugu-Kwangcheon Au-Ag metallogenic province, Republic of Korea, *Environ. Technol.*, 19, 291–298.

Kline, E.R. 1998. Biological impacts and recovery from marine disposal of metal mining waste. Ph.D. thesis. Univ. Alaska, Fairbanks, 167 pp.

Kline, E.R. and M.S. Stekoll. 2000a. Relative sensitivity of marine species to an effluent with total dissolved solids, *Environ. Toxicol. Chem.*, 19, 228–233.

Kline, E.R. and M.S. Stekoll. 2000b. The role of calcium and sodium in toxicity of an effluent to mysid shrimp (*Mysidopsis bahia*), *Environ. Toxicol. Chem.*, 19, 234–241.

Krause, B. 1996. *Mineral Collector's Handbook*. Sterling, New York, 192 pp.

LaPerriere, J.D. and J.B. Reynolds. 1997. Gold placer mining and stream ecosystems of interior Alaska, *Ecol. Stud.: Anal. Synth.*, 119, 265–280.

Lau, S., M. Mohamed, A.T.C. Yen, and S. Su'ut. 1998. Accumulation of heavy metals in freshwater molluscs, *Sci. Total Environ.*, 214, 113–121.

Leis, A.L. and M.G. Fox. 1994. Effect of mine tailings on the *in situ* survival of walleye (*Stizostedion vitreum*) eggs in a northern Ontario River, *Ecoscience*, 1, 215–222.

Leis, A.L. and M.G. Fox. 1996. Feeding, growth, and habitat associations of young-of-year walleye (*Stizostedion vitreum*) in a river affected by a mine tailings spill, *Canad. Jour. Fish. Aquat. Sci.*, 53, 2408–2417.

Matschullat, J., R.P. Borba, E. Deschamps, B.R. Figueiredo, T. Gabrio, and M. Schwenk. 2000. Human and environmental contamination in the iron quadrangle, Brazil, *Appl. Geochem.*, 15, 181–190.

May, T.W., R.H. Wiedmeyer, J. Gober, and S. Larson. 2001. Influence of mining-related activities on concentrations of metals in water and sediment from streams of the Black Hills, South Dakota, *Arch. Environ. Contam. Toxicol.*, 40, 1–9.

McLeay, D.J., I.K. Birtwell, G.F. Hartman, and G.L. Ennis. 1987. Responses of Arctic grayling (*Thymallus arcticus*) to acute and prolonged exposure to Yukon placer mining sediment, *Canad. Jour. Fish. Aquat. Sci.*, 44, 658–673.

Nriagu, J. and H.K.T. Wong. 1997. Gold rushes and mercury pollution, in *Mercury and Its Effects on Environment and Biology*, A. Sigal and H. Sigal, (Eds.), Marcel Dekker, New York, 131–160.

Ogola, J.S., W.V. Mitullah, and M.A. Omulo. 2002. Impact of gold mining on the environment and human health: a case study in the Migori gold belt, Kenya, *Environ. Geochem. Health*, 24, 141–158.

Pain, S. 1987. After the goldrush, *New Scientist*, 115, 36–40.

Petralia, J.F. 1996. *Gold! Gold! A Beginner's Handbook and Recreational Guide: How & Where to Prospect for Gold!* Sierra Outdoor Products, San Francisco, 143 pp.

Ripley, E.A., R.E. Redmann, and A.A. Crowder. 1996. *Environmental Effects of Mining*. St. Lucie Press, Delray Beach, FL, 356 pp.

Scheuhammer, A.M. and K.M. Dickson. 1996. Patterns of environmental lead exposure in waterfowl in eastern Canada, *Ambio*, 25, 14–20.

Skidmore, P.B. 1995. Restoration of a placer mined trout stream. *Land and Water*, July/August, 39, 14–18.

Somer, W.L., and T.J. Hassler. 1992. Effects of suction-dredge gold mining on benthic invertebrates in a northern California stream, *North Amer. Jour. Fish. Manage.*, 12, 244–252.

Stone, G.V. 1975. *Prospecting for Lode Gold*. Dorrance Publ., Pittsburgh, 50 pp.

Stone, R.P. and S.W. Johnson. 1997. Survival, growth, and bioaccumulation of heavy metals by juvenile tanner crabs (*Chionoecetes bairdi*) held on weathered mine tailings, *Bull. Environ. Contam. Toxicol.*, 58, 830–837.

Stone, R.P. and S.W. Johnson. 1998. Prolonged exposure to mine tailings and survival and reproductive success of ovigerous tanner crabs (*Chionoecetes bairdi*), *Bull. Environ. Contam. Toxicol.*, 61, 548–556.

U.S. National Academy of Sciences (USNAS), National Research Council, Committee on Hardrock Mining on Federal Lands. 1999. *Hardrock Mining on Federal Lands*. National Academy Press, Washington, D.C., 247 pp.

West, J.M. 1971. *How to Mine and Prospect for Placer Gold*. U.S. Dept. Interior, Bur. Mines Inform. Circ. 8517, 43 pp.

Wong, H.K.T., A. Gauthier, and J.O. Nriagu. 1999. Dispersion and toxicity of metals from abandoned gold mine tailings at Goldenville, Nova Scotia, Canada, *Sci. Total Environ.*, 228, 35–47.

Yeend, W. 1991. Gold placers of the Circle district, Alaska — past, present, and future, U.S. Geol. Surv. Bull. 1943, 42 pp.

Zaranyika, M.F., T.T. Mukono, N. Jayatissa, and M.T. Dube. 1997. Effect of seepage from a gold mine slime dam on the trace heavy metal levels of a nearby receiving stream and dam in Zimbabwe, *Jour. Environ. Sci. Health*, A32, 2155–2168.

CHAPTER 11

Cyanide Hazards to Plants and Animals from Gold Mining and Related Water Issues

Highly toxic sodium cyanide (NaCN) is used increasingly by the international mining community to extract gold and other precious metals through milling of high-grade ores and heap leaching of low-grade ores. The process to concentrate gold using cyanide was developed in Scotland in 1887 and used almost immediately in the Witwatersrand gold fields of the Republic of South Africa. Heap leaching with cyanide was proposed by the U.S. Bureau of Mines in 1969 as a means of extracting gold from low-grade ores. The gold industry adopted the technique in the 1970s, soon making heap leaching the dominant technology in gold extraction (Da Rosa and Lyon 1997). The heap leach and milling processes, which involve dewatering of gold-bearing ores, spraying of dilute cyanide solutions on extremely large heaps of ores containing low concentrations of gold, or milling of ores with the use of cyanide and subsequent recovery of the gold–cyanide complex, have created a number of serious environmental problems affecting wildlife and water management. This chapter reviews the history of cyanide use in gold mining with emphasis on heap leach gold mining, cyanide hazards to plants and animals, water management issues associated with gold mining, and proposed mitigation and research needs.

11.1 HISTORY OF CYANIDE USE IN GOLD MINING

About 100 million kg cyanide (CN) are consumed annually in North America, of which 80% is used in gold mining (Eisler et al. 1999; Fields 2001). In Canada, more than 90% of the mined gold is extracted from ores with the cyanidation process, which consists of leaching gold from the ore as a gold–cyanide complex and recovering the gold by precipitation. The process involves the dissolution of gold from the ore in a dilute cyanide solution and in the presence of lime and oxygen according to the following reactions (Hiskey 1984; Gasparrini 1993; Korte and Coulston 1998):

(1) $2Au + 4NaCN + O_2 + 2H_2O \rightarrow 2NaAu(CN)_2 + 2NaOH + H_2O_2$

(2) $2Au + 4NaCN + H_2O_2 \rightarrow 2NaAu(CN)_2 + 2NaOH$

Depending on solution pH, free cyanide concentrations, and other factors, gold is recovered from the eluate of the cyanidation process using either activated carbon, zinc, or ion-exchange resins (Adams et al. 1999). Using zinc dust, for example, gold along with silver is precipitated according to the reaction (Hiskey 1984; Gasparrini 1993):

(3) $2NaAu(CN)_2 + Zn \rightarrow Na_2Zn(CN)_4 + 2Au$

The process known as carbon in pulp controls the gold precipitation from the cyanide solution using activated charcoal. It is used on low-grade gold and silver ores in several processing operations in the western United States, mainly to control slime-forming organisms. After precipitation, the product is treated with dilute sulfuric acid to dissolve residual zinc and almost all copper present. The residue is washed, dried, and melted with fluxes. The remaining gold and silver alloy is cast into molds for assay. Refining is accomplished via electrolysis, during which silver and platinum group elements are separated and recovered. Another method of separating gold from silver is by parting, wherein hot concentrated sulfuric or nitric acid is used to differentially dissolve the silver, and the gold is recovered from the residue (Hiskey 1984; Gasparrini 1993).

Milling and heap leaching require cycling of millions of liters of alkaline water containing high concentrations of NaCN, free cyanide, and metal cyanide complexes that are available to the biosphere (Eisler 2000). Some milling operations result in tailings ponds 150 ha in area and larger. Heap leach operations that spray or drip cyanide solution onto the flattened top of the ore heap require solution processing ponds of about 1 ha surface area. Puddles of various sizes may occur on the top of heaps where the highest concentrations of NaCN are found. Solution recovery channels are usually constructed at the base of leach heaps; sometimes, these are buried or covered with netting to restrict access of vertebrates.

All these cyanide-containing water bodies are hazardous to natural resources and human health if not properly managed (Eisler 1991, 2000; Henny et al. 1994). For example, cyanide-laced sludges from gold mining operations stored in diked lagoons have regularly escaped from these lagoons. Major spills occurred in Guyana in 1995 and in Latvia and Kyrgyzstan in the 1990s (Koenig 2000). Failure of gold mine tailings ponds killed one child in Zimbabwe in 1978 and 17 people in South Africa in 1994 after a heavy rainfall, and contaminated streams and rivers in New Zealand in 1995 (Garcia-Guinea and Harffy 1998) and elsewhere (Leduc et al. 1982; Alberswerth et al. 1989; Koenig 2000; Kovac 2000).

In September 1980, the price of gold had increased to $750 per troy ounce (1 Troy ounce = 31.1035 g) from $35 a decade earlier (Gasparrini 1993). This economic incentive resulted in improved cyanide processing technologies to permit cost-effective extraction of small amounts of gold from low-grade ores (Henny et al. 1994). The state of Nevada is a major global gold-producing area, with at least 40 active operations. Increased gold mining activity is also reported in other western states, Alaska, the Carolinas, and northern plains states. Where relatively high-grade ores (>0.09 troy ounce Au/t ore) are found, milling techniques are used, but heap leaching of low-grade ores (0.006 to 0.025 troy ounce Au/t) is the most commonly employed extraction technique (Henny et al. 1994). Heap leach facilities usually produce gold for less than $200 US/troy ounce (Greer 1993).

The amount of gold produced in the United States by heap leaching rose 20-fold throughout the 1980s, accounting for 6% of the supply at the beginning of the decade and more than 33% at the end (Greer 1993). In 1980, there were approximately 24 heap leach facilities in the U.S.; by 1991, there were 265, of which 151 were active. The rise in domestic gold production in this period from 31 tons in 1980 to 295 tons in 1990 is attributable mainly to cyanide heap leaching (Greer 1993). Although more tons of gold ore are heap leached than vat leached in the U.S. today, a greater quantity of gold is actually produced by vat leaching because that method is used on higher-grade ores and has a higher gold recovery rate (Da Rosa and Lyon 1997). In 1989, cyanide heap leaching produced 3.7 million troy ounces from 129.8 million tons of ore, and cyanide vat leaching produced 4.3 million troy ounces of gold from 40.6 million tons of ore (Da Rosa and Lyon 1997).

Heap leaching occurs when ore, stacked on an impermeable liner at the ground surface, is sprayed or dripped with a dilute (usually about 0.05%) NaCN solution on the flattened top for a period of several months. Large leach heaps may include 1 to 25 million tons of ore, tower 100 meters high or more, and occupy several hundred hectares. As the solution percolates through the heap, gold is complexed and dissolved. For best results, heap-leached ores need to be porous, contain fine-grained clean gold particles, have low clay content, and have surfaces accessible to leach solutions. After the gold-containing solution is collected in a drainage pond, the gold is chemically precipitated, and the remaining solution is adjusted for pH and cyanide concentration and recycled to precipitate more gold. Eventually the remaining solution is treated to recycle the cyanide or to destroy it to prevent escape into the environment.

Cyanide and other contaminants may be released through tears and punctures in pad liners; leaks in liners carrying the cyanide solution; open ponds, piles, and solution ponds that can overflow; nitrogen compounds released during cyanide degradation; and release of lead, cadmium, copper, arsenic, and mercury, present in ore, that can be mobilized during crushing or leaching (Hiskey 1984; Alberswerth et al. 1989; Greer 1993; Wilkes and Spence 1995; Mosher and Figueroa 1996; Korte and Coulston 1998; White and Schnabel 1998; Korte et al. 2000; Tarras-Wahlberg et al. 2000) (Table 11.1). The amount of hydrogen cyanide that escapes into the atmosphere from gold mining operations is estimated at 20,000 tons annually, where it is quite stable; the half-time persistence of HCN in the atmosphere is about 267 days (Korte and Coulston 1998).

Cyanide is also used in agitation leaching on ores that require finer grinding than those subjected to heap leaching, and in pressure leaching and pressure cyanidation, in which cyanide penetrates at high temperature and pressure into compact ores where the gold occurs in fine fractures (Gasparrini 1993).

Table 11.1   Cyanide and Metals Concentrations in Water and Sediments Downstream of Portovela-Zaruma Cyanide-Gold Mining Area, Ecuador; Dry Season, 1988

Component and Toxicant	Observed vs. Recommended Safe Value
	Water
Free cyanide	6–13 µg/L vs. 24-hr maximum safe level of <3.5 µg/L
Total cyanide	220–2600 µg/L vs. chronic exposure value of <5.2 µg/L
Arsenic	2–264 µg/L vs. chronic exposure value of <190 µg/L
Cadmium	<0.005-0.7 µg/L vs. chronic exposure value of <0.4 µg/L
Copper	0.3–23.2 µg/L vs. chronic exposure value of <3.6 µg/L
Lead	0.04–2.5 µg/L vs. chronic exposure value of <2.5 µg/L
Mercury	<0.0022–1.1 µg/L vs. chronic exposure value of <0.1 µg/L
	Sediments
Arsenic	403–7700 mg/kg dry weight (DW) vs. no adverse effect level of <17 mg/kg DW
Cadmium	1–48 mg/kg DW vs. no probable effect level of <3.5 mg/kg DW
Copper	303–5360 mg/kg DW vs. no probable effect level of <197 mg/kg DW
Lead	9–4470 mg/kg DW vs. no probable effect level of <91 mg/kg DW
Mercury	0.1–5.8 mg/kg DW vs. no probable effect level of <0.45 mg/kg DW

Source: Modified from Tarras-Wahlberg et al. 2000.

Individual mines often cover thousands of hectares, and mining companies sometimes lease additional thousands of hectares for possible mining (Clark and Hothem 1991). Ultimately, mining converts the site into large flat-topped hills of crushed ores, waste rock, or extracted tailings and large open pits. This alteration may result in permanent damage to wildlife habitat, although most areas, with the general exception of open pits, are reclaimed through revegetation. Between 1986 and 1991, cyanide in heap leach solutions and mill tailings ponds at gold mines in Nevada alone killed at least 9500 birds, mammals, reptiles, and amphibians. Dead birds representing 91 species, especially species of migratory waterfowl, shorebirds, and gulls, comprised about 90% of the total number of animals found dead, mammals 7% (28 species), and amphibians and reptiles together 3% (6 species; Henny et al. 1994). In more recent years, the Nevada Division of Wildlife, through its toxic pond permit program (Nevada Administrative Code 502.460 through 502.495) and cooperative work with mining companies, significantly reduced the number of cyanide-related deaths of vertebrate wildlife.

Heap leaching operations are closely monitored by regulatory agencies. In California, for example, at least six permits are necessary before cyanide extraction may commence: (1) a water use permit, obtained from the California Water Board; (2) a waste discharge permit, obtained from the California Regional Water Quality Board; (3) an air quality permit, from the California Air Pollution Control District; (4) a conditional use permit, from the local county; (5) an operations plan permit, from the U.S. Bureau of Land Management; and (6) a radioactive material license, from the California Department of Health Sciences (Hiskey 1984).

Under certain alkaline conditions, cyanide may persist for at least a century in groundwater, mine tailings, and abandoned leach heaps (Da Rosa and Lyon 1997). Cyanide destruction by natural reaction with the ore, soil, clay, and microorganisms has been advanced as the major mechanism for returning a site to an environmentally safe condition. To legally shut down the operation, concentrations <0.2 mg/L of weak acid dissociable cyanide (metal-bound cyanide dissociable in weak acids, WAD) are required (White and Schnabel 1998). The use of cyanide to extract gold was banned in Turkey by the Turkish Supreme Court in 1999 because of accidental releases into the environment of untreated cyanide wastes stored in open ponds and resultant harm to human and ecosystem health (Korte et al. 2000). In Turkey, where more than 250,000 tons of crushed rocks with mean gold content of 3 g/t were subjected to 125,000 tons of sodium cyanide in 365,000 m^3 water every year, more than 2 million m^3 untreated cyanide/heavy metals solution had accumulated in waste ponds. Other countries that are considering prohibition of the cyanide leaching gold recovery process include the Czech Republic, Greece, and Romania (Korte et al. 2000).

Alkaline chlorination of wastewaters is one of the more widely used methods of treating cyanide wastes. In this process, cyanogen chloride (CNCl) is formed, which is hydrolyzed to the cyanate (CNO^-) at alkaline pH. If free chlorine is present, CNO^- can be further oxidized (Simovic and Snodgrass 1985; Marrs and Ballantyne 1987). The use of sulfur dioxide in a high-dissolved-oxygen environment with a copper catalyst reportedly reduces total cyanide in high-cyanide rinse waters from metal plating shops to less than 1 mg/L; this process may have application in cyanide detoxification of tailings ponds (Robbins 1996).

Other methods used in cyanide waste management include lagooning for natural degradation, evaporation, exposure to ultraviolet radiation, aldehyde treatment, ozonization, acidification–volatilization–neutralization, ion exchange, activated carbon absorption, electrolytic decomposition, catalytic oxidation, treatment with hydrogen peroxide, and biological treatment with cyanide-metabolizing bacteria (Towill et al. 1978; Way 1981; Marrs and Ballantyne 1987; Smith and Mudder 1991; Mosher and Figueroa 1996; Ripley et al. 1996; Dictor et al. 1997; Adams et al. 1999). Additional cyanide detoxification treatments include the use of $FeSO_4$; $FeSO_4$ plus CO_2, H_2O_2, and $Ca(OCl)_2$; dilution with water; and $FeSO_4$ plus H_2O_2, and $(NH_4)HSO_3$ (Adams et al. 1999; Eisler et al. 1999). In Canadian gold mining operations, the main treatment for cyanide removal is to retain wastewaters in impoundments for several days to months; removal occurs through volatilization, photodegradation, chemical oxidation, and secondarily through microbial oxidation (Simovic and Snodgrass 1985).

In general, because chemical treatments do not degrade all cyanide complexes, biological treatments are used (Figueira et al. 1996). Biological treatments include (1) oxidation of cyanide compounds and thiocyanate by *Pseudomonas paucimobilis* with 95% to 98% reduction of cyanides in daily discharges of 15 million L; (2) metabolism of cyanides by strains of *Pseudomonas*, *Acinetobacter*, *Bacillus*, and *Alcaligenes* involving oxygenase enzymes; and (3) bacterial cyanide degraders involving cyanide oxygenase, cyanide nitrilase, and cyanide hydratase (Figueira et al. 1996).

Microbial oxidation of cyanide is reportedly not significant in mine tailings ponds because of the high pH (>10), low number of microorganisms, low nutrient levels, large quiescent zones, and cyanide concentrations >10 mg/L (Simovic and Snodgrass 1985). However, cyanide-resistant strains of microorganisms are now used routinely to degrade cyanide. Biological degradation of cyanide in which CN⁻ is converted to CO_2, NH_3, and OH^- by bacteria, when appropriate, is considered the most cost-effective method in cyanide detoxification and has been used in cyanide detoxification of heap leaches containing more than 1.2 million tons (Mosher and Figueroa 1996). Concentrations of 10^5 cells of *Pseudomonas alcaligenes*/mL can reduce cyanide from 100 to <8 mg/L in 4 days at elevated pH (Zaugg et al. 1997). Strains of *Escherichia coli* isolated from gold extraction liquids metabolically degrade cyanide at concentrations up to 50 mg HCN/L in the presence of a glucose-cyanide complex (Figueira et al. 1996). Ammonia accumulated as the sole nitrogen by-product and was used for growth of *E. coli* involving a dioxygenase enzyme that converted cyanide directly to ammonia without cyanate formation (Figueira et al. 1996).

Removal of free cyanide, thiocyanate, and various metallocyanides from a synthetic gold milling effluent was accomplished using biologically acclimatized sludge; the adapted microbial consortium removed >95% of free cyanide, thiocyanate, copper, and zinc from the original effluent in about 8 hours (Granato et al. 1996). Biological treatment of a leachate containing cyanide was accomplished with a mixed culture of microorganisms, *Pseudomonas* and other species isolated from waste-activated sludge of the Fairbanks, Alaska, municipal wastewater treatment plant, provided with cyanide as the sole carbon and nitrogen source (White and Schnabel 1998). Microorganisms consumed cyanide and produced ammonia in an approximate 1:1 molar yield, reducing initial concentrations of 20.0 mg CN/L to <0.5 mg/L. When supplied with glucose, excess ammonia was readily consumed. This process may have application as a mobile system in the treatment of leachate from cyanidation extraction of gold from ores (White and Schnabel 1998).

Cyanide degradation has also been reported in various strains of cyanide-resistant yeasts isolated from wastewaters of gold mining operations. One strain of *Rhodotorula rubra* was able to use ammonia generated from abiotic cyanide degradation as its sole nitrogen source in the presence of a reducing sugar in aerobic media at pH 9.0 (Linardi et al. 1995). Similar results are reported for strains of *Cryptococcus* sp., *Rhodotorula glutinis*, *R. mucilaginosa*, and *Cryptococcus flavus* isolated from samples of Brazilian gold ores and industrial effluents (Gomes et al. 1999; Rezende et al. 1999).

In soils, cyanide seldom remains biologically available because it is either complexed by trace metals, microbially metabolized, or lost through volatilization (Towill et al. 1978: Marrs and Ballantyne 1987). Cyanide ions are not strongly adsorbed or retained on soils, and leaching into the surrounding groundwater will probably occur. Under aerobic conditions, cyanide salts in the soil are microbially degraded to nitrites or form complexes with trace metals. Under anaerobic conditions, cyanides denitrify to gaseous nitrogen compounds that enter the atmosphere. Mixed microbial communities that can metabolize cyanide and were not previously exposed to cyanide are adversely affected at 0.3 mg HCN/kg; however, these communities can become acclimatized to cyanide and then degrade wastes with higher cyanide concentrations.

Acclimatized microbes in activated sewage sludge can often convert nitriles to ammonia at concentrations as high as 60.0 mg total CN/kg (Towill et al. 1978).

In regard to cyanide use and toxicity on the recovery of gold and other precious metals, most authorities (as summarized in Eisler 1991, 2000; Eisler et al. 1999) currently agree on nine points:

1. Metal mining operations consume most of the current cyanide production.
2. The greatest source of cyanide exposure to humans and range animals is cyano-genic food plants and forage crops, not mining operations.
3. Cyanide is ubiquitous in the environment, with gold mining facilities only one of many sources of elevated concentrations.
4. Many chemical forms of cyanide are present in the environment, including free cyanide, metallocyanide complexes, and synthetic organocyanides, but only free cyanide (the sum of molecular hydrogen cyanide [HCN] and the cyanide anion [CN⁻]) is the primary toxic agent, regardless of origin.
5. Cyanides are readily absorbed through inhalation, ingestion, or skin contact, and are readily distributed throughout the body via blood. Cyanide is a potent and rapid-acting asphyxiant; it induces tissue anoxia through inactivation of cytochrome oxi-dase, causing cytotoxic hypoxia in the presence of normal hemoglobin oxygenation.
6. At sublethal doses, cyanide reacts with thiosulfate in the presence of rhodanese to produce the comparatively harmless thiocyanate, most of which is excreted in the urine. Rapid detoxification enables animals to ingest high sublethal doses of cyanide over extended periods without adverse effects.
7. Cyanides are not mutagenic or carcinogenic.
8. Cyanide does not biomagnify in food webs or cycle extensively in ecosystems, probably because of its rapid breakdown.
9. Cyanide seldom persists in surface waters owing to complexation or sedimenta-tion, microbial metabolism, and loss from volatilization.

11.2 CYANIDE HAZARDS

Cyanide hazards to aquatic plants and animals, terrestrial vegetation, birds, and mammals from heap leach and milling gold mining operations are briefly reviewed.

11.2.1 Aquatic Ecosystems

Fish kills from accidental discharges of cyanide-containing gold mining wastes are common (Eisler et al. 1999; Eisler 2000). In one case, mine effluents containing cyanide from a Canadian tailings pond released into a nearby creek killed more than 20,000 steelhead (*Oncorhynchus mykiss*; Leduc et al. 1982). In Colorado, overflows of 760,000 L NaCN-contaminated water from storage ponds into natural waterways killed all aquatic life along 28 km of the Alamosa River (Alberswerth et al. 1989). In 1990, 40 million L of cyanide wastes from a gold mine spilled into the Lynches River in South Carolina from a breached containment pond after heavy rains, killing an estimated 11,000 fish (Greer 1993; Da Rosa and Lyon 1997). In 1995, 160,000 L cyanide solution from a gold mine tailings pond near Jefferson City, Montana, were released into a nearby creek with loss of all fish and greatly reduced populations of aquatic insects (Da Rosa and Lyon 1997). In August 1995, in Guyana, South America, a dam failed with the release of more than 3.3 billion L cyanide-containing gold mine wastes into the Essequibo River, the nations' primary waterway, killing fish for about 80 km and contaminating drinking and irrigation water (Da Rosa and Lyon 1997).

On January 30, 2000, a dike holding millions of liters of cyanide-laced wastewater gave way at a gold extraction operation in northwestern Romania (owned jointly by Australian and Romanian firms), sending a waterborne plume into a stream that flows into the Somes, a Tisza tributary that crosses into Hungary (Koenig 2000). At least 200 tons of fish were killed, and endangered European otters (*Lutra lutra*) and white-tailed sea eagles (*Haliaeetus albicilla*) that ate the tainted fish were threatened. After devastating the upper Tisza, the 50-km-long pulse of cyanide and heavy metals spilled into the Danube River in northern Yugoslavia, killing more fish before the now-dilute plume filtered into the Danube delta at the Black Sea, more than 1000 km and 3 weeks after the spill. This entire ecosystem was previously heavily contaminated by heavy metals from mining activities (Kovac 2000). Villages close to the accident were provided with alternate water sources. Hungarian officials were most concerned that heavy metals in the Tisza River might enter flooded agricultural areas, with subsequent accumulation by crops and entry into the human food chain (Kovac 2000).

In Zimbabwe, where gold mining is the primary mining activity, tailings from the cyanidation process are treated to ensure that cyanide concentrations in the receiving waters are <5 µg CN⁻/L (Zaranyika et al. 1994). Effluents from two gold mines in Zimbabwe, where gold is extracted by the cyanide process, contained 210 and 2600 mg CN⁻/L, respectively. However, cyanide levels in the receiving stream were much lower at 2.1 µg CN⁻/L and <0.2 µg/L at 500 and 1000 meters, respectively, downstream from the point where effluents entered the receiving body of water (Zaranyika et al. 1994).

Data on the recovery of poisoned ecosystems were scarce. In one case, a large amount of cyanide-containing slag entered a stream from the reservoir of a Japanese gold mine as a result of an earthquake (Yasuno et al. 1981). The slag covered the stream bed for about 10 km from the point of rupture, killing all stream biota; cyanide was detected in the water column for only 3 days after the spill. Within 1 month, flora was established on the silt covering the above-water stones, but there was little underwater growth. After 6 to 7 months, populations of fish, algae, and invertebrates had recovered, although the species composition of algae was altered (Yasuno et al. 1981).

Fish are the most cyanide-sensitive group of aquatic organisms tested. Under conditions of continuous exposure, adverse effects on swimming and reproduction usually occurred between 5.0 and 7.2 µg free CN/L and on survival between 20 and 76 µg/L (Eisler 1991, 2000). Reproductive impairment in adult bluegills (*Lepomis macrochirus*) occurred following exposure to 5.2 µg CN/L for 289 days (USEPA 1989). Concentrations of 10 µg HCN/L caused developmental abnormalities in embryos of Atlantic salmon (*Salmo salar*) after extended exposure (Leduc 1978). These abnormalities, which were absent in controls, included yolk sac dropsy and malformations of eyes, mouth, and vertebral column (Leduc 1984). Exposure of naturally reproducing female rainbow trout (*Oncorhynchus mykiss*) to 10 µg HCN/L for 12 days during the onset of the reproductive cycle produced a reduction in plasma vitellogenin levels and a reduction in ovary weight; vitellogenin is a major source of yolk (Ruby et al. 1986). Oocyte growth was reduced in female rainbow trout (Ruby et al. 1993a) and spermatocyte numbers decreased in males (Ruby et al. 1993b) following exposure to 10 µg HCN for 12 days. Free cyanide concentrations as low as 10 µg/L can rapidly and irreversibly impair the swimming ability of salmonids in well-aerated water (Doudoroff 1976). Exposure of fish to 10 µg HCN/L for 9 days was sufficient to induce extensive necrosis in the liver, although gill tissue showed no damage. Intensification of liver histopathology was evident at dosages of 20 and 30 µg HCN/L and exposure periods up to 18 days (Leduc 1984). Other adverse effects on fish of chronic cyanide exposure included susceptibility to predation, disrupted respiration, osmoregulatory disturbances, and altered growth patterns. Free cyanide concentrations between 50 and 200 µg/L were fatal to sensitive fish species over time, and concentrations >200 µg/L were rapidly lethal to most species of fish (USEPA 1989). The high tolerance of mudskippers (*Boleophthalmus boddaerti*; 96-hour LC50 of 290 µg/L) and perhaps other species of teleosts is attributed to a surplus of cytochrome oxidase and inducible cyanide-detoxifying mechanisms and not to a reduction in metabolic rate or an enhanced anaerobic metabolism (Chew and Ip 1992).

Fish retrieved from cyanide-poisoned environments, dead or alive, can probably be consumed by humans because muscle cyanide residues were considered to be lower than the currently recommended value of 50 mg/kg diet for human health protection (Leduc 1984; Eisler 2000). Cyanide concentrations in fish from streams poisoned with cyanide ranged between 10 and 100 µg total CN/kg whole-body fresh weight (FW) (Wiley 1984). Gill tissues of cyanide-exposed salmonids contained from 30 to >7000 µg/kg FW under widely varying conditions of temperature, nominal water concentrations of free cyanide, and duration of exposure (Holden and Marsden 1964). Unpoisoned fish usually contained <1 µg total CN/kg FW in gills, although values up to 50 µg/kg FW occurred occasionally. Lowest cyanide concentrations in gill occurred at elevated (summer) water temperatures; at lower temperatures, survival was greater and residues were higher (Holden and Marsden 1964).

Among aquatic invertebrates, adverse nonlethal effects occurred between 18 and 43 µg/L, and lethal effects between 30 and 100 µg/L although some deaths occurred between 3 and 7 µg/L for the amphipod *Gammarus pulex* (Eisler 2000). Aquatic plants are comparatively tolerant to cyanide; adverse effects occurred at >160 µg free CN/L (Eisler 2000). Adverse effects of cyanide on aquatic plants are unlikely at concentrations that cause acute effects to most species of freshwater and marine fishes and invertebrates (USEPA 1980).

Biocidal properties of cyanide in aquatic environments are modified by water pH, temperature, and oxygen content; life stage, condition, and species assayed; previous exposure to cyanides; presence of other chemicals; and initial dose tested (Eisler et al. 1999; Eisler 2000). There is general agreement that cyanide is more toxic to freshwater fishes under conditions of low dissolved oxygen; that pH levels within the range 6.8 to 8.3 have little effect on cyanide toxicity but enhance toxicity at more acidic pH; that juveniles and adults are the most sensitive life stages and embryos and sac fry the most resistant; and that substantial interspecies variability exists in sensitivity to free cyanide (Eisler et al. 1999; Eisler 2000). Initial dose and water temperature modify the biocidal properties of HCN to freshwater teleosts. At low lethal concentrations near 10 µg HCN/L, cyanide is more toxic at lower temperatures; at high, rapidly lethal HCN concentrations, cyanide is more toxic at elevated temperatures (Kovacs and Leduc 1982a, 1982b; Leduc et al. 1982; Leduc 1984). By contrast, aquatic invertebrates are most sensitive to HCN at elevated water temperatures, regardless of dose (Smith et al. 1979).

Season and exercise modify the lethality of HCN to juvenile rainbow trout; higher tolerance to cyanide was associated with higher activity induced by exercise and higher temperatures, suggesting a faster detoxification rate or higher oxidative and anaerobic metabolism (McGeachy and Leduc 1988). Low levels of cyanide that are harmful when applied constantly may be harmless under seasonal or other variations that allow the organism to recover and detoxify (Leduc 1981). Acclimatization by fish to sublethal levels of cyanide through continuous exposure was thought to enhance their resistance to potentially lethal concentrations, but studies with Atlantic salmon and rainbow trout were inconclusive (Kovacs and Leduc 1982a; Alabaster et al. 1983).

Cyanides seldom persist in aquatic environments (Leduc 1984). In small, cold oligotrophic lakes treated with NaCN (1 mg/L), acute toxicity to aquatic organisms was negligible within 40 days. In warm shallow ponds, no toxicity was evident to aquatic organisms after application of 1 mg NaCN/L. In rivers and streams, cyanide toxicity fell rapidly on dilution (Leduc 1984). Cyanide was not detectable in water and sediments of Yellowknife Bay, Canada, between 1974 and 1976 despite the continuous input of cyanide-containing effluents from an operating gold mine. Nondetection was attributed to rapid oxidation (Moore 1981).

Several factors contribute to the rapid disappearance of cyanide from water: bacteria and protozoans may degrade cyanide by converting it to carbon dioxide and ammonia; chlorination of water supplies can result in conversion to cyanate; an alkaline pH favors oxidation by chlorine; and an acidic pH favors volatilization of HCN into the atmosphere (USEPA 1980).

Cyanide interacts with other chemicals, and knowledge of these interactions is important in evaluating risk to living resources. Additive, or more than additive, toxicity of free cyanide to aquatic fauna may occur in combination with ammonia (Smith et al. 1979; Alabaster et al. 1983) or arsenic (Leduc 1984). Formation of the nickel-cyanide complex markedly reduced the toxicity of both cyanide and nickel at high concentrations in alkaline pH; at lower concentrations and acidic pH, nickel-cyanide solutions increased in toxicity by more than 1000 times, owing to dissociation of the metallocyanide complex to form hydrogen cyanide (Towill et al. 1978). In 96-hour bioassays with fathead minnows, *Pimephales promelas*, lethality of mixtures of sodium cyanide and nickel sulfate were influenced by water alkalinity and pH. LC50 values decreased with increasing alkalinity and increasing pH, being 0.42 mg CN/L at 5 mg $CaCO_3$/L and pH 6.5, to 730 mg CN/L at 192 mg $CaCO_3$/L and pH 8.0 (Doudoroff 1976).

11.2.2 Birds

Cyanide waste solutions following gold extraction are released into the environment to form ponds, sometimes measuring hundreds of hectares in surface area. In the U.S., these ponds are often located in arid regions of western states and attract wildlife including migratory birds (Pritsos and Ma 1997). Between 1983 and 1992, at least 1018 birds representing 47 species were killed when they drank cyanide-poisoned water from heap leach solution ponds at a gold mine in the Black Hills of South Dakota (Da Rosa and Lyon 1997); in 1995, heap leach ponds from this site overflowed after heavy rains, spilling into a nearby creek with fatal results to all resident fishes (Da Rosa and Lyon 1997). Many species of migratory birds, including waterfowl, shorebirds, passerines, and raptors, were found dead in the immediate vicinity of gold mine heap leach extraction facilities and tailings ponds, presumably as a result of drinking the cyanide-contaminated waters (Clark and Hothem 1991; Henny et al. 1994; Hill and Henry 1996; Da Rosa and Lyon 1997). About 7000 dead birds, mostly waterfowl and songbirds, were recovered from cyanide extraction gold mine leach ponds in the western U.S. between 1980 and 1989; no gross pathological changes related to cyanide were observed in these birds at necropsy (Allen 1990; Clark and Hothem 1991). No gross pathology was evident in cyanide-dosed birds (Wiemeyer et al. 1986), which is consistent with laboratory studies with cyanide and other animal groups tested and examined (Eisler 2000). In one case, waterfowl deaths were recorded in cyanide-containing ponds of an operating gold mine located in western Arizona shortly after the mine began operations in 1987 (Sturgess et al. 1989). Deaths ranged from single birds to flocks of more than 70. At least 33 species of birds, including waterfowl, wading birds, gulls, raptors, and songbirds, and three species of mammals (bats, fox) were found dead in these ponds. Most of the waterfowl deaths were located in desert areas where the nearest water was 8 to 80 km distant.

To protect wildlife, various techniques were used including cyanide recovery, cyanide destruction, physical barriers, hazing, and establishment of decoy ponds. Techniques that were 92% successful (i.e., 8% mortality) cost mine owners about $8.58 per dead bird. This 92% survival was considered unsatisfactory by the U.S. Bureau of Land Management, and mine owners were forced to spend $295 for each dead bird found to reach 99% protection. Under existing legislation, however, zero mortality (100% survival) is the only acceptable solution (Sturgess et al. 1989). It is probable that 100% protection may not be possible using the best available technology. Songbird deaths were associated with hardrock gold mining in the Black Hills, South Dakota (Parrish 1989). This operation used the cyanide heap leaching process. Exposed collection ditches resembled small streams and were particularly attractive to songbirds, mostly red crossbills (*Loxia curvirostrata*) and pine siskins (*Carduelis pinus*), with fatal results. These ditches are now covered to prevent wildlife contact. Ponds containing cyanide solution were found to attract migrant waterfowl, and flagging devices were installed to dissuade waterfowl from landing, with partial success (Parrish 1989).

Free cyanide levels associated with high avian death rates have included 0.12 mg/L in air, 2.1 to 4.6 mg/kg body weight (BW) via acute oral exposure, and

Table 11.2   Single Oral Dose Toxicity of Sodium Cyanide (mg NaCN/kg body weight)
Fatal to 50% of Selected Birds and Mammals (Listed from Most
Sensitive to Most Tolerant)

Species	Oral LD50 (95% Confidence Limits)	Reference[a]
Mallard, *Anas platyrhynchos*	2.7 (2.2–3.2)	1
Human, *Homo sapiens*	3.0 estimated	2
American kestrel, *Falco sparverius*	4.0 (3.0–5.3)	3
Coyote, *Canis latrans*	4.1 (2.1–8.3)	4
Black vulture, *Coragyps atratus*	4.8 (4.4–5.3)	3
Laboratory rat, *Rattus norvegicus*	5.1–6.4	5,6
Little brown bat, *Myotis, lucifugus*	8.4 (5.9–11.9)	7
Eastern screech-owl, *Otus asio*	8.6 (7.2–10.2)	3
House mouse, *Mus musculus*	8.7 (8.2–9.3)	7
Japanese quail, *Coturnix japonica*	9.4 (7.7–11.4)	3
European starling, *Sturnus vulgaris*	17 (14–22)	3
Domestic chicken, *Gallus domesticus*	21 (12–36)	3
White-footed mouse, Peromyscus leucopus	28 (18–43)	7

[a] 1, Henny et al. 1994; 2, Way 1981; 3, Wiemeyer et al. 1986; 4, Sterner 1979; 5, Ballantyne 1987; 6, Egekeze and Oehme 1980; 7, Clark et al. 1991.

1.3 mg/kg BW administered intravenously. In cyanide-tolerant species, such as the domestic chicken (*Gallus domesticus*), dietary levels of 135 mg total CN/kg ration resulted in growth reduction of chicks, but 103 mg total CN/kg ration had no measurable effect on these chicks (Eisler 1991; Hill and Henry 1996). First signs of cyanide toxicosis in sensitive birds appeared between 0.5 and 5 minutes post-exposure, and included panting, eye blinking, salivation, and lethargy (Wiemeyer et al. 1986). In more tolerant species, signs of toxicosis began 10 minutes post-exposure. At higher doses, breathing in all species tested became increasingly deep and labored, followed by gasping and shallow intermittent breathing. Death usually followed in 15 to 30 minutes, although birds alive at 60 minutes frequently recovered (Wiemeyer et al. 1986). The rapid recovery of some cyanide-exposed birds may be due to the rapid metabolism of cyanide to thiocyanate and its subsequent excretion. Species sensitivity to cyanide seems to be associated with diet, with birds that feed predominantly on flesh being more sensitive to NaCN than species that feed mainly on plant materials, with the possible exception of mallards (*Anas platyrhynchos*), as judged by acute oral LD50 values (Table 11. 2).

Some birds may not die immediately after drinking lethal cyanide solutions. Sodium cyanide rapidly forms free cyanide in the avian digestive tract (pH 1.3 to 6.5), whereas formation of free cyanide from metal cyanide complexes is comparatively slow (Huiatt et al. 1983). A high rate of cyanide absorption is critical to acute toxicity, and absorption may be retarded by the lower dissociation rates of metal–cyanide complexes (Henny et al. 1994). In Arizona, a red-breasted merganser (*Mergus serrator*) was found dead 20 km from the nearest known source of cyanide, yet its pectoral muscle tissue tested positive for cyanide (Clark and Hothem 1991). A proposed mechanism to account for this phenomenon involves weak acid dissociable (WAD) cyanide compounds. Cyanide bound to certain metals, usually copper, is dissociable in weak acids such as stomach acids. Clark and Hothem (1991) suggested that drinking of lethal cyanide solutions by animals may not result in immediate death if the cyanide level is sufficiently low; these animals may die later when additional cyanide is liberated by stomach acid. In Canada, regulations typically require measurement of total cyanide and WAD cyanide in mine effluents (Ripley et al. 1996). More research seems needed on WAD cyanide compounds and delayed mortality.

Cyanide is a respiratory poison because of its affinity for the cytochrome oxidase complex of the mitochondrial respiratory chain (Keilin 1929; Nicholls et al. 1972). High dosages of cyanide are lethal through inhibition of cytochrome oxidase via cessation of mitochondrial respiration and depletion of ATP (Jones et al. 1984). Mallards given single oral doses of KCN (1.0 mg KCN/kg BW) at cyanide concentrations and amounts similar to those found at gold mine tailings ponds (40 mg CN/L), although it is NaCN that is used almost exclusively in mining, had elevated concentrations of creatine kinase in serum, suggesting tissue damage (Pritsos and Ma 1997). At 0.5 mg KCN/kg BW, mitochondrial function, an indicator of oxygen consumption, and ATP concentrations were significantly depressed in heart, liver, and brain (Ma and Pritsos 1997). Rhodanese and 3-mercaptopyruvate sulfurtransferase, two enzymes associated with cyanide detoxification, were induced in brain but not in liver or heart of KCN-dosed mallards. Although cyanide concentrations as high as 2.0 mg KCN/kg BW (at 80 mg CN/L) were not acutely toxic to mallards, the long-term effects of such exposures were not determined and may have serious consequences for migratory birds exposed sublethally to cyanide at gold mine tailings ponds.

Under the Migratory Bird Treaty Act, cyanide-containing ponds must be maintained at a level that does not result in deaths of migratory birds (Pritsos and Ma 1997). At present, there is negligible mortality of most avian species at ponds maintained at 50 mg CN⁻/L. However, some deaths of migratory birds have been recorded at <50 mg CN⁻/L, and sublethal effects have been demonstrated in mallards in water containing 20 mg CN⁻/L. These effects include significant decreases in excised liver and brain tissue ATP levels and significant decreases in mitochondrial respiration rates in heart, liver, and brain tissues. It is clear that water containing <50 mg CN⁻/L can cause generalized tissue damage in birds (Pritsos and Ma 1997), and this needs to be addressed in future regulatory actions.

11.2.3 Mammals

Gold and silver mining are probably the most widespread sources of anthropogenic cyanides in critical wildlife habitat, such as deserts in the western United States (Hill and Henry 1996). Between 1980 and 1989, 519 mammals, mostly rodents (35%) and bats (34%), were found dead at cyanide extraction gold mine mill tailings and heap leach ponds in California, Nevada, and Arizona (Clark and Hothem 1991). The list also included coyote (*Canis latrans*), badger (*Taxidea taxus*), beaver (*Castor canadensis*), mule deer (*Odocoileus hemionus*), blacktail jackrabbit (*Lepus californicus*), and kit fox (*Vulpes macrotis*), as well as skunks, chipmunks, squirrels, and domestic dogs, cats, and cattle. Also found dead at these same ponds were 38 reptiles, 55 amphibians, and 6997 birds. At the time of this study (1980 to 1989), there were approximately 160 cyanide extraction gold mines operating in California, Arizona, and Nevada, and these mines were operating within the geographic ranges of 10 endangered, threatened, or otherwise protected species of mammals. Bats comprised 6 of the 10 listed species. Because bats were not identified to species, members of these six protected species could have been among the 174 reported dead bats (Clark and Hothem 1991). A population of Townsend's big-eared bats (*Plecotus townsendii*), one of the 10 protected species, may have been extirpated by cyanide at a nearby mine in California, as quoted in Eisler et al. (1999). Badgers were another of the 10 protected species; 6 were counted among the 519 mammals found dead.

A vat leach gold mine in South Carolina with a large tailings pond reported 271 dead vertebrates found in the immediate vicinity between December 1988 and the end of 1990; 86% were birds, 13% mammals (29 of the 35 dead mammals were bats) and the rest reptiles and amphibians (Clark 1991). Bighorn sheep (*Ovis canadensis*) were found dead in August 1983 on a cyanide heap leach pile in Montana; in 1991, gulls died after landing on an unnetted cyanide pond, and deer died after consuming cyanide solution that had trickled beneath a fence (Da Rosa and Lyon 1997).

In Nevada, the state with the most heap leach sites, cyanide spills occurred weekly during the 1980s (Greer 1993). In South Dakota, a company's state-of-the-art leach pond was leaking cyanide solution at the rate of 19,000 L daily. Also, some companies allegedly punched holes in the heap leach liner when mining ended to allow drainage for more than 1 billion L of cyanide solution (Greer 1993).

In 1983, the drinking water supply of a Montana community was contaminated with 600,000 L cyanide-containing wastes from a gold mine tailings pond (Da Rosa and Lyon 1997). In 1986, an additional 7500 L leached from this same site and allegedly was responsible for the death of five cows. In 1994, cyanide was discovered in a residential drinking water supply near a gold ore processing facility in Montana. The cyanide had leaked from the mill's wastewater ponds located upgradient of the community (Da Rosa and Lyon 1997). In 1989, 350,000 L of cyanide solution spilled from a leach unit in California and polluted a reservoir used for municipal, recreational, and agricultural purposes (Greer 1993). In 1986, one mine operator in Montana dumped 76 million liters of treated cyanide solution onto 8 ha of land when a solution pond threatened to overflow after a rainstorm, with resultant contamination of a nearby creek (Da Rosa and Lyon 1997). And on February 22, 1994, 14 people were drowned when a tailings dam collapsed during a rainstorm in South Africa, releasing a wave of tailings and mine sediments into housing occupied by gold mine workers (Da Rosa and Lyon 1997).

Signs of acute cyanide poisoning in livestock usually occur within 10 minutes and include initial excitability with muscle tremors, salivation, lacrimation, defecation, urination, and labored breathing, followed by muscular incoordination, gasping, and convulsions; death may occur quickly, depending on the dose administered (Towill et al. 1978; Cade and Rubira 1982). Acute oral LD50 values for representative species of mammals ranged between 4.1 and 28.0 mg HCN/kg BW and overlapped those of birds (see Table 11.2). Despite the high lethality of large single exposures, repeated sublethal doses, especially in diets, are tolerated by many species for extended periods, perhaps indefinitely (Eisler 1991). Livestock found dead near a cyanide disposal site had been drinking surface water runoff that contained up to 365 mg HCN/L (USEPA 1980). Rats exposed for 30 days to 100 or 500 mg KCN/L drinking water had mitochondrial dysfunction, depressed ATP concentrations in liver and heart, and a depressed growth rate; little effect was observed at 50 mg KCN/L (Pritsos 1996). The adverse effect on growth is consistent with the biochemical indicators of energy depletion. However, the concentrations should be viewed with caution as CN may have volatilized from the water solutions before ingestion by the rats, due to presumed neutral pH.

Hydrogen cyanide in the liquid state can readily penetrate the skin (Homan 1987). Skin ulceration has been reported from splash contact with cyanides among workers in the electroplating and gold extraction industries, although effects in those instances were more likely due to the alkalinity of the aqueous solutions (Homan 1987). In one case, liquid HCN ran over the bare hand of a worker wearing a fresh air respirator; he collapsed into unconsciousness in 5 minutes, but ultimately recovered (USEPA 1980). No human cases of illness or death caused by cyanide in water supplies are known (USEPA 1980). Accidental acute cyanide poisonings in humans are rare (Towill et al. 1978); however, a male accidentally splashed with molten sodium cyanide died about 10 hours later (Curry 1963).

11.2.4 Terrestrial Flora

Mixed microbial populations capable of metabolizing cyanide and not previously exposed to cyanide were adversely affected at 0.3 mg HCN/kg substrate; however, these populations can become acclimatized to cyanide and can then degrade wastes containing cyanide concentrations as high as 60 mg/kg (Towill et al. 1978).

Cyanide metabolism in higher plants involves amino acids, N-hydroxyamino acids, aldoximes, nitriles, and cyanohydrins (Halkier et al. 1988). Cyanide is a weak competitive inhibitor of green bean (*Phaseolus vulgaris*) lipooxygenase, an enzyme that catalyzes the formation of hydroperoxides from polyunsaturated acids (Adams 1989). In higher plants, elevated cyanide concentrations inhibited respiration through iron complexation in cytochrome oxidase, and ATP production and other processes dependent on ATP (Towill et al. 1978). At lower concentrations, effects include inhibition of germination and growth, although sometimes cyanide enhances seed germination by stimulating the pentose phosphate pathway and inhibiting catalase (Solomonson 1981). The detoxification mechanism of cyanide is mediated by rhodanese, an enzyme widely distributed in plants (Solomonson 1981; Leduc 1984). The rate of production and release of cyanide by plants to the environment through death and decomposition is unknown (Towill et al. 1978).

11.3 CYANIDE MITIGATION AND RESEARCH NEEDS

Aquatic birds are naturally attracted to large open ponds, and efforts to deter or chemically repel them have been generally ineffective (Hill and Henry 1996). However, some chemical repellents showed promise at reducing consumption of dump leachate pond water when tested on European starlings (*Sturnus vulgaris*), especially *o*-aminoacetophenone and 4-ketobenztriazine (Clark and Shah 1993). Exclusion from cyanide solutions or reductions of cyanide concentrations to nontoxic levels are the only certain methods of protecting avian and mammalian wildlife from cyanide poisoning (Henny et al. 1994). Mortality of migratory birds from cyanide toxicosis may be curtailed at small ponds associated with leach heaps by screening birds from toxic solutions (Hallock 1990). Fencing and covering of small solution ponds with polypropylene netting have proved effective for excluding most birds, bats, and larger mammals, provided that the fencing and netting are properly maintained (Henny et al. 1994). Fences installed around cyanide-containing ponds at a heap leach gold mine in South Dakota successfully prevented deer and elk from entering; this, and other practices, reduced overall wildlife mortality at this site by about 95% (Parrish 1989). Reclamation of leach heaps to establish suitable wildlife areas is ongoing by mining corporations and involves mechanical creation of new wildlife habitats of slopes, ledges, and crevices, and revegetation of these habitats, usually with native plants, but sometimes with introduced species (Parrish 1989).

A few mines in Nevada are now covering surfaces of small ponds with 4-inch (10.2-cm)-diameter, high-density polyethylene balls (Eisler et al. 1999); birds are no longer attracted to these ponds as water sources. Although initial costs of the balls are higher than installation of netting, there are no maintenance expenses for the balls, whereas netting needs continual maintenance. Gold mine operators in southern California and Nevada used plastic sheeting to cover the cyanide leach pond, resulting in a cessation of wildlife mortality. The comparatively high cost of this process was soon recouped through reduced evaporation of water and cyanide (Eisler et al. 1999).

Cyanide concentrations in the water column of mill tailings ponds (160 to 207 mg/L) were reduced at one Nevada site using naturally detoxified recycled tailings water (Henny et al. 1994). Lowering the cyanide concentrations in tailings ponds with hydrogen peroxide has been successful at a few mines in Nevada (Allen 1990), but this procedure was still preliminary (Clark and Hothem 1991). To reduce the potential for puddling on ore heaps, ores should be less compacted; this can be accomplished by reducing the clay content of the ores and stacking ores using conveyer belts rather than trucks (Henny et al. 1994). Puddling can also be reduced by careful monitoring of solution application rates and maintenance of solution distribution systems. Wildlife have been excluded from leaching solution on the heaps by substituting drip lines for sprinklers and covering the drip lines with a layer of gravel (Henny et al. 1994; Hill and Henry 1996). Some mines use small net panels over areas of puddling to exclude birds (Henny et al. 1994).

Water hyacinth (*Eichorinia crassipes*) has been proposed as the basis of a cyanide removal technology, although large-scale use has not been implemented. Hyacinths can survive for at least 72 hours in a nutrient solution containing up to 300 mg CN/L and can accumulate up to 6700 mg CN/kg DW plant material. On this basis, 1 ha of hyacinths has the potential to absorb 56.8 kg cyanide in 72 hours, and this property may be useful in reducing the level of CN in untreated wastewaters where concentrations generally exceed 200 mg CN/L (Low and Lee 1981). Large-scale use of water hyacinths for this purpose has not yet been implemented, possibly due to disagreement over appropriate disposal mechanisms (Eisler et al. 1999).

A model biological treatment plant was established to remove cyanides and toxic metals from a daily discharge of 15,000,000 L wastewater of an underground gold mine in South Dakota (Whitlock 1990). Wastes from this mine, along with wastes from other mines, industry, mining camps, and municipalities had been discharged for more than 114 years into a common drainage system, which became devoid of all life. In the 1970s, all tailings were impounded and, in cooperation with the U.S. Environmental Protection Agency (USEPA), the mine owners developed a technology to reduce cyanide and heavy metals by 95 to 98%. The formerly lifeless receiving stream now supports an established trout fishery. In the process, cyanide was degraded to carbon and nitrogen; ammonia to nitrate; and heavy metals removed by adsorption or absorption. Indigenous species of *Pseudomonas* bacteria were used in rotating biological contactors to accomplish the process. Effluent total and WAD cyanide steadily declined between 1984 (0.45 mg total CN/L; 0.09 WAD cyanide) and 1990 (0.06 mg total CN/L; <0.01 mg WAD cyanide/L), and the cost per cubic meter of effluent declined from $0.20 to $0.10 (Whitlock 1990).

Free cyanide criteria currently proposed for the protection of natural resources include <3 µg/L medium for aquatic life and <100 mg/kg diet for birds and livestock (Eisler 2000). For human health protection, free cyanide values are <10 µg/L drinking water, <50 mg/kg diet, and <5 mg/m^3 air (Eisler 2000). Additional research is needed to establish legally enforceable standards and threshold limit values for potentially toxic cyanides in various forms, including HCN and inorganic cyanide. More research is merited on low-level, long-term cyanide intoxication in birds and mammals by oral and inhalation routes in the vicinities of high cyanide concentrations, especially on the incidence of nasal lesions, thyroid dysfunction, and urinary thiocyanate concentrations (Towill et al. 1978; Egekeze and Oehme 1980). Research is also needed on threshold limits in water where birds and mammals may be exposed, including the role of CN–metal complexes, and on sublethal effects of free cyanide on vertebrate wildlife (Eisler et al. 1999). In aquatic systems, research is needed on (1) long-term effects of low concentrations of cyanide on growth, survival, metabolism, and behavior of a variety of aquatic organisms (Towill et al. 1978; Leduc et al. 1982; Eisler 1991); (2) adaptive resistance to cyanide and the influence, if any, of oxygen, pH, temperature, and other environmental variables (Leduc 1981, 1984); and (3) usefulness of various biochemical indicators of cyanide poisoning, such as cytochrome oxidase inhibition (Gee 1987) and vitellogenin levels in fish plasma (Ruby et al. 1986).

Analytical methodologies need to be developed that differentiate between free cyanide (HCN and CN⁻) and other forms of cyanide, and that are simple, sensitive (i.e., in the µg/L range), and accurate (Smith et al. 1979; Leduc et al. 1982). Procedures need to be standardized that ensure prompt refrigeration and analysis of all samples for cyanide determination because some stored samples generate cyanide while others show decreases (Gee 1987). Periodic monitoring of cyanide in waterways is unsatisfactory for assessing potential hazards because of cyanides's rapid action, high toxicity, and low environmental persistence. A similar case is made for cyanide in the atmosphere. Information is needed on the fate of cyanide compounds in natural waters, relative contributions of natural and anthropogenic sources, and critical exposure routes for aquatic organisms (as summarized in Eisler 2000).

Finally, the use of leaching solutions (lixiviants) other than cyanide should be considered as alternatives. Thiourea was first mentioned as a lixiviant at least 60 years ago (Fields 2001). In the presence of a suitable oxidizing agent, thiourea will solubilize gold to produce a positively charged Au^+ complex (Savvaidis 1998; Adams et al. 1999; Saleh et al. 2001) as follows:

$$Au + 2CS(NH_2)_2 \rightarrow Au[CS(NH_2)_2]^+ + e^-$$

The only reported full-scale thiourea leach is at a mine in Australia (Adams et al. 1999). Products in the effluent include thiourea, formamidine, disulfide, cyanamide, sulfur, nitrate, urea, carbon dioxide, ammonia, and sulfide (Adams et al. 1999). However, thiourea is listed by USEPA as a potential carcinogen and is also relatively long-lived in the environment (Adams et al. 1999; Fields 2001). Other commercial alternatives to cyanide for gold dissolution are various halogens, thiocyanate, and thiosulfate (Adams et al. 1999). All are considered less cost-effective than cyanide, however, and collectively present a variety of environmental problems. Other lixiviants for gold under consideration include polysulfide, various nitriles, and cyanamide. However, it does not seem likely that cyanide will be replaced in the near future as a lixiviant for gold. The environmental impact of cyanide is short term; it is the release and stabilization in solution of base metals that may have a longer-term impact (Adams et al. 1999).

11.4   WATER MANAGEMENT ISSUES

To provide the quantities of ore needed for heap leach facilities, large pits are dug. One prospective open-pit mine is expected to measure 0.9 km deep, 1.44 km wide, 2.4 km long, and involves more than 1 billion tons of rock (Greer 1993). Many mining pits intrude below the water table and must be continually pumped dry. After the mine closes, the pumping ceases and the pit fills to become a small lake. Pit lakes have the potential to become acidic and may eventually contain elevated concentrations of various elements. As the level of potentially toxic water rises, it can begin to infiltrate into groundwater (Fields 2001).

Open-pit mining allegedly disturbs 50 times more earth than underground mining to produce the same amount of gold. Open-pit mining is frequently associated with stripped vegetation and topsoil and potential loss of breeding, wintering, and feeding habitat for wildlife. If the pits are left unfilled, the ecological damage may be irreversible (Greer 1993). In Bolivia, the cyanidation method of gold extraction uses 3 million liters of water daily to treat 17,000 tons of minerals, with resultant destruction of agricultural and grazing lands and loss of at least two lakes (Garcia-Guinea and Harffy 1998).

The combination of open-pit mining and heap leaching and milling generates large quantities of waste soil and rock overburden and residual tailings water from ore concentration. Surface mining of gold may generate two to five times more waste as it does ore, and up to 90% of this ore ends up as tailings. The wastes, especially the tailings, may contain residual cyanide, acids, organic toxicants, nitrogen compounds, and oils, as well as iron, copper, zinc, lead, arsenic, nickel, mercury, and cadmium (Greer 1993).

Discharge of this waste into freshwater and marine ecosystems is an inexpensive but ecologically ill-advised disposal method. Suspended solids from mining and from waste processing cause turbidity, reducing light penetration and inhibiting photosynthesis. Fish gills are sensitive to the sharp-edged, irregular pieces of rock and become vulnerable to secondary invaders, such as fungi (Greer 1993). Disposal of tailings and overburden into marine environments is no longer practiced in the United States (Greer 1993). Australia forbids the dumping of tailings and overburden into rivers, and marine disposal in that country is being phased out. The situation is different in the less-industrialized countries, where the least expensive waste disposal route is used. In Papua New Guinea, for example, where gold production exceeds 100 tons annually (up from 34 tons in 1990), more than 600 million tons of tailings were discharged into the Kawerong/Jaba River system over a period of 20 years. All aquatic life in the Jaba River was destroyed, and this disaster was one of the reported causes of the island's civil war in 1989 (Da Rosa and Lyon 1997). Other river systems in Papua New Guinea are more severely affected, with subsequent contamination of the Gulf of Papua and the Torres Strait, with major impact on fish and shrimp stocks (Greer 1993).

Gold mine operators in Nevada and elsewhere are digging large open pits to reach extensive, deep deposits of low-grade gold ore (Plume 1995; Maurer et al. 1996; Da Rosa and Lyon 1997; Eisler and Wiemeyer 2004). To prevent flooding in the mine pits and to permit efficient earth moving of surface soils, it is necessary to withdraw groundwater and use it for irrigation, discharge it to rapid infiltration basins or, in some cases, discharge it into a nearby watercourse (USBLM 2000). Surface waters are diverted around surface mining operations. After cessation of mining operations and pumping, lakes will form in the open-pit mines dug to levels below that of the surrounding groundwater (Da Rosa and Lyon 1997). At least six of Nevada's open-pit mines have filled with water that does not meet federal criteria for the protection of human drinking water and aquatic life for heavy metals and acidity (Da Rosa and Lyon 1997).

11.4.1 Affected Resources

Miners usually pump water by using a combination of in-pit wells and perimeter wells. In-pit wells pump out water that has entered the mine site, and perimeter wells intercept groundwater before it can seep into the pit. The lowering of the water table decreases groundwater elevation kilometers away from the mining site (Da Rosa and Lyon 1997). In the Humboldt River basin, Nevada, gold miners divert pumped water to irrigate fields, to create wetlands in what is naturally desert, or to supply water to another user. Only a small fraction of the pumped water is restored to the original aquifer. Under Nevada law, miners must replace the lost water from dried-out springs or wells, deepen the well that has been dried out, or reduce pumping so that prior water levels are restored (Da Rosa and Lyon 1997). This subsection briefly reviews the potential effects of water management actions on resources in gold mining communities in northern Nevada beginning in 1992. Specific resources addressed include geological structures, groundwater and surface water resources, riparian areas and wetlands, terrestrial wildlife, aquatic habitat and fisheries, special status species, livestock grazing, socioeconomics, and Native American religious concerns (USNAS 1999; USBLM 2000).

Geological impacts were demonstrated by the formation of at least three sink-holes within 5 km of groundwater drawdown and discharge into the Humboldt River. The areas that could be susceptible to sinkhole development are generally undeveloped areas underlain by carbonate rock. Critical mine-related facilities such as waste rock storage facilities, heap leach pads, and mill and tailings facilities are not located within these areas (USBLM 2000). Ground subsidence may also occur as a result of groundwater withdrawals (USBLM 1996). Changes in groundwater levels caused by mining activities show that faults, mineralized zones, and differences in fracture permeability and bedrock lithology control groundwater flow in the mountain blocks, creating complex flow paths; effects of geological structures on groundwater flow become apparent when large stresses are placed on the system (Maurer et al. 1996).

As a result of mine dewatering operations, groundwater levels at the end of 1998 had decreased 110 to 466 meters in the vicinity of the three mines examined (Maurer et al. 1996; USBLM 2000). Several springs near the mines have dried up or shown a reduction in flow as a result of mine dewatering and drought (Plume 1995). The flow and vegetation in a nearby creek were significantly reduced (USNAS 1999; USBLM 2000). Reductions in the baseflow of perennial springs and streams could affect surface water rights within the drawdown area, directly impacting rights for irrigation, stock watering, domestic use, mining and milling, municipal, and other uses (Plume 1995; USBLM 2000). Surface discharge of excess mine dewatering water and other waters to the Humboldt River was initiated in 1992, with major cumulative discharges predicted during the years 1999 through 2006, and significant discharges in 2007 through 2011 (USBLM 2000). The largest percentage increase in discharge would occur during the late summer and fall months when flows were low. The increased Humboldt River flows would probably not create additional flooding upstream, with negligible long-term impacts on surface water rights within the Humboldt River Basin, an interior terminal basin with no outlet. Mine discharges into the Humboldt Sink at the terminus of the river on occasion have exceeded water quality regulations mandated for arsenic (since corrected) and have contributed increasing, but allowable, amounts of total dissolved solids, boron, copper, fluoride, and zinc. These increased loads may potentially result in increased concentrations of these chemicals in the sink wetlands where concentration by evaporation is of concern (USBLM 2000).

Changes in regional hydrology may affect wetlands, especially in arid regions (USNAS 1999). Many arid region wetlands develop at spring orifices, and the spring pools sometimes support threatened or endangered species. In Nevada, small changes in the hydrologic head may lower the water table several meters, causing destruction of springs and their associated wetlands (USNAS 1999). About 281 ha (618 acres) of riparian vegetation occur within the drainage areas where perennial waters could be reduced by groundwater drawdown (USBLM 2000). Elevated water tables from mine dewatering discharges in low-lying areas adjacent to the Humboldt River would be conducive to the establishment of riparian vegetation and wetlands plants; this situation would reverse when mine dewatering discharges cease (USBLM 2000). Metal-contaminated water and sediments that reach wetlands may create a contaminated substrate for plants that accumulate metals (USNAS 1999).

Contaminated soils and sediments from mine sites have the potential to affect bed, bank, and floodplain sediments, as well as riparian areas and wetlands downgradient from the mine. Terrestrial vegetation diversity may be altered by construction of roads or use of offroad vehicles related to development of mines (USNAS 1999). Small alterations of topography by mining exploration activities may create new habitats for hydrophytes in low areas and xerophytes on elevated terrain. Offroad vehicles alter the stability of wetlands by creating ruts that drain the water. The use of soft-tired vehicles can produce linear depressions that create pools and tend to dry up the remaining wetlands. Offroad vehicles used for exploration may carry propagules of nonnative plant species into relatively pristine areas. Invasion of nonnative species is enhanced through mining exploration, which may be the first mechanical intrusion into these areas (USNAS 1999).

Reduction in surface or subsurface flows could result in a reduction or loss of vegetation cover for wildlife. This, in turn, could reduce breeding, foraging, and cover habitats for wildlife; increase animal displacement and loss; reduce prey availability; reduce overall biological diversity; produce possible genetic isolation in Lahontan cutthroat trout (*Oncorhynchus clarki henshawi*); reduce carrying capacity for terrestrial wildlife; and might result in population declines (USBLM 2000). Incremental habitat loss would affect a variety of big game species, upland gamebirds, waterfowl, shorebirds, raptors, songbirds, nongame mammals, and area reptiles and amphibians. The eventual reduction in flows within artificially created wetlands in the study area would result in a transition back into an upland plant community with reduction in use by waterfowl. Increased leaching of minerals and salts from saturated soils into the soil surface and subsurface layers may, however, result in a plant community of salt-tolerant species that will eventually affect wildlife composition in the area (USBLM 2000). Water discharges into the Humboldt River would produce a net increase in water flows, increasing overall water availability for consumption, supporting riparian and wetland vegetation, restoring portions of wetland and marsh habitats, and producing additional breeding, foraging, and resting habitat for resident and migratory wildlife (USBLM 2000). These effects would be most pronounced during the late summer and fall months associated with low-flow periods. Seasonal flooding may cause habitat loss for nesting and foraging of some species, but would be offset by enhancement of existing backwater and slough areas. Increased annual flows may result in additional open water during winter, with increased foraging opportunities. After mine dewatering discharges cease, the effects on terrestrial wildlife are expected to be minimal.

Reduction in stream flows as a result of mine dewatering would directly impact habitat for aquatic resources, including periphyton and invertebrates that support native fish species (USBLM 2000). Habitat reductions could result in decreased biodiversity and biomass in these communities. If stream segments become dry as a result of reduced flows, aquatic habitat and associated biota would disappear

(USBLM 2000). Increased flows to the Humboldt River would increase the habitat for fish and their food organisms. However, the possible elimination of shallow pools and channels could decrease nursery ground habitat. Increased flows may alter the diversity of fish populations because introduced species may be able to disperse and use wider areas of the river and likely compete with native species (USBLM 2000). Increased sediment levels associated with increased flows may affect aquatic biota near outfalls.

Potential reductions in available water and riparian habitat may adversely affect sensitive species of shrews, bats, eagles, hawks, owls, grouse, ibis, and terns (USBLM 2000). Possible adverse effects include habitat loss, death of individuals, reduced prey availability, reduced diversity, genetic isolation, and population declines. However, mine dewatering activities would have no measurable effect on ospreys (*Pandion haliaetus*) and pelicans. Increased flows to the Humboldt River and the Humboldt Sink would provide additional nesting, brooding, foraging, and resting habitat for species that occur along the river, including sensitive species of wildlife and ospreys. The greater availability of open water areas could provide additional foraging habitat for wintering bald eagles (*Haliaeetus leucocephalus*) along the river. Reductions in flow to upgradient feeder streams may adversely affect habitat and survival of the Lahontan cutthroat trout, the Columbia spotted frog (*Rana luteiventris*), and the California floater (*Anodonata californiensis*), a freshwater mussel. Water level reductions may also occur in springs occupied by springsnails, some of which may be endemic (USBLM 2000).

The potential long-term loss of water sources may cause long-term loss of permitted active grazing use or affect livestock distribution and forage use within grazing allotments. In addition, perennial creeks located within these allotments could be affected by groundwater drawdown, which could affect grazing management operations (USBLM 2000). Slightly increased water levels within the Humboldt River during the mine dewatering period would probably increase the areal extent of wetlands adjacent to the river channel producing increased forage and increased water availability for livestock; this would reverse after mine dewatering ceases.

Reductions in groundwater levels may impose economic hardship on those using water for domestic, industrial, commercial, agricultural, and husbandry purposes (Plume 1995; USBLM 2000). Excess mine water discharged into the Humboldt River would constitute a positive effect to water rights holders in the basin, although adverse effects from increased flow may include flooding of irrigated fields during periods of high flow (USBLM 2000). If additional mine discharges occur during periods of high flow, the storage capacity of a reservoir may be exceeded, with flood damage expected to the reservoir gates and agricultural fields downstream from the reservoir. Predicted decreases in Humboldt River flow (about 2 to 3% of the average annual flow) from groundwater drawdown could extend to the year 2019. This decrease may limit some agricultural operations to irrigate late season hay or to water livestock. Specific irrigators with junior water rights may be the most seriously affected; however, mine operators have committed to augmenting low flows if necessary using senior water rights that they own or control (USBLM 2000).

Native Americans in the vicinity, primarily members of the Western Shoshone, believe that disruption of water resources would impact their lives and the spirit forces associated with these waters, plants, and animals to the extent that Shoshone cultural traditions could not be maintained (USBLM 2000).

11.4.2 Pit Lakes

In Nevada, very low concentrations of gold (0.7 g/metric ton) can be mined economically from bulk mineral deposits, and large open-pit mines have been and are currently being developed for this purpose (Shevenell 2000; Eisler and Wiemeyer 2004). Many of these mining operations in Nevada are extracting ore from below the water table and are withdrawing large volumes of water to maintain dry operating conditions. A single gold mine in Eureka County, Nevada, allegedly has pumped as much groundwater from its open mining pit in 1 year as the entire annual consumption of a city with a population of 500,000 (DaRosa and Lyon 1997). In 1994, gold mining operations in the Humboldt River basin withdrew enough water to supply all domestic water users in the greater Seattle area (population 1.1 million) over that same period (Da Rosa and Lyon 1997). After mining is completed and dewatering activities stopped, the pits will begin to fill with water with the ultimate lake surface approaching the elevation of the premining groundwater level (Shevenell 2000). Between 1992 and 2002, about 35 mines in Nevada had a lake in their open-pit mines after dewatering and cessation of mining. Of the existing pit lakes at eight different gold mines, most had near-neutral pH but exceeded drinking water standards for arsenic, sulfate, or total dissolved solids for at least one sampling period. Pit lakes have lower evaporation rates than natural lakes because they are at higher elevations, and the surface-to-depth ratio of pit lakes is more than 1000 times smaller than that of natural lakes. Shevenell (2000) concludes that the ultimate water quality and limnology of the pit lakes and their potential impact on wildlife have not been adequately evaluated.

Open-pit mining is common in the gold mining industry where the ore bodies are large and overburden depths are limited (Braun 2002). Open-pit depths typically extend below the groundwater and may exceed 250 meters beneath the ground surface; they usually require dewatering to access the ore body. As economic development of the pit continues, dewatering systems are continually expanded. At mine closure, dewatering ceases and groundwater inflow to the excavation begins. Pit lakes form when groundwater, surface water, and other postmining drainage accumulate inside inactive open pits below the groundwater level. Pit lake water quality varies considerably owing to variations in groundwater inflow, direct precipitation, and contact with pit wall precipitation or runoff water. Interactions between wall rock and groundwater, evaporation, and geochemical processes operating within the pit lake also affect water quality (Braun 2002). A potential worst-case scenario may occur where both inflow and evaporation are moderate and there is an outflow of evapoconcentrated water containing high concentrations of toxic constituents (Atkinson 2002).

Questions that commonly arise about pit lake hydrology concern the number and amounts of contaminants from groundwater and surface water retained in pit lakes over time and management of stormwater, other surface waters, and groundwater inputs into the lake (Moreno and Sinton 2002). Managing water resources at open-pit mines frequently involves modeling based on long-term analyses of flow rates, water levels, and geochemistry of pit lake waters. Modeling analysis of ground-water flow, for example, was helpful in predicting the range of flow rates, lake levels, and surrounding groundwater levels at different stages in a mine's development (Moreno and Sinton 2002). Risk assessment of pit lakes is recommended on a case-by-case basis because each pit lake is unique, depending on local hydrogeology, the size of the pit lake, and climatic conditions (Atkinson 2002).

When pit lakes exceed local or federal surface water or groundwater quality standards, three main options for mitigation are recommended: neutralize the pit lake in place through treatment; prevent the formation of a pit lake by pumping groundwater; and regulate the pit lake level at a certain height to prevent commin-gling with other aquifers (Kuipers 2002). Treatment options include physical treat-ment processes that consist mainly of screening and filtration techniques to remove particulate matter; chemical treatment, the most common, to raise pH through lime precipitation of sulfates, and other methods to remove contaminants, such as arsenic; and biological processes, including sulfate reduction processes to treat acid drainage and remove sulfates as metals. These treatment options are preferable to the use of clean water for dilution or relying on faulty land application disposal systems (Kuipers 2002).

Gold mining pit lakes in Nevada, when filled, will contain more water than all reservoirs combined within the borders of this arid state (Miller 2002). Pit lakes are important to Nevada, and their water quality will determine their future use, as well as their effects on the aquifer, wildlife, and ecosystems. Oxidation reactions of groundwater with walls and host rock of the pit release sulfate, acid, and metals into the lake. Although oxidation of the pit wall rock releases sulfuric acid, calcium carbonates present in the wall rock also dissolve and neutralize the acid. In many Nevada pit lakes, the water that enters the pit lake will be near neutral, while in others the pit lake pH may be as low as 2.9. Metal solubilities of cadmium, zinc, and copper are high at acidic pH; at acidic pit lakes with high metal loadings, concentrations as high as 2 mg Cd/L, 172 mg Cu/L, and 550 mg Zn/L are reported. Oxidation rates decline steeply as the pit fills and the water covers the reactive surfaces. Aggressive refilling of a pit lake by groundwater inflow and additional pumping may be useful in limiting oxidation, but the acidic water created in the surrounding host rock during dewatering will eventually flush into the lake, with long-term management required. Backfilling with waste or other rock is a costly management option, but should be considered if the infill material is calcareous, as is the case with many desert soils. If no neutralization capacity is available in the backfill material, low pH problems will present additional management problems. "Pit lakes represent an in-perpetuity commitment of groundwater resources and their management needs further study, particularly in arid climates where water is the limiting resource for agricultural and municipal development" (Miller 2002).

State positions on pit water quality issues vary; however, the death of migratory birds at a pit lake is of concern to the U.S. Fish and Wildlife Service, the lead agency in enforcing violations of the Migratory Bird Treaty Act (Braun 2002). Waters of the Berkeley pit in Butte, Montana, were lethal to lesser snow geese (*Chen caer-ulescens caerulescens*) that used the lake in 1995 (USNAS 1999). Nevada regulates pit lake water quality standards on a case-by-case basis, enforcing a state regulation that loosely states that pit lake water cannot degrade surrounding groundwater quality or adversely affect the health of human or terrestrial life (Bolen 2002). Other states may require that pit lake water conform with aquifer water quality standards (Arizona); be suitable for human drinking, cooking, and food processing purposes after conventional treatment (Montana); and meet livestock, agriculture, or domestic water quality standards, depending on geographic location (Wyoming). Most states now require a bond from mine operators, with amounts ranging from nominal to millions of dollars, depending on the state (Bolen 2002). Aquatic communities may also form in pit lakes; these organisms have the potential to biomagnify various compounds and trace elements from pit lake waters. If pit lakes become attractive to migratory birds or other species, these species will be exposed to the contaminants that may be present. The introduction of fish to pit lakes may, in some cases, present unacceptable risks of contaminant exposure to fish-eating birds.

"Perpetual management of mine pit lakes is an unavoidable component of future U.S. land management" (Kempton 2002). This goal is best accomplished in a research-focused, adaptive management framework based on trust funds, a central repository for reports, and formation of a technical management group to assimilate prediction and remediation information (Kempton 2002).

11.5   WATER QUALITY AND MANAGEMENT RESEARCH NEEDS

The long-term environmental impacts of pit lakes are poorly known at this time, and long-term predictions are currently made on the basis of short-term data. Accordingly, the relation between predicted and actual outcomes needs to be evaluated (USNAS 1999). Research is needed on the chemistry, hydrology, and biology of pit lakes and their surroundings to minimize the environmental influence of future pit lakes. Pit lakes now filling need to be monitored over time to evaluate lake chemistry changes. Studies are also needed on the potential development of biological communities in pit lakes and their influence on aquatic biota, and on avian and terrestrial wildlife (USNAS 1999).

Food chain effects in pit lakes should be determined before stable biological habitats are artificially established. Additional investigations are recommended to establish water quality criteria on selected metals for protection of aquatic life and for establishment of acceptable metal burdens in aquatic prey of migratory birds; this is of particular interest in pit lakes located in arid areas with increasing metals concentrations attributed to high evaporation rates.

More research is recommended on mine area dewatering and the discharge of surplus water, especially to surface waters. Water balance models for different hydrogeological settings need to be developed to address local and regional interrelations among surface flow, pit lake hydrology, and hydraulic head of shallow and deep aquifers (USNAS 1999). These models may permit long-term predictions of the consequences of alteration of surface waters and the interruption, use, and withdrawal of groundwater. Results of studies on mine water discharges into nearby watercourses, for example, mine water discharges into the Humboldt River in Nevada, will aid in providing long-term predictions of the response of riverine ecosystems to hydrologic and biological changes (USNAS 1999). Finally, a better understanding is needed of the risks associated with increased loads of various contaminants from mine watering discharges to surface waters that flow to important wetlands in terminal systems where evapoconcentration is a factor.

11.6   SUMMARY

Cyanide extraction of gold through milling of high-grade ores and heap leaching of low-grade ores requires cycling of millions of liters of alkaline water containing high concentrations of potentially toxic sodium cyanide (NaCN), free cyanide, and metal–cyanide complexes. Some milling operations result in tailings ponds of 150 ha and larger. Heap leach operations that spray or drip cyanide onto the flattened top of the ore heap require solution processing ponds of about 1 ha in surface area. Puddles of various sizes occur on the top of heaps, where the highest concentrations of NaCN are found. Exposed solution recovery channels are usually constructed at the base of leach heaps. All these cyanide-containing water bodies are hazardous to wildlife, especially migratory waterfowl and bats, if not properly managed. Accidental spills of cyanide solutions into rivers and streams have produced massive kills of fish and other aquatic biota. Freshwater fish are the most cyanide-sensitive group of aquatic organisms tested, with high mortality documented at free cyanide concentrations >20 μg/L and adverse effects on swimming and reproduction at >5 μg/L.

Exclusion from cyanide solutions or reductions of cyanide concentrations to nontoxic levels are the only certain methods of protecting vertebrate wildlife from cyanide poisoning; a variety of exclusion/cyanide reduction techniques are presented and discussed. Additional research is recommended on: (1) effects of low-level, long-term cyanide intoxication in birds and mammals by oral and inhalation routes in the vicinity of high cyanide concentrations; (2) long-term effects of low concentrations of cyanide on aquatic biota; (3) adaptive resistance to cyanide; and (4) usefulness of various biochemical indicators of cyanide poisoning.

To prevent flooding in mine open pits and to enable earth moving on a large scale, it is often necessary to withdraw groundwater and use it for irrigation, discharge it to rapid infiltration basins, or, in some cases, discharge it to surface waters. Surface waters are diverted around surface mining operations. Adverse effects of groundwater drawdown include formation of sinkholes within 5 km of groundwater drawdown; reduced stream flows with reduced quantities of water available for irrigation, stock watering, and domestic, mining and milling, and municipal uses; reduction or loss of vegetation cover for wildlife, with reduced carrying capacity for terrestrial wildlife; loss of aquatic habitat for native fishes and their prey; and disruption of Native American cultural traditions. Surface discharge of excess mine dewatering water and other waters to main waterways may contain excess quantities of arsenic, total dissolved solids, boron, copper, fluoride, and zinc. When mining operations cease and the water pumps are dismantled, these large open pits may slowly fill with water, forming lakes. The water quality of pit lakes may present a variety of pressing environmental problems.

LITERATURE CITED

Adams, J.B. 1989. Inhibition of green bean lipoxygenase by cyanide, *Food Chem.*, 31, 243–250.

Adams, M.D., M.W. Johns, and D.W. Dew. 1999. Recovery of gold from ores and environmental aspects, in *Gold: Progress in Chemistry, Biochemistry and Technology*, H. Schmidbaur, (Ed.), John Wiley & Sons, New York, 66–104.

Alabaster, J.S., D.G. Shurben, and M.J. Mallett. 1983. The acute lethal toxicity of mixtures of cyanide and ammonia to smolts of salmon, *Salmo salar* L. at low concentrations of dissolved oxygen, *Jour. Fish Biol.*, 22, 215–222.

Albersworth, D., C. Carlson, J. Horning, S. Elderkin, and S. Mattox. 1989. *Poisoned Profits: Cyanide Heap Leach Mining and Its Impacts on the Environment.* National Wildlife Federation, Vienna, VA.

Allen, C.H. 1990. Mitigating impacts to wildlife at FMC Gold Company's Paradise Peak mine. Proceedings of the Nevada wildlife/mining workshop, March 27–29, 1990, Reno, NV. Nevada Mining Association, Reno, 67–71.

Atkinson, L.C. 2002. The hydrology of pit lakes, *Southwest Hydrology*, 1(3), 14–15.

Ballantyne, B. 1987. Toxicology of cyanides, in *Clinical and Experimental Toxicology of Cyanides*, B. Ballantyne, and T.C. Marrs, (Eds.), Wright, Bristol, 41–126.

Bolen, A. 2002. Regulating the unknown. Pit lake policies state by state, *Southwest Hydrology*, 1(3), 22–23.

Braun, T. 2002. Introduction to pit lakes in the Southwest, *Southwest Hydrology*, 1(3), 12–13.

Cade, J.W. and R.J. Rubira. 1982. Cyanide poisoning of livestock by forage sorghum, *Govt. Victoria, Dept. Agricult., Agnote,* 1960/82, 1–2.

Chew, S.F. and Y.K. Ip. 1992. Cyanide detoxification in the mud skipper, *Boleophthalmus boddaerti, Jour. Exper. Zool.,* 261, 1–8.

Clark, D.R., Jr. 1991. Bats, cyanide, and gold mining, *Bats*, 9, 17–18.

Clark, D.R., Jr., E.F. Hill, and P.F.P. Henry. 1991. Comparative sensitivity of little brown bats (*Myotis lucifugus*) to acute dosages of sodium cyanide, *Bat Res. News*, 32(4), 68.

Clark, D.R., Jr. and R.L. Hothem. 1991. Mammal mortality in Arizona, California, and Nevada gold mines using cyanide extraction, *Calif. Fish Game*, 77, 61–69.

Clark, L. and P.S. Shah. 1993. Chemical bird repellents: possible use in cyanide ponds, *Jour. Wildl. Manage.*, 57, 657–664.

Curry, A.S. 1963. Cyanide poisoning, *Acta Pharmacol. Toxicol.*, 20, 291–294.

Da Rosa, C.D. and J.S. Lyon (Eds.). 1997. *Golden Dreams, Poisoned Streams.* Mineral Policy Center, Washington, D.C., 269 pp.

Dictor, M.C., F. Battaglia-Brunet, D. Morin, A. Bories, and M. Clarens. 1997. Biological treatment of gold ore cyanidation wastewater in fixed bed reactors, *Environ. Pollut.*, 97, 287–294.

Doudoroff, P. 1976. Toxicity of fish to cyanides and related compounds — a review, *U.S. Environ. Protect. Agen. Rep.,* 600/3-76-038.

Egekeze, J.O. and F.W. Oehme. 1980. Cyanides and their toxicity: a literature review, *Veterin. Quart.*, 2, 104–114.

Eisler, R. 1991. Cyanide hazards to fish, wildlife, and invertebrates: a synoptic review, U.S. Fish Wildl. Serv. Biol Rep., 85 (1.23).

Eisler, R. 2000. Cyanide, in *Handbook of Chemical Risk Assessment: Health Hazards to Humans, Plants, and Animals. Volume 2. Organics.* Lewis Publishers, Boca Raton, FL, 903–959.

Eisler, R., D.R. Clark Jr., S.N. Wiemeyer, and C.J. Henny. 1999. Sodium cyanide hazards to fish and other wildlife from gold mining operations, in *Environmental Impacts of Mining Activities: Emphasis on Mitigation and Remedial Measures*, J.M. Azcue, (Ed.), Springer-Verlag, Berlin, 55–67.

Eisler, R and S.N. Wiemeyer. 2004. Cyanide hazards to plants and animals from gold mining and related water issues, *Rev. Environ. Contam. Toxicol.*, 182, 21–54.

Fields, S. 2001. Tarnishing the earth: gold mining's dirty secret, *Environ, Health Perspec.*, 109, A474–A482.

Figueira, M.M., V.S.T. Ciminelli, M.C. de Andrade, and V.R. Linardi. 1996. Cyanide degradation by an *Escherichia coli* strain, *Canad. Jour. Microbiol.*, 42, 519–523.

Garcia-Guinea, J. and M. Harffy. 1998. Bolivian mining pollution: past, present and future, *Ambio*, 27, 251–253.

Gasparrini, C. 1993. *Gold and Other Precious Metals. From Ore to Market*. Springer-Verlag, Berlin, 336 pp.

Gee, D.J. 1987. Cyanides in murder, suicide, and accident, in *Clinical and Experimental Toxicology of Cyanides*, B. Ballantyne and T.C. Marrs, (Eds.), Wright, Bristol, 209–216.

Gomes, N.C.M., C.A. Rosa, P.F. Pimental, V.R. Linardi, and L.C.S. Mendonca-Hagler. 1999. Uptake of free and complexed silver ions by yeast from a gold mining industry in Brazil, *Jour. Gen. Appl. Microbiol.*, 45, 121–124.

Granato, M., M.M.M. Goncalves, R.C.V. Boas, and G.L.S. Anna Jr. 1996. Biological treatment of a synthetic gold milling effluent, *Environ. Pollut.*, 91, 343–350.

Greer, J. 1993. The price of gold: environmental costs of the new gold rush, *The Ecologist*, 23 (3), 91–96.

Halkier, B.A., H.V. Scheller, and B.L.Moller. 1988. Cyanogenic glucosides: the biosynthetic pathway and the enzyme system involved, in *Cyanide Compounds in Biology*, D. Evered and S. Harnett, (Eds.), Ciba Foundation Symposium 140, John Wiley & Sons, Chichester, 49–66.

Hallock, R.J. 1990. Elimination of migratory bird mortality at gold and silver mines using cyanide extraction. Proceedings of the Nevada Wildlife/Mining workshop, March 27–29, 1990, Nevada Mining Assoc., Reno, 9–17.

Henny, C.J., R.J. Hallock, and E.F. Hill. 1994. Cyanide and migratory birds at gold mines in Nevada, USA, *Ecotoxicology*, 3, 45–58.

Hill, E.F. and P.F.P. Henry. 1996. Cyanide, in *Noninfectious Diseases of Wildlife*, 2nd ed. A. Fairbrother, S.N. Locke, and G.L. Hoff, (Eds.), Iowa State University Press, Ames, 99–107.

Hiskey, J.B. (Ed.). 1984. Au & Ag heap and dump leaching practice. Proceedings 1983 Meeting of the Society of Mining Engineers, Salt Lake City, UT, Oct. 19–21, 1983. Guinn Printing, Hoboken, NJ.

Holden, A.V. and K. Marsden. 1964. Cyanide in salmon and brown trout. *Dept. Agricul. Fish. Scotland, Freshwater Salmon Res. Ser.*, 33.

Homan, E.R. 1987. Reactions, processes and materials with potential for cyanide exposure. In *Clinical and Experimental Toxicology of Cyanides*, B. Ballantyne, and T.C. Marrs, (Eds.), Wright, Bristol, 1–21.

Huiatt, J.L., J.L. Kerrigan, F.A. Oslo, and G.L. Potter. 1983. *Cyanide from Mineral Processing*. Utah Mining Mineral Resour. Inst., Salt Lake City.

Jones, M.G., D. Bickar, M.T. Wilson, M. Brunori, A. Colosimo, and P. Sarti. 1984. A re-examination of the reactions of cyanide with cytochrome *c* oxidase, *Biochem. Jour.*, 220, 57–66.

Keilin, D. 1929. Cytochrome and respiratory enzymes, *Proc. Roy. Soc. Lond. Biol. Sci.*, 104, 206–252.

Kempton, H. 2002. Dealing with the legacy of mine pit lakes, *Southwest Hydrology*, 1(3), 24–26.

Koenig, R. 2000. Wildlife deaths are a grim wake-up call in eastern Europe, *Science*, 287, 1737–1738.

Korte, F. and F. Coulston. 1998. Some considerations on the impact of ecological chemical principles in practice with emphasis on gold mining and cyanide, *Ecotoxicol. Environ. Safety*, 41, 119–129.

Korte, F., M. Spiteller, and F. Coulston. 2000. The cyanide leaching gold recovery process is a nonsustainable technology with unacceptable impacts on ecosystems and humans: the disaster in Romania, *Ecotoxicol. Environ. Safety*, 46, 241–245.

Kovac, C. 2000. Cyanide spill could have long term impact, *Brit. Med. Jour.*, 320, 1294.

Kovacs, T.G. and G. Leduc. 1982a. Sublethal toxicity of cyanide to rainbow trout (*Salmo gairdneri*) at different temperatures, *Canad. Jour. Fish. Aquat. Sci.*, 39, 1389–1395.

Kovacs, T.G. and G. Leduc. 1982b. Acute toxicity of cyanide to rainbow trout acclimated at different temperatures, *Canad. Jour. Fish. Aquat. Sci.*, 39, 1426–1429.

Kuipers, J.R. 2002. Water treatment as a mitigation method for pit lakes, *Southwest Hydrology*, 1(3), 18–19.

Leduc, G. 1978. Deleterious effects of cyanide on early life stages of Atlantic salmon (*Salmo salar*), *Jour Fish. Res. Bd. Canada*, 35, 166–174.

Leduc, G. 1981. Ecotoxicology of cyanides in freshwater, in *Cyanide in Biology*, B. Vennesland, E.E. Conn, C.J. Knowles, J. Westley, and F. Wissing, (Eds.), Academic Press, New York, 487–494.

Leduc, G. 1984. Cyanides in water: toxicological significance, in *Aquatic Toxicology*, Volume 2. L.J. Weber, (Ed.), Raven Press, New York, 153–224.

Leduc, G., R.C. Pierce, and I.R. McCracken. 1982. The effects of cyanides on aquatic organisms with emphasis upon freshwater fishes, Nat. Res. Coun. Canada, Ottawa. Publ. NRCC 19246, 139 pp.

Linardi, V.R., M.C. Amancio, and N.C.M. Gomes. 1995. Maintenance of *Rhodotorula rubra* isolated from liquid samples of gold mine effluents, *Folia Microbiol.*, 40, 487–489.

Low, K.S. and C.K. Lee. 1981. Cyanide uptake by water hyacinths, *Eichornia crassipes* (Mart) Solms, *Pertanika*, 42, 122–128.

Ma, J. and C.A. Pritos. 1997. Tissue-specific bioenergetic effects and increased enzymatic activities following acute sublethal peroral exposure to cyanide in the mallard duck, *Toxicol. Appl. Pharmacol.*, 142, 297–302.

Marrs, T.C. and B. Ballantyne. 1987. Clinical and experimental toxicology of cyanides: an overview, in *Clinical and Experimental Toxicology of Cyanides*, B. Ballantyne, and T.C. Marrs, (Eds.), Wright, Bristol, 473–495.

Maurer, D.K., R.W. Plume, J. M. Thomas, and A.K. Johnson. 1996. Water resources and effects of changes in ground-water use along the Carlin Trend, north-central Nevada. U.S. Geol. Surv., Boulder, CO, Water Resour. Invest. Rep., 96-4134.

McGeachy, S.M. and G. Leduc. 1988. The influence of season and exercise on the lethal toxicity of cyanide of rainbow trout (*Salmo gairdneri*), *Arch. Environ. Contam. Toxicol.*, 17, 313–318.

Miller, G.C. 2002. Precious metals pit lakes: controls on eventual water quality, *Southwest Hydrology*, 1(3), 16–17.

Moore, J.W. 1981. Influence of water movements and other factors on distribution and transport of heavy metals in a shallow bay (Canada), *Arch. Environ. Contam. Toxicol.*, 10, 715–724.

Moreno, J. and P. Sinton. 2002. Modeling mine pit lakes, *Southwest Hydrology*, 1(3), 20–21, 31.

Mosher, J.B. and L. Figueroa. 1996. Biological oxidation of cyanide: a viable treatment option for the minerals processing industry, *Minerals Engin.*, 9, 573–581.

Nicholls, P., K.J.H. Van Buren, and B.F. Van Gelder 1972. Biochemical and biophysical studies on cytochrome aa$_3$, *Biochim. Biophys. Acta*, 275, 279–287.

Parrish, B. 1989. Wildlife impact mitigation and reclamation in open pit, cyanide heap leach gold mining, in *Issues and Technology in the Management of Impacted Wildlife*, Proceedings, National Symposium, Glenwood Springs, CO, February 6–8, 1989, P.R. Davis, J.C. Emerick, D.M. Finch, S.Q. Foster, J.W. Monarch, S. Rush, O. Thorne, and J. Todd, (Eds.), 103-106.

Plume, R.W. 1995. Water resources and potential effects of ground-water development in Maggie, Marys, and Susie Creek basins, Elko and Eureka counties, Nevada. U.S. Geol. Surv., Water Resour. Invest. Rep., 94-4222.

Pritsos, C.A. 1996. Mitochondrial dysfunction and energy depletion from subchronic and peroral exposure to cyanide using the Wistar rat as a mammalian model, *Toxic Subst. Mechan.*, 15, 219–229.

Pritsos, C.A. and J. Ma. 1997. Biochemical assessment of cyanide-induced toxicity in migratory birds from gold mining hazardous waste ponds, *Toxicol. Indus. Health*, 13, 203–209.

Rezende, R.P., J.C.T. Dias, C.A. Rosa, F. Carazza, and V.R. Linardi. 1999. Utilization of nitriles by yeasts isolated from a Brazilian gold mine, *Jour. Gen. Appl. Microbiol.*, 45, 185–192.

Ripley, E.A., R.E. Redmann, and A.A. Crowder. 1996. *Environmental Effects of Mining*. St. Lucie Press, Delray Beach, FL, 356 pp.

Robbins, G.H. 1996. Historical development of the INCO SO$_2$/AIR cyanide destruction process, *Canad. Mining Metallur. Bull.*, 89, (1003), 63–69.

Ruby, S.M., D.R. Idler, and Y.P. So. 1986. The effect of sublethal cyanide exposure on plasma vitellogenin levels in rainbow trout (*Salmo gairdneri*) during early vitellogenesis, *Arch. Environ. Contam. Toxicol.*, 15, 603–607.

Ruby, S.M., D.R. Idler, and Y.P. So. 1993a. Plasma vitellogenin, 17b-estradiol, T$_3$ and T$_4$ levels in sexually maturing rainbow trout *Oncorhynchus mykiss* following sublethal HCN exposure, *Aquat. Toxicol.*, 26, 91–102.

Ruby, S.M., P. Jaroslawski, and R. Hull. 1993b. Lead and cyanide toxicity in sexually maturing rainbow trout, *Oncorhynchus mykiss* during spermatogenesis, *Aquat. Toxicol.*, 26, 225–238.

Saleh, S.M., S.A. Said, and M.S. El-Shahawi. 2001. Extraction and recovery of Au, Sb and Sn from electrorefined solid waste, *Anal. Chim. Acta*, 436, 69–77.

Savvaidis, I. 1998. Recovery of gold from thiourea solutions using microorganisms, *Biometals*, 11, 145–151.

Shevenell, L.A. 2000. Water quality in pit lakes in disseminated gold deposits compared to two natural, terminal lakes in Nevada, *Environ. Geol.*, 39, 807–815.

Simovic, L. and W.J. Snodgrass. 1985. Natural removal of cyanides in gold milling effluents — evaluation of removal kinetics, *Water Pollut. Res. Jour. Canada*, 20, 120–135.

Smith, A. and T. Mudder. 1991. *The Chemistry and Treatment of Cyanide Wastes*. Mining Journal Books, London.

Smith, L.L. Jr., S.J. Broderius, D.M. Oseid, G.L. Kimball, W.M. Koenst, and D.T. Lind. 1979. Acute and chronic toxicity of HCN to fish and invertebrates, U.S. Environ. Protect. Agen. Rep., 600/3-79-009.

Solomonson, L.P. 1981. Cyanide as a possible metabolic inhibitor, in *Cyanide in Biology*, B. Vennesland, E.E. Conn, C.J. Knowles, J. Westley, and F. Wissing, (Eds.), Academic Press, New York, 11–28.

Sterner, R.T. 1979. Effects of sodium cyanide and diphacinome in coyotes (*Canis latrans*): applications as predacides in livestock toxic collars, *Bull. Environ. Contam. Toxicol.*, 23, 211–217.

Sturgess, J.A., D.C. Robertson, L. Sharp, and G. Stephan. 1989. Mitigating duck losses at cyanide ponds: methods, costs and results at an operating gold mine, in *Issues and Technology in the Management of Impacted Wildlife*, Proceedings, National Symposium, Glenwood Springs, CO, February 6–8, 1989, P.R. Davis, J.C. Emerick, D.M. Finch, S.Q. Foster, J.W. Monarch, S. Rush, O. Thorne, and J. Todd, (Eds.), 98–102.

Tarras-Wahlberg, N.H., A. Flachier, G. Fredriksson, S. Lane, B. Lundberg, and O. Sangfors. 2000. Environmental impact of small-scale and artisanal gold mining in southern Ecuador, *Ambio*, 29, 484–491.

Towill, L.E., J.S. Drury, B.L. Whitfield, E.B. Lewis, E.L. Galyan, and A.S. Hammons. 1978. Reviews of the environmental effects of pollutants: V. Cyanide. U.S. Environ. Protect. Agen. Rep. 600/1-78-027.

U.S. Bureau of Land Management (USBLM). 2000. *Cumulative Impact Analysis of Dewatering and Water Management Operations for the Betze Project, South Operations Area Amendment, and Leevile Project.* USBLM, Elko, NV.

U.S. Environmental Protection Agency (USEPA). 1980. Ambient water quality criteria for cyanides, U.S. Environ. Protect. Agen. Rep., 440/5-80-037.

U.S. Environmental Protection Agency (USEPA). 1989. Cyanide. *Rev. Environ. Contam. Toxicol.*, 107, 53–64.

U.S. National Academy of Sciences (USNAS), National Research Council, Committee on Hardrock Mining on Federal Lands. 1999. *Hardrock Mining on Federal Lands.* National Academy Press, Washington, D.C., 247 pp.

Way, J.L. 1981. Pharmacologic aspects of cyanide and its antagonism. In *Cyanide in Biology*, B. Vennesland, E.E. Conn, C.J. Knowles, J. Westley, and F. Wissing, (Eds.), Academic Press, New York, 29–40.

White, D.M. and W. Schnabel. 1998. Treatment of cyanide waste in a sequencing batch biofilm reactor, *Water Res.*, 32, 254–257.

Whitlock, J.L. 1990. Biological detoxification of precious metal processing wastewaters, *Geomicrobiol. Jour.*, 8, 241–290.

Wiemeyer, S.N., E.F. Hill, J.W. Carpenter, and A.J. Krynitsky. 1986. Acute oral toxicity of sodium cyanide in birds, *Jour. Wildl. Dis.*, 22, 538–546.

Wiley, R.W. 1984. A review of sodium cyanide for use in sampling stream fishes, *North Amer. Jour. Fish Manage.*, 4, 249–256.

Wilkes, B.D. and J.M.B. Spence. 1995. Assessing the toxicity of surface waters downstream from a gold mine using a battery of bioassays, in Proceedings of the 21st Annual Aquatic Toxicity Workshop: October 3–5, 1994, Sarnia, Ontario. Canad. Tech. Rep. Fish. Aquat. Sci. 2050, G.F. Westlake, J.L. Parrott, and A.J. Niimi, (Eds.), 38–44.

Yasuno, M., S. Fukushima, F. Shioyama, J. Hasegawa, and S. Kasuga. 1981. Recovery processes of benthic flora and fauna in a stream after discharge of slag containing cyanide, *Verhandl. Intern. Verein. Theoret. Ange. Limnologie*, 21, 1154–1164.

Zaugg, S.E., R.A. Davidson, J.C. Walker, and F.B. Walker. 1997. Electrophoretic analysis of cyanide depletion by *Pseudomonas alcaligenes*, *Electrophoresis*, 18, 202–204.

Zaranyika, M.F., L. Mudungwe, and R.C. Gurira. 1994. Cyanide ion concentration in the effluent from two gold mines in Zimbabwe and in a stream receiving effluent from one of the goldmines, *Jour. Environ. Sci. Health*, A29, 1295–1303.

CHAPTER 12

Arsenic Hazards from Gold Mining for Humans, Plants, and Animals

Arsenic contamination of the biosphere from various gold mining and refining operations jeopardizes the health and well-being of biological communities. This section documents the sources and extent of arsenic discharges to the environment associated with gold mining operations; arsenic risks to human health, with emphasis on gold miners, gold refinery workers, and children residing near gold mining and refining activities; arsenic concentrations in biota and abiotic materials near gold extraction and refining facilities; lethal and sublethal effects of different chemical forms of arsenic to representative species of flora and fauna; and proposed arsenic criteria for the protection of human health and selected natural resources.

12.1 ARSENIC SOURCES TO THE BIOSPHERE FROM GOLD MINING

Gold-bearing ores worldwide contain variable quantities of sulfide and arsenic compounds that interfere with efficient gold extraction using current cyanidation technology. Arsenic occurs in many types of Canadian gold ore deposits, mainly as arsenopyrite (FeAsS), niccolite (NiAs), cobaltite (CoAsS), tennantite ($[Cu,Fe]_{12}As_4S_{13}$), enargite (Cu_3AsS_4), orpiment (As_2S_3), and realgar (AsS) (Azcue et al. 1994). Some gold-containing ores in Colombia, South America, contain up to 32% of arsenic-bearing minerals, and surrounding sediments may hold as much as 6300 mg As/kg DW (Grosser et al. 1994).

Arsenic enters the environment from a variety of sources associated with gold mining, including waste soil and rocks, tailings, atmospheric emissions from ore roasting, and bacterially enhanced leaching. The combination of open-cast mining and heap leaching generates large quantities of waste soil and rock (overburden) and residual water from ore concentrations (tailings). The wastes, especially the tailings, are rich sources of arsenic (Greer 1993; Lim et al. 2003). In Nova Scotia, for example, about 3 million tons of tailings — containing 20,700 kg of arsenic — were left from gold mining activities between 1860 and 1945; tailings tend to diffuse into the surrounding environment over time, with subsequent spread of arsenic contamination (Wong et al. 1999).

Discharges from gold mines into the Humboldt Sink, Nevada, sometimes exceed water quality regulations mandated for arsenic (U.S. Bureau of Land Management [USBLM] 2000). In the Black Hills of South Dakota, a cluster of 11 abandoned gold mines discharged up to 10,000 kg of arsenopyrites daily into nearby creeks (Rahn et al. 1996). The present treatment of gold mine tailings to reduce arsenic availability to the environment involves peroxide addition to oxidize cyanide to cyanate, ferric sulfate and lime addition to precipitate arsenic as ferric arsenate ($FeAsO_4$), and polyacrylamide flocculent addition to enhance sedimentation (Bright et al. 1994, 1996). Bioremediation of arsenic from mine tailings containing 3290 mg As/kg and sediments containing 339 mg As/kg from a Korean gold mine using introduced strains of sulfur-oxidizing bacteria in a bioleaching process is possible under acidic (<pH 4.0) conditions; however, costs were excessive (Lee et al. 2003). A cost-effective alternative is the use of indigenous bacteria under anaerobic conditions and various carbon sources (Lee et al. 2003).

As will be discussed later, roasting of some types of gold-containing ores to remove sulfur resulted in significant atmospheric emissions of arsenic trioxide (As_2O_3) and sulfur oxides (Ripley et al. 1996). Arsenic previously used to be extracted as a by-product in many gold mines and sold mainly for the manufacture of pesticides; however, this use is no longer profitable (Azcue et al. 1994). In Fairbanks, Alaska, some groundwaters are still contaminated with arsenic originating from gold mining activities 30 years earlier and are considered unsafe for drinking; bacteria associated with arsenic in mine drainage may accelerate the rate at which arsenic leaches from the sediment into groundwater (Pain 1987).

Refractory gold ores are those that are not free milling and require pretreatment prior to cyanide leaching (Adams et al. 1999). In most refractory ores, gold is locked in sulfides or is substituted in the sulfide mineral lattice. Commercial treatment of these ores involves roasting to destroy the sulfide minerals and liberate the gold, the calcine being treated by conventional cyanidation. In the treatment of ores containing arsenopyrite, environmental contamination may occur due to release of sulfur dioxide and arsenic trioxide:

$$2FeAsS + 5O_2 \rightarrow 2SO_2 + Fe_2O_3 + As_2O_3$$

In Canada, roasting has been largely discontinued; however, at least three operating facilities in that country were still using this practice in 1992 (Ripley et al. 1996). In Ghana, arsenic trioxides and other arsenic oxides from roasting of gold ores that were lost to the atmosphere were subsequently deposited in rainfall, causing extensive arsenic contamination of soil, vegetation, crops, humans, rivers, and livestock (Golow et al. 1996). Despite pollution aspects, roasting is still recommended as the most cost effective method for the treatment of refractory gold ores (Adams et al. 1999). To reduce arsenic emissions, new processes have been developed for the treatment of refractory ores. These include pressure-oxidation, bio-oxidation, whole ore roasting, ultra-fine grinding, nitric acid oxidation, and fine milling combined with low pressure oxidation. In whole ore roasting, pressure oxidation, and bio-oxidation, arsenic is fixed as basic ferric arsenate instead of As_2O_3 (Adams et al. 1999). Other operations have extracted the arsenic through flotation, cycloning, alkaline chlorination, ferric ion precipitation, bioleaching and bacterial oxidation, and pressure oxidation using an autoclave (Ripley et al. 1996).

Bacterial decomposition of arsenopyrite assists in opening the molecular mineral structure, allowing access of the gold to cyanide. Arsenic can become a limiting factor in the bioleaching of arsenopyrite for the recovery of gold at high temperatures owing to the formation of soluble As^{+3} and As^{+5}, and their toxicity, especially that of As^{+3}, to strains of bacteria that were not resistant to arsenic (Hallberg et al. 1996). Bio-oxidation of difficult to treat gold-bearing arsenopyrite ores is now done commercially in aerated, stirred tanks and with rapidly growing, arsenic-resistant bacterial strains of *Thiobacillus* spp., *Sulfolobus* sp., and *Leptospirullium* sp. (Ngubane and Baecker 1990; Agate 1996; Rawlings 1998). These obligate chemoautolithotrophic strains of bacteria obtain their energy through the oxidation of ferrous to ferric iron or through the reduction of inorganic sulfur compounds to sulfate.

Arsenic is often found as a mineral in combination with iron and sulfur. Oxidation of these insoluble forms results in the formation of arsenite (As^{+3}). In environments such as acid mine drainage of abandoned gold mines, As^{+3} concentrations ranged from 2 to 13 mg/L (Santini et al. 2000). The As^{+3} can then be oxidized to arsenate (As^{+5}). Both these soluble forms of arsenic are toxic to living organisms, especially inorganic arsenite. The chemical oxidation of arsenite to arsenate is slow compared with microbiological processes (Santini et al. 2000). Some species of bacteria protect against arsenic by reducing As^{+5} that has entered the cell to As^{+3} and then transporting As^{+3} out of the cell; however, arsenate reduction does not seem to support growth.

12.2 ARSENIC RISKS TO HUMAN HEALTH

Beneficial uses of arsenic compounds in medicine have been known for at least 2400 years. Inorganic arsenicals have been used for centuries, and organoarsenicals for at least a century in the treatment of syphilis, yaws, amoebic dysentery, asthma, tuberculosis, leprosy, dermatoses, and trypanosomiasis (Asperger and Ceina-Cizmek 1999; Eisler 2000). The advent of penicillin and other newer drugs nearly eliminated the use of organic arsenicals as human therapeutic agents, although arsenical drugs are still used in treating African sleeping sickness and amoebic dysentery and in veterinary medicine to treat filariasis in dogs and blackhead in poultry (Eisler 2000).

By contrast, arsenic contamination of the environment, even at low levels of exposure, has potential human health hazards, including skin cancer, stomach cancer, respiratory tract cancer, hearing and vision impairment, melanosis, leucomelanosis, keratosis, hyperkeratosis, edema, gangrene, and extensive liver damage (Kabir and Bilgi 1993; Kusiak et al. 1993; Simonato et al. 1994; Huang and Dasgupta 1999; Matschullat et al. 2000). Arsenic-contaminated drinking water is a major health problem in Bangladesh and other parts of the Indian subcontinent as a result of arsenic-bearing sediments in contact with the aquifer. Ironically, the use of groundwater for drinking water was implemented to eliminate waterborne pathogens; this effort was initiated by international organizations led by the United Nations (Huang and Dasgupta 1999; Eisler 2000).

Canadian gold miners had an excess of mortality from carcinoma of the stomach and respiratory tract when compared with other miners. The increased frequency of stomach cancer appeared 5 to 19 years after they began gold mining in Ontario (Kusiak et al. 1993). A number of explanations are offered to account for the high death rate, including exposure to arsenic (Kusiak et al. 1993). Gold miners in Ontario with 5 or more years of gold mining experience before 1945 had a significantly increased risk of primary cancer of the trachea, bronchus, and lung (Kabir and Bilgi 1993). A minimum latency period of 15 years was recorded between first employment and diagnosis of lung cancer. Underground miners were exposed to air concentrations of 2.4 to 5.6 µg As/m^3 and had significantly elevated concentrations of arsenic in urine. For purposes of work-relatedness, it was concluded that arsenic exposure was one of several causes of primary lung cancer in Ontario gold miners (Kabir and Bilgi 1993).

In France, a high incidence of neoplasms of the respiratory system among gold extraction and refinery workers was first reported in 1977, and again in 1985, and appears related to occupational exposure (Simonato et al. 1994). Statistics showed that mine and smelter workers at this very same site were twice as likely as the general population to die of lung cancer. The lung cancer excess was strongly associated with exposure to soluble and insoluble forms of arsenic (Simonato et al. 1994). In Zimbabwe, arsenic exposure was implicated in the increase of lung cancer among gold miners (Boffetta et al. 1994).

Active gold mining in the state of Minas Gerais, Brazil, has been documented since the early 1700s (Matschullat et al. 2000). Three major gold deposits can be discerned within the volcanic sedimentary sequence of the Nova Lima group near the city of Belo Horizonte. In the 1990s, yearly gold production was around 6 metric tons extracted from about 1 million tons of ore. Most of the ores contained arsenopyrites with high potential for arsenic contamination. Although arsenic emissions from ore processing should be minimal because of modern control facilities, this was not the case here due to the overall poverty in the area. In addition, the local population used surface waters not only for fishing and gardening, but frequently as their drinking water. Sources of arsenic to the biosphere included weathering of mine wastes via erosion, dissolution of arsenic-contaminated soils and tailings into surface waters and sediments, and smelting activities that released arsenic into the air through oxidation of arsenopyrites. In April 1998, 126 school children of mean age 9.8 years (range 8.7 to 10.9) in this southeastern Brazilian mining district had low urinary levels of cadmium (mean 0.13, range 0.04 to 0.35 µg/L), partly elevated concentrations of mercury (mean 1.1, range 0.1 to 16.5 µg/L), and generally elevated to high concentrations of arsenic (mean 25.7, range 2.2 to 106.0 µg/L). Of the total population, 20% showed elevated arsenic concentrations associated with future adverse health effects. Arsenic concentrations were high in local surface waters, soils, sediments, and mine tailings (Table 12.1), with arsenic-contaminated drinking water as the probable causative factor of elevated arsenic in urine (Matschullat et al. 2000).

Residents of La Oraya, Peru, experienced respiratory problems caused by arsenic and sulfur dioxide emissions released from an area smelter that processed gold and other ores; a soil sample collected 4 km downwind of the smelter contained 12,600 mg/kg of surface arsenic as well as 22,000 mg/kg of lead and 305 mg/kg of cadmium (Da Rosa and Lyon 1997).

12.3   ARSENIC CONCENTRATIONS IN ABIOTIC MATERIALS AND BIOTA NEAR GOLD EXTRACTION FACILITIES

Arsenic is a relatively common element that occurs in air, water, soil, and all living tissues (Eisler 2000). It ranks 20th in abundance in the Earth's crust, 14th in seawater, and 12th in the human body. Arsenic is a teratogen and carcinogen that can traverse placental barriers and produce fetal death and malformations in many species of mammals. It is carcinogenic in humans, but evidence for arsenic-induced carcinogenicity in other mammals is scarce. Arsenic concentrations are usually low (<1.0 mg/kg FW) in most living organisms, but they are frequently elevated in marine biota, in which arsenic occurs as arsenobetaine and poses little risk to organisms or their consumers, and in plants and animals from areas that are naturally arseniferous or near anthropogenic sources (Eisler 2000).

Arsenic concentrations in samples collected near gold mining and processing facilities worldwide were elevated in sediments, sediment pore waters, water column, mine tailings, mine tailing drainage waters, soils, terrestrial plants (including edible plants used in human diets), aquatic plants, aquatic bivalve molluscs, terrestrial and aquatic insects, fishes, bird tissues, and human urine (Table 12.1; Eisler 2004). Inorganic arsenicals are considered more toxic than organic arsenicals and trivalent arsenite (As^{+3}) compounds more toxic than pentavalent arsenate (As^{+5}) compounds. Total arsenic, As^{+3}, and As^{+5} can now be measured under field conditions at a detection limit of 1 μg/L with a portable stripping voltammetric instrument using a gold film electrode (Huang and Dasgupta 1999).

Gold mining has been a major activity in Canada for more than a century (Azcue et al. 1994). Since 1921, Canada has ranked among the top three gold-producing nations. Abandoned gold mine tailings and waste rock contain large quantities of arsenic with high potential for adverse environmental effects. In one case, gold was extracted by underground mining between 1933 and 1964 near a lake located in northeastern British Columbia leaving tailings and waste rock 4.5 meters thick over 25 ha of land adjacent to the lake. The tailings contained >2000 mg As/kg, the lake sediments up to 1104 mg As/kg, and lake water up to 556 μg/L. The greatest proportion of arsenic in the sediment cores is associated with iron oxides and sulfides. Under aerobic conditions, the high concentrations of iron in the tailings were effective at limiting arsenic migration (Azcue et al. 1994).

Abnormally high concentrations of arsenic in sediment (max. 3090 mg As/kg DW) and water samples were documented in 1990–1991 from a watershed receiving gold mine effluent near Yellowknife, Northwest Territories, Canada (Bright et al.

Table 12.1   Arsenic Concentrations in Biota and Abiotic Materials Collected near Gold
Mining and Processing Facilities

Location and Sample	Concentration (mg total arsenic/kg Dry Weight [DW] or Fresh Weight [FW])[a]	Ref[b]
South America		
Brazil: April 1998; southeastern gold mining districts		
Schoolchildren, age 8–11 yr; urine	0.026 (0.002–0.106) FW	1
Surface waters	0.031 (0.004–0.35) FW	1
Soils	200–800 DW	1
Sediments	350 (22–3200) DW	1
Tailings	10,500 (300–21,000) DW	1
Columbia, stream sediments	Max. 6300 DW	2
Ecuador: 1988; dry season, downstream of cyanide-gold mining area		
Water: measured vs. recommended	0.002–0.264 FW vs. <0.19 FW	2
Sediments: measured vs. recommended	403–7700 DW vs. <17 DW	3
Peru: surface soils 4 km downwind of gold smelter	12,600 DW	4
North America		
British Columbia, Canada: site of underground gold mine; 1933–1964 (northeast shore of Jack of Clubs Lake)		
Tailings	>2000 DW	5
Lake sediments	Max. 1,104 DW	5
Lake water	Max. 0.56 FW	5
Nova Scotia, Canada: stream waters at Goldenville mine; upstream vs. at mine discharge	0.03–0.05 FW vs. 0.23–0.25 FW	6
Yellowknife, NWT, Canada: 1990–1991; subarctic lakes; watershed contaminated with arsenic from effluent of two gold mines over several decades		
Surface sediments (gold content maximum 6.75 mg/kg DW)	2,186 (22–3090) DW	7,8
Sediment pore waters	Max. 5.2 FW	8
Overlying water column	Max. 0.53–0.55 FW	7,8
United States		
Whitewood Creek, South Dakota (recipient of gold mine tailings 1876–1977) vs. reference site; 1987		
Sediments	764 DW vs. 18 DW	9
Aquatic insects, 4 species	73, 77, 278, and 625 DW vs. 1–16 DW	9
Whitewood Creek (arsenic impacted from gold tailings containing an estimated 270,000 t arsenic between 1920 and 1977) vs. reference site in Casper, Wyoming		
Sediments, 1989	1920 DW vs. 9 DW	10
House wren, *Troglodytes aedon*; 1997		
Eggs	<0.5 DW vs. <0.5 DW	10
Chicks		
Livers	2.9 (1.8–5.6) DW vs. <0.5 DW	10
Diet (benthic insects)	103.0 DW vs. <0.5 DW	10

Table 12.1 (continued)   Arsenic Concentrations in Biota and Abiotic Materials Collected Near Gold Mining and Processing Facilities

Location and Sample	Concentration (mg total arsenic/kg Dry Weight [DW] or Fresh Weight [FW])[a]	Ref[b]
Africa		
Ghana		
Near gold ore processing facility vs. reference sites; topsoil		
Total arsenic	50 DW vs. 3–10 DW	11
As^{+5}	35 DW vs. no data	11
As^{+3}	15 DW vs. 1–2 DW	11
Near gold ore-roasting facility (17 t arsenic discharged to atmosphere/d) vs. reference site		
Cooked foods, edible portions		
Cassava, *Manihot esculenta*	2.7 DW vs. 1.9 DW	12
Plantain, *Musa paradisiaca*	3.4 DW vs. 3.0 DW	12
Other cooked foods	2.4 DW vs. 1.4 DW	12
Oil palm fruit, *Elaeis guineensis*	Max. 5.9 DW vs. Max. 3.7 DW	12
Stargrass, *Eleusine indica*	11.3 DW vs. 6.7 DW	12
Water	5.2 (2.8–10.4) DW vs. no data (USEPA drinking water criterion, <0.01 FW)	12
Active gold mining town and environs; 14 sites; 1992–1993		
Soil	12.9 (2.1–48.9) DW	13
Plantain, edible portions	Max. 4.3 DW	13
Water fern, *Ceratopterus cornuta*; whole	9.1 (0.5–78.7) DW	13
Elephant grass, *Pennisetum purpureum*; whole	Max. 27.4 DW	13
Cassava, edible portions	Max. 2.6 DW	13
Mudfish, *Heterobranchus bidorsalis*; whole	Max. 2.7 DW	13
Tanzania; Serengeti National Park; drainage water from Lake Victoria gold field tailings	324 FW	14
Europe		
Poland and Czech Republic; 5 species of aquatic bryophytes collected spring-summer		
Ten sites draining an area with high arsenic mineralization	3.4 DW	15
Two sites as above in areas of former gold mining activities	19.4 DW	15
Twenty-two reference sites	0.8 DW	15
Korea		
Abandoned Au-Ag-Mo mine; maximum concentrations; Songcheon		
Tailings	20,140 DW	16
Farmland soil	496 DW	16
Cabbage	3.2 DW	16
Stream waters	0.64 FW	16

Table 12.1 (continued)   Arsenic Concentrations in Biota and Abiotic Materials Collected Near Gold Mining and Processing Facilities

Location and Sample	Concentration (mg total arsenic/kg Dry Weight [DW] or Fresh Weight [FW])[a]	Ref[b]
Abandoned Au-Ag-Cu-Zn mine; Dongil		
Tailings	8720 DW	17
Farm Soils	40 DW	17
Paddy soils	31 DW	17
Abandoned Au-Ag mine; Myungbong		
Tailings	5810 DW	17
Farm soils	92 DW	17
Paddy soils	129 DW	17
Malaysia		
Tributary that received gold mine effluents for at least 10 yr		
Sediments	147 DW	18
Bivalve molluscs; 3 species; soft parts; from sediments containing 6.3 mg As/kg DW (plus, in mg/kg DW, 3.4 Cu, 0.02 Hg, 0.7 Pb, and 27 Zn); no bivalves found in more heavily contaminated sediments	Max. 225 DW (plus 115 mg Cu/kg DW, 127 mg Zn/kg DW, and negligible concentrations of Cd, Pb, and Hg)	18

[a] Ranges in parentheses.
[b] References: 1, Matschullat et al. 2000; 2, Grosser et al. 1994; 3, Tarras-Wahlberg et al. 2000; 4, Da Rosa and Lyon 1997; 5, Azcue et al. 1994; 6, Wong et al. 1999; 7, Bright et al. 1994; 8, Bright et al. 1996; 9, Cain et al. 1992; 10, Custer et al. 2002; 11, Golow et al. 1996; 12, Amonoo-Neizer and Amekor 1993; 13, Amonoo-Neizer et al. 1996; 14, Bowell et al. 1995; 15, Samecka-Cymerman and Kempers 1998; 16, Lim et al. 2003; 17, Lee and Chon 2003; 18, Lau et al. 1998.

1994, 1996). Inorganic arsenic concentrations were maximal in water column, sediment particulates, and sediment pore water about 4 to 6 km downstream of the gold mine input. Arsenite (As^{+3}) was the predominant arsenical in sediment pore water, and arsenate (As^{+5}) was the primary dissolved arsenic species in water column samples. Water samples also contained a variety of methylated arsenicals; methylation of As^{+3} and As^{+5} compounds through biological and other processes reduces their toxicity. Particulate concentrations of arsenic comprised up to 70% of the total arsenic in the water column downstream of the gold mine discharge. The high concentrations of arsenicals in sediment pore water (max. 5.16 mg/L) and the overlying water (max. 547 µg/L) in dissolved form in areas distant from the input are attributable to remobilization from sediments through redox-related dissolution (Bright et al. 1994, 1996).

Soil contamination by gold mining operations tends to be localized and because of the phytotoxic effects of arsenic, not easily overlooked (O'Neill 1990). At Yellowknife, Canada, high concentrations of arsenic were measured in soils near a gold smelter: >21,000 mg/kg DW soil at 0.28 km from the smelter and 600 mg As/kg DW at a site 1 km distant. The tailings deposit also led to contamination of surrounding soils. Vegetation that grew in these contaminated areas usually contained low concentrations of arsenic except when soil levels were >1000 mg As/kg, which produced either phytotoxic effects in sensitive species or growth in a few tolerant genotypes.

Maximum acceptable concentrations of arsenic in soils used for food production or for soil in parks range between 10 and 40 mg As/kg DW in Europe and the United Kingdom (O'Neill 1990). Galbraith et al. (1995) state that soil arsenic concentrations in excess of 20 to 50 mg/kg are injurious to plant growth and development, and sensitive species may be affected by concentrations as low as 5 mg/kg; greater levels of these concentrations can lead to toxic responses that include root plasmolysis, necrosis of leaf tips, and seed germination failure. In arsenic-enriched areas, evergreen forests were replaced with bare ground devoid of vegetation, grasslands were dominated by weeds, and there was overall species impoverishment, including wildlife species (Galbraith et al. 1995). Phytoremediation of gold mining sites contaminated by arsenic using arsenic-tolerant plants, such as *Equisetum* spp., is recommended (Wong et al. 1999).

Arsenic contamination in Whitewood Creek, South Dakota, from a gold mine was assessed in aquatic insects and bed sediments over a 40-km reach (Cain et al. 1992). From 1876 to 1977, about 100 million tons of finely ground gold mine tailings were discharged via a small tributary into Whitewood Creek; the main contaminant was arsenic derived from arsenopyrites (May et al. 2001). Transport and deposition of the discharged tailings led to extensive downstream arsenic contamination of sediments and biota (Cain et al. 1992). In spring 1987, the maximum arsenic concentration in Whitewood Creek sediments was 764 mg/kg DW compared with 18 mg/kg at a reference site. For four species of aquatic insects, the maximum value was 625 mg As/kg DW (versus 16 for a reference site), with most arsenic concentrated in the exoskeleton (Cain et al. 1992). Insectivorous birds (house wren, *Troglodytes aedon*) feeding on these same species of aquatic insects near Whitewood Creek in 1997 had elevated arsenic concentrations in liver (maximum 5.6 mg As/kg DW) when compared to a reference site in Wyoming (<0.5 mg As/kg DW) (Custer et al. 2002).

In Ghana, where gold accounts for the largest proportion of foreign exchange, large quantities (17 tons daily) of arsenic are discharged into the atmosphere from a single roasting/smelting facility (Amonoo-Neizer and Amekor 1993). Total arsenic, pentavalent arsenate (As^{+5}), and trivalent arsenite (As^{+3}), were usually highest in soils near the gold ore processing facility (Table 12.1), with background levels attained 7 to 15 km from the site, depending on wind direction and velocity (Golow et al. 1996). Freshwaters in the vicinity of the smelter had grossly elevated concentrations of arsenic (mean 5.2 mg As/L; range 2.8 to 10.4 mg/L; Table 12.1), and were considered unfit for aquatic life, irrigation, and for human consumption.

In Korea, tailings from a gold–silver–molybdenum mine is the primary source of arsenic contamination in the soil–water system of the Songcheon mine area (Lim et al. 2003). In Malaysia, edible clams and mussels from a tributary receiving gold mine wastes contained up to 225 mg As/kg DW soft parts, a level that exceeded mandatory levels for arsenic set by the Malaysian Food Act of 1983 (Lau et al. 1998). Because arsenic enhances the toxicity of free cyanide to aquatic fauna (Leduc 1984), this knowledge needs to be incorporated into future arsenic risk assessments.

12.4 ARSENIC EFFECTS ON SENSITIVE SPECIES

Adverse effects of various arsenicals on sensitive species of organisms are documented (Table 12.2; Eisler 2000). The most sensitive of the aquatic species tested showing adverse effects were three species of marine algae, with reduced growth evident in the range of 19 to 22 μg As^{+3}/L; developing embryos of the narrow-mouthed toad (*Gastrophryne carolinensis*), of which 50% were dead or malformed in 7 days at 40 μg As^{+3}/L; and a freshwater alga (*Scenedesmus obliquis*), in which growth was inhibited 50% in 14 days at 48 μg As^{+5}/L. Adverse biological effects have also been documented at 75 to 100 μg As/L: growth reduction in freshwater and marine algae at 75 μg As^{+5}/L; 10% to 32% mortality in 28 days of a freshwater amphipod (*Gammarus pseudolimnaeus*) at 85 to 88 μg/L of As^{+5} or various methylated arsenicals; inhibition of sexual reproduction of marine algae at 95 μg As^{+3}/L; and death of marine copepods and impaired swimming ability of goldfish at 100 μg As^{+5}/L (Table 12.2; Eisler 2000).

Juvenile tanner crabs (*Chionoecetes bairdi*) held for 502 days on weathered gold mine tailings with elevated arsenic concentrations (29.7 mg As/kg DW) or reference sediments (2.5 mg As/kg DW) showed the same concentrations of arsenic in gill (8.9 vs. 9.8 mg As/kg DW) and muscle (8.9 vs. 8.1 mg As/kg DW) tissues (Stone and Johnson 1997). Female tanner crabs may initially avoid areas affected by submarine tailings but later recolonize the altered sea floor and incorporate lead, but not arsenic, into their tissues (Stone and Johnson 1998). In a 90-day study of ovigerous tanner crabs in forced contact with fresh gold mine tailings, survival and reproduction were normal, although egg survival was lower than among crabs held on control sediments, which was attributed to the action of lead; arsenic concentrations in muscle and ova were similar for those held on control and tailings sediments (Stone and Johnson 1998). Reduced food availability to ovigerous females due to smothering of the sea floor could result in reduced fecundity, poor larval survival, and increased susceptibility to disease (Johnson et al. 1998b).

Juvenile yellowfin sole (*Pleuronectes asper*) avoid fresh tailings (15 mg As/kg DW) in favor of natural marine sediments (7 mg As/kg DW), but when tailings are covered with 2 cm of control sediments, there is no significant avoidance of the covered fresh tailings (Johnson et al. 1998a). Growth was inhibited for sole held on fresh tailings for 30 days but not during days 30 to 60; survival was similar (90 to 93% survival) for fish held on all sediments (Johnson et al. 1998a).

Among terrestrial plants and invertebrates, yields of most crops decreased at soil arsenic levels of 3 to 28 mg water-soluble arsenic/L and 25 to 85 mg/kg of total arsenic; yields of peas (*Pisum sativum*) were decreased at 1 mg/L of water-soluble arsenic or 25 mg/kg total soil arsenic; soybeans (*Glycine max*) grew poorly when plant residues exceeded 1 mg As/kg DW; and earthworms (*Lumbricus terrestris*) held in soils containing 40 to 100 mg As^{+5}/kg DW soil for 23 days showed reduced survival, especially among worms held in soils <70 mm in depth when compared with worms held at 500 to 700 mm (Table 12.2; Eisler 2000, 2004).

Signs of inorganic trivalent arsenite poisoning in birds (muscular incoordination, debility, slowness, jerkiness, falling, hyperactivity, fluffed feathers, drooped eyelid, huddled position, unkempt appearance, loss of righting reflex, immobility, seizures)

Table 12.2   Lethal and Sublethal Effects of Various Arsenicals on Humans and Selected Species of Plants and Animals

Ecosystem, Species, Arsenic Compound, Dose, and Other Variables	Effect	Ref[a]
Freshwater Plants		
Algae; 4 species; As^{+3} (inorganic trivalent arsenite); 1.7–2.3 mg/L	95–100% fatal in 2–4 weeks	1,2
Algae; As^{+5} (inorganic pentavalent arsenate); 2 species; 0.048–0.26 mg/L	50% growth inhibition in 14 days	2
Freshwater Invertebrates		
Cladocerans		
Bosmina longirostris; As^{+5}; 0.85 mg/L	50% immobilization in 96 h	3
Daphnia magna		
As^{+5}; 0.52 mg/L	16% reproductive impairment in 3 weeks	2
As^{+3}; 0.63–1.32 mg/L	MATC[b]	2
As^{+5}; 7.4 mg/L	50% dead in 96 h	1
Daphnia pulex		
As^{+3}; 1.3 mg/L	50% dead in 96 h	1,2
As^{+3}; 3.0 mg/L	50% immobilized in 48 h	4
As^{+5}; 49.6 mg/L	50% immobilized in 48 h	3
Simocephalus serrulatus; As^{+3}; 0.81 mg/L	50% dead in 96 h	2
Amphipod, *Gammarus pseudolimnaeus*		
DSMA = disodium methylarsenate [$CH_3AsO(ONa)_2$]; 0.086 mg/L	10% dead in 28 days	5
As^{+3}; 0.088 mg/L	20% dead in 28 days	5
SDMA = sodium dimethylarsenate [$(CH_3)_2As(ONa)$]; 0.85 mg/L	No deaths in 28 days	5
As^{+3}; 0.96 mg/L	All dead in 28 days	5
As^{+5}; 0.97 mg/L	20% dead in 28 days	5
DSMA; 0.97 mg/L	40% dead in 28 days	5
Snail, *Helisoma campanulata*		
SDMA; 0.085 mg/L	No deaths in 28 days	5
As^{+3}; 0.96 mg/L	10% dead in 28 days	5
As^{+5}; 0.97 mg/L	No deaths in 28 days; maximum bioconcentration factor of 99	5
DSMA; 0.97 mg/L	No deaths in 28 days	5
Red crayfish, *Procambarus clarki*		
MSMA = monosodium methanearsonate [CH_4AsNaO_3]; 100 mg/L, equivalent to 46.3 mg As/L	No effect on growth or survival during exposure for 24 weeks but hatching success reduced to 17 vs. 78% for controls	6
MSMA; 1000 mg/L	50% dead in 96 h	6
Stoneflys		
Pteronarcys californica; As^{+3}; 38.0 mg/L	50% dead in 96 h	4
Pteronarcys dorsata		
DSMA, SDMA, or As^{+3}; 0.85–0.97 mg/L	No deaths in 28 days	5
As^{+5}; 0.97 mg/L	20% dead in 28 days	5
Freshwater Vertebrates		
Fishes		
Goldfish, *Carassius auratus*; As^{+5}; 0.1 mg/L	15% behavioral impairment in 24 h; 30% impairment in 48 h	7

Table 12.2 (continued)   Lethal and Sublethal Effects of Various Arsenicals on Selected
Species of Plants, Animals, and Humans

Ecosystem, Species, Arsenic Compound, Dose, and Other Variables	Effect	Ref[a]
Flagfish, *Jordanella floridae*		
As^{+3}; 2.1–4.1 mg/L	MATC[b]	2
As^{+3}; 14.4 mg/L	50% dead in 96 h	8
Fathead minnow, *Pimephales promelas*		
As^{+5}; 0.53–1.50 mg/L	MATC[b]	2
As^{+3}; 2.1–4.8 mg/L	MATCb[b]	8
As^{+3}; 14.1 mg/L	50% dead in 96 h	8
As^{+5}; 25.6 mg/L	50% dead in 96 h	2
Rainbow trout, *Oncorhynchus mykiss*		
As^{+3}; 0.54 mg/L	Embryos: 50% dead in 28 days	1
DSMA or SDMA; 0.85–0.97 mg/L	No deaths in 28 days	5
As^{+3}; 0.96 mg/L	50% dead in 28 days	4,9
As^{+3}; 23.0–26.6 mg/L	Adults: 50% dead in 28 days	5
Sodium cacodylate (SC); 1000 mg/L	No deaths in 28 days	11
As^{+5}; 10–90 mg/kg diet for 16 weeks	No effect level at about 10 mg/kg diet. Some adaptation to 90 mg/kg diet as initial negative growth gave way to slow positive growth over time	10
DSA = Disodium arsenate heptahydrate; 13–33 mg As as DSA/kg ration for 12–24 weeks (0.28–0.52 mg As/kg body weight [BW] daily)	MATC[b]	5
As^{+3} or As^{+5}; 120–1600 mg/kg diet for 8 weeks	Growth depression, food avoidance, and impaired feed efficiency at all levels	10
DMA = dimethyl arsinic acid, or ABA = *p*-amino-benzenearsonic acid; 120–1600 mg/kg diet for 8 weeks	No toxic response at any level tested	10
Amphibians		
Marbled salamander, *Ambystoma opacum*; As^{+3}; 4.5 mg/L	Developing embryos: 50% dead or malformed in 8 days	2
Narrow-mouthed toad, *Gastrophryne carolenisis*; As^{+3}; 0.04 mg/L	Developing embryos: 50% dead or malformed in 7 days	2
Marine Plants		
Algae, 3 species; As^{+3}; 0.019–0.022 mg/L	Reduced growth	2
Red alga, *Champia parvula*		
As^{+3}; 0.065 mg/L	Normal sexual reproduction	12
As^{+3}; 0.095 mg/L	No sexual reproduction	12
As^{+3}; 0.300 mg/L	Death	12
As^{+5}; 10.0 mg/L	Normal growth but no sexual reproduction	12
Phytoplankton; As^{+5}; 0.075 mg/L	Reduced biomass of populations in 4 days	2
Alga, *Skeletonema costatum*; As^{+5}; 0.13 mg/L	Growth inhibition	2
Marine Invertebrates		
Copepod, *Acartia clausi*; As^{+3}; 0.51 mg/L	50% dead in 96 h	2
Copepod, *Eurytermora affinis*		
As^{+5}; 0.1 mg/L	Reduced juvenile survival	13
As^{+5}; 1.0 mg/L	Reduced adult survival	13

Table 12.2 (continued)   Lethal and Sublethal Effects of Various Arsenicals on Selected Species of Plants, Animals, and Humans

Ecosystem, Species, Arsenic Compound, Dose, and Other Variables	Effect	Ref[a]
Dungeness crab, *Cancer magister*; As^{+3}; 0.23 mg/L	Zoea: 50% dead in 96 h	2
Pacific oyster, *Crassostrea gigas*; As^{+3}; 0.33 mg/L	Embryos: 50% dead in 96 h	2
Mysid, *Mysidopsis bahia*		
As^{+3}; 0.63–1.27 mg/L	MATC[b]	2
As^{+5}; 2.3 mg/L	50% dead in 96 h	2

Marine Fishes

Marine fishes, 3 species; As^{+3}; 12.7–16.0 mg/L	50% dead in 96 h	2
Pink salmon, *Oncorhynchus gorbuscha*		
As^{+3}; 2.5 mg/L	No deaths in 10 days	9
As^{+3}; 3.8 mg/L	54% dead in 10 days	2
As^{+3}; 7.2 mg/L	All dead in 7 days	2

Terrestrial Plants

Crops		
Total water soluble arsenic; 3–28 mg/kg soil	Depressed crop yield	7
Total arsenic; 25–85 mg/kg soil	Depressed crop yield	7
Common bermudagrass, *Cynodon dactylon*; As^{+3}; arsenic-amended soils containing up to 90 mg/kg soil	Arsenic residues were up to 17 mg/kg dry weight [DW] in stems, 20 in leaves, and 304 in roots	14
Soybean, *Glycine max*; total arsenic; >1 mg/kg DW plant	Toxic signs	7
Rice, *Oryza sativa*; DSMA; 50 mg/kg soil	75% decrease in yield	7
Scots pine, *Pinus sylvestris*		
As^{+5}; >62 mg/kg shoots DW	Toxic	15
As^{+5}; >250 mg/kg soil DW	Seedlings die	15
As^{+5}; >3300 mg/kg shoots DW	Fatal	15
Pea, *Pisum sativum*; As^{+3}; 15 mg/L	Inhibition of light activation and photosynthetic CO_2 fixation in chloroplasts	19
Grasslands; CA = cacodylic acid [$(CH_3)_2AsO(OH)$]; 17 kg/ha	75–90% of all species killed; recovery modest	11
Sandhill plant communities		
CA; 2.25 kg/ha	No lasting effect	11
CA; 6.8 kg/ha	Some species defoliated	11
CA; 34.0 kg/ha	75% defoliation of oaks and death of all pine trees	11

Terrestrial Invertebrates

Beetles; CA; dietary levels of 100–1000 mg/kg	Fatal to certain pestiferous species	16
Western spruce budworm, *Christoneura occidentalis*; sixth instar larvae		
As^{+3}; 99.5 mg/kg ration fresh weight [FW]	Fatal to 10%	17
As^{+3}; 2250 mg/kg ration FW	Fatal to 50%	17
As^{+3}; 65,300 mg/kg ration FW	Fatal to 90%	17

Table 12.2 (continued)   Lethal and Sublethal Effects of Various Arsenicals on Selected Species of Plants, Animals, and Humans

Ecosystem, Species, Arsenic Compound, Dose, and Other Variables	Effect	Ref[a]
As[+3]; 100–65,300 mg/kg ration FW	Newly-molted pupae and adults of As-exposed larvae had reduced weight. Regardless of dietary levels, concentrations of As ranged up to 2640 mg/kg DW in dead pupae and 1708 mg/kg DW in adults	
Earthworm, *Lumbricus terrestris*		
As[+5]; 40 mg/kg DW soil; exposure for 23 days	No accumulations in first 12 days, with bioconcentration factor [BCF] of 3 by day 23	18
As[+5]; 100 mg/kg DW soil	Fatal to 50% in 8 days	18
As[+5]; 400 mg/kg DW soil	Fatal to 50% in 2 days	18

Birds

Mallard, *Anas platyrhynchos*		
Adult breeding pairs; As[+5]; fed diets with 0, 25, 100, or 400 mg/kg ration for up to 173 days. Ducklings produced were fed the same diet as their parents for 14 days	Dose-dependent increase in liver arsenic from 0.23 mg As/kg DW in controls to 6.6 in the 400 mg/kg group and in eggs from 0.23 in controls to 3.6 mg/kg DW in the 400 mg/kg group. Dose-dependent adverse effects on growth, onset of egg laying, and eggshell thinning. In ducklings, arsenic accumulated in the liver from 0.2 mg As/kg DW in controls to 33.0 in the 400 mg/kg group and caused a dose-dependent decrease in growth rate of whole body and liver	20
Ducklings; As[+5]; fed 30, 100, or 300 mg/kg diet for 10 weeks	All treatments produced elevated hepatic glutathione and ATP concentrations and decreased overall weight gain and rate of growth in females. Arsenic concentrations were elevated in brain and liver of ducklings fed 100 or 300 mg/kg diet; all ducklings had altered behavior, e.g., increased resting time; males had reduced growth	21
Day-old ducklings; As[+5]; fed diets containing 200 mg/kg ration for 4 weeks	When protein was adequate (22%), some growth reduction resulted. With only 7% protein in diet, growth and survival was reduced and frequency of liver histopathology increased	22
Adult males; As[+5]; fed rations containing 300 mg/kg	Equilibrium reached in 10–30 days; 50% loss from liver in 1–3 days on transfer to an uncontaminated diet	23
As[+3]; 323 mg As[+3]/kg BW	Acute oral LD50	7, 9, 24
As[+3]; 500 mg/kg diet	Fatal to 50% in 32 days	2
As[+3]; 1000 mg/kg diet	Fatal to 50% in 6 days	2
CA; 1740–5000 mg/kg diet	Fatal to 50% in 5 days	11
California quail, *Callipepla californica*; As[+3]; 47.6 mg/kg BW	Acute oral LD50	24

Table 12.2 (continued)   Lethal and Sublethal Effects of Various Arsenicals on Selected Species of Plants, Animals, and Humans

Ecosystem, Species, Arsenic Compound, Dose, and Other Variables	Effect	Ref[a]
Common bobwhite, *Colinus virginianus*		
SC; 1740 mg/kg diet for 5 days	No effects on behavior, no signs of intoxication, negative necropsy	11
MSMA; 3300 mg/kg BW	Acute oral LD50	11
Chicken, *Gallus gallus*		
As^{+3}; 0.01–1.0 mg/embryo	Up to 34% dead; malformation threshold at 0.03–0.3 mg/embryo	7
As^{+5}; 0.01–1.0 mg/embryo	Up to 8% dead	7
As^{+5}; 0.3–3.0 mg/embryo	Malformation threshold	7
DSMA; 1–2 mg/egg	Teratogenic when injected	7,11
SC; 1–2 mg/egg	Developmental abnormalities when injected	11
DC = dodecylamine p-chlorophenylarsonate; 23.3 mg/kg diet for 9 weeks	Liver residues were 2.9 mg/kg FW at end; no ill effects noted	25
CA; 100 mg/kg BW	No adverse effects at daily oral dosing for 10 days	11
Ring-necked pheasant, *Phasianus colchicus*; As^{+3}; 363 mg/kg BW	Acute oral LD50	24

Nonhuman Mammals

Cattle, *Bos* spp.		
As^{+5}; fed 33 mg daily per animal for 33 months	Elevated levels of arsenic in muscle (0.02 mg/kg FW vs. 0.005 in controls) and liver (0.03 vs. 0.012) but normal levels in milk and kidney	26
As^{+3}; fed 33 mg daily per animal for 15–28 months	Elevated arsenic levels, in mg/kg FW, of 0.002 for milk (vs. 0.001 for controls), 0.03 for muscle (vs. 0.005), 0.1 for liver (vs. 0.012), and 0.16 for kidney (vs. 0.053)	26
As^{+3}; single oral dose of 15–45 g per animal, as arsenic trioxide	Fatal	7
As^{+3}; single oral dose of 1–4 g per animal, as sodium arsenite	Fatal	7
MSMA; 10 mg/kg BW daily for 10 days	Fatal	7
As^{+3}; 33-55 mg/kg BW, or 13.2–22 g for a 400-kg animal; topical application	Arsenic-poisoned cows contained up to 15 mg As/kg FW liver, 23 in kidney, and 45 in urine (vs. <1 for all normal tissues)	27
CA or MAA (methanearsonic acid); calves fed diets containing 4000–4700 mg/kg ration	Appetite loss in 3–6 days	11
CA: adults given oral dose of 10 mg/kg BW daily for 3 weeks, followed by 20 mg/kg BW daily for 5–6 weeks	Lethal	11
CA; adults given oral dose of 25 mg/kg BW daily for 10 days	Adverse effects	11
Dog, *Canis familiaris*		
CA or MMA; 30 mg/kg diet for 90 days	No adverse effects	11
CA; 1000 mg/kg BW	Oral LD50	11
As^{+3}; 50–150 mg	Fatal	7

Table 12.2 (continued)   Lethal and Sublethal Effects of Various Arsenicals on Selected
Species of Plants, Animals, and Humans

Ecosystem, Species, Arsenic Compound, Dose, and Other Variables	Effect	Ref[a]
Guinea pig, *Cavia* sp.; As^{+3} as arsenic trioxide; fed diet containing 50 mg/kg for 21 days	Elevated arsenic residues, in mg/kg FW, of 4 in blood and 15 in heart vs. <1 for controls	25
Hamster, *Cricetus* sp.		
As^{+5}; maternal dose of 5 mg/kg BW	Some fetal deaths, but no malformations	7
As^{+5}; maternal dose of 20 mg/kg BW	54% fetal deaths and malformations	7
As^{+5}, as sodium arsenate; dosed intravenously on day 8 of gestation		
2 mg/kg BW	No measurable effect	28
8 mg/kg BW	Increased incidence of malformation and resorption	28
16 mg/kg BW	All embryos died	28
SC; single intraperitoneal injection of 900–1000 mg/kg BW during mid-gestation	Some maternal deaths and increased incidences of fetal malformations	11
Horse, *Equus caballus*; As^{+3}; 2–6 mg/kg BW daily (1–3 g of sodium arsenite)	Fatal in 14 weeks	7
Cat, *Felis domesticus*; As^{+3} or As^{+5}; 1.5 mg/kg BW daily	Chronic oral toxicity	28
Mammals, representative species		
Calcium arsenate; 35–1000 mg/kg BW	Single oral LD50 range	7
Lead arsenate; 10–50 mg/kg BW	Single oral LD50 range	7
As^{+3}, as arsenic trioxide; 3–250 mg/kg BW	Lethal	9
As^{+3}, as sodium arsenite; 1–25 mg/kg BW	Lethal	9
Mouse, Mus spp.		
As^{+5}; maternal dose of 10 mg/kg BW	Some fetal deaths and malformations	7
As^{+5}; 20–50 mg/kg BW; pregnant mice, day 18 of gestation	No deaths or abortions at lower dose when administered intraperitoneally, or higher dose when given orally. Residue half-life was 10 h regardless of route	32
As^{+3}, as arsenic trioxide		
10.4 mg/kg BW	Oral LD0 in 96 h	9
39.4 mg/kg BW	Oral LD50 in 96 h	9
0.26 mg/m^3 air for 4 h daily on days 9–12 of gestation	3.1% decrease in fetal weight	29
2.9 mg/m^3 air for 4 h daily on days 9–12 of gestation	9.9% decrease in fetal weight	29
28.5 mg/m^3 air for 4 h daily on days 9–12 of gestation	Fetotoxic effects (reduced survival, impaired growth, retarded limb ossification, bone abnormalities) and chromosomal damage to liver cells by day 18	29
As^{+5}, as sodium arsenate; 0.5 mg/L drinking water for up to 26 months, equivalent to 0.07-0.08 mg/kg BW daily	No tumors in controls vs. 41.1% of mice in treated groups with 1 or more tumors, mostly of the lung, liver and GI tract	33
As^{+3}, as sodium arsenite; 5 mg/kg diet for 3 generations	Reduced litter size, but outwardly normal	28
As^{+3}, as sodium arsenite; 9.6–11.3 mg/kg BW via subcutaneous injection	Lower dose is LD50; higher dose is LD90 7 days postexposure	34
As^{+3}, as sodium arsenite; 10–12 mg/kg BW via intraperitoneal route	Lower dose causes damage to bone marrow and sperm; higher dose is LD50	35

Table 12.2 (continued)   Lethal and Sublethal Effects of Various Arsenicals on Selected Species of Plants, Animals, and Humans

Ecosystem, Species, Arsenic Compound, Dose, and Other Variables	Effect	Ref[a]
Single oral dose		
Arsenous oxide; 34 mg As/kg BW	LD50	31
Tetramethylarsonium iodide; 890 mg As/kg BW	LD50	31
Arsenocholine; 6500 mg As/kg BW	LD50	31
Arsenobetaine; >100,000 mg As/kg BW	LD50	31
DMA: 200-600 mg/kg BW daily for 10 days	Fetal and maternal toxicity	30
CA; oral dosages of 400–600 mg/kg BW on days 7–16 of gestation	Fetal malformations (cleft palate), delayed skeletal ossification, and fetal weight reduction	11
SC; 1200 mg/kg BW during mid-gestation via intraperitoneal injection	Increased rates of fetal skeletal malformations	11
Rabbit, *Oryctolagus* sp.; MMA; 50 mg/kg ration for 7–12 weeks	Hepatotoxicity	30
Domestic sheep, *Ovis aries*		
As^{+3}, as sodium arsenite; single oral dose of 5–12 mg/kg BW (0.2–0.5 g)	Acutely toxic	7
As^{+5}, as soluble arsenic; lambs fed diets containing 2 mg As/kg supplemental arsenic for 3 months	Maximum arsenic concentrations, in mg/kg FW, were 2 in brain (vs. 1 in controls), 14 in muscle (2), 24 in liver (4), and 57 in kidney (10)	36
Total arsenic; diets contained lakeweed (*Lagarosiphon major*) (288 mg As/kg DW) at 58 mg total As/kg diet for 3 weeks	No ill effects. Tissue residues increased during feeding, but rapidly declined when lakeweed was removed from diet	25
Rat, *Rattus* spp.		
Arsanilic acid; 17.5 mg/kg diet for 7 generations	No teratogenesis observed; positive effect on litter size and survival	28
As^{+5}; fed diets containing 50 mg/kg for 10 weeks	No effect on serum uric acid levels	37
As^{+3}, as arsenic trioxide; single oral dose of 15.1 mg/kg BW	LD50 (96 h)	9
As^{+3}, as arsenic trioxide; fed diets with 50 mg/kg for 21 days	Tissue arsenic levels elevated in blood (125 mg/L vs. 15 in controls), heart (43.0 mg/kg FW vs. 3.3), spleen (60.0 vs. 0.7), and kidney (25.0 vs. 1.5)	25
As^{+3}; oral administration of 12 mg/kg BW daily for 6 weeks	Serum uric acid levels reduced 67%	37
As^{+3}; 10 mg/L in drinking water for 7 months	Urinary metabolites were mainly methylated arsenic metabolites with about 6% in inorganic form	38
Arsenobetaine; 100 mg As/L drinking water for 7 months	Eliminated in urine unchanged without transformation	38
Cacodylic acid (CA); pregnant rats dosed by gavage at 50-60 mg/kg BW daily during gestation days 6–13	Maternal deaths and fetal deaths and abnormalities noted	11
Dimethylarsinic acid (DMA); 100 mg/L in drinking water for 7 months	Main metabolites in urine were DMA and trimethylarsin oxide (TMAO) with minute amounts of tetramethylarsonium (TMA)	38
DMA; 40–60 mg/kg BW daily for 10 days	Fetal and maternal toxicity	30

Table 12.2 (continued)   Lethal and Sublethal Effects of Various Arsenicals on Selected Species of Plants, Animals, and Humans

Ecosystem, Species, Arsenic Compound, Dose, and Other Variables	Effect	Ref[a]
Monomethylarsonic acid (MMA); 200 mg/L in drinking water for 7 months	Main products in urine were unchanged MMA, DMA, and small amounts of TMA and TMAO	38
Rodents, various species		
Cacodylic acid (CA); 470–830 mg/kg BW	LD50 range by various routes of administration	11
Sodium cacodylate (SC); 600–2600 mg/kg BW	LD50 range, various routes of administration	
Cotton rat, *Sigmodon hispidus*; As[+3] as sodium arsenite; adult males given 0, 5, or 10 mg/L in drinking water for 6 weeks	Dose-dependent decrease in daily food intake. Minimal effects on immune function, tissue weights, and blood chemistry	39
Pig, *Sus* sp.		
As[+3], as sodium arsenite; 500 mg/L in drinking water	Lethal when arsenic residues ranged from 100–200 mg/kg BW	9
3-nitro-4-hydroxyphenylarsonic acid; 100–250 mg/kg diet	Arsenosis documented after 2 months on diets containing 100 mg/kg, or after 3–10 days on diets containing 250 mg/kg	9
Human Health		
As[+5]; 3.5 mg daily for 1 month	12,000 Japanese infants accidentally poisoned (128 deaths) from consumption of dry milk contaminated with arsenic. Post-exposure effects (15 years later) included severe hearing loss, brain wave abnormalities, and other CNS disturbances	28
As[+3] as arsenic trioxide		
1–2.6 mg/kg BW (70–189 mg)	Some deaths	7
7 mg/kg BW	LD50	7
CA; 1350 mg/kg BW	LD50	7
Total arsenic; 1–3 mg/kg BW daily for 3 months in children or 80 mg kg/BW daily for 3 months in adults	Symptoms of chronic arsenic poisoning	7
Total arsenic in drinking and cooking water; prolonged use		
0.29 mg/L	Skin cancer	7
0.6 mg/L	Chronic arsenic intoxication	7
Total inorganic arsenic; 3 mg daily for 2 weeks	May cause severe poisoning in infants and symptoms of toxicity in adults	28

[a] 1, USEPA 1980a; 2, USEPA 1985; 3, Passino and Novak 1984; 4, Johnson and Finley 1980; 5, Spehar et al. 1980; 6, Naqvi and Flagge 1990; 7, NRCC 1978; 8, Lima et al. 1984; 9, NAS 1977; 10, Cockell and Hilton 1985; 11, Hood 1985; 12, Thursby and Steele 1984; 13, Sanders 1986; 14, Wang et al. 1984; 15, Sheppard et al. 1985; 16, Jenkins 1980; 17, Robertson and McLean 1985; 18, Meharg et al. 1998; 19, Marques and Anderson 1986; 20, Stanley et al. 1994; 21, Camardese et al. 1990; 22, Hoffman et al. 1992; 23, Pendleton et al. 1995; 24, Hudson et al. 1984; 25, Woolson 1975; 26, Vreman et al. 1986; 27, Robertson et al. 1984; 28, Pershagen and Vahter 1979; 29, Nagymajtenyi et al. 1985; 30, Hughes and Kenyon 1998; 31, Hamasaki et al. 1995; 32, Hood et al. 1987; 33, Ng et al. 1998; 34, Stine et al. 1984; 35, Deknudt et al. 1986; 36, Veen and Vreman 1986; 37, Jauge and Del-Razo 1985; 38, Yoshida et al. 1998; 39, Savabieasfahani et al. 1998.

Table 12.2 (continued)   Lethal and Sublethal Effects of Various Arsenicals on Selected Species of Plants, Animals, and Humans

[b] MATC = maximum acceptable toxicant concentration. Lower value in each pair indicates highest concentration tested producing no measurable effect on growth, survival, reproduction, or metabolism during chronic exposure; higher value indicates lowest concentration tested producing a measurable effect.

were similar to those induced by many other toxicants and did not seem to be specific for arsenosis. Signs occurred within 1 hour of arsenite administration and deaths within 1 to 6 days postadministration; remission took up to 1 month (Hudson et al. 1984). Internal examination suggested that lethal effects of acute inorganic arsenic poisoning were due to the destruction of the blood vessels lining the gut, which resulted in decreased blood pressure and subsequent shock (Nystrom 1984). Mallard ducklings fed a diet that contained 30 mg As/kg ration had reduced growth and altered physiology, and those fed a diet containing 300 mg As/kg had disrupted brain biochemistry and nesting behavior; decreased energy levels and altered behavior can further decrease duckling survival in a natural environment (Table 12.2; Camardese et al. 1990).

In mammals, arsenic uptake may occur by ingestion (the most likely route), inhalation, and absorption through the skin and mucous membranes. Soluble arsenicals are absorbed more rapidly and completely than are the sparingly soluble arsenicals, regardless of route of administration (National Research Council of Canada [NRCC] 1978). In humans, inorganic arsenic at high concentrations is associated with adverse reproductive outcomes, including increased rates of spontaneous abortion, low birth weight, congenital malformations, and death (Hopenhayn-Rich et al. 1998). However, at environmentally relevant levels and routes of exposure, humans are not at risk for birth defects due to arsenic (Holson et al. 1998). *In vitro* tests with human erythrocytes demonstrate that inorganic As^{+5} as sodium arsenate was as much as 1000 times more effective than inorganic As^{+3} as sodium arsenite after exposure to 750 mg As/L in causing death, morphologic changes, and ATP depletion (Winski and Carter 1998).

Acute episodes of poisoning in warm-blooded organisms by inorganic and organic arsenicals are usually characterized by high mortality and morbidity over a period of 2 to 3 days (National Academy of Sciences [NAS] 1977; Selby et al. 1977). General signs of arsenic toxicosis include intense abdominal pain, staggering gait, extreme weakness, trembling, salivation, vomiting, diarrhea, fast and feeble pulse, prostration, collapse, and death. Gross necropsy shows a reddening of gastric mucosa and intestinal mucosa, a soft yellow liver, and red edematous lungs, Histopathological findings show edema of gastrointestinal mucosa and submucosa, necrosis and sloughing of mucosal epithelium, renal tubular degeneration, hepatic fatty changes and necrosis, and capillary degeneration in the gastrointestinal tract, vascular beds, skin, and other organs.

In subacute episodes, in which animals live for several days, signs of arsenosis include depression, anorexia, increased urination, dehydration, thirst, partial paralysis of rear limbs, trembling, stupor, coldness of extremities, and subnormal body temperatures (NAS 1977; Selby et al. 1977; U.S. Public Health Service [USPHS] 2000). In cases involving cutaneous exposure to arsenicals, a dry, cracked, leathery,

and peeling skin may be a prominent feature (Selby et al. 1977). Nasal discharges
and eye irritation were documented in rodents exposed to organoarsenicals in inhalation toxicity tests (Hood 1985). Subacute effects in humans and laboratory animals
include peripheral nervous disturbances, melanosis, anemia, leukopenia, cardiac
abnormalities, and liver changes. Most adverse signs rapidly disappear after exposure
ceases (Pershagen and Vahter 1979).

Research results on arsenic poisoning in mammals (see Table 12.2; Eisler 2000,
2004) show general agreement on eight points:

1. Arsenic metabolism and effects are significantly influenced by the organism tested,
   the route of administration, the physical and chemical form of the arsenical, and
   the dose.
2. Inorganic arsenic compounds are more toxic than organic arsenic compounds, and
   trivalent species are more toxic than pentavalent species.
3. Inorganic arsenicals can cross the placenta in most species of mammals.
4. Early developmental stages are the most sensitive, and humans appear to be one
   of the more susceptible species.
5. Animal tissues usually contain low levels (<0.3 mg As/kg fresh weight) of arsenic.
   After the administration of arsenicals, these levels are elevated, especially in liver,
   kidney, spleen, and lung; and several weeks later, arsenic is translocated to ecto-
   dermal tissues (hair, nails) because of the high concentration of sulfur-containing
   proteins in these tissues.
6. Inorganic arsenicals are oxidized *in vivo*, biomethylated, and usually excreted
   rapidly in the urine, but organoarsenicals are usually not subject to similar transformations.
7. Acute or subacute arsenic exposure can lead to elevated tissue residues, appetite
   loss, reduced growth, loss of hearing, dermatitis, blindness, degenerative changes
   in liver and kidney, cancer, chromosomal damage, birth defects, and death.
8. Death or malformations have been documented at single oral doses of 2.5 to 33 mg
   As/kg body weight, at chronic doses of 1 to 10 mg As/kg body weight, and at
   dietary levels >5 and <50 mg As/kg diet.

Unlike wildlife, reports of arsenosis in domestic animals are common in cattle
and house cats, less common in sheep and horses, and rare in pigs and poultry (NAS
1977). In practice, the most dangerous arsenic preparations are dips, herbicides, and
defoliants in which the arsenical is in a highly soluble trivalent form, usually as
trioxide or arsenite (Selby et al. 1977). Accidental poisoning of cattle with arsenicals,
for example, is well documented. In one instance, more than 100 cattle died after
accidental overdosing with arsenic trioxide applied topically to control lice. On
necropsy, there were subcutaneous edematous swellings and petechial hemorrhages
in the area of application, and histopathology of the intestine, mucosa, kidney, and
epidermis (Robertson et al. 1984).

When extrapolating animal data from one species to another, the species tested
must be considered. For example, the metabolism of arsenic in the rat (*Rattus* sp.)
is unique and very different from that in humans and other mammals. Rats store
arsenic in blood hemoglobin, excreting it slowly, unlike most mammals, which
rapidly excrete ingested inorganic arsenic in the urine as methylated derivatives
(NAS 1977). Blood arsenic, whether given as As^{+3} or As^{+5}, rapidly clears from

humans, mice, rabbits, dogs, and primates; the half-life is about 6 hours for the fast phase and 60 hours for the slow phase (U.S. Environmental Protection Agency [USEPA] 1980). In the rat, however, blood arsenic is mostly retained in erythrocytes and clears slowly; the half-life is 60 to 90 days (USEPA 1980). In rats, the excretion of arsenic into bile is 40 times faster than in rabbits and up to 800 times faster than in dogs (Pershagen and Vahter 1979). Most researchers agree that the rat is unsatisfactory for use in arsenic research (NAS 1977; NRCC 1978; Pershagen and Vahter 1979; USEPA 1980; Webb et al. 1986).

12.5 PROPOSED ARSENIC CRITERIA

Numerous arsenic criteria have been proposed for the protection of human health and natural resources; some are shown in Table 12.3. Most proposed arsenic criteria have been exceeded, sometimes by orders of magnitude, in samples collected near gold mining extraction and refining facilities (Tables 12.1, 12.2). Arsenic criteria are undergoing constant revision. For example, the criterion of 190 µg As^{+3}/L for freshwater-life protection (USEPA 1985) was reduced over a 5-year period from 440 µg As^{+3}/L (USEPA 1980) but still does not afford adequate protection; many species of freshwater biota are adversely affected at <190 µg/L of As^{+3}, As^{+5}, or various organoarsenicals (Table 12.2). These adverse effects include death and malformations of toad embryos at 40 µg/L, growth inhibition of algae at 48 to 74 µg/L, mortality of amphipods and gastropods at 85 to 88 µg/L, and behavioral impairment of goldfish (*Carassius auratus*) at 100 µg/L. A downward adjustment in the current freshwater aquatic life protection criterion seems merited. A similar scenario exists for saltwater life protection, where the water quality criterion of 36 µg As^{+3}/L had been reduced from 508 µg As^{+3}/L 5 years earlier (USEPA 1980, 1985), with only a few species of algae showing adverse effects at <36 µg As/L (e.g., reduced growth at 19 to 22 µg/L).

Arsenic criteria in marine products of commerce also need to be reexamined because most of the arsenic in seafoods is in the form of arsenobetaine or some other comparatively harmless form and does not pose a threat to the consumer. It is now clear that the formulation of maximum permissible concentrations of arsenic in seafoods for health regulation purposes should recognize the chemical nature of arsenic (Jelinek and Corneliussen 1977; Phillips et al. 1982; Ozretic et al. 1990; McGeachy and Dixon 1990; Eisler 2000).

Various phenylarsonic acids, including arsanilic acid, sodium arsinilate, and 3-nitro-4-hydroxyphenylarsonic acid, have been used as feed additives for disease control and for improvement of weight gain in swine and poultry for more than 40 years (NAS 1977). The arsenic is present as As^{+5} and is rapidly excreted; present regulations require withdrawal of arsenical feed additives 5 days before slaughter for satisfactory feed depuration (NAS 1977). Under these conditions, total arsenic residues in edible tissues do not exceed the maximum permissible limit of 3 mg/kg fresh weight (Jelinek and Corneliussen 1977). Organoarsenicals will probably continue to be used as feed additives until new evidence indicates the contrary.

Table 12.3   Proposed Arsenic Criteria for the Protection of Human Health and Selected Natural Resources

Resource and Other Variables	Criterion or Effective Arsenic Concentration	Ref[a]
Human Health		
Total diet	<0.5 mg As/kg dry weight (DW) diet; 0.0003–0.0008 mg/kg body weight (BW) daily	1,13
Total intake	No observable effect at <0.021 mg arsenic daily based on 0.0003 mg/kg BW daily for 70-kg adult	1
Muscle of poultry and swine, eggs, swine edible by-products	<2 mg As/kg fresh weight (FW)	2
Shellfish Diet		
Crustaceans, edible tissues	<76 mg total As/kg FW tissue	3
Tolerable daily intake	<0.13 mg	4
Maximum allowable	<30 mg total As/kg FW diet	4
90th percentile consumers of shellfish		
Bivalve molluscs	0.057 mg daily	4
Lobsters, shrimp	0.18 mg daily	4
Drinking Water		
Total arsenic, recommended	<10 µg/L	5,6,7,8
Symptoms of arsenic toxicity observed	9% incidence at 50 µg/L, 16% at 50–100 µg/L, 44% incidence at >100 µg/L	9
Cancer frequency	0.01% at 82 µg As/L; 0.17% at 600 µg As/L	9
Tissue Residues		
No observed effect levels	<0.05 mg As/L urine; <0.5 mg/kg liver or kidney; <0.7 mg/L blood; <2 mg/kg hair; <5 mg/kg fingernail	9
Arsenic-poisoned, liver or kidney	2–100 mg As/kg FW	10
Arsenic-poisoned; whole body; children vs. adults	1 mg As/kg BW (equivalent to intake of 10 mg per month for 3 months) vs. 80 mg As/kg BW (intake of 2 g per year for 3 years)	9
Air		
Inorganic arsenic, occupational vs. residential	<2 µg/m^3 vs. <10 µg/m^3	1
Organic arsenic	<500 µg/m^3	1
Increased mortality	>3 µg/m^3 for 1 year	9
Respiratory cancer, increased risk	Lifetime occupational exposure >54.6 µg As/m^3; 50 µg As/m^3 for more than 25 years	9,11
Skin diseases	60–13,000 µg As/m^3	9
Dermatitis	300–81,500 µg As/m^3	9
Soils used for food production or parks in Europe and UK	10–40 mg As/kg DW	12
Terrestrial Vegetation		
No observable effects	<1.0 mg total water-soluble soil As/L, <25.0 mg total As/kg soil, <3.9 µg As/m^3 air	9
Adverse effects, crops and vegetation	3–28 mg water soluble As/L, equivalent to 25–85 mg total As/kg soil; air concentrations >3.9 µg As/m^3	7

Table 12.3 (continued)   Proposed Arsenic Criteria for the Protection of Human Health and Selected Natural Resources

Resource and Other Variables	Criterion or Effective Arsenic Concentration	Ref[a]
Soils, recommended	<20 mg/kg (Germany) to <500 mg/kg elsewhere	8
Phytotoxic or growth inhibition of tolerant genotypes	>1000 mg/kg DW soil	12
Aquatic Biota		
Freshwater biota: medium	96-h average water concentration should not exceed 190 µg total recoverable inorganic As^{+3}/L more than once every 3 years	14
Freshwater biota: tissue residues	Diminished growth and survival in immature bluegills, *Lepomis macrochirus*, when total arsenic residues in muscle are >1.3 mg/kg FW or >5 mg/kg in adults	9
Saltwater biota: medium	96-h average water concentration should not exceed 36 µg As^{+3}/L more than once every 3 years	14
Saltwater biota: tissues	Depending on chemical form of arsenic, certain marine fishes can tolerate muscle loading of 40 mg total As/kg FW	9
Birds		
Single oral dose fatal to 50%, sensitive species	17-48 mg As/kg BW	7
Tissue residues, liver and kidney	Residues of 2–10 mg total As/kg FW are considered elevated and residues >10 mg/kg are indicative of arsenic poisoning	15, 16
Diet	Reduced growth in mallard (*Anas platyrhynchos*) ducklings fed more than 30 mg As/kg diet as sodium arsenate	17
Small Laboratory Mammals		
Adverse effects, sensitive species	Single oral dose of 2.5–33.0 mg As/kg BW; chronic doses of 1–10 mg As/kg BW; 50 mg As/kg diet	7
Domestic Livestock		
Feedstuffs	Usually <2 mg total As/kg FW; <4 mg total As/kg in grasses and <10 in fish meals	18
Tissue residues		
Normal, muscle	<0.3 mg total As/kg FW	19
Poisoned, liver and kidney	5–10 mg total As/kg FW	18, 20

[a] 1, USPHS 2000; 2, Jelinek and Corneliussen 1977; 3, Jewett and Naidu 2000; 4, Adams et al. 1993a; 5, Kurttio et al. 1998; 6, Huang and Dasgupta 1999; 7, Eisler 2000c; 8, Matschullat et al. 2000; 9, NRCC 1978; 10, NAS 1977; 11, Pershagen and Vahter 1979; 12, O'Neill 1990; 13, Sorensen et al. 1985; 14, USEPA 1985; 15, Goede 1985; 16, Custer et al. 2002; 17, Camardese et al. 1990; 18, Vreman et al. 1986; 19, Veen and Vreman 1986; 20, Thatcher et al. 1985.

Many authorities now recognize that current arsenic criteria are not sufficient for adequate protection and that additional data are required for meaningful arsenic standards (NAS 1977; USEPA 1980, 1985; Abernathy et al. 1997; Society for Environmental Geochemistry and Health [SEGH] 1998; Eisler 2000). Specifically, there is general agreement that data are needed on the following subjects:

1. Cancer incidence and other abnormalities in natural resources with elevated arsenic levels, and the relation to potential carcinogenicity of arsenic compounds.
2. Interaction effects of arsenic with other carcinogens, cocarcinogens, promoting agents, inhibitors, and common environmental contaminants.
3. Controlled studies with aquatic and terrestrial indicator organisms on physiological and biochemical effects of long-term, low-dose exposures to inorganic and organic arsenicals, including effects on reproduction and genetic makeup.
4. Methodologies for establishing maximum permissible tissue concentrations for arsenic.
5. Effects of arsenic in combination with infectious agents.
6. Mechanisms of arsenical growth-promoting agents.
7. Role of arsenic in nutrition.
8. Extent of animal adaptation to arsenicals and the mechanisms of action.
9. Identification and quantification of mineral and chemical forms of arsenic in rocks, soils, and sediments that constitute the natural forms of arsenic entering water and the food chain.
10. Physicochemical processes influencing arsenic cycling.

12.6 SUMMARY

Arsenic sources to the biosphere associated with gold mining include waste soil and rocks, residual water from ore concentrations, roasting of some types of gold-containing ores to remove sulfur and sulfur oxides, and bacterially enhanced leaching. Arsenic concentrations near gold mining operations are elevated in abiotic materials and biota: maximum total arsenic concentrations measured were 560 µg/L in surface waters, 5.16 mg/L in sediment pore waters, 5.6 mg/kg dry weight (DW) in bird liver, 27 mg/kg DW in terrestrial grasses, 50 mg/kg DW in soils, 79 mg/kg DW in aquatic plants, 103 mg/kg DW in bird diets, 225 mg/kg DW in soft parts of bivalve molluscs, 324 mg/L in mine drainage waters, 625 mg/kg DW in aquatic insects, 7700 mg/kg DW in sediments, and 21,000 mg/kg DW in tailings.

Single oral doses of arsenicals that were fatal to 50% of tested species ranged from 17 to 48 mg/kg body weight (BW) in birds and from 2.5 to 33 mg/kg BW in mammals. Susceptible species of mammals were adversely affected at chronic doses of 1 to 10 mg As/kg BW or 50 mg As/kg diet. Sensitive aquatic species were damaged at water concentrations of 19 to 48 µg As/L, 120 mg As/kg diet, or tissue residues (in the case of freshwater fish) >1.3 mg/kg fresh weight. Adverse effects to crops and vegetation were recorded at 3 to 28 mg of water-soluble As/L (equivalent to about 25 to 85 mg total As/kg soil) and at atmospheric concentrations >3.9 µg As/m^3. Gold miners had a number of arsenic-associated health problems including excess mortality from cancer of the lung, stomach, and respiratory tract. Miners and school-children in the vicinity of gold mining activities had elevated urine arsenic of 25.7 µg/L (range 2.2 to 106.0 µg/L). Of the total population at this location, 20% showed elevated urine arsenic concentrations associated with future adverse health effects; arsenic-contaminated drinking water is the probable causative factor of elevated arsenic in urine. Proposed arsenic criteria to protect human health and natural resources are listed and discussed. Many of these proposed criteria do not adequately protect sensitive species.

LITERATURE CITED

Abernathy, C.O., R.L. Calderon, and W.R. Chappell (Eds.). 1997. *Arsenic. Exposure and Health Effects.* Chapman & Hall, London, 429 pp.

Adams, M.A., M. Bolger, C.D. Carrington, C.E. Coker, G.M. Cramer, M.J. DiNovi, and S. Dolan. 1993. *Guidance Document for Arsenic in Shellfish.* U.S. Food Drug. Admin., Washington, D.C., 27 pp.

Adams, M.D., M.W. Johns, and D.W. Dew. 1999. Recovery of gold from ores and environmental aspects, in *Gold: Progress in Chemistry, Biochemistry and Technology,* H. Schmidbaur, (Ed.), John Wiley & Sons, New York, 66–104.

Agate, A.D. 1996. Recent advances in microbial mining, *World Jour. Microbiol. Biotechnol.,* 12, 487–495.

Amonoo-Neizer, E.H. and E.M.K. Amekor. 1993. Determination of total arsenic in environmental samples from Kumasi and Obuasi, Ghana, *Environ. Health Perspec.,* 101, 46–49.

Amonoo-Neizer, E.H., D. Nyamah, and S.B. Bakiamoh. 1996. Mercury and arsenic pollution in soil and biological samples around the mining town of Obuasi, Ghana, *Water Air Soil Pollut.,* 91, 363–373.

Asperger, S. and B. Cetina-Cizmek. 1999. Metal complexes in tumour therapy, *Acta Pharmaceut.,* 49, 225–236.

Azcue, J.M., A. Mudroch, F. Rosa, and G.E.M. Hall. 1994. Effects of abandoned gold mine tailings on the arsenic concentrations in water and sediments of Jack of Clubs Lake, B.C., *Environ. Technol.,* 15, 669–678.

Boffetta, P., M. Kogevinas, N. Pearce, and E. Matos. 1994. Cancer, in *Occupational Cancer in Developing Countries,* Int. Agen. Res. Cancer, IARC Sci. Publ. 129, Oxford University Press, New York, 111–126.

Bowell, R.J., A. Warren, H.A. Minjera, and N. Kimaro. 1995. Environmental impact of former gold mining on the Orangi River, Serengeti N.P., Tanzania, *Biogeochemistry,* 28, 131–160.

Bright, D.A., B. Coedy, W.T. Dushenko, and K.J. Reimer. 1994. Arsenic transport in a watershed receiving gold mine effluent near Yellowknife, Northwest Territories, Canada, *Sci. Total Environ.,* 155, 237–252.

Bright, D.A., M. Dodd, and K.J. Reimer. 1996. Arsenic in subArctic lakes influenced by gold mine effluent: the occurrence of organoarsenicals and 'hidden' arsenic, *Sci. Total Environ.,* 180, 165–182.

Cain, D.J., S.N. Luoma, J.L. Carter, and S.V. Fend. 1992. Aquatic insects as bioindicators of trace element contamination in cobble-bottom rivers and streams, *Canad. Jour. Fish. Aquat. Sci.,* 49, 2141–2154.

Camardese, M.B., D.J. Hoffman, L.J. LeCaptain, and G.W. Pendleton. 1990. Effects of arsenate on growth and physiology in mallard ducklings, *Environ. Toxicol. Chem.,* 9, 785–795.

Cockell, K.A. and J.W. Hilton. 1985. Chronic toxicity of dietary inorganic and organic arsenicals to rainbow trout (*Salmo gairdneri* R.), *Feder. Proc.,* 44(4), 938.

Custer, T.W., C.M. Custer, S. Larson, and K.K. Dickerson. 2002. Arsenic concentrations in house wrens from Whitewood Creek, South Dakota, USA, *Bull. Environ. Contam. Toxicol.,* 68, 517–524.

Da Rosa, C.D. and J.S. Lyon (Eds.). 1997. *Golden Dreams, Poisoned Streams*. Mineral Policy Center, Washington, D.C., 269 pp.

Deknudt, G., A. Leonard, J. Arany, G.J. Du Buisson, and E. Delavignette. 1986. *In vivo* studies in male mice on the mutagenic effects of inorganic arsenic, *Mutagenesis*, 1, 33–34.

Eisler, R. 2000. Arsenic, in *Handbook of Chemical Risk Assessment: Health Hazards to Humans, Plants, and Animals. Volume 3, Metalloids, radiation, cumulative index to chemicals and species*. Lewis Publishers, Boca Raton, FL, 1501–1566.

Eisler, R. 2004. Arsenic hazards to humans, plants, and animals from gold mining, *Rev. Environ. Contam. Toxicol.*, 180, 133–165.

Galbraith, H., K. LeJeune, and J. Lipton. 1995. Metal and arsenic impacts to soils, vegetation communities and wildlife habitat in southwest Montana uplands contaminated by smelter emissions: I. Field evaluation, *Environ. Toxicol. Chem.*, 14, 1895–1903.

Goede, A.A. 1985. Mercury, selenium, arsenic and zinc in waders from the Dutch Wadden Sea, *Environ. Pollut.*, 37A, 287–309.

Golow, A.A., A. Schleuter, S. Amihere-Mensah, H.L.K. Granson, and M.S. Tetteh. 1996. Distribution of arsenic and sulphate in the vicinity of Ashanti goldmine at Obuasi, Ghana, *Bull. Environ. Contam. Toxicol.*, 56, 703 –710.

Greer, J. 1993. The price of gold: environmental costs of the new gold rush, *The Ecologist*, 23 (3), 91–96.

Grosser, J.R., V. Hagelgans, T. Hentschel, and M. Priester. 1994. Heavy metals in stream sediments: a gold mining area near Los Andes, southern Colombia S.A., *Ambio*, 23, 46–149.

Hallberg, K.B., H.M. Sehlin, and E.B. Lindstrom. 1996. Toxicity of arsenic during high temperature bioleaching of gold-bearing arsenical pyrite, *Appl. Microbiol. Biotechnol.*, 45, 212–216.

Hamasaki, T., H. Nagase, Y. Yoshioka, and T. Sato. 1995. Formation, distribution, and ecotoxicity of methylmetals of tin, mercury, and arsenic in the environment, *Crit. Rev. Environ. Sci. Technol.*, 25, 45–91.

Hoffman, D.J., C.J. Sanderson, L.J. LeCaptain, E. Cromartie, and G.W. Pendleton. 1992. Interactive effects of arsenate, selenium, and dietary protein on survival, growth, and physiology in mallard ducklings, *Arch. Environ. Contam. Toxicol.*, 22, 55–62.

Holson, J.F., J.M. DeSesso, A.R. Scialli, and C.F. Farr. 1998. Inorganic arsenic and prenatal development: a comprehensive evaluation for human risk assessment. Society for Environmental Geochemistry and Health 3rd Inter. Conf. Arsenic Expos. Health Effects, 23.

Hood, R.D. 1985. *Cacodylic Acid: Agricultural Uses, Biologic Effects, and Environmental Fate*. VA Monograph. U.S. Govt. Printing Off., Washington, D.C., 171 pp.

Hood, R.D., G.C. Vedel-Macrender, M.J. Zaworotko, F.M. Tatum, and R.G. Meeks. 1987. Distribution, metabolism, and fetal uptake of pentavalent arsenic in pregnant mice following oral or intraperitoneal administration, *Teratology*, 35, 19–25.

Hopenhayn-Rich, C., K.D. Johnson,, and J. Hertz-Picciotto. 1998. Reproductive and developmental effects associated with chronic arsenic exposure. Society for Environmental Geochemistry and Health 3rd Inter. Conf. Arsenic Expos. Health Effects, 21.

Huang, H. and P.K. Dasgupta. 1999. A field-deployable instrument for the measurement and speciation of arsenic in potable water, *Anal. Chim. Acta*, 380, 27–37.

Hudson, R.H., R.K. Tucker, and M.A. Haegle. 1984. *Handbook of Toxicity of Pesticides to Wildlife*. U.S. Fish Wildl. Serv. Resour. Publ. 153, 90 pp.

Hughes, M.F. and E.M. Kenyon. 1998. Dose-dependent effects on the disposition of monomethylarsonic acid and dimethylarsinic acid in the mouse after intravenous administration, *Jour. Toxicol. Environ. Health*, 53A, 95–112.

Jauge, P. and L.M. Del-Razo. 1985. Uric acid levels in plasma and urine in rats chronically exposed to inorganic As(III) and As(V), *Toxicol. Lett.*, 26, 31–35.

Jelinek, C.F. and P.E. Corneliussen. 1977. Levels of arsenic in the United States food supply, *Environ. Health Perspec.*, 19, 83–87.

Jenkins, D.W. 1980. Biological monitoring of toxic trace metals. Vol. 2. Toxic trace metals in plants and animals of the world, Part 1. U.S. Environ. Protection Agen. Rep. 600/3-80-090, 30–138.

Jewett, S.C. and S. Naidu. 2000. Assessment of heavy metals in red king crabs following offshore placer gold mining, *Mar. Pollut. Bull.*, 40, 478–490.

Johnson, S.W., S.D. Rice, and D.A. Moles. 1998a. Effects of submarine mine tailings disposal on juvenile yellowfin sole (*Pleuronectes asper*): a laboratory study, *Mar. Pollut. Bull.*, 36, 278–287.

Johnson, S.W., R. P. Stone, and D.C. Love. 1998b. Avoidance behavior of ovigerous tanner crabs *Chionoecetes bairdi* exposed to mine tailings: a laboratory study, *Alaska Fish. Res. Bull.*, 5, 39–45.

Johnson, W.W. and M.T. Finley. 1980. Handbook of acute toxicity of chemicals to fish and aquatic invertebrates. U.S. Fish Wildl. Serv. Resour. Publ. 137.

Kabir, H. and C. Bilgi. 1993. Ontario gold miners with lung cancer, *Jour. Occup. Med.*, 35, 1203–1207.

Kurttio, P., H. Komulainen, E. Hakala, and J. Pekkanen. 1998. Urinary excretion of arsenic species after exposure to arsenic present in drinking water, *Arch. Environ. Contam. Toxicol.*, 34, 297–305.

Kusiak, R.A., A.C. Ritchie, J. Springer, and J. Muller. 1993. Mortality from stomach cancer in Ontario miners, *Brit. Jour. Indus. Med.*, 50, 117–126.

Lau, S., M. Mohamed, A.T.C. Yen, and S. Su'ut. 1998. Accumulation of heavy metals in freshwater molluscs, *Sci. Total Environ.*, 214, 113–121.

Leduc, G. 1984. Cyanides in water: toxicological significance, in *Aquatic Toxicology, Volume 2*, L.J. Weber, (Ed.), Raven Press, New York, 153–224.

Lee, J-S. and H.T. Chon. 2003. Toxic risk assessment of heavy metals on abandoned metal mine areas with various exposure pathways. Sixth International Symposium on Environmental Geochemistry, Edinburgh, Scotland, 7–11 Sept. 2003, Book of Abstracts, 191.

Lee, S.W., J.U. Lee, and K.W. Kim. 2003. Bioremediation of sediment and tailings contaminated with arsenic by indigenous bacteria. Sixth International Symposium on Environmental Biogeochemistry, Edinburgh, Scotland, 7–11 Sept. 2003, Book of Abstracts, 155.

Lim, H-S., J.S. Lee, and H.T. Chon. 2003. Arsenic and heavy metal contamination in the vicinity of abandoned Songcheon Au-Ag-Mo mine, Korea. Sixth International Symposium on Environmental Geochemistry, Edinburgh, Scotland, 7–11 Sept. 2003, Book of Abstracts, 157.

Lima, A.R., C. Curtis, D.E. Hammermeister, T.P. Markee, C.E. Northcutt, and L.T. Brooke. 1984. Acute and chronic toxicities of arsenic (III) to fathead minnows, flagfish, daphnids, and an amphipod, *Arch. Environ. Contam. Toxicol.*, 13, 595–601.

Marques, I.A. and L.E. Anderson 1986. Effects of arsenite, sulfite, and sulfate on photosynthetic carbon metabolism in isolated pea (*Pisum sativum* L., cv Little Marvel) chloroplasts, *Plant Physiol.*, 82, 488–493.

Matschullat, J., R.P. Borba, E. Deschamps, B.R. Figueiredo, T. Gabrio, and M. Schwenk. 2000. Human and environmental contamination in the iron quadrangle, Brazil, *Appl. Geochem.*, 15, 181–190.

May, T.W., R.H. Wiedmeyer, J. Gober, and S. Larson. 2001. Influence of mining-related activities on concentrations of metals in water and sediment from streams of the Black Hills, South Dakota, *Arch. Environ. Contam. Toxicol.*, 40, 1–9.

McGeachy, S.M. and D.G. Dixon. 1990. Effect of temperature on the chronic toxicity of arsenate to rainbow trout (*Oncorhynchus mykiss*), *Canad. Jour. Fish. Aquat. Sci.*, 47, 2228–2234.

Meharg, A.A., R.F. Shore, and K.F. Broadgate. 1998. Edaphic factors affecting the toxicity and accumulation of arsenate in the earthworm *Lumbricus terrestris*, *Environ. Toxicol. Chem.*, 17, 1124–1131.

Nagymajtenyi, L., A. Selypes, and G. Berencsi. 1985. Chromosomal aberrations and fetotoxic effects of atmospheric arsenic exposure in mice, *Jour. Appl. Toxicol.*, 5, 61–63.

Naqvi, S.M. and C.T. Flagge. 1990. Chronic effects of arsenic on American red crayfish, *Procambarus clarki*, exposed to monosodium methanearsonate (MSMA) herbicide, *Bull. Environ. Contam. Toxicol.*, 45, 101–106.

National Academy of Sciences (NAS). 1977. *Arsenic.* NAS, Washington, D.C., 332 pp.

National Research Council of Canada (NRCC). 1978. *Effects of Arsenic in the Canadian Environment.* NRCC Publ. 15391, 349 pp.

Ng, J.C., A.A. Seawright, L. Qi, C.M. Garnett, M.R. Moore, and B. Chiswell. 1998. Tumours in mice induced by chronic exposure of high arsenic concentrations in drinking water. Society for Environmental Geochemistry and Health 3rd Inter. Conf. Arsenic Expos. Health Effects. 28.

Ngubane, W.T. and A.A.W. Baecker. 1990. Oxidation of gold-bearing pyrite and arsenopyrite by *Sulfolobus acidocaldarius* and *Sulfolobus* BC in airlift reactors, *Biorecovery*, 1, 255–259.

Nystrom, R.R. 1984. Cytological changes occurring in the liver of coturnix quail with an acute arsenic exposure, *Drug Chem. Toxicol.*, 7, 587–594.

O'Neill, P. 1990. Arsenic, in *Heavy Metals in Soils.* Halsted Press, Glasgow, 83–99.

Ozretic, B., M. Krajinovic-Ozretic, J. Santin, B. Medjugorac, and M. Kras, 1990. As, Cd, Pd, and Hg in benthic animals from the Kvarber-Rijeka region, Yugoslovia, *Mar. Pollut. Bull.*, 21, 595–597.

Pain, S. 1987. After the goldrush, *New Scientist*, 115 (1574), 36–40.

Passino, D.R.M. and A.J. Novak. 1984. Toxicity of arsenate and DDT to the cladoceran *Bosmina longirostris*, *Bull. Environ. Contam. Toxicol.*, 33, 325–329.

Pendleton, G.W., M.R. Whitworth, and G.H. Olsen. 1995. Accumulation and loss of arsenic and boron, alone and in combination, in mallard ducks, *Environ. Toxicol. Chem.*, 14, 1357–1364.

Pershagen, G. and M. Vahter. 1979. *Arsenic — A Toxicological and Epidemiological Appraisal.* Naturvards-verket Rapp. SNV PM 1128, Liber Tryck, Stockholm.

Phillips, D.J.H., G.B. Thompson, K.M. Gabuji, and C.T. Ho. 1982. Trace metals of toxicological significance to man in Hong Kong seafood, *Environ. Pollut.*, 3B, 27–45.

Rahn, P.H., A.D. Davis, C.J. Webb, and A.D. Nichols. 1996. Water quality impacts from mining in the Black Hills, South Dakota, USA, *Environ. Geol.*, 27, 38–53.

Rawlings, D.E. 1998. Industrial practice and the biology of leaching of metals from ores, *Jour. Indus. Microbiol. Biotechnol.*, 20, 268–274.

Ripley, E.A., R.E. Redmann, and A.A. Crowder. 1996. *Environmental Effects of Mining.* St. Lucie Press, Delray Beach, FL, 356 pp.

Robertson, I.D., W.E. Harms, and P.J. Ketterer. 1984. Accidental arsenic toxicity of cattle. *Austral. Veterin. Jour.*, 61, 366–367.

Robertson, J.L. and J.A. McLean. 1985. Correspondence of the LC 50 for arsenic trioxide in a diet-incorporation experiment with the quantity of arsenic ingested as measured by X-ray, energy-dispersive spectrometry, *Jour. Econ. Entomol.*, 78, 1035–1036.

Samecka-Cymerman, A. and A.J. Kempers. 1998. Bioindication of gold by aquatic bryophytes, *Acta Hydrochim. Hydrobiol.*, 26, 90–94.

Sanders, J.G. 1986. Direct and indirect effects of arsenic on the survival and fecundity of estuarine zooplankton, *Canad. Jour. Fish. Aquat. Sci.*, 43, 694–699.

Santini, J.M., L.I. Sly, R.D. Schnagl, and J.M. Macy. 2000. A new chemolithoautotrophic arsenite-oxidizing bacterium isolated from a gold mine: phylogenetic, physiological and preliminary biochemical studies, *Appl. Environ. Microbiol.*, 66, 92–97.

Savabieasfahani, M., R.L. Lochmiller, D.P. Rafferty, and J.A. Sinclaiar. 1998. Sensitivity of wild cotton rats (*Sigmodon hispidus*) to the immunotoxic effects of low-level arsenic exposure, *Arch. Environ. Contam. Toxicol.*, 34,, 289–296.

Selby, L.A., A.A. Case, G.D. Osweiler, and H.M. Hages, Jr, 1977. Epidemiology and toxicology of arsenic poisoning in domestic animals, *Environ. Health Perspec.*, 19, 183–189.

Sheppard, M.I., D.H. Thibault, and S.C. Sheppard. 1985. Concentrations and concentration ratios of U, As and Co in Scots pine grown in a waste-site soil and an experimentally contaminated soil, *Water Air Soil Pollut.*, 26, 85–94.

Simonato, L., J.J. Moulin, B. Javelaud, G. Ferro, P. Wild, R. Winkelmann, and R. Saracci. 1994. A retrospective mortality study of workers exposed to arsenic in a gold mine and refinery in France, *Amer. Jour. Indus. Med.*, 25, 625–633.

Society for Environmental Geochemistry and Health (SEGH). 1998. Third Int. Conf. on Arsenic Expos. and Health Effects. San Diego, CA, July 12–15, 1998. Book of Abstracts.

Sorensen, E.M.B., R.R. Mitchell, A. Pradzynski, T.L. Bayer, and L.L. Wenz. 1985. Stereological analyses of hepatocyte changes parallel arsenic accumulation in the livers of green sunfish, *Jour. Environ. Pathol. Toxicol. Oncol.*, 6, 195–210.

Spehar, R.L., J.T. Fiandt, R.L. Anderson, and D.L. DeFoe. 1980. Comparative toxicity of arsenic compounds and their accumulation in invertebrates and fish, *Arch. Environ. Contam. Toxicol.*, 9, 53–63.

Stine, E.R., C.A. Hsu, T.D. Hoovers, H.V. Aposhian, and D.E. Carter. 1984. N-(2,3-dimercaptopropyl) phthalamidic acid: protection *in vivo* and *in vitro* against arsenic intoxication, *Toxicol. Appl. Pharmacol.*, 75, 329–336.

Stone, R.P. and S.W. Johnson. 1997. Survival, growth, and bioaccumulation of heavy metals by juvenile tanner crabs (*Chionoecetes bairdi*) held on weathered mine tailings, *Bull. Environ. Contam. Toxicol.*, 58, 830–837

Stone, R.P. and S.W. Johnson. 1998. Prolonged exposure to mine tailings and survival and reproductive success of ovigerous tanner crabs (*Chionoecetes bairdi*), *Bull. Environ. Contam. Toxicol.*, 61, 548–556.

Tarras-Wahlberg, N.H., A. Flachier, G. Fredriksson, S. Lane, B. Lundberg, and O. Sangfors. 2000. Environmental impact of small-scale and artisanal gold mining in southern Ecuador, *Ambio*, 29, 484–491.

Thatcher, C.D., J.B. Meldrum, S.E. Wikse, and W.D. Whittier. 1985. Arsenic toxicosis and suspected chromium toxicosis in a herd of cattle, *Jour. Amer. Vet. Assoc.*, 187, 179–182.

Thursby, G.B. and R.L. Steele. 1984. Toxicity of arsenite and arsenate to the marine macroalgae *Champia parvula* (Rhodophyta), *Environ. Toxicol Chem.*, 52, 641–648.

U.S. Bureau of Land Management (USBLM). 2000. *Cumulative Impact Analysis of Dewatering and Water Management Operations for the Betze Project, South Operations Area Amendment, and Leevile Project.* USBLM, Elko, NV, 403 pp.

U.S. Environmental Protection Agency (USEPA). 1980. Ambient water quality criteria for arsenic, USEPA Rep. 440/5-80-021, 205 pp.

U.S. Environmental Protection Agency (USEPA). 1985. Ambient water quality criteria for arsenic — 1984. USEPA Rep. 440/5-84-033, 66 pp.

U.S. Public Health Service (USPHS). 2000. Toxicological profile for arsenic (update). Draft for public comment. Agen. Toxic Substances Dis. Registry, 446 pp.

Veen, N.G. van der, and K. Vreman. 1986. Transfer of cadmium, lead, mercury and arsenic from feed into various organs and tissues of fattening lambs, *Nether. Jour. Agric. Sci.*, 34, 134–153.

Vreman, K, N.G. van der Veen, E.J. van der Molen, and W.G. de Ruig. 1986. Transfer of cadmium, lead, mercury and arsenic from feed into milk and various tissues of dairy cows: chemical and pathological data, *Nether. Jour. Agric. Sci.*, 34, 129–144.

Wang, D.S., R.W. Weaver, and J.R. Melton. 1984. Microbial decomposition of plant tissue contaminated with arsenic and mercury, *Environ. Pollut.*, 34A, 275–282.

Webb, D.R., S.E. Wilson, and D.E. Carter. 1986. Comparative pulmonary toxicity of gallium arsenide, gallium (III) oxide or arsenic (III) oxide intratracheally instilled into rats, *Toxicol. Appl. Pharmacol.*, 82, 405–416.

Winski, S.L. and D.E. Carter. 1998. Arsenate toxicity in human erythrocytes: characterization of morphologic changes and determination of the mechanism of damage, *Jour. Toxicol. Environ. Health*, 53A, 345–355.

Wong, H.K.T., A. Gauthier, and J.O. Nriagu. 1999. Dispersion and toxicity of metals from abandoned gold mine tailings at Goldenville, Nova Scotia, Canada, *Sci. Total Environ.*, 228, 35–47.

Woolson, E.A. (Ed.). 1975. Arsenical pesticides, *Amer. Chem. Soc. Symp.*, Ser. 7.

Yoshida, K., Y. Inoue, K. Kuroda, H. Chen, H. Wanibuchi, S. Fukushima, and G. Endo. 1998. Urinary excretion of arsenic metabolites after long-term oral administration of various arsenic compounds to rats, *Jour. Toxicol. Environ. Health*, 54A, 179–192.

CHAPTER 13

Mercury Hazards from Gold Mining for Humans, Plants, and Animals

Mercury has no beneficial biological function, and its presence in living organisms is associated with cancer, birth defects, and other undesirable outcomes (Eisler 2000). The use of liquid mercury (Hg^0) to separate microgold (Au^0) particles from sediments through formation of amalgam (Au-Hg) with subsequent recovery and reuse of mercury is a technique that has been in force for at least 4700 years (Lacerda 1997a); however, this process is usually accompanied by massive mercury contamination of the biosphere (Petralia 1996). It is estimated that gold mining currently accounts for about 10% of the global mercury emissions from human activities (Lacerda 1997a).

This chapter documents the history of mercury in gold production; ecotoxicological aspects of the amalgamation process in various geographic regions, with emphasis on Brazil and North America; lethal and sublethal effects of mercury and its compounds to plants and animals; and proposed mercury criteria to protect sensitive natural resources and human health.

13.1 HISTORY OF MERCURY IN GOLD MINING

The use of mercury in the mining industry to amalgamate and concentrate precious metals dates from about 2700 BCE when the Phoenicians and Carthaginians used it in Spain. The technology became widespread by the Romans in 50 CE and is similar to that employed today (Lacerda 1997a; Rojas et al. 2001; Eisler 2004). In 177 CE, the Romans banned elemental mercury use for gold recovery in mainland Italy, possibly in response to health problems caused by this activity (de Lacerda and Salomons 1998). Gold extraction using mercury was widespread until the end of the first millennium (Meech et al. 1998).

In the Americas, mercury was introduced in the 16th century to amalgamate Mexican gold and silver. In 1849, during the California gold rush, mercury was widely used, and mercury poisoning was allegedly common among miners (Meech et al. 1998). In the 30-year period 1854–1884, gold mines in California's Sierra Nevada range released between 1400 and 3600 metric tons to the environment (Fields 2001); dredge tailings from this period still cover more than 73 km^2 in the Folsom-Natomas region of California and continue to represent a threat to current residents (de Lacerda and Salomons 1998). In South America, mercury was used extensively by the Spanish colonizers to extract gold, releasing nearly 200,000 tons of mercury to the environment between 1550 and 1880 as a direct result of this process (Malm 1998). At the height of the Brazilian gold rush in the 1880s, more than 6 million people were prospecting for gold in the Amazon region alone (Frery et al. 2001).

It is doubtful whether there would have been gold rushes without mercury (Nriagu and Wong 1997). Supplies that entered the early mining camps included hundreds of flasks of mercury weighing 34.5 kg each, consigned to the placer diggings and recovery mills. It is alleged that the unit of measure for mercury (the flask) is equivalent to the Phoenician talent, which is indicative of the long association of mercury with gold recovery. Mercury amalgamation provided an inexpensive and efficient process for the extraction of gold, and the process can be learned rapidly by itinerant gold diggers. The mercury amalgamation process absolved the miners from any capital investment in equipment. This was important where riches were obtained instantaneously and ores contained only a few ounces of gold per ton and could not be economically transported elsewhere for processing (Nriagu and Wong 1997).

Mercury released to the biosphere between 1550 and 1930 as a result of gold mining activities, mainly in Spanish colonial America, but also in Australia, southeast Asia, and England, may have exceeded 260,000 tons (Lacerda 1997a). Exceptional increases in gold prices in the 1970s concomitant with worsening socioeconomic conditions in developing regions of the world resulted in a new gold rush in the Southern Hemisphere involving more than 10 million people on all continents. At present, mercury amalgamation is used as the major technique for gold production in South America, China, Southeast Asia, and some African countries. Most of the mercury released to the biosphere through gold mining may still participate in the global mercury cycle through remobilization from abandoned tailings and other contaminated areas (Lacerda 1997a).

From 1860 to 1925, amalgamation was the main technique for gold recovery worldwide, and it was common in the United States until the early 1940s (Greer 1993). The various procedures in current use can be grouped into two categories (de Lacerda and Salomons 1988; Korte and Coulston 1998):

1. Recovery of gold from soils and rocks containing 4 to 20 g gold/t. The metal-rich material is passed through grinding mills to produce a metal-rich concentrate. In colonial America, mules and slaves were used instead of electric mills. This practice is associated with pronounced deforestation, soil erosion, and river silt-ation. The concentrate is moved to small amalgamation ponds or drums, mixed with liquid mercury, squeezed to remove excess mercury, and taken to a retort for roasting. Any residue in the concentrate is returned to the amalgamation pond and reworked until the gold is extracted.
2. Extraction of gold from dredged bottom sediments. Stones are removed by iron meshes. The material is then passed through carpeted riffles for 20 to 30 hours,

which retains the heavier gold particles. The particles are collected in barrels, amalgamated, and treated as described previously. However, residues of the procedure are released into the rivers. Vaporization of mercury and losses also occur due to human error (de Lacerda and Salomons 1998).

The organized mining sector abandoned amalgamation because of economic and environmental considerations. However, small-scale mine operators in South America, Asia, and Africa, often driven by unemployment, poverty, and landlessness, have resorted to amalgamation because they lack affordable alternative technologies. Typically, these operators pour liquid mercury over crushed ore in a pan or sluice. The amalgam, a mixture of gold and mercury (Au-Hg), is separated by hand, passed through a chamois cloth to expel the excess mercury, which will be reused, then heated with a blowtorch to volatilize the mercury. About 70% of the loss of mercury to the environment occurs during the blowtorching. Most of these atmospheric emissions quickly return to the river ecosystem in rainfall and concentrate in bottom sediments (Greer 1993).

Residues from mercury amalgamation remain at many stream sites around the globe. Amalgamation should not be applied because of associated health hazards and is, in fact, forbidden almost everywhere; however, it remains in use today, especially in the Amazon section of Brazil. In Latin America, more than 1 million gold miners collect between 115 and 190 tons of gold annually, emitting more than 200 tons of mercury in the process (Korte and Coulston 1998). The world production of gold is about 225 tons annually, with 65 tons of the total produced in Africa. It is alleged that only 20% of the mined gold is recorded officially.

About 1 million people are employed globally on nonmining aspects of artisanal gold, 40% of them female with an average yearly income of US $600 (Korte and Coulston 1998). The total number of gold miners in the world using mercury amalgamation to produce gold ranges from 3 to 5 million, including 650,000 in Brazil, 250,000 in Tanzania, 250,000 in Indonesia, and 150,000 in Vietnam (Jernelov and Ramel 1994). To provide a living, marginal at best, for this large number of miners, gold production and mercury use would come to thousands of tons annually; however, official figures account for only 10% of the production level (Jernelov and Ramel 1994). At least 90% of the gold extracted by individual miners in Brazil is not registered with authorities for a variety of reasons, some financial. Accordingly, official gold production figures reported in Brazil and probably most other areas of the world are grossly underreported (Porvari 1995).

Cases of human mercury contamination have been reported from various sites around the world ever since mercury was introduced as the major mining technique to produce gold and other precious metals in South America hundreds of years ago (de Lacerda and Salomons 1998). Contamination in humans is reflected by elevated mercury concentrations in air, water, and diet, and in hair, urine, blood, and other tissues. However, only a few studies actually detected symptoms or clinical evidence of mercury poisoning in gold mining communities (Eisler 2003).

After the development of the cyanide leaching process for gold extraction, mercury amalgamation disappeared as a significant mining technology (de Lacerda and Salomons 1998). However, when the price of gold soared from US $58/troy ounce in 1972 to $430 in 1985, a second gold rush was triggered, particularly in Latin America, and later in the Philippines, Thailand, and Tanzania (de Lacerda and Salomons 1998). In modern Brazil, where there has been a gold rush since 1980, at least 2000 tons of mercury were released, with subsequent mercury contamination of sediments, soils, air, fish, and human tissues; a similar situation exists in Colombia, Venezuela, Peru, and Bolivia (Malm 1998). Recent estimates of global anthropogenic total mercury emissions range from 2000 to 4000 t/yr of which 460 tons is from small-scale gold mining (Porcella et al. 1995, 1997). Major contributors of mercury to the environment from recent gold mining activities include Brazil (3000 tons since 1979), China (596 tons since 1938), Venezuela (360 tons since 1989), Bolivia (300 tons since 1979), the Philippines (260 tons since 1986), Colombia (248 tons since 1987), the United States (150 tons since 1969), and Indonesia (120 tons since 1988; Lacerda 1997a).

The most heavily mercury-contaminated site in North America is the Lahontan Reservoir and environs in Nevada (Henny et al. 2002). Millions of kilograms of liquid mercury used to process gold and silver ore mined from Virginia City, Nevada, and vicinity between 1859 and 1890, along with waste rock, were released into the Carson River watershed. The inorganic elemental mercury was readily methylated to water-soluble methylmercury. Over time, much of this mercury was transported downstream into the lower reaches of the Carson River, especially the Lahontan Reservoir and Lahontan wetlands near the terminus of the system, with significant damage to wildlife (Henny et al. 2002).

Most authorities on mercury ecotoxicology now agree on seven points:

1. Mercury and its compounds have no known biological function, and its presence in living organisms is undesirable and potentially hazardous.
2. Forms of mercury with relatively low toxicity can be transformed into forms with very high toxicity through biological and other processes.
3. Methylmercury, the most toxic form, can be bioconcentrated in organisms and biomagnified through food chains, returning mercury directly to humans and other upper-trophic-level consumers in concentrated form.
4. Mercury methylation is dependent on many biological and chemical factors but is usually most rapid in organic substrates as a result of anaerobic microbial activity.
5. Mercury is a mutagen, teratogen, and carcinogen and causes embryocidal, cytochemical, and histopathological effects.
6. High body burdens of mercury normally encountered in some species of fish and wildlife from remote locations emphasize the complexity of natural mercury cycles and human impacts on these cycles.
7. Anthropogenic release of mercury should be curtailed because the difference between tolerable natural background levels of mercury and harmful effects in the environment is exceptionally small.

These concerns, and other aspects of mercury and its compounds in the environment as a result of natural or anthropogenic processes, are the subject of many reviews, including those by Montague and Montague (1971), D'Itri (1972), Friberg and Vostal (1972), Jernelov et al. (1972, 1975), Keckes and Miettinen (1972), Buhler (1973),

Holden (1973), D'Itri and D'Itri (1977), Eisler (1978, 1981, 1987, 2000, 2004), U.S. National Academy of Sciences [USNAS] (1978), Beijer and Jernelov (1979), Birge et al. (1979), Magos and Webb (1979), Nriagu (1979), Clarkson and Marsh (1982), Das et al. (1982), Boudou and Ribeyre (1983), Elhassni (1983), Clarkson et al. (1984), Robinson and Touvinen (1984), U.S. Environmental Protection Agency [USEPA] (1985), Wren (1986), Clarkson (1990), Lindqvist (1991), U.S. Public Health Service [USPHS] (1994), Watras and Huckabee (1994), Hamasaki et al. (1995), Porcella et al. (1995), Heinz (1996), Thompson (1996), Wiener and Spry (1996), de Lacerda and Salomons (1998), Wolfe et al. (1998), Ullrich et al. (2001), and Wiener et al. (2002).

13.2 ECOTOXICOLOGICAL ASPECTS OF AMALGAMATION

Mercury emissions from historic gold mining activities and from present gold production operations in developing countries represent a significant source of local pollution. Poor amalgamation distillation practices account for a significant part of the mercury contamination, followed by inefficient amalgam concentrate separation and gold melting operations (Meech et al. 1998). Ecotoxicological aspects of mercury amalgamation of gold are presented next for selected geographic regions, with special emphasis on Brazil and North America.

13.2.1 Brazil

High mercury levels found in the Brazilian Amazon environment are attributed mainly to gold mining practices, even though elevated mercury concentrations are reported in fish and human tissues in regions far from any anthropogenic mercury source (Fostier et al. 2000). Since the late 1970s, many rivers and waterways in the Amazon have been exploited for gold using mercury in the mining process as an amalgamate to separate the fine gold particles from other components in the bottom gravel (Malm et al. 1990). Between 1979 and 1985, at least 100 tons of mercury were discharged into the Madeira River basin with 45% reaching the river and 55% passing into the atmosphere. As a result of gold mining activities using mercury, elevated concentrations of mercury were measured in bottom sediments from small forest streams [up to 157 mg Hg/kg dry weight (DW)], in stream water (up to 10 μg/L), in fish [up to 2.7 mg/kg fresh weight (FW) muscle], and in human hair (up to 26.7 mg/kg DW) (Malm et al. 1990). Mercury transport to pristine areas by rainwater, water currents, and other vectors could be increased with increasing deforestation, degradation of soil cover from gold mining activities, and increased volatilization of mercury from gold mining practices (Davies 1997; Fostier et al. 2000). Population shifts as a result of gold mining are common in Brazil. For example, from 1970 to 1985, the population of Rondonia, Brazil, increased from about 111,000 to 904,000, mainly due to gold mining and agriculture. One result was a major increase in deforested areas and in gold production from 4 kg Au/yr to 3600 kg Au/yr (Martinelli et al. 1988).

Mercury is lost during two distinct phases of the gold mining process. In the first phase, sediments are aspirated from the river bottom and passed through a series of seines. Metallic mercury is added to the seines to separate and amalgamate the gold. Part of this mercury escapes into the river, with risk to fish and to livestock that drink river water, and to humans from occupational exposure and from ingestion of mercury-contaminated fish, meat, and water. In the second phase, the gold is purified by heating the amalgam, usually in the open air, with mercury vapor lost to the atmosphere. The workers take few precautions to avoid inhalation of the mercury vapor (Palheta and Taylor 1995; Martinelli et al. 1988).

Mercury Sources and Release Rates

All mercury used in Brazil is imported, mostly from the Netherlands, Germany, and England, reaching 340 tons in 1989 (Lacerda 1997b). For amalgamation purposes, mercury in Brazil is sold in small quantities (200 g) to a great number (about 600,000) of individual miners. Serious ecotoxicological damage is likely because much, if not most, of the human population in these regions depends on local natural resources for food (Lacerda 1997b). The amount of gold produced in Brazil was 9.6 tons in 1972 and 218.6 tons in 1988; an equal amount of mercury is estimated to have been discharged into the environment (Camara et al. 1997). In Brazil, industry was responsible for almost 100% of total mercury emissions to the environment until the early 1970s, at which time existing mercury control policies were enforced with subsequent declines in mercury releases (Lacerda 1997b). Mercury emissions from gold mining were insignificant up to the late 1970s, but by the mid 1990s this accounted for 80% of the total mercury emissions. About 210 tons of mercury are now released to the biosphere each year in Brazil: 170 tons from gold mining, 17 tons from the chloralkali industry, and the rest from other industrial sources. Emission to the atmosphere is the major pathway of mercury releases to the environment, with the gold mining industry accounting for 136 tons annually in Brazil (Lacerda 1997b).

At least 400,000, and perhaps as many as 1 million, small-scale gold miners, known as garimpos, are active in the Brazilian Amazon region at more than 2000 sites (Pessoa et al. 1995; Veiga et al. 1995). It is estimated that each garimpo is indirectly responsible for another four to five people, including builders and operators of production equipment, dredges, aircraft (at least 1000), small boats or motor-driven canoes (at least 10,000), and about 1100 pieces of digging and excavation equipment. It is conservatively estimated that this group discharges 100 tons of mercury each year into the environment. There are five main mining and concentration methods used in the Amazon region to extract gold from rocks and soils containing 0.6 to 20 g/t gold (Pessoa et al. 1995):

1. Manual: this method involves the use of primitive equipment, such as shovels and hoes. About 15% of the garimpos use this method, usually in pairs. Gold is recovered in small concentration boxes with crossed riffles. Very few tailings are discharged into the river.
2. Floating dredges with suction pumps: this method is considered inefficient, with large loss of mercury and low recovery of gold.

3. Rafts with underwater divers directing the suction process: this is considered a hazardous occupation, with many fatalities. Incidentally, there is a comparatively large mercury loss using this procedure.
4. Hydraulic disintegration: this involves breaking down steep banks using a high-pressure water jet pump.
5. Concentration mills: gold recovered from underground veins is pulverized and extracted, sometimes by cyanide heap leaching.

The production of gold by garimpos (small- and medium-scale, often clandestine and transitory mineral extraction operations) is from three sources: extraction of auriferous materials from river sediments; from veins where gold is found in the rocks; and alluvial, where gold is found on the banks of small rivers (Camara et al. 1997). The alluvial method is the most common and includes installation of equipment and housing, hydraulic pumping (high pressure water to bring down the pebble embankment), concentration of gold by mercury, and burning the gold to remove the mercury. The latter step is responsible for about 70% of the mercury entering the environment. The gold is sold at specialized stores where it is again fired. Metallic mercury can also undergo methylation in the river sediments and enter the food chain (Camara et al. 1997).

In the Amazon region of Brazil, more than 3000 tons of mercury were released into the biosphere from gold mining activities between 1987 and 1994, especially into the Tapajos River basin (Boas 1995; Castilhos et al. 1998). Local ecosystems receive about 100 tons of metallic mercury yearly, of which 45% enters river systems and 55% the atmosphere (Akagi et al. 1995). Mercury lost to rivers and soils as Hg^0 is comparatively unreactive and contributes little to mercury burdens in fish and other biota (de Lacerda 1997). Mercury entering the atmosphere is redeposited with rainfall at 90 to 120 $\mu g/m^2$ annually, mostly as Hg^{+2} and particulate mercury; these forms are readily methylated in floodplains, rivers, lakes, and reservoirs (de Lacerda 1997). Health hazards to humans include direct inhalation of mercury vapor during the processes of burning the Hg-Au amalgam and consuming mercury-contaminated fish. Methylmercury, the most toxic form of mercury, is readily formed (Akagi et al. 1995).

About 130 tons of Hg^0 are released annually by alluvial gold mining to the Amazonian environment, either directly to rivers or into the atmosphere, after reconcentration, amalgamation, and burning (Reuther 1994). In the early 1980s, the Amazon region in northern Brazil was the scene of the most intense gold rush in the history of Brazil (Hacon et al. 1995). Metallic mercury was used to amalgamate particulate metallic gold. Refining of gold to remove the mercury is considered to be the source of environmental mercury contamination; however, other sources of mercury emissions in Amazonia include tailings deposits and burning of tropical forests and savannahs (Hacon et al. 1995). In 1989 alone, gold mining in Brazil contributed 168 tons of mercury to the environment (Aula et al. 1995).

Lechler et al. (2000) asserted that natural sources of mercury and natural biogeochemical processes contribute heavily to reported elevated mercury concentrations in fish and water samples collected as much as 900 km downstream from local gold mining activities. Based on analysis of water, sediments, and fish samples

systematically collected along a 900-km stretch of the Madeira River in 1997, they
concluded that the elevated mercury concentrations in samples were mainly derived
from natural sources, and that the effects of mercury released from gold mining sites
were localized (Lechler et al. 2000). This conclusion needs to be verified.

Mercury Concentrations in Abiotic Materials and Biota

Since 1980, during the present gold rush in Brazil, at least 2000 tons of mercury
were released into the environment (Malm 1998). Elevated mercury concentrations
are reported in virtually all abiotic materials, plants, and animals collected near
mercury-amalgamation gold mining sites (Table 13.1). Mercury concentrations in
samples show high variability, which may be related to seasonal differences,
geochemical composition of the samples, and species differences (Malm 1998). In
1992, more than 200 tons of mercury were used in the gold mining regions of Brazil
(von Tumpling et al. 1995). One area, near Pocone, has been mined for more than
200 years. In the 1980s, about 5000 miners were working 130 gold mines in this
region. Mercury was used to amalgamate the preconcentrated gold particles for
separation of the gold from the slag. Mercury-contaminated wastes from the separation process were combined with the slag from the reconcentration process and
collected as tailings. The total mercury content in tailings piles in this geographic
locale was estimated at about 1600 kg, or about 12% of all mercury used in the past
10 years. Surface runoff from tropical rains caused extensive erosion of tailings
piles, some 4.5 meters high, with contaminated material reaching nearby streams
and rivers. In the region of Pocone, mercury concentrations in waste tailings material
ranged from 2 to 495 µg/kg, occupied 4.9 km^2, and degraded an estimated 12.3 km^2
(von Tumpling et al. 1995).

Tropical ecosystems in Brazil are under increasing threat of development and
habitat degradation from population growth and urbanization, agricultural expansion,
deforestation, and mining (Lacher and Goldstein 1997). Where mercury has been
released into the aquatic system as a result of unregulated gold mining, subsequent
contamination of invertebrates, fish, and birds was measured and biomagnification
of mercury was documented from gastropod molluscs (*Ampullaria* spp.) to birds
(snail kite, *Rostrhamus sociabilis*) and from invertebrates and fish to water birds and
humans (Lacher and Goldstein 1997). Indigenous peoples of the Amazon living near
gold mining activities have elevated levels of mercury in their hair and blood. Other
indigenous groups are also at risk from mercury contamination as well as from
malaria and tuberculosis (Greer 1993). The miners, mostly former farmers, are also
victims of hard times and limited opportunities. Small-scale gold mining offers an
income and an opportunity for upward mobility (Greer 1993).

Throughout the Brazilian Amazon, about 650,000 small-scale miners are responsible for about 90% of Brazil's gold production and for the discharge of 90 to 120 tons
of mercury to the environment every year. About 33% of the miners had elevated
concentrations in tissues over the tolerable limit set by the World Health Organization
(WHO) (Greer 1993). In Brazil, it is alleged that health authorities are unable to
detect conclusive evidence of mercury intoxication due to difficult logistics and the
poor health conditions of the mining population that may mask evidence of mercury

Table 13.1 Total Mercury Concentrations in Abiotic Materials, Plants, and Animals near Active Brazilian Gold Mining and Refining Sites

Location, Sample, and Other Variables	Concentration[a]	Ref[b]
Amazon Region		
Livestock; Gold Field vs. Reference Site		
Hair		
Cattle	0.2 mg/kg dry weight (DW) vs. 0.1 mg/kg DW	1
Pigs	0.9 mg/kg DW vs. 0.2 mg/kg DW	1
Sheep	0.2 mg/kg DW vs. 0.1 mg/kg DW	1
Blood		
Cattle	12 µg/L vs. 5 µg/L	1
Pigs	18 µg/L vs. 13 µg/L	1
Sheep	3 µg/L vs. 1 µg/L	1
Humans		
Blood		
Miners	(2–29) µg/L	1
Villagers	(3–10) µg/L	1
River dwellers	(1–65) µg/L	1
Reference site	(2–10) µg/L	1
Urine		
From people in gold processing shops vs. maximum allowable level vs. reference site	269 (10–1168) µg/L vs. <50 µg/L vs. 12.0 (1.5–74.3) µg/L	16
Miners	(1–155) µg/L	1
Villagers	(1–3) µg/L	1
Reference site	(0.1–7) µg/L	1
Hair		
Pregnant women vs. maximum allowable for this cohort	3.6 (1.4–8.0) mg/kg fresh weight (FW) vs. <10 mg/kg FW	16
Miners	(0.4–32.0) mg/kg DW	1
Villagers	(0.8–4.6) mg/kg DW	1
River dwellers	(0.2–15.0) mg/kg DW	1
Reference site	<2 mg/kg DW	1
Soils		
Forest soils; 20–100 m from amalgam refining area vs. reference site	2.0 (0.4–10.0) mg/kg DW vs. 0.2 (max 0.3) mg/kg DW	16
Urban soils; 5–350 m from amalgam refining area vs. reference site	7.5 (0.5–64.0) mg/kg DW vs. 0.4 (0.03–1.3) mg/kg DW	16
Alta Floresta and Vicinity		
Air; near mercury emission areas from gold purification vs. indoor gold shop	(0.02–5.8) µg/m^3 vs. (0.25–40.6) µg/m^3	2,3
Fish muscle, carnivorous species	0.3–3.6 mg/kg FW	3
Soil	(0.05–4.1) mg/kg DW	2,4
Madeira River and Vicinity		
Air	(10–296) µg/m^3	4
Aquatic macrophytes		
Leaves; floating vs. submerged	0.9–1.0 mg/kg DW vs. 0.001 mg/kg DW	4
Victoria amazonica	0.9 mg/kg DW	5

Table 13.1 (continued)   Total Mercury Concentrations in Abiotic Materials, Plants, and Animals near Active Brazilian Gold Mining and Refining Sites

Location, Sample, and Other Variables	Concentration[a]	Ref[b]
Eichornia crassipes	(0.04–1.01) mg/kg DW	5
Echinocloa polystacha	<0.008 DW	5
Fish eggs, detritovores	(0.05–3.8) mg/kg FW	5
Fish muscle		
Carnivores vs. omnivores	0.5–2.2 mg/kg FW vs. 0.04–1.0 mg/kg FW	5
Carnivorous species vs. noncarnivorous species	Max. 2.9 mg/kg FW vs. Max. 0.65 mg/kg FW	4
7 species		
Herbivores	0.08 mg/kg FW; Max. 0.2 mg/kg FW	6
Omnivores	0.8 mg/kg FW; Max. 1.7 mg/kg FW	6
Piscivores	0.9 mg/kg FW; Max. 2.2 mg/kg FW	6
Maximum	2.7 mg/kg FW	7, 15
Water	Max. 8.6 to 10.0 µg/L	7, 15
Sediments	19.8 mg/kg DW; Max. 157 mg/kg DW	7, 15

Mato Grosso

Freshwater molluscs (*Ampullaria* spp., *Marisa planogyra*); soft parts	Max. 1.2 mg/kg FW	8
Sediments	Max. 0.25 mg/kg FW	8

Negro River

Fish muscle; fish-eating species vs. herbivores	Max. 4.2 mg/kg FW vs. Max. 0.35 mg/kg FW	4

Pantanal

Clam, *Anodontitis trapesialis*; soft parts	0.35 mg/kg FW	9
Clam, *Castalia* sp.; soft parts	0.64 mg/kg FW	9

Parana River

Water; dry season vs. rainy season	0.41 µg/L vs. 2.95 µg/L	4

Pocone and Vicinity

Air	<0.14–1.68 µg/m^3	4
Fish muscle; carnivores vs. noncarnivores	Max. 0.68 mg/kg FW vs. Max. 0.16 mg/kg FW	4
Surface sediments	0.06–0.08 mg/kg DW	4

Porto Velho

Air	0.1–7.5 µg/m^3	4
Soils; near gold dealer shops vs. reference site	(0.4–64.0) mg/kg DW vs. (0.03–1.3) mg/kg DW	4

Tapajas River

Fish muscle; carnivores vs. noncarnivores	Max. 2.6 mg/kg FW vs. Max. 0.31 mg/kg FW	4
Fish muscle; contaminated site vs. reference site 250 km downstream		

Table 13.1 (continued)   Total Mercury Concentrations in Abiotic Materials, Plants, and Animals near Active Brazilian Gold Mining and Refining Sites

Location, Sample, and Other Variables	Concentration[a]	Ref[b]
Carnivorous fishes	0.42 mg/kg FW vs. 0.23 mg/kg FW	10
Noncarnivorous fishes	0.06 mg/kg FW vs. 0.04 mg/kg FW	10
Sediments; mining area vs. reference site		
Total mercury	0.14 mg/kg FW vs. (0.003–0.009) mg/kg FW	4
Methylmercury	0.8 µg/kg FW vs. 0.07–0.19 µg/kg FW	4
Teles River Mining Site		
Air	(0.01–3.05) µg/m^3	4
Fish muscle	Max. 3.8 mg/kg FW	4
Tucurui Reservoir and Vicinity		
Aquatic macrophytes		
Floating vs. submerged	0.12 mg/kg DW vs. 0.03 mg/kg DW	4
Floating plants; roots vs. shoots	Max.0.098 mg/kg DW vs. Max. 0.046 mg/kg DW	11
Fish muscle, 7 species	0.06–2.6 mg/kg FW; Max. 4.5 mg/kg FW	12
Fish muscle; carnivores vs. noncarnivores	Max. 2.9 mg/kg FW vs. Max. 0.16 mg/kg FW	4
Gastropods; soft parts vs. eggs	0.06 (0.01–0.17) mg/kg FW vs. ND[a]	12
Turtle, *Podocnemis unifilis*; egg	0.01 (0.007–0.02) mg/kg FW	12
Caiman (crocodile), *Paleosuchus* sp.; muscle vs. liver	1.9 (1.2–3.6) mg/kg FW vs. 19.0 (11.0–30.0) mg/kg FW	12
Capybara (mammal), *Hydrochoerus hydrochaeris*		
Hair	0.16 (0.12–0.19) mg/kg DW	12
Liver	0.01 (0.006–0.01) mg/kg FW	12
Muscle	0.02 (0.007–0.03) mg/kg FW	12
Sediments	0.13 (0.07–0.22) mg/kg DW	11
Various Locations, Brazil		
Air		
Mining areas	Max. 296 µg/m^3	4
Rio de Janeiro	(0.02-0.007) µg/m^3	4
Rural areas vs. urban areas	0.001–0.015 µg/m^3 vs. 0.005–0.05 µg/m^3	4
Bromeliad epiphyte (plant), *Tillandsia usenoides*; Exposure for 45 days; Dry Season vs. Rainy Season		
Near mercury emission sources	12.2 (1.9–22.5) mg/kg FW vs. 5.2 (2.5–9.5) mg/kg FW	14
Inside gold shop	4.3 (0.6–26.8) mg/kg FW vs. 1.7 (0.2–5.3) mg/kg FW	14
Local controls	0.2 (<0.08–0.4) mg/kg FW vs. 0.09 (<0.08–0.12) mg/kg FW	14
Rio de Janeiro controls	0.2 (<0.08–0.4) mg/kg FW vs. <0.08 mg/kg FW	14
Fish muscle		
Near gold mining areas	0.21–2.9 mg/kg FW	13
Global, mercury contaminated	1.3–24.8 mg/kg FW	13

Table 13.1 (continued)   Total Mercury Concentrations in Abiotic Materials, Plants, and Animals near Active Brazilian Gold Mining and Refining Sites

Location, Sample, and Other Variables	Concentration[a]	Ref[b]
Reference sites; carnivorous species vs. noncarnivorous species	Max. 0.17 mg/kg FW vs. Max. <0.10 mg/kg FW	4
Lake water; gold mining areas vs. reference sites	0.04–8.6 µg/L vs. <0.03 µg/L	1
River water; mining areas vs. reference sites	0.8 µg/L vs. <0.2 µg/L	1
Sediments; gold mining areas vs. reference sites	0.05–19.8 mg/kg DW vs. <0.04 mg/kg DW	13
Soils (forest); gold mining areas vs. reference sites	0.4-10.0 mg/kg DW vs. 0.03–0.34 mg/kg DW	4

[a] Concentrations are shown as means, range (in parentheses), maximum (Max.), and nondetectable (ND).
[b] Reference: 1, Palheta and Taylor 1995; 2, Hacon et al. 1995; 3, Hacon et al. 1997; 4, de Lacerda and Salomons 1998; 5, Martinelli et al. 1988; 6, Dorea et al. 1998; 7, Pfeiffer et al. 1989; 8, Vieira et al. 1995; 9, Callil and Junk 1999; 10, Castilhos et al. 1998; 11, Aula et al. 1995; 12, Aula et al. 1994; 13, Pessoa et al. 1995; 14, Malm et al. 1995a; 15, Malm et al. 1990; 16, Malm et al. 1995b.

poisoning. There is a strong belief that a silent outbreak of mercury poisoning has the potential for regional disaster (de Lacerda and Salomons 1998).

In the Madeira River Basin, mercury levels in certain sediments were 1500 times higher than similar sediments from nonmining areas, and dissolved mercury concentrations in the water column were 17 times higher than average for rivers throughout the world (Greer 1993). High concentrations of mercury were measured in fish and sediments from a tributary of the Madeira River affected by alluvial small-scale mining (Reuther 1994). The local safety limit of 0.1 mg Hg/kg DW sediment was exceeded by a factor of 25 and the safety level for fish muscle of 0.5 mg Hg/kg FW muscle was exceeded by a factor of 4. Both sediments and fish act as potential sinks for mercury because existing physicochemical conditions in these tropical waters (low pH, high organic load, high microbial activity, elevated temperatures) favor mercury mobilization, methylation, and availability (Reuther 1994).

In Amazonian river sediments, mercury methylation accounts for less than 2.2% of the total mercury in sediments (de Lacerda and Salomons 1998). In soils, mercury mobility is low, in general (de Lacerda and Salomons 1998). There is an association between the distribution of mercury-resistant bacteria in sediments and the presence of mercury compounds (Cursino et al. 1999). Between 1995 and 1997, mercury concentrations were measured in sediment along the Carmo stream, Minas Gerais, located in gold prospecting areas. Most sediments contained more than the Brazilian allowable limit of 0.1 mg Hg/kg DW. Mercury-resistant bacteria were present in sediments at all sites and ranged from 27 to 77% of all bacterial species, with a greater percentage of species showing resistance at higher mercury concentrations (Cursino et al. 1999).

The Pantanal is one of the largest wetlands in the world and extends over 300,000 km^2 along the border area of Brazil, Bolivia, Argentina, and Paraguay (Guimaraes et al. 1998). Half this surface is flooded annually. Since the 18th century, gold has been extracted from quartz veins in Brazil using amalgamation as a concentration process, resulting in metallic mercury releases to the atmosphere, soils, and sediments.

The availability to aquatic biota of Hg^0 released by gold mining activities is limited to its oxidation rate to Hg^{+2} and then by conversion to methylmercury (CH_3Hg^+), which is readily soluble in water (Guimaraes et al. 1998).

The Pantanal in Brazil, at 140,000 km², is an important breeding ground for storks, herons, egrets, and other birds, as well as a refuge for threatened or endangered mammals including jaguars (*Panthera onca*), giant anteaters (*Myrmecophaga tridactyla*), and swamp deer (*Cervus duvauceli*) (Alho and Viera 1997). Gold mining is common in the northern Pantanal. There are approximately 700 operating gold mining dredges along the Cuiba River. Unregulated gold mines have contaminated the area with mercury, and 35 to 50% of all fishes collected from this area contain more than 0.5 mg Hg/kg FW muscle, the current Brazilian and international WHO standard for fish consumed by humans (Alho and Viera 1997). Gastropod molluscs that are commonly eaten by birds contained 0.02 to 1.6 mg Hg/kg FW soft tissues. Mercury concentrations in various tissues of birds that ate these molluscs were highest in the anhinga (*Anhinga anhinga*) at 0.4 to 1.4 mg/kg FW and the snail kite at 0.3 to 0.6 mg/kg FW, and lower in the great egret [*Casmerodius* (formerly *Ardea*) *albus*] at 0.02 to 0.04 mg/kg FW and in the limpkin (*Aramus guarauna*) at 0.1 to 0.5 mg/kg FW. The high mercury levels detected, mainly in fishes, show that the mercury used in gold mining and released into the environment has reached the Pantanal and spread throughout the ecosystem with potential biomagnification (Alho and Viera 1997).

Floating plants accumulate small amounts of mercury (see Table 13.1), but their sheer abundance makes them likely candidates for mercury phytoremediation. For example, in the Tucurui Reservoir in the state of Para, it is estimated that 32 tons of mercury are stored in floating plants, mostly *Scurpus cubensis* (Aula et al. 1995). Mercury methylation rates in sediments and floating plants were evaluated in Fazenda Ipiranga Lake, 30 km downstream from gold mining fields near the Pantanal during the dry season of 1995 (Guimaraes et al. 1998). Sediments and roots of dominant floating macrophytes (*Eichornia azurea*, *Salvina* sp.) were incubated *in situ* for 3 days with about 43 µg Hg^{+2}/kg DW added as $^{203}HgCl_2$. Net methylation was about 1% in sediments under floating macrophytes, being highest at temperatures in the 33° to 45°C range and high concentrations of sulfate-reducing bacteria. Methylation was inhibited above 55°C, under saline conditions, and under conditions of low sulfate. Mercury-203 was detectable to a depth of 16 cm in the sediments, coinciding with the depth reached by chironomid larvae. Methylation was up to nine times greater in the roots of floating macrophytes than in the underlying surface sediments: an average of 10.4% of added Hg^{+2} was methylated in *Salvina* roots in 3 days and 6.5% in *Eichornia* roots (Guimaraes et al. 1998). Using radiomercury-203 tracers (^{203}Hg), no methylation was observed under anoxic conditions in organic-rich, flocculent surface sediments due to the formation of HgS, a compound that is much less available for methylation than is Hg^{+2} (Guimaraes et al. 1995). The authors concluded that floating macrophytes should be considered in evaluation of mercury methylation rates in tropical ecosystems (Guimaraes et al. 1998).

Clams collected near gold mining operations had elevated concentrations of mercury (up to 0.64 mg Hg/kg FW) in soft tissues (see Table 13.1). Laboratory studies suggest that mercury adsorbed to suspended materials in the water column is the most likely route for mercury uptake by filter-feeding bivalve molluscs (Callil and Junk 1999).

Mercury concentrations in fish collected near gold mining activities in Brazil were elevated and decreased with increasing distance from mining sites (see Table 13.1). In general, muscle is the major tissue of mercury localization in fishes, and concentrations are higher in older, larger, predatory species (Aula et al. 1994; Eisler 2000; Lima et al. 2000). In the Tapajos River region, which receives between 70 and 130 tons of mercury annually from gold mining activities, mercury concentrations in fish muscle were highest in carnivorous species, lowest in herbivores, and intermediate in omnivores (Lima et al. 2000). However, only 2% of fish collected in 1988 (vs. 1% in 1991) from the Tapajos region exceeded the Brazilian standard of 0.5 mg total mercury/kg FW muscle, and all violations were from a single species of cichlid (tucunare/speckled pavon, *Cichla temensis*) (Lima et al. 2000). In a 1991 survey of 11 species of fishes collected from a gold mining area (Cachoeira de Teotonio), it was found that almost all predatory species had >0.5 mg Hg/kg FW muscle versus <0.5 mg/kg FW in conspecifics collected from Guajara, a distant reference site (Padovani et al. 1995). Limits on human food consumption were set for individual species on the basis of mercury concentrations in muscle. Specifically, no restrictions were placed on some species, mostly herbivores and omnivores, some restrictions on some omnivores and small predators, and severe restrictions on larger predators (Padovani et al. 1995).

Fish that live near gold mining areas have elevated concentrations of mercury in their flesh and are at high risk of reproductive failure (Oryu et al. 2001). Mercury concentrations of 10 to 20 mg/kg FW in fish muscle are considered lethal to the fish, and 1 to 5 mg/kg FW sublethal; predatory fishes frequently contain 2 to 6 mg Hg/kg FW muscle. Mercury-contaminated fish pose a hazard to humans and other fish consumers, including the endangered giant otter (*Pteronura brasiliensis*) and the jaguar. Giant otters eat mainly fish and are at risk from mercury intoxication: 1 to 2 mg Hg/kg FW diet is considered lethal. Jaguars consume fish and giant otters; however, no data are available on the sensitivity of this top predator to mercury (Oryu et al. 2001). Caiman crocodiles are also threatened by gold mining and related mercury contamination of habitat, increased predation by humans, extensive agriculture, and deforestation (Brazaitis et al. 1996). Caiman crocodiles (*Paleosuchus* spp.) from the Tucurui Reservoir area had up to 3.6 mg Hg/kg FW muscle and 30.0 mg Hg/kg FW liver (Aula et al. 1994). Extensive habitat destruction and mercury pollution attributed to mining activity were observed at 19 localities in Mato Grosso and environs. Crocodiles (*Caiman* spp., *Melanosuchus niger*, *Crocodilus crocodilus*) captured from these areas were emaciated, algae covered, in poor body condition, and heavily infested with leeches (Brazaitis et al. 1996).

Mitigation

There are two populations at significant risk from mercury intoxication in Brazilian gold mining communities: riverine populations, which routinely eat mercury-contaminated fish and have high levels of mercury in hair, and gold dealers in indoor shops exposed to Hg^0 vapors (de Lacerda and Salomons 1998). These two critical groups should receive special attention regarding exposure risks. Riverine populations, especially children and women of childbearing age, should avoid consumption of carnivorous fishes, and in gold dealer shops, adequate ventilation and treatment systems for mercury vapor retention should be installed (de Lacerda and Salomons 1998).

An occupational health and safety program was launched to educate Brazilian adolescents to industrial hazards including health risks associated with mercury amalgamation of gold (Camara et al. 1997). Adolescents, together with young adults, constitute a large proportion of the garimpos and are in a critical physical and psychological growth phase. The number of adolescents participating in gold extraction increases with decreasing family income and with the structure of the labor market as approved by public policy and the courts. The program was designed to transfer information on risks associated with mining and other dangerous occupations, to apprise them of their choices, and to promote training. The method had been successfully tested earlier in selected urban and rural schools of different economic strata.

The community selected was Minas Gerais, where gold was discovered more than 300 years ago and is the main source of community income. The municipality does not have a sewer system, and the water supply is not treated. The program was initiated in April 1994 over a 5-week period. The 70 students were from the 5th to 8th grade. Almost all worked, most in mining, where exposure to mercury was illegal. Students successfully improved safety habits and recognition of potential accident sites. A generalized occupational safety program, such as this one, is recommended for other school districts (Camara et al. 1997).

13.2.2 South America Other than Brazil

Mercury pollution is now recognized as one of the main environmental problems in tropical South America, as judged by increasingly elevated concentrations of mercury in water, sediments, aquatic macrophytes, freshwater snails, fish, and fish-eating birds. Human health is compromised through occupational mercury exposures from mining and by ingestion of methylmercury-contaminated fish (Mol et al. 2001).

A major portion of the mercury contamination noted in Central and South America originated from amalgamation of silver by the Spanish for at least 300 years starting in 1554 (Nriagu 1993). Until the middle of the 18th century, about 1.5 kg mercury were lost to the environment for every kilogram of silver produced. Between 1570 and 1820, mercury loss in this geographic region averaged 527 tons annually, or a total of about 126,000 tons during this period. Between 1820 and 1900, another 70,000 tons of mercury was lost to the environment through silver production for a total of 196,000 tons of mercury during the period 1570 to 1900. By comparison, the input of mercury into the Brazilian Amazon associated with gold mining is 90 to 120 tons annually. Under the hot tropical conditions typical of Mexico and parts of South America, mercury in abandoned mine wastes or deposited in aquatic sediments is likely to be methylated and released to the atmosphere where it cycles for considerable periods. It now seems reasonable to conclude that the Spanish American silver mines were responsible, in part, for the high background concentrations of mercury in the global environment now being reported (Nriagu 1993).

In Colombia, mercury intoxication was reported among fishermen and miners living in the Mina Santa Cruz marsh, possibly from ingestion of mercury-contaminated fish. This marsh received an unknown amount of mercury-contaminated gold mine wastes (Olivero and Solano 1998). However, mercury concentrations in marsh sediments, fish muscle, and macrophytes were low. Examination of abiotic and biotic materials from this location in 1996 showed maximum mercury concentrations of 0.4 mg/kg FW in sediment (range, 0.14 to 0.4), 0.4 mg/kg DW in roots of *Eichornia crassipes*, an aquatic macrophyte (range, 0.1 to 0.4), and 1.1 mg/kg FW in fish muscle. Among eight species of fishes examined, the mean mercury concentrations in muscle ranged between 0.03 and 0.38 mg/kg FW, and the range for all observations extended from 0.01 to 1.1 mg/kg (Olivero and Solano 1998).

In Peru, mercury in scat of the giant otter and in the otter's fish diet was measured in samples collected between 1990 and 1993 near a gold mine (Gutleb et al. 1997). Total mercury in fish muscle ranged between 0.05 and 1.54 mg/kg FW. In 68% of fish muscle samples analyzed, the total mercury levels exceeded the maximum tolerated level proposed by Gutleb et al. (1997) of 0.1 mg/kg FW in fish for the European otter *Lutra lutra*, and 17.6% exceeded 0.5 mg/kg FW, the recommended maximum level for human consumption. Most (61 to 97%) of the mercury in fish muscle was in the form of methylmercury. In otter scat, no methylmercury and a maximum of 0.12 mg total Hg/kg DW were measured. The authors concluded that the concentrations of mercury in fish flesh may pose a threat to humans and wildlife feeding on the fish and that all mercury discharges into tropical rainforests should cease immediately to protect human health and endangered wildlife (Gutleb et al. 1997).

In Suriname (population 400,000), gold mining has existed on a small scale since 1876, producing about 200 kg gold yearly through the 1970s and 1209 kg in 1988 using amalgamation as the extraction procedure of choice (de Kom et al. 1998). In the 1990s, gold mining activities increased dramatically because of hyperinflation and poverty; up to 15,000 workers produced 10,000 kg crude gold in 1995 using mercury methodology unchanged for at least a century (de Kom et al. 1998). Riverine food fishes from the vicinity of small-scale gold mining activities in Suriname were contaminated with mercury, and concentrations in muscle of fish-eating fishes collected there exceeded the maximum permissible concentration of 0.5 mg total Hg/kg FW in 57% of the samples collected (Mol et al. 2001). Total mercury and methylmercury concentrations were elevated in sediments and water from artisanal gold mines in Suriname (Gray et al. 2002). Maximum concentrations recorded in mine waters were 0.93 µg total Hg/L and 0.0038 µg methylmercury/L; for reference sites, these values were 0.01 µg total Hg/L and 0.00028 µg methylmercury/L. Maximum concentrations in stream sediments below the mines were 0.15 mg total Hg/kg (vs. 0.048 mg total Hg/kg from a reference site) and 0.0014 µg methylmercury/kg (vs. 0.00008 µg methylmercury/kg from reference site). Increasing total mercury contents in discharged mine water correlated positively with increasing water turbidity, indicating that most mercury transport is on suspended particulates (Gray et al. 2002).

In Venezuela, miners exploring the region for gold over a 10-year period cut and burned virgin forests, excavated large pits through the floodplain with pumps and high-pressure hoses, hydraulically dredged stream channels, and used mercury (called exigi by miners) to isolate gold from sediments (Nico and Taphorn 1994). The use of mercury in Venezuelan gold mining is common and widespread, with annual use estimated at 40 to 50 tons. Between 1979 and 1985, at least 87 tons of mercury were discharged into river systems (Nico and Taphorn 1994). Forest soils in Venezuela near active gold mines contained as much as 129.3 mg Hg/kg DW compared with 0.15 to 0.28 mg/kg from reference sites (Davies 1997; de Lacerda and Salomons 1998). Maximum mercury concentrations measured in nine species of fish collected in 1992 from gold mining regions in the upper Cuyuni River system, Venezuela, were (in mg total Hg/kg FW) 2.6 in liver, 0.9 in muscle, 0.8 in stomach contents, and 0.1 in ovary (Nico and Taphorn 1994). The highest value, 8.9 mg/kg FW muscle, was from a 63-cm-long aimara (*Hoplias macrophthalmus*), a locally important food fish (Nico and Taphorn 1994).

13.2.3 Africa

Gold mining in Tanzania dates back to 1884 during Germany's colonial rule, with >100,000 kg gold produced officially since 1935. Between 150,000 and 250,000 Tanzanians are now involved in small-scale gold mining, with extensive use of mercury in gold recovery; about 6 tons of mercury are lost to the environment annually from gold mining activities. The current gold rush, which began in the 1980s and reached a peak in 1983, was stimulated, in part, by the relaxation of mining regulations by the state-controlled economy. Gold ore processing involves crushing and grinding the ore, gravity separation involving simple panning or washing, amalgamation with mercury of the gold-rich concentrate, and firing in the open air. Additional firing is conducted at the gold ore processing sites, in residential compounds, and inside residences. Further purification is done in the national commercial bank branches before the bank purchases the gold from the miners (Ikingura and Akagi 1996; Ikingura et al. 1997).

Mercury concentrations recorded in water at the Lake Victoria gold fields in Tanzania averaged 0.68 (0.01 to 6.8) µg/L versus 0.02 to 0.33 µg/L at an inland water reference site and 0.02 to 0.35 µg/L at a coastal water reference site (Ikingura et al. 1997). Mercury concentrations in sediments, forest soils, and tailings at the Lake Victoria gold fields were 0.02 to 136.0 mg/kg in sediments, 3.4 (0.05 to 28.0) mg/kg in soils versus 0.06 mg/kg at a reference site, and 16.2 (0.3 to 31.2) mg/kg in tailings (Ikingura and Akagi 1996; Ikingura et al. 1997; de Lacerda and Salomons 1998). The highest mercury concentrations recorded in any fish species from Lake Victoria or in rivers draining from gold processing sites were in the marbled lungfish *Protopterus aethiopicus*, with 0.24 mg Hg/kg FW in muscle and 0.56 mg Hg/kg FW in liver; however, a catfish (*Clarias* sp.) from a tailings pond contained 2.6 mg Hg/kg FW muscle and 4.5 in liver (van Straaten 2000).

In Obuasi, Ghana, elevated mercury concentrations were recorded in various samples collected from 14 sites near active gold mining towns and environs during 1992 to 1993 (Amonoo-Neizer et al. 1996). Mercury concentrations (in mg/kg DW) were 0.9 (0.3 to 2.5) in soil, 2.0 maximum in whole mudfish (*Heterobranchus bidorsalis*), 2.3 maximum in edible portions of plantain (*Musa paradisiaca*), 3.3 maximum in edible portions of cassava (*Manihot esculenta*), 9.1 maximum in whole elephant grass (*Pennisetum purpureum*), and 12.4 maximum in whole water fern (*Ceratopterus cornuta*).

In Kenya, Ogola et al. (2002) reported elevated mercury concentrations near gold mines in tailings (maximum 1920 mg/kg DW) and surficial stream sediments (maximum 348 mg/kg DW).

Tailings also contained elevated concentrations of arsenic (76 mg/kg DW) and lead (510 mg/kg DW); stream sediments contained up to 1.9 mg As/kg DW and up to 11,075 mg Pb/kg DW.

13.2.4 People's Republic of China

Dexing County in Jiangxi Province was the site of about 200 small-scale gold mines using mercury amalgamation to extract gold in 1990 to 1995 (Lin et al. 1997). Gold firing was usually conducted in private residences. One result of this activity was mercury contamination of air (1.0 mg/m^3 in workshop, 2.6 mg/m^3 in workroom vs. 10 to 20 ng/m^3 at reference sites), wastewater (0.71 mg/L), and solid tailings (maximum 189.0 mg/kg), with serious implications for human health. Since September 1996, however, most small-scale mining activities have been prohibited through passage of national environmental legislation (Lin et al 1997).

13.2.5 The Philippines

The gold rush began in eastern Mindanao in the 1980s, resulting in the development of several mining communities with more than 100,000 residents. Between 1986 and 1988, about 140 tons of mercury were released into the environment from 53 mining communities (Appleton et al. 1999). In the early 1990s, about 200,000 small-scale gold miners produced about 15 tons of gold each year and, in the process, released 25 tons of mercury annually, with concomitant contamination of aquatic ecosystems (Greer 1993). Much of the discharged mercury is in the river sediments, where it can be recycled through flash flooding, consumption by bottom-feeding fish, or microbial digestion and methylation (Greer 1993). In mercury-contaminated areas of Davao del Norte on Mindanao, mercury concentrations were as high as 2.6 mg/kg FW in muscle of carnivorous fishes and 136.4 µg/m^3 air (de Lacerda and Salomons 1998). In 1995, drainage downstream of gold mines in Mindanao was characterized by extremely high levels of mercury in solution (2.9 mg/L) and in bottom sediments (>20 mg Hg/kg; Appleton et al. 1999).

The Environment Canada sediment quality mercury toxic effect threshold for the protection of aquatic life (<1 mg Hg/kg DW) was exceeded for a downstream distance of 20 km. In sections of stream used for fishing and potable water supply, the surface water mercury concentrations, for at least 14 km downstream, exceeded the WHO international standard for drinking water (set at <1 µg Hg/L) and the U.S. Environmental Protection Agency water quality criterion for the protection of aquatic life (<2.1 µg/L; Appleton et al. 1999). At present, about 1.6 tons of mercury are released for every ton of gold produced in the Philippines, with releases usually highest to the atmosphere and lower to rivers and soils. However, mercury releases can be reduced by at least an order of magnitude through use of closed amalgamation systems and use of retorts in the roasting process (de Lacerda and Salomons 1998).

13.2.6 Siberia

Measurements of air, soil, surface water, and groundwater near gold mines in eastern Transbaikalia, Siberia, where mercury amalgamation was used to extract gold from ores, showed no evidence of significant mercury contamination in any compartment measured (Tupyakov et al. 1995). Mean mercury concentrations, in ng/m^3, in air near the gold mines versus a reference site were 2.0 versus 2.2. For soils it was 0.03 mg/kg versus 0.02. And for surface water and groundwater, these values were 0.01 µg/L versus 0.006, and 0.008 µg/L versus 0.005, respectively.

13.2.7 Canada

Gold was discovered in Nova Scotia in 1861 (Nriagu and Wong 1997). Since then, more than 1.2 million ounces of gold were produced from about 65 mines in the province, with about 20% from mines located in Goldenville. Processing of the 3.3 million tons of ores from Goldenville mills between 1862 and 1935 released an estimated 63 tons of mercury to the biosphere, mostly before 1910. Mine tailings at one Goldenville site still contain large quantities of mercury (up to 2600 mg Hg/kg) as unreacted elemental mercury, unrecovered Au-Hg amalgams, and mercury compounds formed by side reactions during the amalgamation process. Current mercury levels in muds and sediments at the mine site range from 100 to 250 mg/kg, well above the 0.4 to 2.0 µg/kg typically found in uncontaminated stream sediments (Nriagu and Wong 1997).

Concentrations of other elements in Goldenville mine tailings were also high: up to 2000 mg As/kg, 1400 mg Pb/kg, 500 mg Cu/kg, and 80 mg Se/kg. Except for some mosses, no vegetation grows on the tailings-covered areas. Stream sediments located 200 meters downstream of the mine were 10 to 1000 times higher than upstream concentrations for Hg, As, Cd, Cu, Pb, and Se. Little is known about the forms of mercury in contaminated mine sites, which can affect the loss rate of mercury to the environment from the waste materials (Nriagu and Wong 1997). Aquatic macrophytes from gold mining areas in Nova Scotia, where mercury was used extensively, accumulated significant quantities of mercury (de Lacerda and Salomons 1998). Mercury concentrations were higher in emergent species (16.3 mg/kg DW) when compared with floating and submerged species (0.54 to 0.56 mg/kg DW). Also, aquatic macrophytes growing on tailings had higher concentrations of mercury in roots than in shoots (de Lacerda and Salomons 1998).

In Quebec, gold mines at Val d'Or used mercury amalgamation techniques throughout most of the 20th century to produce gold (Meech et al. 1998). Today, most abandoned sites and environs show high mercury concentrations in sediments (up to 6 mg/kg) and in fish muscle (up to 2.6 mg/kg FW). As alluvial gold deposits have become exhausted in North America, cyanidation has become the primary method for primary gold production, and amalgamation is now limited to individual prospectors (Meech et al. 1998).

In Yellowknife, NWT, about 2.5 tons of mercury were discharged between the mid-1940s and 1968, together with tailings, into Giauque Lake. This lake is now listed as a contaminated site under the Environment Canada National Contaminated Site Program (Meech et al. 1998). In 1977 to 1978, about 70% of the bottom sediments contained >0.5 mg Hg/kg DW (Moore and Sutherland 1980). Also in 1977 to 1978, lake trout (*Salvelinus namaycush*) were found to contain 3.8 mg Hg/kg FW muscle; for northern pike (*Esox lucius*) and round whitefish (*Prosopium cylindraceum*), these values were 1.8 and 1.2 mg/kg FW, respectively. The authors concluded that contamination of only a small part of the lake results in high levels in fish throughout the lake, probably due to movement of fish from mercury-contaminated to uncontaminated areas (Moore and Sutherland 1980). In 1995, sediments from Giauque Lake were dated using lead-210 (^{210}Pb) and cesium-137 (^{137}Cs), and analyzed for mercury (Lockhart et al. 2000). The peak mercury concentrations in sediment cores ranged between 2 and 3 mg Hg/kg DW. The history of mercury deposition derived from the dated sediment cores agreed well with the known history of input from gold mining (Lockhart et al. 2000).

13.2.8 The United States

Gold mining in the U.S. is ubiquitous; however, persistent mercury hazards to the environment were considered most severe from activities conducted during the latter portion of the 19th century, especially in Nevada. In the 50-year period 1850 to 1900, gold mining in the U.S. consumed 268 to 2820 tons of mercury yearly, or about 70,000 tons during that period (de Lacerda and Salomons 1998).

A recent study that measured total mercury accumulations in predatory fishes collected nationwide in 1998 showed that mercury levels in muscle were significantly correlated with methylmercury concentrations in water, pH of the water, percent wetlands in the basin, and the acid volatile content of the sediment (Brumbaugh et al. 2001). These four variables, especially methylmercury levels in water, accounted for 45% of the variability in mercury concentrations of fish, normalized by total length. A methylmercury water concentration of 0.12 ng/L was, on average, associated with a fish fillet concentration of 0.3 mg Hg/kg FW for an age 3 (3-year-old) fish when all species were considered. Sampling sites with the highest overall mercury concentrations in water, sediment, and fish were highest in the Nevada Basins and environs (from historic gold mining activities), followed, in order, by the South Florida Basin, Santee River Basin, Sacramento River Basin in California, and drainages in South Carolina, and the Long Island and New Jersey coastal drainages (Table 13.2). Elevated mercury concentrations in fish, except for Nevada, are not necessarily a result of gold mining activities. The mercury criterion for human health protection set by the U.S. Environmental Protection Agency in 2001 is now 0.3 mg Hg/kg FW diet, down from 0.5 mg/kg FW previously (Brumbaugh et al. 2001).

Beginning in 1859, gold from the Comstock Lode near Virginia City, Nevada, was processed at 30 sites using a crude mercury amalgamation process, discharging

Table 13.2 Total Mercury Concentrations in Abiotic Materials, Plants, and Animals near Historic Gold Mining and Refining Sites in the United States

Location, Sample, and Other Variables	Concentration[a]	Ref[b]
Alaska; Northeastern Bering Sea; Sediments		
80 m below sea floor; typical vs. anomalies	0.03 mg/kg DW vs. 0.2–1.3 mg/kg DW	1
Modern beach; mean vs. max.	0.22 mg/kg DW vs. 1.3 mg/kg DW	1
Nearshore subsurface gravels; mean vs. max.	0.06 mg/kg DW vs. 0.6 mg/kg DW	1
Seward Peninsula: mean vs. max.	0.1 mg/kg DW vs. 0.16 mg/kg DW	1
Georgia; Former Gold Mining Areas		
Historical values (1829–1940)		
Sediments	0.12–12.0 mg/kg DW	14
Soils	0.2–0.6 mg/kg DW	14
Samples collected in 1990s		
Surface waters	Max. 0.013 µg/L	12
Sediments	Max. 4.0 mg/kg DW near source mines; <0.1 mg/kg DW 10–15 km downstream	13
Clams and mussels, soft parts	0.7 mg/kg DW	13
Nationwide; 1998; 106 Sites; Freshwater Fish; Muscle		
Lahontan Reservoir, Nevada (near historic gold mining sites)	3.34 mg/kg FW	2
South Florida Basin	0.95 mg/kg FW	2
Santee River, South Carolina	0.70 mg/kg FW	2
Sacramento River Basin, California	0.46 mg/kg FW	2
Yellowstone River Basin	0.44 mg/kg FW	2
Acadian-Ponchartrain Basin	0.39 mg/kg FW	2
Others	<0.3 mg/kg FW[c]	2
Nevada; Migratory Waterbirds; 1997–1998; Nesting on Lower Carson River vs. Reference Site (Humboldt River)		
Double Crested Cormorant, *Phalacrocorax auritus*; Adults		
Blood	17.1 (11.6–22.0) mg/kg FW vs. 3.1 mg/kg FW	3
Brain	11.3 (8.9–14.9) mg/kg FW vs. 1.2 mg/kg FW	3
Kidney	69.4 (36.2–172.9) mg/kg FW vs. 9.0 mg/kg FW	3
Liver	134 (82.4–222.2) mg/kg FW vs. 18.0 mg/kg FW	3
Snowy Egret, *Egretta thula*; Adults		
Blood	5.9 (2.8–10.3) mg/kg FW vs. 3.1 mg/kg FW	3
Brain	2.3 (2.0–3.8) mg/kg FW vs. 0.9 mg/kg FW	3
Kidney	11.1 (5.7–21.7) mg/kg FW vs. 3.1 mg/kg FW	3
Liver	43.7 (16.0–109.9)mg/kg FW vs. 7.9 mg/kg FW	3

Table 13.2 (continued)   Total Mercury Concentrations in Abiotic Materials, Plants, and
                         Animals near Historic Gold Mining and Refining Sites in the United States

Location, Sample, and Other Variables	Concentration[a]	Ref[b]
Black-Crowned Night-Heron, *Nycticorax nycticorax*; Adults		
Blood	6.6 (3.2–14.2) mg/kg FW vs. 2.5 mg/kg FW	3
Brain	1.7 (0.8–5.5) mg/kg FW vs. 0.5 mg/kg FW	3
Kidney	6.1 (3.4–15.1) mg/kg FW vs. 1.8 mg/kg FW	3
Liver	13.5 (5.0–49.6) mg/kg FW vs. 2.1 mg/kg FW	3
Nevada; Migratory Waterbirds; 1997–1998; Lahontan Reservoir		
Stomach Contents; Adults; Total Mercury vs. Methylmercury		
Black-crowned night-heron	0.5 mg/kg FW vs. 0.48 mg/kg FW	3
Snowy egret	1.0 mg/kg FW vs. 0.0 mg/kg FW	3
Double crested cormorant	1.4 (0.8–2.2) mg/kg FW vs. 1.2 (0.8–1.6) mg/kg FW	3
Nestlings/Fledglings; Lahontan Reservoir vs. Reference Site		
Black-Crowned Night-Heron		
Blood	3.2 (1.2–5.7) mg/kg FW vs. 0.6 mg/kg FW	3
Brain	1.1 (0.6–1.7) mg/kg FW vs. 0.1 mg/kg FW	3
Feathers	32.3 (21.9–67.1) mg/kg FW vs. 5.5 (3.4–16.6) mg/kg FW	3
Kidney	3.4 (1.8–6.3) mg/kg FW vs. 0.5 mg/kg FW	3
Liver	4.0 (1.1–7.8) mg/kg FW vs. 0.7 mg/kg FW	3
Snowy Egret		
Blood	2.7 mg/kg FW vs. 0.4 mg/kg FW	3
Brain	0.7 mg/kg FW vs. 0.1 mg/kg FW	3
Feathers	30.6 (14.7–59.8) mg/kg FW vs. 7.8 mg/kg FW	3
Kidney	2.2 mg/kg FW vs 0.3 mg/kg FW	3
Liver	2.7 mg/kg FW vs. 0.4 mg/kg FW	3
Double Crested Cormorant		
Blood	5.4 (3.4–7.2) mg/kg FW vs. 0.8 (0.6–1.2) mg/kg FW	3
Brain	2.7 (2.1–3.3) mg/kg FW vs. 0.5 (0.4–0.5) mg/kg FW	3
Feathers	66.3 (54.0–87.3) mg/kg FW vs. 11.8 (10.9–13.5) mg/kg FW	3
Kidney	6.2 (2.1–9.8) mg/kg FW vs. 1.2 (1.1–1.3) mg/kg FW	3
Liver	10.9 (8.7–14.9) mg/kg FW vs. 1.8 (1.4–2.6) mg/kg FW	3
Nevada; Abiotic Materials		
Air; over Comstock Lode tailings vs. reference site	0.23 μg/m^3 vs. 0.01–0.04 μg/m^3	4

Table 13.2 (continued) Total Mercury Concentrations in Abiotic Materials, Plants, and Animals near Historic Gold Mining and Refining Sites in the United States

Location, Sample, and Other Variables	Concentration[a]	Ref[b]
Carson River–Lahontan Reservoir; sampled in early 1990s		
Surface waters: total mercury vs. methylmercury	Max. 7.6 mg/L vs. Max. 0.007 mg/L	5
Surface waters: vs. downstream	0.004 mg/L vs. 1.5–2.1 mg/L	6
Water; filtered vs. unfiltered	0.113 mg/L vs. max. 0.977 mg/L	8
Unfiltered reservoir water vs. reference site	0.053–0.391 mg/L vs. 0.001–0.0033 mg/L	7
Mill tailings	3–1610 mg/kg FW	7
Atmospheric vapor	0.002–0.294 mg/m^3 vs. 0.001–0.004 mg/m^3	7
Sediments	27 (2–156) mg/kg FW vs. 0.01–0.05 mg/kg FW	7
Sediments; Carson River; below mines vs. above mines	Max. 881 mg/kg DW vs. 0.03–6.1 mg/kg DW	8
Sediments; Lahontan Reservoir; before dam vs. after dam	0.011 mg/kg DW vs. 1–60 mg/kg DW	8
Sediments; 1990s; 60 km downstream of Comstock Lode discharges	>100 mg/kg FW	11
Alluvial Fan Soils; Six Mile Canyon		
Premining (1859)	0.5 (0.07–1.9) mg/kg DW	11
Post mining		
Fan deposits	103.7 (1.3–368.9) mg/kg DW	11
Modern channel	4.8 (1.1–9.3) mg/kg DW	11

Nevada; Carson River vs. Reference Site; Sampled in Early 1990s

Fish muscle	Max. 5.5 mg/kg FW vs. Max. <0.5 mg/kg FW	7
Duck muscle	>0.5 mg/kg FW vs. <0.5 mg/kg FW	7

North Carolina; Near Marion (Gold Mining Activity Using Mercury in the 1800s to Early 1900s); 1991; Maximum Concentrations

Heavy metal concentrates	784 mg/kg	9
Gold grains	Max. 448 mg/kg	9
Sediments	7.4 mg/kg	9
Moss (max. gold content of 1.85 mg/kg)	4.9 mg/kg	9
Fish muscle	<0.2 mg/kg FW	9
Stream and well waters	<2 µg/L (limit of detection)	9

South Dakota; Lake Oahe; Mold Mining Discontinued in 1970; Sampled 1970–1971

Water	<0.2 µg/L	10
Sediments	0.03–0.62 mg/kg DW; <0.03–0.33 mg/kg FW	10
Fish muscle	0.02–1.05 mg/kg FW; 13% of samples had >0.5 mg/kg FW, especially older, larger northern pike, *Esox lucius*	10

Table 13.2 (continued)   Total Mercury Concentrations in Abiotic Materials, Plants, and Animals near Historic Gold Mining and Refining Sites in the United States

[a] Concentrations are shown as means, range (in parentheses), maximum (Max.), and non-detectable (ND).

[b] Reference: 1, Nelson et al. 1975; 2, Brumbaugh et al. 2001; 3, Henny et al. 2002; 4, de Lacerda and Salomons 1998; 5, Bonzongo et al. 1996a; 6, Bonzongo et al. 1996b; 7, Gustin et al. 1994; 8, Wayne et al. 1996; 9, Callahan et al. 1994; 10, Walter et al. 1973; 11, Miller et al. 1996; 12, Mastrine et al. 1999; 13, Leigh 1994; 14, Leigh 1997.

[c] Current mercury criterion to protect human health is <0.3 mg/kg FW, down from <0.5 mg/kg FW (Brumbaugh et al. 2001).

about 6.75 million kg of mercury to the environment during the first 30 years of mine operation (Miller et al. 1996). Over time, mercury-contaminated sediments were eroded and transported downstream by fluvial processes. The most heavily contaminated wastes, with a total estimated volume of 710,700 m^3, contained 31,500 kg mercury, 248 kg gold, and 37,000 kg silver. If site remediation is conducted, extraction of the gold and silver, worth about US \$12 million, would defray a significant portion of the cleanup costs (Miller et al. 1996).

In the Carson River–Lahontan Reservoir (Nevada) watershed, approximately 7100 tons of metallic mercury were released into the watershed between 1859 and 1890 as a by-product of silver and gold ore refining (Wayne et al. 1996; Lawrence 2003). Mercury-contaminated tailings were dispersed throughout the lower Carson River, Lahontan Reservoir, and the Carson Sink by floods that occurred 19 times between 1861 and 1997 (Hunerlach and Alpers 2003; Lawrence 2003). During the past 130 years, mercury has been redistributed throughout 500 km^2 of the basin, and mercury concentrations at this site rank among the highest reported in North America (Gustin et al. 1994). Mercury contamination was still severe in this region in 1993 (see Table 13.2). Nevada authorities have issued health advisories against fish consumption from the Carson River; an 80-km stretch of the river has been declared a superfund site (Da Rosa and Lyon 1997). Low levels of methylmercury in surface waters of the Carson River–Lahontan Reservoir are attributed to increasing pH and increasing concentrations of various anions (SeO_4^{-2}, MoO_4^{-2}, and WO_4^{-2}), both of which are inhibitory to sulfate-reducing bacteria known to play a key role in methylmercury production in anoxic sediments (Bonzongo et al. 1996a). Methylmercury concentrations ranged from 0.1 to 0.7 ng/L in this ecosystem and were positively associated with total suspended solids (Bonzongo et al. 1996b). Removal of mercury from the water column was attributed to binding to particles, sedimentation, and volatilization of dissolved gaseous mercury.

The Lahontan Reservoir supports game and commercial fisheries, and the Lahontan Valley wetlands, home to many species of birds, is considered to be the most mercury-contaminated natural system in the United States (Henny et al. 2002). Mercury concentrations in sediments from this site and environs ranged from 10 to 30 mg/kg, or about 80 times higher than uncontaminated sediments; elevated concentrations of mercury were documented from the site in annelid worms, aquatic insects, fishes, frogs, toads, and birds (Da Rosa and Lyon 1997). In 1997–1998, three species of fish-eating birds nesting along the lower Carson River were examined for mercury contamination: double-crested cormorants (*Phalacrocorax auritus*), snowy egrets (*Egretta thula*), and black-crowned night-herons (*Nycticorax nycticorax*) (Henny et al. 2002). The high concentrations of total mercury observed in livers [means (in mg/kg FW): 13.5 in herons, 43.7 in egrets, and 69.4 in cormorants] and kidneys (6.1 in herons, 11.1 in egrets, 69.4 in cormorants) of adult birds (see Table 13.2) were possibly due to a threshold-dependent demethylation coupled with sequestration of the resultant inorganic mercury (Henny et al. 2002).

Demethylation and sequestration processes also appeared to have reduced the amount of methylmercury distributed to eggs, although the short time spent by adults in the contaminated area was a factor in the lower-than-expected mercury concentrations in eggs (Henny et al. 2002; see Table 13.2). Most eggs had mercury concentrations, as methylmercury, below 0.8 mg/kg FW, the threshold concentration at which reproductive problems may be expected (Heinz 1979; Heinz and Hoffman 2003). After hatching, young birds were fed diets by parent birds averaging 0.36 to 1.18 mg methylmercury/kg FW for 4 to 6 weeks through fledging (Henny et al. 2002). Eisler (2000) recommends avian dietary intakes <0.1 mg Hg/kg FW ration. Mercury concentrations in organs of the fledglings were much lower than those found in adults, but evidence was detected of toxicity to immune, detoxicating, and nervous systems (Henny et al. 2002). Immunodeficiencies and neurological impairment of fledglings may affect survival when these are burdened with stress associated with learning to forage and ability to complete the first migration. Oxidative stress was noted in young cormorants containing the highest concentrations of mercury, as evidenced by increasing thiobarbituric acid-reactive substances, and altered glutathione metabolism (Henny et al. 2002). Henny et al. (2002), based on their studies with fish-eating birds, concluded the following:

1. Adults tolerate relatively high levels of mercury in critical tissues through demethylation processes that occur above threshold concentrations.
2. Adults demethylate methylmercury to inorganic mercury, which is excreted or complexed with selenium and stored in liver and kidney. This change in form and the sequestering process reduce the amount of methylmercury circulating in tissues and the amount available for deposition in eggs.
3. The low concentrations of methylmercury in eggs are attributed to the short duration of time spent in the area before egg laying, and to the demethylation and sequestering processes within the birds.
4. Young of these migratory fish-eating birds experience neurological and histological damage associated with exposure to dietary mercury.

In the southeastern United States, gold mining was especially common between 1830 and 1849, and again during the 1880s. Historical gold mining activities contributed significantly to mercury problems in this region, as evidenced by elevated mercury concentrations in surface waters (Mastrine et al. 1999). Mercury in surface waters was positively correlated with total suspended solids and with bioavailable iron. Vegetation in the Southeast is comparatively heavy and by controlling erosion may reduce the total amount of mercury released from contaminated mining sites to the rivers. Mercury concentrations in surface waters of southeastern states were significantly lower, by orders of magnitude, than those from western states where amalgamation extraction techniques were also practiced, and this may reflect the higher concentrations of mercury used and the sparser vegetation of the western areas (Mastrine et al. 1999).

In northern Georgia, gold was discovered in 1829 and mined until about 1940 (Leigh 1994, 1997). Extensive use of mercury probably began in 1838 when stamp mills (ore crushers) were introduced to help recover gold from vein ore, with about 38% of all mercury used in gold mining escaping into nearby streams. Mercury concentrations in historical floodplain sediments near the core of the mining district were as high as 4.0 mg/kg DW, but decreased with increasing distance downstream to <0.1 mg/kg at 10 to 15 km from the source mines. Near Dahlonega, Georgia, historical sediments contained as much as 12.0 mg Hg/kg, and mean values in streambanks near the mining district were 0.2 to 0.6 mg Hg/kg DW.

Mercury-contaminated floodplain sediments in Georgia pose a potential hazard to wildlife. Clams and mussels, for example, collected from these areas more than 50 years after mining had ceased contained elevated (0.7 mg/kg DW) mercury concentrations in soft tissues. Erosion of channel banks and croplands, with subsequent transport of mercury-contaminated sediments from the mined watersheds, is likely to occur for hundreds or thousands of years. Mercury was the only significant trace metal contaminant resulting from former gold mining activities in Georgia, exceeding the USEPA "heavily polluted" guideline for sediments of >1.0 mg Hg/kg. Other metals examined did not exceed the "heavily polluted" sediment guidelines, which are >50 mg/kg for copper, >200 mg/kg for lead, and >200 mg/kg for zinc (Leigh 1994, 1997).

In South Dakota, most of the fish collected from the Cheyenne River arm of Lake Oahe in 1970 contained elevated (>0.5 mg/kg FW muscle) concentrations of mercury (Walter et al. 1973). Elemental mercury was used extensively in this region between 1880 and 1970 to extract gold from ores and is considered to be the source of the contamination. In 1970, when the use of mercury in the gold recovery process was discontinued at this site, liquid wastes containing 5.5 to 18 kg of mercury were being discharged daily into the Cheyenne River arm (Walter et al. 1973).

In Nome, Alaska, gold mining is responsible, in part, for the elevated mercury levels (maximum 0.45 mg/kg DW) measured in modern beach sediments (Nelson et al. 1975). However, higher concentrations (maximum 0.6 mg/kg) routinely occur in buried Pleistocene sediments immediately offshore and in modern nearby unpolluted beach sediments (1.3 mg Hg/kg), which suggests that the effects of mercury contamination from mining are less than natural concentration processes in the Seward Peninsula region of Alaska (Nelson et al. 1975).

13.3   LETHAL AND SUBLETHAL EFFECTS OF MERCURY

Mercury is a known mutagen, teratogen, and carcinogen. At comparatively low concentrations in birds and mammals, it adversely affects reproduction, growth and development, behavior, blood chemistry, motor coordination, vision, hearing, histology, and metabolism. It has a high potential for bioaccumulation and biomagnification and is slow to depurate. Organomercury compounds were more effective in producing adverse effects than were inorganic mercury compounds, although effects were enhanced or ameliorated by many biotic and nonbiological modifiers (Eisler 2000). Lethal concentrations of mercury to sensitive organisms ranged from 0.1 to 2.0 µg/L medium for aquatic fauna, from 2.2 to 31.1 mg/kg body weight (acute oral) and 4.0 to 40.0 mg/kg (dietary) for birds, and from 0.1 to 0.5 mg/kg body weight (daily dose) and from 1.0 to 5.0 mg/kg (dietary) for mammals (Table 13.3). Adverse sublethal effects were documented for sensitive species of aquatic organisms at water concentrations of 0.03 to 0.1 µg/L. For sensitive species of birds, harmful levels were 640 µg Hg/kg body weight daily, or 50 to 500 µg Hg/kg in the diet. For sensitive mammals, these levels were 250 µg Hg/kg body weight daily, or 1100 µg Hg/kg diet (Eisler 2000).

13.3.1 Aquatic Organisms

Signs of acute mercury poisoning in fish included flaring of gill covers, increased frequency of respiratory movements, loss of equilibrium, excessive mucous secretion, darkening coloration, and sluggishness (Armstrong 1979; Hilmy et al. 1987). Signs of chronic mercury poisoning include emaciation (due to appetite loss), brain lesions, cataracts, diminished response to change in light intensity, inability to capture food, abnormal motor coordination, and various erratic behaviors (Armstrong 1979; Hawryshyn et al. 1982). Mercury residues in severely poisoned fish that died soon after exposure were (in mg/kg FW) from 26 to 68 in liver, 16 to 20 in brain, and 5 to 7 in the whole body (Armstrong 1979).

Sublethal concentrations of mercury are known to adversely affect sensitive species of aquatic biota through inhibition of reproduction (Dave and Xiu 1991; Kanamadi and Saidapur 1991; Kirubagaran and Joy 1992; Khan and Weis 1993; Punzo 1993), reduction in growth rate (Kanamadi and Saidapur 1991; Punzo 1993), increased frequency of tissue histopathology (Kirubagaran and Joy 1988, 1989; Handy and Penrice 1993; Voccia et al. 1994), impairment in ability to capture prey (Weis and Weis 1995) and olfactory receptor function (Baatrup et al. 1990; Baatrup and Doving 1990), alterations in blood chemistry (Allen 1994) and enzyme activities (Nicholls et al. 1989; Kramer et al. 1992), disruption of thyroid function (Kirubagaran and Joy 1989), chloride secretion (Silva et al. 1992), and other biochemical and metabolic functions (Nicholls et al. 1989; Angelow and Nicholls 1991). In general, the accumulation of mercury by aquatic biota is rapid and depuration is slow (Newman and Doubet 1989; Angelow and Nicholls 1991; Wright et al. 1991; Handy and Penrice 1993; Pelletier and Audet 1995; Geffen et al. 1998). It is emphasized that organomercury compounds, especially methylmercury, were significantly more effective than inorganic mercury compounds in producing adverse effects and accumulations (Baatrupet et al. 1990; Wright et al. 1991; Kirubagaran and Joy 1992; Odin et al. 1995).

Reproduction was inhibited among sensitive species of aquatic organisms at water concentrations of 0.03 to 1.6 µg Hg/L. In the planarian (*Dugesia dorotocephala*), asexual fission was suppressed at 0.03 to 0.1 µg organomercury/L (Best et al. 1981); in the slipper limpet (*Crepidula fornicata*), spawning was delayed and fecundity was decreased at 0.25 Hg^{+2}/L (Thain 1984); in the zebrafish (*Brachydanio rerio*), hatching success was reduced at 0.1 µg Hg^{+2}/L and egg deposition was reduced at

Table 13.3   Lethal Effects of Mercury to Sensitive Species of Aquatic Organisms, Birds, and Mammals

Ecosystem, Species, and Other Variables	Mercury Concentration and Effect	Ref[a]
Aquatic Organisms, via Medium		
Rainbow Trout, *Oncorhynchus mykiss*; embryo-Larvae		
Flowthrough test	0.1 µg/L; LC50 (28 d)	1
Static test	4.7 µg/L; LC50 (28 d)	1
Channel catfish, *Ictalurus punctatus*; embryo-larvae; flowthrough test	0.3 µg/L; LC50 (10 d)	1
Brook trout, *Salvelinus fontinalis*	0.3–0.9 µg/L; LC50 (lifetime exposure)	2
Narrow-mouthed toad, *Gastrophryne carolinensis*; embryo-larvae	1.3 µg/L; LC50 (96 h)	1
Daphnid, *Daphnia magna*	1.3–1.8 µg/L; LC50 (lifetime exposure)	2
Crayfish, *Orconectes limosus*	2.0 µg/L; LC50 (30 d)	2
Birds, Single Oral Dose		
Mallard, *Anas platyrhynchos*	2.2–23.5 mg/kg body weight (BW); LD50 within 14 d after treatment	3
Ring-necked pheasant, *Phasianus colchicus*	11.5–26.8 mg/kg BW; LD50 within 14 d	3
Prairie chicken, *Tympanuchus cupido*	11.5 mg/kg BW; LD50 within 14 d	3
Coturnix, *Coturnix japonica*	11.0–31.1 mg/kg BW; LD50 within 14 d	3,4,5
House sparrow, *Passer domesticus*	12.6-37.8 mg/kg BW; LD50 within 14 d	3
Gray partridge, *Perdix perdix*	17.6 mg/kg BW; LD50 within 14 d	3
Rock dove, *Columba livia*	22.8 mg/kg BW; LD50 within 14 d	3
Northern bobwhite, *Colinus virginianus*	23.8 mg/kg BW; LD50 within 14 d	3
Birds, via Diet		
Mallard	3.0 mg/kg diet for two reproductive seasons; reduced duckling survival	6
Zebra finch, *Poephila guttata*	5.0 mg/kg diet for 77 d; LD25	7
Coturnix	8 mg/kg diet for 5 d; some deaths	5
Ring-necked pheasant	12.5 mg/kg diet for 70 d; LD50	8
Ring-necked pheasant	37.4 mg/kg diet for 28 d; LD50	8
Birds, 3 species	33.0 mg/kg diet for 35 d; LD8 to LD90	10
Birds, 4 species	40.0 mg/kg diet for 6 to 11 d; LD33	9
Mammals, Oral Dose		
Domestic dog, *Canis familiaris*	0.1–0.25 mg/kg BW daily during pregnancy; high incidence of stillbirths	11
Pig, *Sus* spp.	0.5 mg/kg BW daily during pregnancy; high incidence of stillbirths	11
Rhesus monkey, *Macaca mulatta*	0.5 mg/kg BW daily during d 20–30 of pregnancy; maternally toxic and abortient	11
Mule deer, *Odocoileus hemionus hemionus*	17.9 mg/kg BW, single oral dose; LD50	3
Harp seal, *Pagophilus groenlandica*	25.0 mg/kg BW daily; death in 20–26 d. Blood Hg levels just before death were 26.8 to 30.3 mg/L	16

Table 13.3 (continued)   Lethal Effects of Mercury to Sensitive Species of Aquatic Organisms, Birds, and Mammals

Ecosystem, Species, and Other Variables	Mercury Concentration and Effect	Ref[a]
Mammals, Dietary Route		
Domestic cat, *Felis domesticus*	0.25 mg/kg BW daily for 90 d for total of about 85 mg Hg; LD50 (78 d); convulsions starting at day 68, all with signs by day 90. Liver residues of survivors were 40.2 and 18.1 mg/kg FW for total mercury and inorganic mercury, respectively	12
Mink, *Mustela vison*	1.0 mg/kg diet; all dead in about 2 months	13
River otter, *Lutra canadensis*	>2.0 mg/kg diet; fatal within 7 months	14, 15

[a] Reference: 1, Birge et al. 1979; 2, USEPA 1985; 3, Hudson et al. 1984; 4, Hill 1981; 5, Hill and Soares 1984; 6, Heinz and Locke 1976; 7, Scheuhammer 1988; 8, Spann et al. 1972; 9, Finley et al. 1979; 10, Hamasaki et al. 1995; 11, Khera 1979; 12, Eaton et al. 1980; 13, Sheffy and St. Amant 1982; 14, Kucera 1983; 15, Ropek and Neely 1993; 16, Ronald et al. 1977.

0.8 µg/L (Armstrong 1979); fathead minnows (*Pimephales promelas*) exposed to 0.12 µg methylmercury/L for 3 months failed to reproduce (Birge et al. 1979); the leopard frog (*Rana pipiens*) did not metamorphose during exposure to 1.0 µg methylmercury/L for 4 months (USEPA 1985); and in the mysid shrimp (*Mysidopsis bahia*), the abortion rate increased and population size decreased after lifetime (i.e., 28 days) exposure to 1.6 µg/L of mercury as mercuric chloride (Gentile et al. 1983). For sensitive species of marine invertebrates such as hydroids, protozoans, and mysid shrimp, reproduction was inhibited at concentrations between 1.1 and 2.5 µg Hg^{+2}/L; this range was 5 to 71 µg/L for more resistant species of marine invertebrates (Gentile et al. 1983).

Reduced growth of sensitive species of aquatic organisms has been recorded at water concentrations of 0.04 to 1.0 µg Hg/L. The rainbow trout (*Oncorhynchus mykiss*) was the most sensitive species tested; growth reduction was observed after 64 days in 0.04 µg Hg/L as methylmercury or 0.11 µg Hg/L as phenylmercury (USEPA 1985). In adults of the marine mollusc *Crepidula fornicata*, growth was reduced after 16 weeks in 0.25 µg Hg^{+2}/L (Thain 1984). Growth inhibition was recorded in freshwater algae after exposure of 24 hours to 10 days to 0.3 to 0.6 µg organomercury/L, in brook trout (*Salvelinus fontinalis*) alevins after exposure for 21 days to 0.79 µg organomercury/L (USEPA 1985), and in the marine alga *Scripsiella faeroense* exposed to 1.0 µg Hg^{+2}/L for 24 hours (Kayser 1976). Adverse effects of mercury on aquatic organisms, in addition to those listed for reproduction and growth, are documented at water concentrations of 0.88 to 5.0 µg/L (reviewed by Eisler 2000). These effects include enzyme disruption in fish embryos, increased incidence of frustule abnormalities in marine algae, arrested development of sea urchin larvae, and, in adult fishes, decreased rate of intestinal transport of glucose and fructose, altered blood chemistry, decreased respiration, and elevated liver metallothioneins.

At lower trophic levels, the efficiency of mercury transfer was low through natural aquatic food chains; however, in animals of higher trophic levels, such as predatory teleosts and fish-eating birds and mammals, the transfer was markedly amplified (Eisler 1978, 1981, 1987). High uptake and accumulation of mercury from the medium by representative species of marine and freshwater fishes and invertebrates are extensively documented (Kopfler 1974; Eisler 1978, 1981; Birge et al. 1979; Huckabee et al. 1979; USEPA 1985; Stokes et al. 1981; Rodgers and Beamish 1982; Hirota et al. 1983; Clarkson et al. 1984; McClurg 1984; Niimi and Lowe-Jindi 1984; Ramamorthy and Blumhagen 1984; Ribeyre and Boudou 1984; Thain 1984; Newman and Doubet 1989; Angelow and Nicholls 1991; Wright et al. 1991; Handy and Penrice 1993). Accumulation patterns were enhanced or significantly modified by a variety of biological and abiotic factors (USNAS 1978; Eisler 1978, 1981, 1984, 1985; USEPA 1985; Stokes et al. 1981; Rodgers and Beamish 1982; Clarkson et al. 1984; Ramamorthy and Blumhagen 1984; Ribeyre and Boudou 1984; Ponce and Bloom 1991; Odin et al. 1995; Choi et al. 1998).

In general, the accumulation of mercury was markedly enhanced at elevated water temperatures, reduced salinity or water hardness, reduced water pH, increased age of the organism, and reduced organic matter content of the medium; in the presence of zinc, cadmium, or selenium in solution; after increased duration of exposure; and in the presence of increased nominal concentrations of protein-bound mercury. Uptake patterns were significantly modified by sex, sexual condition, prior history of exposure to mercury salts, the presence of complexing and chelating agents in solution, dietary composition, feeding niche, tissue specificity, and metabolism. However, trends were not consistent between species and it is difficult to generalize.

In one example, Ribeyre and Boudou (1984) immersed rainbow trout in solutions containing 0.1 μg Hg/L as methylmercury: after 30 days, bioconcentration factors (BCF) ranged from 28,300 for brain to 238,000 for spleen; values were intermediate for muscle (30,000), whole fish (36,000), blood (102,000), liver (102,000), kidney (137,000), and gill (163,000). These values may have been higher if exposure had extended beyond 30 days. Rodgers and Beamish (1982) showed that whole-body mercury residues in rainbow trout subjected to mercury insult continued to increase for 66 days before stabilizing. When mercury was presented as inorganic mercury ion at 0.1 μg/L for 30 days, BCF values were usually lower than in trout exposed to methylmercury: 2300 for muscle; 6800 for brain; 7000 for whole trout; 14,300 for blood; 25,000 for liver; 53,000 for kidney; 68,600 for gill; and 521,000 for spleen (Ribeyre and Boudou 1984). The high BCF values recorded for rainbow trout were probably due to efficient uptake from water coupled with slow depuration (Rodgers and Beamish 1982).

Total mercury concentrations (mg/kg FW) in tissues of adult freshwater fishes with signs of methylmercury intoxication ranged from 3 to 42 in brain, from 6 to 114 in liver, from 5 to 52 in muscle, and from 3 to 35 in whole body (Wiener and Spry 1996). Whole-body levels up to 100 mg Hg/kg were reportedly not lethal to rainbow trout, although a level of 20 to 30 mg/kg was associated with reduced appetite, loss of equilibrium, and hyperplasia of gill epithelium (Niimi and Lowe-Jinde 1984). However, brook trout showed toxic signs and death at whole body residues of only 5 to 7 mg Hg/kg (Armstrong 1979).

In another example, the marine copepod *Acartia clausi*, when subjected to 0.05 µg/L of mercury and higher, reached equilibrium with the medium in only 24 hours. In that study (Hirota et al. 1983), BCF values for whole *Acartia* after 24-hour exposures were 14,360 for inorganic mercury ion (0.05 µg/L) and, for methylmercury, 179,200 (0.05 µg/L) and 181,000 (0.1 µg/L). Time to eliminate 50% of biologically assimilated mercury and its compounds (Tb 1/2) is variable. Among various species of freshwater fishes, Tb 1/2 values (in days) ranged from 20 in guppies (*Poecilia reticulatus*), to 23 for the goldfish (*Carassius auratus*), 100 for northern pike (*Esox lucius*), to about 1000 for rainbow trout and brook trout (Huckabee et al. 1979). Tb 1/2 values for marine fishes and invertebrates were 297 days for decapod crustaceans, 435 to 481 days for bivalve molluscs, and 1030 to 1200 days for eels and flounders (USNAS 1978).

Mercury tolerance or resistance has been reported in bacteria (Colwell et al. 1976; Baldi et al. 1991), protozoans (Berk et al. 1978), crustaceans (Green et al. 1976; Weis 1976), and fish (Weis 1984), but the mechanisms of action are not fully understood.

13.3.2 Birds

Signs of mercury poisoning in birds include muscular incoordination, falling, slowness, fluffed feathers, calmness, withdrawal, hyporeactivity, hypoactivity, and eyelid drooping. In acute oral exposures, signs appeared as early as 20 minutes postadministration in mallards and 2.5 hours in pheasants. Deaths occurred between 4 and 48 hours in mallards and 2 and 6 days in pheasants; remission required as long as 7 days (Hudson et al. 1984). In studies with coturnix (*Coturnix coturnix coturnix*), Hill (1981) found that methylmercury was always more toxic than inorganic mercury and that young birds were usually more sensitive than older ones. In addition, some birds poisoned by inorganic mercury recovered after treatment was withdrawn, but chicks that were fed methylmercury and later developed toxic signs usually died, even if the treated feed was removed. The withdrawal syndrome in coturnix poisoned by inorganic mercury was usually preceded by intermittent, nearly undetectable tremors, coupled with aggressiveness toward cohorts; time from onset to remission was usually 3 to 5 days, but sometimes was extended to 7 days. Coturnix poisoned by methylmercury appeared normal until 2 to 5 days posttreatment; then, ataxia and low body carriage with outstretched neck were often associated with walking. In advanced stages, coturnix lost locomotor coordination and did not recover; in mild to moderate clinical signs, recovery usually took at least 1 week (Hill 1981).

Residues of mercury in experimentally poisoned passerine birds usually exceeded 20 mg/kg FW and were similar to concentrations reported in wild birds that died of mercury poisoning (Finley et al. 1979). In one study with the zebra finch (*Poephila guttata*), adults were fed methylmercury in the diet for 76 days at dietary levels <0.92 (controls), 1, 2.5, or 5 mg Hg/kg DW ration (Scheuhammer 1988). There were no signs of mercury intoxication in any group except the high-dose group, which had 25% dead and 40% neurological impairment. Dead birds from the high-dose group had 73 mg Hg/kg FW in liver, 65 in kidney, and 20 in brain; survivors without signs had 30 mg Hg/kg FW in liver, 36 in kidney, and 14 in brain; and impaired birds had 43 mg Hg/kg FW in liver, 55 in kidney, and 20 in brain (Scheuhammer 1988).

Mercury levels in tissues of poisoned wild birds were highest (45 to 126 mg/kg FW) in red-winged blackbirds (*Agelaius phoeniceus*), intermediate in starlings (*Sturnus vulgaris*) and cowbirds (*Molothrus ater*), and lowest (21 to 54) in grackles (*Quiscalus quiscula*). In general, mercury residues were highest in the brain, followed by the liver, kidney, muscle, and carcass. Some avian species are more sensitive than passerines (Solonen and Lodenius 1984; Hamasaki et al. 1995): liver residues in birds experimentally killed by methylmercury ranged from 17 mg Hg/kg FW in red-tailed hawks (*Buto jamaicensis*) to 70 in jackdaws (*Corvus monedula*); values were intermediate in ring-necked pheasants, kestrels (*Falco tinnunculus*), and magpies (*Pica pica*). Experimentally poisoned grey herons (*Ardea cinerea*) seemed to be unusually resistant to mercury; lethal doses produced residues of 415 to 752 mg Hg/kg DW in liver (Van der Molen et al. 1982). However, levels of this magnitude were encountered in livers from grey herons collected during a massive die-off in the Netherlands during a cold spell in 1976; the interaction effects of cold stress, mercury loading, and poor physical condition of the herons are unknown (Van der Molen et al. 1982).

Sublethal effects of mercury on birds, administered by a variety of routes, included histopathology, accumulation, and adverse effects on growth, development, reproduction, blood and tissue chemistry, metabolism, and behavior. The dietary route of administration is the most extensively studied pathway of avian mercury intake.

Domestic chickens fed diets containing as little as 50 μg/kg mercury, as methylmercury, contained elevated total mercury (2.0 mg/kg FW) residues in liver and kidney after 28 weeks; at 150 μg/kg diet, residues ranged from 1.3 to 3.7 mg/kg in heart, muscle, brain, kidney, and liver, in that general order. At 450 μg/kg in diets, residues in edible chicken tissues (3.3 to 8.2 mg/kg) were considered hazardous to human consumers, although no overt signs of mercury toxicosis were observed in the chickens (March et al. 1983). The dietary concentration of 0.5 mg Hg/kg DW (equivalent to about 0.1 mg/kg FW) in the form of methylmercury was fed to three generations of mallards (Heinz 1979). Females laid a greater percentage of their eggs outside nest boxes than did controls, and also laid fewer eggs and produced fewer ducklings. Ducklings from parents fed methylmercury were less responsive than controls to tape-recorded maternal calls but were hyperresponsive to a fright stimulus in avoidance tests (Heinz 1979). Adult female mallards fed diets containing 1 or 5 mg Hg/kg ration, as methylmercury chloride, produced eggs that contained 1.4 mg Hg/kg FW (1.0 mg/kg ration) or 8.7 mg Hg/kg FW (5 mg Hg/kg diet); breast muscle had 1.0 or 5.3 mg Hg/kg FW (Heinz 1987).

Lesions in the spinal cord were the primary effect in adult female mallards fed diets containing 1.5 or 2.8 mg Hg/kg DW ration as methylmercury (Pass et al. 1975). The tissues and eggs of ducks and other species of birds collected in the wild have sometimes contained levels of mercury equal to, or far exceeding, those associated with reproductive and behavioral deficiencies in domestic mallards (e.g., 9 to 11 mg/kg FW in feathers; >2.0 mg/kg FW in other tissues); therefore, it is possible that reproduction and behavior of wild birds can be modified by methylmercury contamination (Heinz 1979).

Interaction effects of mercury with other metals need to be considered when evaluating avian dietary mercury risks. In some cases, mercury effects were exacerbated when diets contained lead and cadmium (Rao et al. 1989). Other metals or metalloids may protect ducks and other species against the toxic effects of methylmercury, as was the case for selenium in the form of seleno-DL-methionine, by restoring glutathione status involved in antioxidative defense mechanisms (Heinz and Hoffman 1998; Hoffman and Heinz 1998).

Dietary concentrations of 1.1 mg total Hg/kg have been associated with kidney lesions in juvenile starlings and with elevated residues in the liver (5.5 mg/kg DW) and kidney (36.3 mg/kg DW), after exposure for 8 weeks (Nicholson and Osborn 1984). In American black ducks (*Anas rubripes*) fed diets containing 3.0 mg Hg/kg as methylmercury for 28 weeks, reproduction was significantly inhibited; tissue residues were elevated in kidney (16.0 mg/kg FW) and liver (23.0 mg/kg); and brain lesions characteristic of mercury poisoning were present (Finley and Stendell 1978). Japanese quail (*Coturnix japonica*) fed diets containing 8 mg Hg/kg of inorganic mercury for 3 weeks had depressed gonad weights; those fed 3 mg/kg ration inorganic mercury or 1 mg/kg ration methylmercury for 9 weeks showed alterations in brain and plasma enzyme activities (Hill and Soares 1984). Grossly elevated tissue mercury residues of 400 mg/kg in feathers and 17 to 130 mg/kg in other tissues were measured in gray partridge (*Perdix perdix*) after dietary exposure of 20 to 25 mg total Hg/kg for 4 weeks (McEwen et al. 1973).

Reduced reproductive ability was noted in gray partridges ingesting 640 µg Hg (as organomercury)/kg BW daily for 30 days (McEwen et al. 1973); similar results were observed in ring-necked pheasants (Spann et al. 1972; Mullins et al. 1977). Behavioral alterations were noted in rock doves (*Columba livia*) given 3.0 mg inorganic Hg/kg BW daily for 17 days (Leander et al. 1977) or 1.0 mg/kg BW methylmercury for 5 weeks (Evans et al. 1982). Observed behavioral changes in posture and motor coordination of pigeons were permanent after the brain accumulated >12 mg Hg/kg FW and were similar to the "spastic paralysis" observed in wild crows during the Minamata, Japan, outbreak of the 1950s, although both species survived for years with these signs (Evans et al. 1982).

Tissue concentrations >15,000 µg Hg/kg FW brain and >30,000 to 40,000 µg/kg FW liver or kidney are associated with avian neurological impairment (Scheuhammer 1988). Mercury residues of 790 to 2000 µg/kg in egg and 5000 to 40,000 µg/kg in feathers are linked to impaired reproduction in various bird species (Spann et al. 1972; USNAS 1978; Fimrite 1979; Heinz 1979; Solenen and Lodenius 1984). Residues in eggs of 1300 to 2000 µg Hg/kg FW were associated with reduced hatching success in white-tailed sea-eagles (*Haliaeetus albicilla*), the common loon (*Gavia immer*), and in several seed-eating species (Fimreite 1979). This range was 900 to 3100 µg/kg egg FW for ring-necked pheasant (Spann et al. 1972) and 790 to 860 µg/kg for mallards (Heinz 1979). Residues of 5000 to 11,000 µg Hg/kg in feathers of various species of birds have been associated with reduced hatch of eggs and with sterility (USNAS 1978). Sterility was observed in the Finnish sparrow hawk (*Accipiter nisus*) at mercury concentrations of 40,000 µg/kg in feathers (Solenen and Lodenius 1984).

13.3.3 Mammals

Mercury is easily transformed into stable and highly toxic methylmercury by microorganisms and other vectors (de Lacerda and Salomons 1998; Eisler 2000). Methylmercury has long residence times in aquatic biota. Consumption of methyl-mercury-contaminated fish is implicated in more than 150 deaths and more than 1000 birth defects in Minamata, Japan, between 1956 and 1960. By 1987, more than 17,000 people had been affected by methylmercury poisoning in Japan, with 999 deaths. Worldwide, de Lacerda and Salomons (1998) estimated that mercury poi-soning from ingestion of contaminated food is responsible for more than 1400 deaths and 200,000 sublethal cases. Methylmercury affects the central nervous system in humans, especially the sensory, visual, and auditory areas concerned with coordi-nation. The most severe effects lead to widespread brain damage, resulting in mental derangement, coma, and death (Clarkson and Marsh 1982; USPHS 1994).

In mule deer (*Odocoileus hemionus hemionus*), after acute oral mercury poison-ing was induced experimentally, additional signs included belching, bloody diarrhea, piloerection (hair more erect than usual), and loss of appetite (Hudson et al. 1984). The kidney is the probable critical organ in adult mammals because of the rapid degradation of phenylmercurials and methoxyethylmercurials to inorganic mercury compounds and subsequent translocation to the kidney (Suzuki 1979), whereas in the fetus the brain is the principal target (Khera 1979).

Organomercury compounds, especially methylmercury, were the most toxic mer-cury species tested. Among sensitive species of mammals, death occurred at daily organomercury concentrations of 0.1 to 0.5 mg/kg BW or 1 to 5 mg/kg in the diet (see Table 13.3). Larger animals such as mule deer and harp seals (*Pagophilus groenlandica*) appear to be more resistant to mercury than smaller animals such as mink (*Mustela vison*), domestic cats (*Felis domesticus*) and dogs (*Canis familiaris*), pigs (*Sus* sp.), monkeys, and river otters (*Lutra canadensis*); the reasons for this difference are unknown, but may be related to differences in metabolism and detox-ification rates. Tissue residues in fatally poisoned mammals (mg Hg/kg FW) were 6.0 in brain, 10.0 to 55.6 in liver, 17.0 in whole body, about 30.0 in blood, and 37.7 in kidney (Eisler 2000; see Table 13.3).

Mercury has no known physiological function (USEPA 1985). In humans and other mammals, it causes teratogenic, mutagenic, and carcinogenic effects; the fetus is the most sensitive life stage (USNAS 1978; Chang 1979: Khera 1979; USEPA 1985; Elhassni 1983; Greener and Kochen 1983; Clarkson et al. 1984). Methylmercury irreversibly destroys the neurons of the central nervous system. Frequently, a substantial latent period intervenes between the cessation of exposure to mercury and the onset of signs and symptoms. This interval is usually measured in weeks or months, but some-times in years (Clarkson et al. 1984). At high sublethal doses in humans, mercury causes cerebral palsy, gross motor and mental impairment, speech disturbances, blind-ness, deafness, microcephaly, intestinal disturbances, tremors, and tissue pathology (Chang 1979; USEPA 1985; Elhassni 1983; Clarkson et al. 1984; USPHS 1994).

Pathological and other effects of mercury may vary from organ to organ, depending on factors such as the effective toxic dose in the organ, the compound involved and its metabolism within the organ, the duration of exposure, and the other contaminants to which the animal is concomitantly exposed (Chang 1979). Many compounds, especially salts of selenium, protect humans and other animals against mercury toxicity, although their mode of action is not clear (USNAS 1978; Chang 1979; USEPA 1985; Eisler 1985).

Mercury transfer and biomagnification through mammalian food chains are well documented (Galster 1976; USNAS 1978; Eaton et al. 1980; Eisler 1981; Huckabee et al. 1981; Sheffy and St. Amant 1982; Kucera 1983; Clarkson et al. 1984; Wren 1986), but there is considerable variation. Among terrestrial mammals, for example, herbivores such as mule deer, moose (*Alces alces*), caribou (*Rangifer tarandus*), and various species of rabbits usually contained <1.0 mg Hg/kg FW in liver and kidney, but carnivores such as the marten (*Martes martes*), polecat (*Mustela putorius*), and red fox (*Vulpes vulpes*) frequently contained more than 30 mg/kg (USNAS 1978). The usually higher mercury concentrations in fish-eating furbearers than in herbivorous species seemed to reflect the amounts of fish and other aquatic organisms in the diet. In river otter and mink from the Wisconsin River (U.S.) drainage system, mercury levels paralleled those recorded in fish, crayfish, and sediments at that location. Highest mercury levels in all samples were found about 30 km downstream from known industrial discharges of mercury into the Wisconsin River. Residues were highest in the fur, followed by the liver, kidney, muscle, and brain (Sheffy and St. Amant 1982).

In marine mammals, more than 90% of the mercury content is inorganic; however, enough methylmercury occurs in selected tissues to result in the accumulation of high tissue concentrations of methylmercury in humans and wildlife consuming such meat (Clarkson et al. 1984). The liver of the ringed seal (*Phoca hispida*) normally contains 27,000 to 187,000 µg Hg/kg FW and is a traditional and common food of the coastal Inuit (Canada) people (Eaton et al. 1980). Although levels of mercury in hair (109,000 µg/kg) and blood (37 µg/L) of Inuits were grossly elevated, no symptoms of mercury poisoning were evident in the coastal Inuits. Similar high concentrations have been reported for Alaskan Inuit mothers who, during pregnancy, ate seal oil twice a day, and seal meat or fish from the Yukon–Kuskokwim Coast every day (Galster 1976). Despite the extremely high total Hg content of seal liver, only the small organomercury component was absorbed and appeared in the tissues. Cats fed a diet of seal liver (26,000 µg Hg/kg FW) for 90 days showed no neurologic or histopathologic signs (Eaton et al. 1980). It seems that the toxic potential of seal liver in terms of accumulated tissue levels in cats (up to 862 µg total Hg/L blood and 7600 µg total Hg/kg hair) is better indicated by the organomercury fraction in seal liver than by the concentration of total Hg (Eaton et al. 1980).

Retention of mercury by mammalian tissues is longer for organomercury compounds, especially methylmercury, than for inorganic mercury compounds (USNAS 1978; Clarkson and Marsh 1982; Elhassni 1983; Clarkson et al. 1984). Excretion of all mercury species follows a complex, variable, multicompartmental pattern; the longer-lived chemical mercury species have a biological half-life that ranges from about 1.7 days in human lung to 1.36 years in whole body of various pinnipeds. In humans, increased urinary excretion and blood levels of mercury were observed in volunteers who used phenylmercuric borate solutions or lozenges intended for treatment of throat infections (USPHS 1994).

13.3.4 Other Groups

Methylmercury compounds at concentrations of 25.0 mg Hg/kg soil were fatal to all tiger worms (*Eisenia foetida*) in 12 weeks; at 5.0 mg/kg, however, only 21% died in a similar period (Beyer et al. 1985). Inorganic mercury compounds were also toxic to earthworms (*Octochaetus pattoni*); in 60 days, 50% died at soil mercury levels of 0.79 mg/kg, and all died at 5.0 mg/kg (Abbasi and Soni 1983). Seedlings of rice (*Oryza sativa*) grown on mercury-contaminated waste soil from a chloralkali factory (and applicable to soil mercury levels near gold mining activities) for 75 days showed increasing mercury concentrations over time with increasing soil mercury content. At 2.5% waste soil in garden soil, rice seedlings contained 8.0 mg Hg/kg FW; at 10%, 15.2 mg Hg/kg; and at 17.5% waste soil in garden soil, seedlings contained 19.1 mg Hg/kg FW (Nanda and Mishra 1997). Seedlings of spruce (*Picea abies*) exposed to solutions containing up to 0.2 mg Hg/L, as inorganic mercury or methylmercury, showed a dose-dependent inhibition in root growth, especially for methylmercury (Godbold 1991).

Methylmercury compounds have induced abnormal sex chromosomes in the fruit fly (*Drosophila melanogaster*) (USNAS 1978; Khera 1979). Earthworms (*Eisenia foetida*) exposed to soil containing 5.0 mg Hg/kg showed a significant reduction in the number of segments regenerated after 12 weeks and contained 85 mg Hg/kg FW whole body. Regeneration was normal at soil mercury levels of 1.0 mg/kg, although body burdens up to 27 mg/kg were recorded with potential damage effects to earthworm predators (Beyer et al. 1985). Studies with a different species of earthworm (*Octochaetus pattoni*) and mercuric chloride demonstrated a progressive initial increase in reproduction as soil mercury levels increased from 0.0 to the 60-day lethal level of 5.0 mg/kg (Abbasi and Soni 1983).

13.4 PROPOSED MERCURY CRITERIA

Proposed mercury criteria for the protection of sensitive crops, aquatic organisms, birds, mammals, and human health are shown in Table 13.4. In almost every instance, these criteria are listed as concentrations of total mercury, with most of, if not all, the mercury present as an organomercury species. For freshwater aquatic life protection, the USEPA mercury criteria were <0.012 µg/L, 4-day average (not to be exceeded more than once every 3 years) and <2.4 µg/L, 1-hour average (not to be exceeded more than once every 3 years), based on mercury measurements of samples that passed through a 0.45-µm membrane filter after the sample had been acidified to pH 1.5 to 2.0 with nitric acid (USEPA 1985).

However, as documented earlier, mercury concentrations in freshwater as low as 0.1 to 2.0 µg/L were fatal to sensitive aquatic species, and concentrations as low as 0.03 to 0.1 µg/L were associated with significant sublethal effects. Safety factors

Table 13.4 Proposed Mercury Criteria for the Protection of Selected Natural Resources and Human Health

Resource and Other Variables	Criterion or Effective Mercury Concentration	Ref[a]
Crops		
Irrigation water, Brazil	<0.2 µg/L	1
Land Application of Sludge and Solid Waste; Maximum Permissible Concentration		
Iowa, Maine, Vermont	10.0 mg/kg waste	2
California	<20.0 mg/kg waste	2
Soils, Germany	<2.0 mg/kg dry weight (DW)	3
Aquatic Life		
Freshwater		
Total mercury	<0.1 µg/L	4
Methylmercury	<0.01 µg/L	5
Total mercury	<0.012 µg/L, 4-d average; <2.4 µg/L, 1-h average	6
Inland surface waters, India	<10.0 µg/L from point source discharge	7
Public water supply, Wisconsin	<0.079 µg/L	2
Saltwater	<0.025 µg/L, 4-d average; <2.1 µg/L, 1-h average	6
Sediments		
California		
Acceptable	<0.51 mg/kg DW	8
Bivalve mollusc larval abnormalities	>0.51 mg/kg DW	8
Hazardous	>1.2–.3 mg/kg DW	8
Canada, Marine and Freshwater		
Safe	<0.14 mg/kg DW	5
Adverse effects expected	>2.0 mg/kg DW	8
Washington State, safe	<0.41 mg/kg DW	
Tissue Residues		
Rainbow Trout, *Oncorhynchus mykiss*		
Lethal		
Eggs	>70 µg/kg fresh weight (FW)	9
Muscle, adults	>10.0 mg/kg FW	9
Adverse effects probable	1.0–5.0 mg/kg FW	10
Various Species of Freshwater Adult Fishes; Adverse Effects Expected		
Brain, whole body	>3.0 mg/kg FW	9
Muscle	>5.0 mg/kg FW	9
Birds		
Tissue residues		
Safe		
Brain, muscle	<15.0 mg/kg FW	11
Feather	<5.0 mg/kg FW	11
Kidney, seabirds	<30.0 mg/kg FW	13
Kidney, not seabirds	<20.0 mg/kg FW	11
Liver		
Normal	1.0–10.0 mg/kg FW	14
Toxic to sensitive species	>5.0–6.0 mg/kg FW	14, 15
Hazardous, possibly fatal	>20.0 mg/kg FW	14, 16

Table 13.4 (continued)   Proposed Mercury Criteria for the Protection of Selected Natural Resources and Human Health

Resource and Other Variables	Criterion or Effective Mercury Concentration	Ref[a]
Egg		
Mallard, *Anas platyrhynchos*; safe	<0.8–<1.0 mg/kg FW	17, 55
Ring-necked pheasant, *Phasianus colchicus*; safe	0.5–<0.9 mg/kg FW	18, 20
Common tern, *Sterna hirundo*; normal reproduction vs. reduced hatching and fledging success	<1.0 mg/kg FW vs. 2.0–4.7 mg/kg FW	19
Various species; safe	<0.5 to <2.0 mg/kg FW	13, 20
Waterbirds; adverse effects	1.0 to 3.6 mg/kg FW	15
Feather, acceptable	<9.0 mg/kg FW	22
Diet, fish-eating birds	<20 µg Hg/kg FW ration, as methylmercury	23
Diet, non fish-eating birds	50–<100 µg Hg/kg FW ration, as methylmercury	17, 24
Diet, loon	<300 µg/kg FW ration[b]	25, 26
Diet	<1.0–<3.0 mg/kg DW	13, 27
Daily intake	<640 µg/kg body weight (BW)	18, 28, 29, 30
Daily intake	<32 µg/kg BW[c]	30
Mammals		
Daily intake	<250 µg/kg BW	31
Diet; fish-eating mammals	<100 µg Hg/kg FW ration, as methylmercury	23
Diet	<1.1 mg/kg FW ration	13
Drinking Water		
Feral and domestic animal water supply, Wisconsin	<0.002 µg/L	2
Terrestrial vertebrate wildlife	<0.0013 µg/L[d]	30
Soils; terrestrial ecosystem protection; agricultural and residential land use vs. commercial and industrial use	<2.0 mg/kg DW vs. <30.0 mg/kg DW	5
Tissue Residues		
Kidney	<1.1 mg/kg FW	22
Liver, kidney	<30 mg/kg FW	13
Blood	<1.2 mg/kg FW	34
Brain	<1.5 mg/kg FW	34
Hair	<2.0 mg/kg FW	34
Florida Panther, *Felis concolor coryi*		
Reproduction normal (1.46 kittens per female annually)	<250 µg/kg FW blood	35
Reproduction inhibited (0.167 kittens per female annually)	>500 µg/kg FW blood	35
European Otter, *Lutra lutra*; Liver		
Normal	<4.0 mg/kg FW	36
Adverse sublethal effects possible	>10.0 mg/kg FW	36
Wildlife protection, Slovak Republic		
Fat	<1.0 µg/kg FW	21
Muscle	<50 µg/kg FW	21
Liver, kidney	<100 µg/kg FW	21

Table 13.4 (continued)   Proposed Mercury Criteria for the Protection of Selected Natural Resources and Human Health

Resource and Other Variables	Criterion or Effective Mercury Concentration	Ref[a]
Human Health		
Air; safe		
California	<0.00 µg/m^3	2
North Dakota	<0.0005 µg/m^3 for 8 h	2
Montana	<0.008 µg/m^3 for 24 h	2
Texas	<0.05 µg/m^3 for 1 year	2
New York	<0.167 µg/m^3 per year	2
Connecticut	<1.0 µg/m^3 for 8 h	2
Arizona	<1.5 µg/m^3 for 1 h	2
Virginia	<1.7 µg/m^3 for 24 h	2
Workplace		
Organic mercury	<10.0 µg/m^3	2
Metallic mercury vapor	<50.0 µg/m^3	2
Air; Adverse Effects Possible		
Skin	>30.0 µg/m^3	2
Emissions at individual sites	>2300–3200 g mercury daily	2
Drinking Water		
Brazil	<0.2 µg/L	1
International	<1.0 µg/L	2
U.S. (most states)	<2.0 µg/L	2
Effluent Limitations from Wastewater Treatment Plants		
Delaware, Oklahoma, Texas	<5.0 µg/L	2
Illinois, Wisconsin	<0.5 µg/L	2
New Jersey	<2.0 µg/L	2
Tennessee	<50.0 µg/L	2
Diet		
Australia	10–100 µg/kg FW ration	12
Benelux countries	<30 µg/kg FW ration	12
Brazil	<50 µg/kg FW ration	12
Canada	<500 µg/kg FW ration	49
U.S.	<1000 µg/kg FW ration	49, 50
Permissible Tolerable Weekly Intake		
Total mercury	<5 µg/kg BW	2
Total mercury	Maximum of 4.28 µg/kg BW	42
Methylmercury	<3.3 µg/kg BW	2, 51
Fish consumption advisory; Florida vs. most states	>0.5 mg total Hg/kg FW edible aquatic product vs. >1.0 mg total Hg/kg FW edible aquatic product	37
Fish and Seafood, Edible Parts		
U.S.	<300 µg/kg FW	53, 54
Acceptable intake		
60-kg adult	25 µg daily	38
70-kg adult	200 µg weekly	38
Adult	500 µg weekly	39
Pregnant Women		
All	<250 µg/kg FW	38
Japan	<400 µg/kg FW	12

Table 13.4 (continued) **Proposed Mercury Criteria for the Protection of Selected Natural Resources and Human Health**

Resource and Other Variables	Criterion or Effective Mercury Concentration	Ref[a]
Canada, Germany, U.S., Brazil	<500 µg/kg FW	2, 12, 40, 41
Italy	<700 µg/kg FW	42
Israel	<1000 µg/kg FW	43
Foods of Animal Origin		
Livestock tissues	<500 µg/kg FW	44
Wildlife tissues	<50 µg/kg FW	45
Breast muscle		
Domestic poultry	<500 µg/kg FW	12
Ducks (wildlife)	<1000 µg/kg FW	46
All Foods; Adult Weekly Intake		
As methylmercury	<100 µg	12
As total mercury	<150 µg	12
As methylmercury	<200 µg	23, 47
As total mercury	<300 µg	42, 47
Adult daily intake; nonpregnant vs. pregnant	<4.3 µg/kg BW vs. <0.6–1.1 µg/kg BW[e]	48
Tissue Residues		
Blood	<200 µg/kg FW	52
Hair	<6 mg/kg FW	39

[a] Refererence: 1, Palheta and Taylor 1995; 2, U.S. Public Health Service (USPHS) 1994; 3, Zumbroich 1997; 4, Dave and Xiu 1991; 5, Gaudet et al. 1995; 6, U.S. Environmental Protection Agency (USEPA) 1985; 7, Abbasi and Soni 1983; 8, Gillis et al. 1993; 9, Wiener and Spry 1996; 10, Niimi and Kissoon 1994; 11, Heinz 1996; 12, National Academy of Sciences (NAS) 1978; 13, Thompson 1996; 14, Wood et al. 1996; 15, Zillioux et al. 1993; 16, Littrell 1991; 17, Heinz 1979; 18, Spann et al. 1972; 19, Mora 1996; 20, Fimreite 1979; 21, Zilincar et al. 1992; 22, Beyer et al. 1997; 23, Yeardley et al. 1998; 24, March et al. 1983; 25, Scheuhammer et al. 1998; 26, Gariboldi et al. 1998; 27, Scheuhammer 1988; 28, McEwen et al. 1973; 29, Mullins et al. 1977; 30, Wolfe and Norman 1998; 31, Ramprashad and Ronald 1977; 32, Kucera 1983; 33, USEPA 1985; 34, Suzuki 1979; 35, Roelke et al. 1991; 36, Mason and Madsen 1992; 37, Facemire et al. 1995; 38, Khera 1979; 39, Lodenius et al. 1983; 40, Lathrop et al. 1991; 41, Hylander et al. 1994; 42, Barghigiani and De Ranieri 1992; 43, Krom et al. 1990; 44, Best et al. 1981; 45, Krynski et al. 1982; 46, Lindsay and Dimmick 1983; 47, Buzina et al. 1989; 48, Clarkson 1990; 49, Bodaly et al. 1984; 50, Kannan et al. 1998; 51, Petruccioli and Turillazzi 1991; 52, Galster 1976; 53, Brumbaugh et al. 2001; 54, Fields 2001; 55, Heinz and Hoffman 2003.

[b] Reproduction declined in loons, *Gavia imer*, when mercury in prey exceeded 300 µg total mercury/kg FW

[c] No observed adverse effect level with uncertainty factor of 20.

[d] Based on food chain biomagnification in aquatic webs.

[e] Assuming continuous exposure until steady-state balance for methylmercury is achieved. This process will take up to one year in most cases (Clarkson 1990).

for mercury and acute toxicities are now as low as 8 (8 to 167), based on the 96-hour average of 0.012 µg/L, and as low as 0.04 (0.04 to 0.8) based on the maximum permissible concentration of 2.4 µg/L. For protection against sublethal effects, these values were as low as 2 (2 to 8) based on the 4-day average and only 0.01 to 0.04 based on the maximum permissible concentration, or essentially no significant protection. A similar case is made against the proposed saltwater criteria (USEPA 1985; see Table 13.4). It seems that some downward modification is needed in the proposed mercury saltwater criteria if marine and estuarine biota are to be provided even minimal protection.

The significance of elevated mercury residues in tissues of aquatic organisms is not fully understood. Induction of liver metallothioneins and increased translatability of mRNA are biochemical indicators of the response of fish to mercury exposure (Angelow and Nicholls 1991; Schlenk et al. 1995), and more research seems needed on these and other indicators of mercury stress. Concentrations exceeding 1 mg Hg/kg FW can occur in various tissues of selected species of fish and aquatic mammals eaten by humans, but not all mercury incorporated in aquatic food chains is a result of anthropogenic activities (Barber et al. 1984). Some organisms, however, contain mercury tissue residues associated with known adverse effects to the organism and its predators. Thus, whole body residues of 5 to 7 mg Hg/kg FW in brook trout eventually proved fatal to that species (USEPA 1985).

To protect sensitive species of mammals and birds that regularly consume fish and other aquatic organisms, total mercury concentrations in these food items should probably not exceed 100 µg/kg for avian protection or 1.1 mg/kg for small mammals (Eisler 2000; see Table 13.4). For human health protection, proposed mercury levels in fish and seafood should not exceed 250 µg/kg FW for expectant mothers and 400 to 1000 µg/kg FW for adults worldwide (see Table 13.4). In humans, methylmercury concentration in scalp hair during pregnancy is considered to be the most reliable indicator for predicting the probability of psychomotor retardation in the child. The minimum toxic intake for humans is estimated to range between 0.6 and 1.1 µg methylmercury/kg BW daily (Clarkson 1990).

Because long-lived, slow-growing, high-trophic-position aquatic organisms usually contain the highest tissue mercury residues (Eisler 1981), some fisheries managers have proposed a legal maximum limit based on fish length or body weight (Lyle 1984; Chvojka 1988). Other strategies to control mercury concentrations in predatory fish include control of forage fish (Futter 1994), overfishing (Verta 1990), and various chemical treatments, including selenium (Paulsson and Lundbergh 1989), liming (Lindqvist et al. 1991), and various group VI derivatives, including sulfur, selenium and tellurium compounds, and thiamine (Siegel et al. 1991). Among sensitive avian species, adverse effects, primarily on reproduction, have been reported at mercury concentrations (µg/kg FW) of 5000 in feathers, 900 in eggs, 50 to 100 in diet, and daily administered doses of 640 on a body weight basis (Table 13.4; Eisler 2000).

Although low mercury concentrations (e.g., 50 µg/kg in the diets of domestic chickens) sometimes produced no adverse effects in the chickens, the tissue residues of mercury were sufficiently elevated to pose a hazard to human consumers (March et al. 1983). Mammals such as the domestic cat and the harp seal showed birth defects, histopathology, and elevated tissue residues at doses of 250 µg Hg/kg BW (Table 13.4). The mink, at dietary levels of 1100 µg Hg/kg, had signs of mercury poisoning; mercury residues in mink brain at this dietary level ranged from 7100 to 9300 µg/kg FW (Kucera 1983). Tissue residues in kidney, blood, brain, and hair in excess of 1100 µg/kg FW in nonhuman mammals are usually considered presumptive evidence of significant mercury contamination (Table 13.4). Tissue residues of mercury, as methylmercury, considered harmful to adult birds and terrestrial mammals ranged from 8 mg/kg FW in brain to 15 in muscle and 20 mg/kg FW in liver and kidney (Heinz 1996). In Canadian aboriginal peoples, a 20-year follow-up study on methylmercury levels has been initiated, with emphasis on age, sex, location, relation between maternal and fetal levels, and a reassessment of potential risk in communities where the highest known methylmercury levels have been found (Wheatley and Paradis 1995). Similar studies are recommended for avian and mammalian wildlife.

In the specialized case of environmental mercury contamination from historic and current gold mining activities, more research is recommended on physical and biological mercury removal technologies; development of nonmercury technologies to extract gold with minimal environmental damage; measurement of loss rates of mercury through continued periodic monitoring of fishery and wildlife resources in mercury-contaminated areas; and mercury accumulation and detoxification rates in comparatively pristine ecosystems. In view of the demonstrable adverse effects of uncontrolled mercury releases into the biosphere from gold production, it is imperative that all use of liquid mercury in gold amalgamation should cease at the earliest opportunity and that the ban be made permanent.

13.5 SUMMARY

Mercury contamination of the environment from historical and ongoing mining practices that rely on mercury amalgamation for gold extraction is widespread. Contamination was particularly severe in the immediate vicinity of gold extraction and refining operations; however, mercury, especially in the form of water-soluble methylmercury, may be transported to pristine areas by rainwater, water currents, deforestation, volatilization, and other vectors. Examples of gold mining-associated mercury pollution have been shown for Canada, the U.S., Africa, China, the Philippines, Siberia, and South America. In parts of Brazil, for example, mercury concentrations in all abiotic materials, plants, and animals, including endangered species of mammals and reptiles, collected near ongoing mercury amalgamation gold mining sites were far in excess of allowable mercury levels promulgated by regulatory agencies for the protection of human health and natural resources. Although health authorities in Brazil are unable to detect conclusive evidence of human mercury intoxication, the potential exists in the absence of mitigation for epidemic mercury poisoning of the mining population and environs. In the U.S., environmental mercury contamination is mostly from historical gold mining practices, and portions of Nevada remain sufficiently mercury-contaminated to pose a hazard to reproduction of carnivorous fishes and fish-eating birds.

Concentrations of total mercury lethal to sensitive representative natural resources range from 0.1 to 2.0 µg/L of medium for aquatic organisms; from 2200 to 31,000 µg/kg BW (acute oral) and from 4000 to 40,000 µg/kg (dietary) for birds; and from 100 to 500 µg/kg BW (daily dose) and from 1000 to 5000 µg/kg diet for mammals. Significant adverse sublethal effects were observed among selected aquatic species at water concentrations of 0.03 to 0.1 µg Hg/L. For some birds, adverse effects, mainly on reproduction, have been associated with total mercury concentrations (µg/kg FW) of 5000 in feathers, 900 in eggs, and 50 to 100 in diet, and with daily intakes of 640 µg/kg BW. Sensitive nonhuman mammals showed significant adverse effects of mercury when daily intakes were 250 µg/kg BW, when dietary levels were 1100 µg/kg, or when tissue concentrations exceeded 1100 µg/kg. Proposed mercury criteria for protection of aquatic life range from 0.012 µg/L for freshwater life to 0.025 µg/L for marine life; for birds, less than 100 µg/kg diet FW; and for small mammals, less than 1100 µg/kg FW diet. All of these proposed criteria provide, at best, minimal protection.

LITERATURE CITED

Abbasi, S.A. and R. Soni. 1983. Stress-induced enhancement of reproduction in earthworms *Octochaetus pattoni* exposed to chromium (VI) and mercury (II) — implications in environmental management, *Int. Jour. Environ. Stud.*, 22, 43–47.

Akagi, H., O. Malm, F.J.P. Branches, Y. Kinjo, Y. Kashima, J.R.D. Guimaraes, R.B. Oliveira, K. Haraguchi, W.C. Pfeiffer, Y. Takizawa, and H. Kato. 1995. Human exposure to mercury due to goldmining in the Tapajos River Basin, Amazon, Brazil: speciation of mercury in human hair, blood and urine, *Water Air Soil Pollut.*, 80, 85–94.

Alho, C.J.R. and L.M. Viera. 1997. Fish and wildlife resources in the Pantanal wetlands of Brazil and potential disturbances from the release of environmental contaminants, *Environ. Toxicol. Chem.*, 16, 71–74.

Allen, P. 1994. Changes in the haematological profile of the cichlid *Oreochromis aureus* (Steindachner) during acute inorganic mercury intoxication, *Comp. Biochem. Physiol.*, 108C, 117–121.

Amonoo-Neizer, E.H., D. Nyamah, and S.B. Bakiamoh. 1996. Mercury and arsenic pollution in soil and biological samples around the mining town of Obuasi, Ghana, *Water Air Soil Pollut.*, 91, 363–373.

Angelow, R.W. and D.M. Nicholls. 1991. The effect of mercury exposure on liver mRNA translatability and metallothionein in rainbow trout, *Comp. Biochem. Physiol.*, 100C, 439–444.

Appleton, J.D., T.M. Williams, N. Breward, A. Apostol, J. Miguel, and C. Miranda. 1999. Mercury contamination associated with artisanal gold mining on the island of Mindanao, the Philippines, *Sci. Total Environ.*, 228, 95–109.

Armstrong, F.A.J. 1979. Effects of mercury compounds on fish, in *The Biogeochemistry of Mercury in the Environment*, J.O. Nriagu, (Ed.), Elsevier/North-Holland, New York, 657–670.

Aula, I., H. Braunschweiler, T. Leino, I. Malin, P. Porvari, T. Hatanaka, M. Lodenius, and A. Juras. 1994. Levels of mercury in the Tucurui reservoir and its surrounding area in Para, Brazil, in *Mercury Pollution: Integration and Synthesis*, C.J. Watras and J.W. Huckabee, (Eds.), Lewis Publishers, CRC Press, Boca Raton, FL, 21–40.

Aula, I., H. Braunschweiler, and I. Malin. 1995. The watershed flux of mercury examined with indicators in the Tucurui reservoir in Para, Brazil, *Sci. Total Environ.*, 175, 97–107.

Baatrup, E. and K.B. Doving. 1990. Histochemical demonstration of mercury in the olfactory system of salmon (*Salmo salar* L.) following treatment with dietary methylmercuric chloride and dissolved mercuric chloride, *Ecotoxicol. Environ. Safety*, 20, 277–289.

Baatrup, E., K.B. Doving, and S. Winberg. 1990. Differential effects of mercurial compounds on the electroolfactogram (EOG) of salmon (*Salmo salar* L.), *Ecotoxicol. Environ. Safety*, 20, 269–276.

Baldi, F., F. Semplici, and M. Filippelli. 1991. Environmental applications of mercury resistant bacteria, *Water Air Soil Pollut.*, 56, 465–475.

Barber, R.T., P.J. Whaling, and D.M. Cohen. 1984. Mercury in recent and century-old deep-sea fish, *Environ. Sci. Technol.*, 18, 552–555.

Barghigiani, C. and S. De Ranieri. 1992. Mercury content in different size classes of important edible species of the northern Tyrrhenian Sea, *Mar. Pollut. Bull.*, 24, 114–116.

Beijer, K. and A. Jernelov. 1979. Methylation of mercury in natural waters, in *The Biogeochemistry of Mercury in the Environment*, J.O. Nriagu, (Ed.), Elsevier/North-Holland, New York, 201–210.

Berk, S.G., A.L. Mills, D.L. Henricks, and R.R. Colwell. 1978. Effects of ingesting mercury containing bacteria on mercury tolerance and growth rate of ciliates, *Microb. Ecol.*, 4, 319–330.

Best, J.B., M. Morita, J. Ragin, and J. Best, Jr. 1981. Acute toxic responses of the freshwater planarian, *Dugella dorotocephala*, to methylmercury, *Bull. Environ. Contam. Toxicol.*, 27, 49–54.

Beyer, W.N., E. Cromartie, and G.B. Moment. 1985. Accumulation of methylmercury in the earthworm, *Eisenia foetida*, and its effect on regeneration, *Bull. Environ. Contam. Toxicol.*, 35, 157–162.

Beyer W.N., M. Spalding, and D. Morrison. 1997. Mercury concentrations in feathers of wading birds from Florida, *Ambio*, 26, 97–100.

Birge, W.J., J.A. Black, A.G. Westerman, and J.E. Hudson. 1979.The effect of mercury on reproduction of fish and amphibians, in *The Biogeochemistry of Mercury in the Environment*, J.O. Nriagu, (Ed.), Elsevier/North-Holland, New York, 629–655.

Boas, R.C.V. 1995. Mineral extraction in Amazon and the environment. The mercury problem, in *Chemistry of the Amazon: Biodiversity, Natural Products, and Environmental Issues*. ACS Symposium Series 588. American Chemical Society, Washington, D.C., 295–303.

Bodaly, R.A., R.E. Hecky, and R.J.P. Fudge. 1984. Increases in fish mercury levels in lakes flooded by the Churchill River diversion, northern Manitoba, *Canad. Jour. Fish. Aquat. Sci.*, 41, 682–691.

Bonzongo, J.C. J, K.J. Heim, Y. Chen, W.B. Lyons, J.J. Warwick, G.C. Miller, and P.J. Lechler. 1996a. Mercury pathways in the Carson River-Lahontan Reservoir system, Nevada, USA, *Environ. Toxicol. Chem.*, 15, 677–683.

Bonzongo, J.C., K.J. Heim, J.J Warwick, and W.B. Lyons. 1996b. Mercury levels in surface waters of the Carson River-Lahontan Reservoir system, Nevada: influence of historic mining activities, *Environ. Pollut.*, 92, 193–201.

Boudou, A. and F. Ribeyre. 1983. Contamination of aquatic biocenoses by mercury compounds: an experimental toxicological approach, in *Aquatic Toxicology*, J.O. Nriagu, (Ed.), John Wiley & Sons, New York, 73–116.

Brazaitis, P., G.H. Rebelo, C. Yamashita, E.A Odierna, and M.E. Watanabe. 1996. Threats to Brazilian crocodile populations, *Oryx*, 30, 275–284.

Brumbaugh, W.G., D.P. Krabbenhoft, D.R. Helsel, J.G. Wiener, and K.R. Echols. 2001. A national pilot study of mercury contamination of aquatic ecosystems along multiple gradients: bioaccumulation in fish, U.S. Geol. Surv., Biol. Sci. Rep. USGS/BRD/BSR 2001-0009, 25 pp.

Buhler, D.R. (Ed.). 1973. Mercury in the western environment. Proceedings of a workshop, Portland, Oregon, February 25–26, 1971. Oregon State Univ., Corvallis, 360 pp.

Buzina, R., K. Suboticanec, J. Vukusic, J. Sapunar, and M. Zorica. 1989. Effect of industrial pollution on seafood content and dietary intake of total and methylmercury, *Sci. Total Environ.*, 78, 45–57.

Callahan, J.E., J.W. Miller, and J.R. Craig. 1994. Mercury pollution as a result of gold extraction in North Carolina, U.S.A., *Appl. Geochem.*, 9, 235–241.

Callil, C.T. and W.J. Junk. 1999. Concentration and incorporation of mercury by mollusc bivalves *Anodontites trapesialis* (Lamarck, 1819) and *Castalia ambigua* (Lamarck, 1819) in Pantanal of Pocone-MT, Brasil, *Biociecias*, 7(1), 3–28. (Portuguese, English abstract).

Camara, V. de M., M.I.D.F. Filhote, M.I.M. Lima, F.V. Aleira, M.S. Martins, T.O. Dantes, and R.R. Luiz. 1997. Strategies for preventing adolescent mercury exposure in Brazilian gold mining areas, *Toxicol. Indus. Health*, 13, 285–297.

Castilhos, Z.C., E.D. Bidone, and L.D. Lacerda. 1998. Increase of the background human exposure to mercury through fish consumption due to gold mining at the Tapajos River region, Para State, Amazon, *Bull. Environ. Contam. Toxicol.*, 61, 202–209.

Chang, L.W. 1979. Pathological effects of mercury poisoning, in *The Biogeochemistry of Mercury in the Environment*, J.O. Nriagu, (Ed.), Elsevier/North-Holland, New York, 519–580.

Choi, M.H., J.J. Cech, Jr., and M.C. Lagunas-Solar. 1998. Bioavailability of methylmercury to Sacramento blackfish (*Orthodon microlepidotus*): dissolved organic carbon effects, *Environ. Toxicol. Chem.*, 17, 695–701.

Chvojka, R. 1988. Mercury and selenium in axial white muscle of yellowtail kingfish from Sydney, Australia, *Mar. Pollut. Bull.*, 19, 210–213.

Clarkson, T.W. 1990. Human health risks from methylmercury in fish, *Environ. Toxicol. Chem.*, 9, 957–961.

Clarkson, T.W., R. Hamada, and L. Amin-Zaki. 1984. Mercury, in *Changing Metal Cycles and Human Health*, J.O. Nriagu, (Ed.), Springer-Verlag, Berlin, 285–309.

Clarkson, T.W. and D.O. Marsh. 1982. Mercury toxicity in man, in *Clinical, Biochemical, and Nutritional Aspects of Trace Elements, Vol. 6*, A.S. Prasad, (Ed.), Alan R. Liss, Inc., New York, 549–568.

Colwell, R.R., G.S. Sayler, J.D. Nelson, Jr., and A. Justice. 1976. Microbial mobilization of mercury in the aquatic environment, in *Environmental Biogeochemistry, Vol. 2, Metals Transfer and Ecological Mass Balances*, J.O. Nriagu, (Ed.), Ann Arbor Science Publ., Ann Arbor, MI, 437–487.

Cursino, L., S.M. Olberda, R.V. Cecilio, R.M. Moreira, E. Chartone-Souza, and A.M.A. Nascimento. 1999. Mercury concentrations in the sediment at different gold prospecting sites along the Carmo stream, Minas Gerais, Brazil, and frequency of resistant bacteria in the respective aquatic communities, *Hydrobiologia*, 394, 5–12.

Da Rosa, C.D. and J.S. Lyon (Eds.). 1997. *Golden Dreams, Poisoned Streams*. Mineral Policy Center, Washington, D.C., 269 pp.

Das, S.K., A. Sharma, and G. Talukder. 1982. Effects of mercury on cellular systems in mammals — a review, *Nucleus (Calcutta)*, 25, 193–230.

Dave, G. and R. Xiu. 1991. Toxicity of mercury, copper, nickel, lead, and cobalt to embryos and larvae of zebrafish, *Brachydanio rerio*, *Arch. Environ. Contam. Toxicol.*, 21, 126–34.

Davies, B.E. 1997. Deficiencies and toxicities of trace elements and micronutrients in tropical soils: limitations of knowledge and future research needs, *Environ. Toxicol. Chem.*, 16, 75–83.

de Kom, J.F.M., G.B. van der Voet, and F.A. de Wolff. 1998. Mercury exposure of maroon workers in the small scale gold mining in Suriname, *Environ. Res.*, 77A, 91–97.

de Lacerda, L.D. 1997. Atmospheric mercury and fish contamination in the Amazon, *Cien. Cult. Jour. Brazil. Assoc. Adv. Sci.*, 49, 54–57.

de Lacerda, L.D. and W. Salomons. 1998. *Mercury from Gold and Silver Mining: A Chemical Time Bomb?* Springer, Berlin, 146 pp.

D'Itri, F.M. 1972. Mercury in the aquatic ecosystem, *Tech. Rep. 23*, Institute of Water Research, Michigan State University, East Lansing, MI, 101 pp.

D'Itri, P. and F.M. D'Itri. 1977. *Mercury Contamination: A Human Tragedy.* John Wiley & Sons, New York, 311 pp.

Dorea, J.G., M.B. Moreira, G. East, and A.C. Barbosa. 1998. Selenium and mercury concentrations in some fish species of the Madeira River, Amazon Basin, Brazil, *Biol. Trace Elem. Res.,* 65, 211–220.

Eaton, R.D.P., D.C. Secord, and P. Hewitt. 1980. An experimental assessment of the toxic potential of mercury in ringed-seal liver for adult laboratory cats, *Toxicol. Appl. Pharmacol.,* 55, 514–521.

Eisler, R. 1978. Mercury contamination standards for marine environments, in *Energy and Environmental Stress in Aquatic Systems,* J.H. Thorp and J.W. Gibbons, (Eds.), U.S. Dept. Energy Symposium Series 48, CONF-771114, Natl. Tech. Infor. Serv., U.S. Dept. Commerce, Springfield, VA, 241–272 .

Eisler. R. 1981. *Trace Metal Concentrations in Marine Organisms.* Pergamon, New York, 687 pp.

Eisler, R. 1984. Trace metal changes associated with age of marine vertebrates, *Biol. Trace Elem. Res.,* 6, 165–180.

Eisler, R. 1985. Selenium hazards to fish, wildlife, and invertebrates: a synoptic review, U.S. Fish Wildl. Serv., Biol. Rep. 85 (1.5).

Eisler, R. 1987. Mercury hazards to fish, wildlife, and invertebrates: a synoptic review, U.S. Fish Wildl. Serv. Biol. Rep. 85 (1.10).

Eisler, R. 2000. Mercury, in *Handbook of Chemical Risk Assessment: Health Hazards to Humans, Plants, and Animals. Volume 1. Metals.* Lewis Publishers, Boca Raton, FL, 313–409.

Eisler, R. 2003. Health risks of gold miners: a synoptic review, *Environ. Geochem. Health,* 25, 325–345.

Eisler, R. 2004. Mercury hazards from gold mining to humans, plants, and animals, *Rev. Environ. Contam. Toxicol.,* 181, 139–198.

Elhassni, S.G. 1983. The many faces of methylmercury poisoning, *Jour. Toxicol.,* 19, 875–906.

Evans, H.L., R.H. Garman, and V.G. Laties. 1982. Neurotoxicity of methylmercury in the pigeon, *Neurotoxicology,* 3(3), 21–36.

Facemire, C., T. Augspurger, D. Bateman, M. Brim, P. Conzelmann, S. Delchamps, E. Douglas, L. Inmon, K. Looney, F. Lopez, G. Masson, D. Morrison, N. Morse, and A. Robison. 1995. Impacts of mercury contamination in the southeastern United States, *Water Air Soil Pollut.,* 80, 923–926.

Fields, S. 2001. Tarnishing the earth: gold mining's dirty secret, *Environ. Health Perspect.,* 109, A474–A482.

Fimreite, M. 1979. Accumulation and effects of mercury on birds, in *The Biogeochemistry of Mercury in the Environment,* J.O. Nriagu, (Ed.), Elsevier/North-Holland, New York, 601–627.

Finley, M.T. and R.C. Stendell. 1978. Survival and reproductive success of black ducks fed methylmercury, *Environ. Pollut.,* 16, 51–64.

Finley, M.T., W.H. Stickel, and R.E. Christensen. 1979. Mercury residues in tissue of dead and surviving birds fed methylmercury, *Bull. Environ. Contam. Toxicol.,* 21, 105–110.

Fostier, A.H., M.C. Forti, J.R.D. Guimaraes, A.J. Melfi, R. Boulet, C.M.E. Santo, and F.J. Krug. 2000. Mercury fluxes in a natural forested Amazonian catchment (Serra do Navio, Amapa State, Brazil), *Sci. Total Environ.,* 260, 201–211.

Frery, N., R. Maury-Brachet, E. Maillot, M. Deheeger, B. de Merona, and A. Boudou. 2001. Gold-mining activities and mercury contamination of native Amerindian communities in French Guiana: key role of fish in dietary uptake, *Environ. Health Perspect.,* 109, 449–456.

Friberg, L., and J. Vostal (Eds.). 1972. *Mercury in the Environment.* CRC Press, Cleveland, OH, 215 pp.

Futter, M.N. 1994. Pelagic food-web structure influences probability of mercury contamination in lake trout (*Salvelinus namaycush*), *Sci. Total Environ.*, 145, 7–12.

Galster, W.A. 1976. Mercury in Alaskan Eskimo mothers and infants, *Environ. Health Perspect.*, 15, 135–140.

Gariboldi, J.C., C.H. Jagoe, and A.L. Bryan, Jr. 1998. Dietary exposure to mercury in nestling wood storks (*Mycteria americana*) in Georgia, *Arch. Environ. Contam. Toxicol.*, 34, 398–405.

Gaudet, C., S. Lingard, P. Cureton, K. Keenleyside, S. Smith, and G. Raju. 1995. Canadian environmental guidelines for mercury, *Water Air Soil Pollut.*, 80, 1149–1159.

Geffen, A.J., N.J.G. Pearce, and W.T. Perkins. 1998. Metal concentrations in fish otoliths in relation to body composition after laboratory exposure to mercury and lead, *Mar. Ecol. Prog. Ser.*, 165, 235–245.

Gentile, J.H., S.M. Gentile, G. Hoffman, J.F. Heltshe, and N. Hairston, Jr. 1983. The effects of a chronic mercury exposure on survival, reproduction and population dynamics of *Mysidopsis bahia*, *Environ. Toxicol. Chem.*, 2, 61–68.

Gillis, C.A., N.L. Bonnevie, and R.J. Wenning. 1993. Mercury contamination in the Newark Bay estuary, *Ecotoxicol. Environ. Safety*, 25, 214–226.

Godbold, D.L. 1991. Mercury-induced root damage in spruce seedlings, *Water Air Soil Pollut.*, 56, 823–831.

Gray, J.E., V.F. Labson, J.N. Weaver, and D.P. Krabbenhoft. 2002. Mercury and methylmercury contamination related to artisanal gold mining, Suriname, *Geophysical Research. Lett.*, 29 (23/2105), 20/1–20/4.

Green, F.A., Jr., J.W. Anderson, S.R. Petrocelli, B.J. Presley, and R. Sims. 1976. Effect of mercury on the survival, respiration, and growth of postlarval white shrimp, *Penaeus setiferus*, *Mar. Biol.*, 37, 75–81.

Greener, Y. and J. Kochen. 1983. Methyl mercury toxicity in the chick embryo, *Teratology*, 28, 23–28.

Greer, J. 1993. The price of gold: environmental costs of the new gold rush, *The Ecologist*, 23 (3), 91–96.

Guimaraes, J.R.D., O. Malm, and W.C. Pfeiffer. 1995. A simplified technique for measurements of net mercury methylation rates in aquatic systems near goldmining areas, Amazon, Brazil, *Sci. Total Environ.*, 175, 151–162.

Guimaraes, J.R.D., M. Meili, O. Malm, and E.M.S. Brito. 1998. Hg methylation in sediments and floating meadows of a tropical lake in the Pantanal floodplain, Brazil, *Sci. Total Environ.*, 213, 165–175.

Gustin, M.S., G.E. Taylor Jr., and T.L. Leonard. 1994. High levels of mercury contamination in multiple media of the Carson River drainage system of Nevada: implications for risk assessment, *Environ. Health Perspect.*, 102, 772–778.

Gutleb, A.C., C. Schenck, and E. Staib. 1997. Giant otter (*Pteronura brasiliensis*) at risk? Total mercury and methylmercury levels in fish and otter scats, Peru, *Ambio*, 26, 511–514.

Hacon, S., P. Artaxo, F. Gerab, M.A. Yamasoe, R.C. Campos, L.F. Conti, and L.D. de Lacerda. 1995. Atmospheric mercury and trace elements in the region of Alta Floresta in the Amazon basin, *Water Air Soil Pollut.*, 80, 273–283.

Hacon, S., E.R. Rochedo, R. Campos, G. Rosales, and L.D. de Lacerda. 1997. Risk assessment of mercury in Alta Floresta. Amazon Basin — Brazil, *Water Air Soil Pollut.*, 97, 91–105.

Hamasaki, T., H. Nagase, Y. Yoshioka, and T. Sato. 1995. Formation, distribution, and ecotoxicity of methylmetals of tin, mercury, and arsenic in the environment, *Crit. Rev. Environ. Sci. Technol.*, 25, 45–91.

Handy, R.D. and W.S. Penrice. 1993. The influence of high oral doses of mercuric chloride on organ toxicant concentrations and histopathology in rainbow trout, *Oncorhynchus mykiss*, *Comp. Biochem. Physiol.*, 106C, 717–724.

Hawryshyn, C.W., W.C. Mackay, and T.H.Nilsson. 1982. Methyl mercury induced visual deficits in rainbow trout, *Canad. Jour. Zool.*, 60, 3127–3133.

Heinz, G.H. 1979. Methylmercury: reproductive and behavioral effects on three generations of mallard ducks, *Jour. Wildl. Manage.*, 43, 394–401.

Heinz, G.H. 1987. Mercury accumulations in mallards fed methylmercury with or without added DDE, *Environ. Res.*, 42, 372–376.

Heinz, G.H. 1996. Mercury poisoning in wildlife, in *Noninfectious Diseases of Wildlife*, 2nd ed., A. Fairbrother, L.N. Locke, and G.L. Hoff, (Eds.), Iowa State University Press, Ames, 118–127.

Heinz, G.H. and D.J. Hoffman. 1998. Methylmercury chloride and selenomethionine interactions on health and reproduction in mallards, *Environ. Toxicol. Chem.*, 17, 139–145.

Heinz, G.H. and D.J. Hoffman. 2003. Embryotoxic thresholds of mercury: estimates from individual mallard eggs, *Arch. Environ. Toxicol. Chem.*, 44, 267–264.

Heinz, G.H. and L.N. Locke. 1976. Brain lesions in mallard ducklings from parents fed methylmercury, *Avian Dis.*, 20, 9–17.

Henny, C.J., E.F. Hill, D.J. Hoffman, M.G. Spalding, and R.A. Grove. 2002. Nineteenth century mercury: hazard to wading birds and cormorants of the Carson River, Nevada, *Ecotoxicology*, 11, 213–231.

Hill, E.F. 1981. Inorganic and organic mercury toxicity to coturnix: sensitivity related to age and quantal assessment of physiological responses. Ph.D. thesis, University of Maryland, College Park.

Hill, E.F. and J.H. Soares, Jr. 1984. Subchronic mercury toxicity in *Coturnix* and a method of hazard evaluation, *Environ. Toxicol. Chem.*, 3, 489–502.

Hilmy, A.M., N.A. El Domiaty, A.Y. Daabees, and F.I. Moussa. 1987. Short-term effects of mercury on survival, behaviour, bioaccumulation and ionic pattern in the catfish (*Clarias lazera*), *Comp. Biochem. Physiol.*, 87C, 303–308.

Hirota, R., J. Asada, S. Tajima, and M. Fujiki. 1983. Accumulation of mercury by the marine copepod *Acartia clausi*, *Bull. Japan. Soc. Fish.*, 49, 1249–1251.

Hoffman, D.J. and G.H. Heinz 1998. Effects of mercury and selenium on glutathione metabolism and oxidative stress in mallard ducks, *Environ. Toxicol. Chem.*, 17, 161–166.

Holden, A.V. 1973. Mercury in fish and shellfish, a review, *Jour. Food Technol.*, 8, 1–25.

Huckabee, J.W., J.W. Elwood, and S.G. Hildebrand. 1979. Accumulation of mercury in freshwater biota, in *The Biogeochemistry of Mercury in the Environment*, J.O. Nriagu, (Ed.), Elsevier/North-Holland, New York, 277–301.

Huckabee, J.W., D.M. Lucas, and J.M. Baird. 1981. Occurrence of methylated mercury in a terrestrial food chain, *Environ. Res.*, 26, 174–181.

Hudson, R.H., R.K. Tucker, and M.A. Haegle. 1984. Handbook of toxicity of pesticides to wildlife. *U.S. Fish Wildl. Serv. Resour. Publ. 153*.

Hunerlach, M.P. and C.N. Alpers. 2003. Mercury contamination from hydraulic gold mining in the Sierra Nevada, California, in *Geologic Studies of Mercury by the U.S. Geological Survey*, J.E. Gray (Ed.). USGS Circular 1248, USGS Information Services, Denver, 23–28.

Hylander, L.D., E.C. Silva, L.J. Oliveira, S.A. Silva, E.K. Kuntze, and D.X. Silva. 1994. Mercury levels in Alto Pantanal: a screening study, *Ambio*, 23, 478–484.

Ikingura, J.R. and H. Akagi. 1996. Monitoring of fish and human exposure to mercury due to gold mining in the Lake Victoria goldfields, Tanzania, *Sci. Total Environ.*, 191, 59–68.

Ikingura, J.R., M.K.D. Mutakyahwa, and J.M.J. Kahatano. 1997. Mercury and mining in Africa with special reference to Tanzania, *Water Air Soil Pollut.*, 97, 223–232.

Jernelov, A. R. Hartung, P.B. Trost, and R.E. Bisque. 1972. Environmental dynamics of mercury, in *Environmental Mercury Contamination*, R. Hartung and B.D. Dinman, (Eds.), Ann Arbor Science Publ., Ann Arbor, MI, 167–201.

Jernelov, A., L. Landner, and T. Larsson. 1975. Swedish perspectives on mercury pollution, *Jour. Water Pollut. Control Fed.*, 47, 810–822.

Jernelov, A. and C. Ramel. 1994. Mercury in the environment, *Ambio*, 23, 166.

Kanamadi, R.D. and S.K. Saidapur. 1991. Effect of sublethal concentration of mercuric chloride on the ovary of the frog *Rana cyanophlyctis*, *Jour. Herpetol.*, 25, 494–497.

Kannan, K, R.J. Smith, Jr., R.F. Lee, H.L. Windom, P.T. Heitmuller, J.M. Macauley, and J.K. Summers. 1998. Distribution of total mercury and methyl mercury in water, sediment, and fish from south Florida estuaries, *Arch. Environ. Contam. Toxicol.*, 34, 109–118.

Kayser, H. 1976. Waste-water assay with continuous algal cultures: the effect of mercuric acetate on the growth of some marine dinoflagellates, *Mar. Biol.*, 36, 61–72.

Keckes, S. and J.J. Miettinen. 1972. Mercury as a marine pollutant, in *Marine Pollution and Sea Life*, M. Ruivo, (Ed.), Fishing Trading News Ltd., London, 279–289.

Khan, A.T. and J.S. Weis. 1993. Differential effects of organic and inorganic mercury on the micropyle of the eggs of *Fundulus heteroclitus*, *Environ. Biol. Fish.*, 37, 323–327.

Khera, K.S. 1979. Teratogenic and genetic effects of mercury toxicity, in *The Biogeochemistry of Mercury in the Environment*, J.O. Nriagu, (Ed.), Elsevier/North-Holland, New York, 501–518.

Kirubagaran R. and K.P. Joy. 1988. Toxic effects of three mercurial compounds on survival and histology of the kidney of the catfish *Clarias batrachus* (L.), *Ecotoxicol. Environ. Safety*, 15, 171–179.

Kirubagaran R. and K.P. Joy. 1989. Toxic effects of mercurials on thyroid function of the catfish, *Clarias batrachus* (L.), *Ecotoxicol. Environ. Safety*, 17, 265–271.

Kirubagaran R. and K.P. Joy. 1992. Toxic effects of mercury on testicular activity in the freshwater teleost, *Clarias batrachus* (L.), *Jour. Fish Biol.*, 41, 305–315.

Kopfler, F.C. 1974. The accumulation of organic and inorganic mercury compounds by the eastern oyster (*Crassostrea virginica*), *Bull. Environ. Contam. Toxicol.*, 11, 275–280.

Korte, F. and F. Coulston. 1998. Some considerations on the impact of ecological chemical principles in practice with emphasis on gold mining and cyanide, *Ecotoxicol. Environ. Safety*, 41, 119–129.

Kramer, V.J., M.C. Newman, M. Mulvey, and G.R. Ultsch. 1992. Glycolysis and Krebs cycle metabolites in mosquitofish, *Gambusia holbrooki*, Girard 1859, exposed to mercuric chloride: allozyme genotype effects, *Environ. Toxicol. Chem.*, 11, 357–364.

Krom, M.D., H. Hornung, and Y. Cohen. 1990. Determination of the environmental capacity of Haifa Bay with respect to the input of mercury, *Mar. Pollut. Bull.*, 21, 349–354.

Krynski, A., J. Kaluzinski, M. Wlazelko, and A. Adamowski. 1982. Contamination of roe deer by mercury compounds, *Acta Theriol.*, 27, 499–507.

Kucera, E. 1983. Mink and otter as indicators of mercury in Manitoba waters, *Canad. Jour. Zool.*, 61, 2250–2256.

Lacerda, L.D. 1997a. Global mercury emissions from gold and silver mining, *Water Air Soil Pollut.*, 97, 209–221.

Lacerda, L.D. 1997b. Evolution of mercury contamination in Brazil, *Water Air Soil Pollut.*, 97, 247–255.

Lacher, T.E., Jr. and M.I. Goldstein. 1997. Tropical ecotoxicology: status and needs, *Environ. Toxicol. Chem.*, 16, 100–111.

Lathrop, R.C., P.W. Rasmussem, and D.R. Knauer. 1991. Mercury concentrations in walleyes from Wisconsin (USA) lakes, *Water Air Soil Pollut.*, 56, 295–307.

Lawrence, S.J. 2003. Mercury in the Carson River Basin, Nevada, in *Geologic Studies of Mercury by the U.S. Geological Survey,* J.E. Gray (Ed.). USGS Circular 1248, USGS Information Services, Denver, 29–34.

Leander, J.D., D.E. McMillan, and T.S. Barlow. 1977. Chronic mercuric chloride: behavioral effects in pigeons, *Environ. Res.*, 14, 424–435.

Lechler, P.J., J.R. Miller, L.D. Lacerda, D. Vinson, J.C. Bonzongo, W.B. Lyons, and J.J. Warwick. 2000. Elevated mercury concentrations in soils, sediments, water, and fish of the Madeira River basin, Brazilian Amazon: a function of natural enrichments? *Sci. Total Environ.*, 260, 87–96.

Leigh, D.S. 1994. Mercury contamination and floodplain sedimentation from former gold mines in north Georgia, *Water Resour. Bull.*, 30, 739–748.

Leigh, D.S. 1997. Mercury-tainted overbank sediment from past gold mining in north Georgia, USA, *Environ. Geol.*, 30, 244–251.

Lima, A.R., C. Curtis, D.E. Hammermeister, T.P. Markee, C.E. Northcutt, and L.T. Brooke. 1984. Acute and chronic toxicities of arsenic (III) to fathead minnows, flagfish, daphnids, and an amphipod, *Arch. Environ. Contam. Toxicol.*, 13, 595–601.

Lin, Y., M. Guo, and W. Gan. 1997. Mercury pollution from small gold mines in China, *Water Air Soil Pollut.*, 97, 233–239.

Lindsay, R.C. and R.W. Dimmick. 1983. Mercury residues in wood ducks and wood duck foods in eastern Tennessee, *Jour. Wildl. Dis.*, 19, 114–117.

Lindqvist, O. (Ed.). 1991. Mercury as an environmental pollutant, *Water Air Soil Pollut.*, 56, 1–847.

Lindqvist, O., K. Johansson, M. Aastrup, A. Andersson, L. Bringmark, G. Hovsenius, L. Hakanson, A. Iverfeldt, M. Meili, and B. Timm. 1991. Mercury in the Swedish environment — recent research on causes, consequences and corrective methods, *Water Air Soil Pollut.*, 55, 1–261.

Littrell, E.E. 1991. Mercury in western grebes at Lake Berryessa and Clear Lake, California, *Calif. Fish Game*, 77, 142–144,

Lockhart, W.L., R.W. Macdonald, P.M. Outridge, P. Wilkinson, J.B. DeLaronde, and J.W.M. Rudd. 2000. Tests of the fidelity of lake sediment core records of mercury deposition to known histories of mercury contamination, *Sci. Total Environ.*, 260, 171–180.

Lodenius, M., A. Seppanen, and M. Herrnanen. 1983. Accumulation of mercury in fish and man from reservoirs in northern Finland, *Water Air Soil Pollut.*, 19, 237–246.

Lyle, J.M. 1984. Mercury concentrations in four carcharinid and three hammerhead sharks from coastal waters of the Northern Territory, *Austral. Jour. Mar. Freshwater Res.*, 35, 441–451.

Magos, L. and M. Webb. 1979. Synergism and antagonism in the toxicology of mercury, in *The Biogeochemistry of Mercury in the Environment*, J.O. Nriagu, (Ed.), Elsevier/North-Holland, New York, 581–599.

Malm, O. 1998. Gold mining as a source of mercury exposure in the Brazilian Amazon, *Environ. Res.*, 77A, 73–78.

Malm, O., F.J.P. Branches, H. Akagi, M.B. Castro, W.C. Pfeiffer, M. Harada, W.R. Bastos, and H. Kato. 1995a. Mercury and methylmercury in fish and human hair from the Tapajos river basin, Brazil, *Sci. Total Environ.*, 175, 141–150.

Malm, O., M.B. Castro, W.R. Bastos, F.J.P. Branches, J.R.D. Guimaraes, C.E. Zuffo, and W.C. Pfeiffer. 1995b. An assessment of Hg pollution in different goldmining areas, Amazon Brazil, *Sci. Total Environ.*, 175, 127–140.

Malm, O., W.C. Pfeiffer, C.M.M. Souza, and R. Reuther. 1990. Mercury pollution due to gold mining in the Madeira River Basin, Brazil, *Ambio*, 19, 11–15.

March, B.E., R.E. Poon, and S. Chu. 1983. The dynamics of ingested methyl mercury in growing and laying chickens, *Poult. Sci.*, 62, 1000–1009.

Martinelli, L.A., J.R. Ferreira, B.R. Forsberg, and R.L. Victoria. 1988. Mercury contamination in the Amazon: a gold rush consequence, *Ambio*, 17, 252–254.

Mason, C.F. and A.B. Madsen. 1992. Mercury in Danish otters (*Lutra lutra*), *Chemosphere*, 34, 1845–1849.

Mastrine, J.A., J-C.J. Bonzongo, and W.B. Lyons. 1999. Mercury concentrations in surface waters from fluvial systems draining historical precious metals mining areas in southeastern U.S.A, *Appl. Geochem.*, 14, 147–158.

McClurg, T.P. 1984. Effects of fluoride, cadmium and mercury on the estuarine prawn *Penaeus indicus*, *Water SA*, 10, 40–45.

McEwen, L.C., R.K. Tucker, J.O. Ells, and M. Haegle. 1973. Mercury-wildlife studies by the Denver Wildlife Research Center, in *Mercury in the Western Environment*, D.R. Buhler, (Ed.), Oregon State University, Corvallis, 146–156.

Meech, J.A., M.M. Veiga, and D. Tromans. 1998. Reactivity of mercury from gold mining activities in darkwater ecosystems, *Ambio*, 27, 92–98.

Miller, J.R., J. Rowland, P.J. Lechler, M. Desilets, and L.C. Hsu. 1996. Dispersal of mercury-contaminated sediments by geomorphic processes, Sixmile Canyon, Nevada, USA: implications to site characterization and remediation of fluvial environments, *Water Air Soil Pollut.*, 86, 373–388.

Mol, J.H., J.S. Ramlal, C. Lietar, and M. Verloo. 2001. Mercury contamination in freshwater, estuarine, and marine fishes in relation to small-scale gold mining in Suriname, South America, *Environ. Res.*, 86A, 183–197.

Montague, K. and P. Montague. 1971. *Mercury.* Sierra Club, New York, 158 pp.

Moore, J.W. and D.J. Sutherland. 1980. Mercury concentrations in fish inhabiting two polluted lakes in northern Canada, *Water Res.*, 14, 903–907.

Mora, M.A. 1996. Organochlorines and trace elements in four colonial waterbird species nesting in the lower Laguna Madre, Texas, *Arch. Environ. Contam. Toxicol.*, 31, 533–537.

Mullins, W.H., E.G. Bizeau, and W.W. Benson. 1977. Effects of phenyl mercury on captive game farm pheasants, *Jour. Wildl. Manage.*, 41, 302–308.

Nanda, D.E. and B.B. Mishra. 1997. Effect of solid waste from a chlor-alkali factory on rice plants; mercury accumulation and changes in biochemical variables, in *Ecological Issues and Environmental Impact Assessment*, P.N. Cheremisinoff, (Ed.), Gulf Publishing Co., Houston, 601–612.

National Academy of Sciences (NAS). 1978. *An Assessment of Mercury in the Environment.* Natl. Acad. Sci., Washington, D.C., 185 pp.

Nelson, C.H., D.E. Pierce, K.W. Leong, and F.F.H. Wang. 1975. Mercury distribution in ancient and modern sediment of northeastern Bering Sea, *Mar. Geol.*, 18, 91–104.

Newman, M.C. and D.K. Doubet. 1989. Size-dependence of mercury (II) accumulation kinetics in the mosquitofish, *Gambusia affinis* (Baird and Girard), *Arch. Environ. Contam. Toxicol.*, 18, 819–825.

Nicholls, D.M., K. Teichert-Kuliszewska, and G.R. Girgis. 1989. Effect of chronic mercuric chloride exposure on liver and muscle enzymes in fish, *Comp. Biochem. Physiol.*, 94C, 265–270.

Nicholson, J.K. and D. Osborn. 1984. Kidney lesions in juvenile starlings *Sturnus vulgaris* fed on a mercury-contaminated synthetic diet, *Environ. Pollut.*, 33A, 195–206.

Nico, L.G. and D.C. Taphorn. 1994. Mercury in fish from gold-mining regions in the upper Cuyuni River system, Venezuela, *Fresenius Environ. Bull.*, 3, 287–292.

Niimi, A.J. and G.P. Kissoon. 1994. Evaluation of the critical body burden concept based on inorganic and organic mercury toxicity to rainbow trout (*Oncorhynchus mykiss*), *Arch. Environ. Contam. Toxicol.*, 26, 169–178.

Niimi, A.J. and L. Lowe-Jinde. 1984. Differential blood cell ratios of rainbow trout (*Salmo gairdneri*) exposed to methylmercury and chlorobenzenes, *Arch. Environ. Contam. Toxicol.*, 13, 303–311.

Nriagu, J.O. (Ed.). 1979. *The Biogeochemistry of Mercury in the Environment.* Elsevier/North-Holland, New York, 696 pp.

Nriagu, J.O. 1993. Legacy of mercury pollution, *Nature*, 363, 589.

Nriagu, J. and H.K.T. Wong. 1997. Gold rushes and mercury pollution, in *Mercury and Its Effects on Environment and Biology*, A. Sigal and H. Sigal, (Eds.), Marcel Dekker, New York, 131–160.

Odin, M., A. Fuertet-Mazel, F. Ribeyre, and A. Boudou. 1995. Temperature, pH and photo-period effects on mercury bioaccumulation by nymphs of the burrowing mayfly *Hexagenia rigida*, *Water Air Soil Pollut.*, 80, 1003–1006.

Ogola, J.S., W.V. Mitullah, and M.A. Omulo. 2002. Impact of gold mining on the environment and human health: a case study in the Migori gold belt, Kenya, *Environ. Geochem. Health*, 24, 141–158.

Olivero, J. and B. Solano. 1998. Mercury in environmental samples from a waterbody con-taminated by gold mining in Columbia, South America, *Sci. Total Environ.*, 217, 83–89.

Oryu, Y., O. Malm, I. Thornton, I. Payne, and D. Cleary. 2001. Mercury contamination of fish and its implications for other wildlife of the Tapajos Basin, Brazilian Amazon, *Conserv. Biol.*, 15, 438–446.

Padovani, C.R., B.R. Forsberg, and T.P. Pimental. 1995. Contaminacao mercurial em peixes do Rio Madeira: resultados e recomendacoes para consumo humano, *Acta Amazonica*, 25, 127–136 (Portuguese, English summary).

Palheta, D. and A. Taylor. 1995. Mercury in environmental and biological samples from a gold mining area in the Amazon region of Brazil, *Sci. Total Environ.*, 168, 63–69.

Pass, D.A., P.B. Little, and L.A. Karstad. 1975. The pathology of subacute and chronic methyl mercury poisoning of the mallard duck (*Anas platyrhynchos*), *Jour. Comp. Pathol.*, 85, 7–21.

Paulsson, K. and K. Lundbergh. 1989. The selenium method for treatment of lakes for elevated levels of mercury in fish, *Sci. Total Environ.*, 87/88, 495–507.

Pelletier, E. and C. Audet. 1995. Tissue distribution and histopathological effects of dietary methylmercury in benthic grubby *Myoxocephalus aenaeus*, *Bull. Environ. Contam. Toxicol.*, 54, 724–730.

Pessoa, A., G.S. Albuquerque, and M.L. Barreto. 1995. The "garimpo" problem in the Amazon region. In *Chemistry of the Amazon: Biodiversity, Natural Products, and Environ-mental Issues.* ACS Symposium Series 588, American Chemical Society, Washington, D.C., 281–294

Petralia, J.F. 1996. *Gold! Gold! A Beginner's Handbook and Recreational Guide: How & Where to Prospect for Gold!* Sierra Outdoor Products, San Francisco, 143 pp.

Petruccioli, L. and P. Turillazzi 1991. Effect of methylmercury on acetylcholinesterase and serum cholinesterase activity in monkeys, *Macaca fascicularis*, *Bull. Environ. Con-tam. Toxicol.*, 46, 769–773.

Pfeiffer, W.C., L.D. de Lacerda, O. Malm, C.M.M. Souza, E.G. da Silveira, and W.R. Bastos. 1989. Mercury concentrations in inland waters of gold-mining areas in Rondonia, Brazil, *Sci. Total Environ.*, 87/88, 233–240.

Ponce, R.A. and N.S. Bloom. 1991. Effects of pH on the bioaccumulation of low level, dissolved methylmercury by rainbow trout (*Oncorhynchus mykiss*), *Water Air Soil Pollut*, 56, 631–640.

Porcella, D., J. Huckabee, and B. Wheatley (Eds.). 1995. Mercury as a global pollutant, *Water Air Soil Pollut.*, 80, 3–1336.

Porcella, D.B., C. Ramel, and A. Jernelov. 1997. Global mercury pollution and the role of gold mining: an overview, *Water Air Soil Pollut.*, 97, 205 –207.

Porvari, P. 1995. Mercury levels of fish in Tucurui hydroelectric reservoir and in River Moju in Amazonia, in the state of Para, Brazil, *Sci. Total Environ.*, 175, 109–117.

Punzo, F. 1993. Ovarian effects of a sublethal concentration of mercuric chloride in the river frog, *Rana heckscheri* (Anura: Ranidae), *Bull. Environ. Contam. Toxicol.*, 50, 385–391.

Ramamorthy, S. and K. Blumhagen. 1984. Uptake of Zn, Cd, and Hg by fish in the presence of competing compartments, *Canad. Jour. Fish. Aquat. Sci.*, 41, 750–756.

Ramprashad, F. and K. Ronald. 1977. A surface preparation study on the effect of methyl-mercury on the sensory hair cell population in the cochlea of the harp seal (*Pagophilus groenlandicus* Erxleben, 1977), *Canad. Jour. Zool.*, 55, 223–230.

Rao, R.V.V.P., S.A. Jordan, and M.K. Bhatnager. 1989. Ultrastructure of kidney of ducks exposed to methylmercury, lead and cadmium in combination, *Jour. Environ. Pathol. Toxicol. Oncol.*, 9, 19–44.

Reuther, R. 1994. Mercury accumulation in sediment and fish from rivers affected by alluvial gold mining in the Brazilian Madeira River basin, Amazon, *Environ. Monitor. Assess.*, 32, 239–258.

Ribeyre, F. and A. Boudou. 1984. Bioaccumulation et repartition tissulaire du mercure — $HgCl_2$ et CH_3HgCl_2 — chez *Salmo gairdneri* apres contamination par voie directe, *Water Air Soil Pollut.*, 23, 169–186.

Robinson, J.B. and O. H. Touvinen. 1984. Mechanisms of microbial resistance and detoxifi-cation of mercury and organomercury compounds: physiological, biochemical and genetic analysis, *Microbiol. Rev.*, 48, 95–124.

Rodgers, D.W. and F.W.H. Beamish. 1982. Dynamics of dietary methylmercury in rainbow trout, *Salmo gairdneri*, *Aquat. Toxicol.*, 2, 271–290.

Roelke, M.E., D.P. Schultz, C.F. Facemire, and S.F. Sundlof. 1991. Mercury contamination in the free-ranging endangered Florida panther (*Felis concolor coryi*), *Proc. Amer. Assoc. Zoo Veterin.*, 1991, 277–283.

Rojas, M., P.L. Drake, and S.M. Roberts. 2001. Assessing mercury health effects in gold workers near El Callao, Venezuela, *Jour. Occup. Environ. Med.*, 43, 158–165.

Ronald, K., S.V. Tessaro, J.F. Uthe, H.C. Freeman, and R. Frank. 1977. Methylmercury poisoning in the harp seal (*Pagophilus groenlandicus*), *Sci. Total Environ.*, 8, 1–11.

Ropek, R.M. and R.K. Neely. 1993. Mercury levels in Michigan river otters, *Lutra canadensis*, *Jour. Freshwat. Ecology*, 8, 141–147.

Scheuhammer, A.M. 1988. Chronic dietary toxicity of methylmercury in the zebra finch, *Poephila guttata*, *Bull. Environ. Contam. Toxicol.*, 40, 123–130.

Scheuhammer, A.M., C.M. Atchison, A.H.K. Wong, and D.C. Evers. 1998. Mercury exposure in breeding common loons (*Gavia imer*) in central Ontario, Canada, *Environ. Toxicol. Chem.*, 17, 191–196.

Schlenk, D., Y.S. Zhang, and J. Nix. 1995. Expression of hepatic metallothionein messenger RNA in feral and caged fish species correlates with muscle mercury levels, *Ecotoxicol. Environ. Safety*, 31, 282–286.

Sheffy, T.B. and J.R. St. Amant. 1982. Mercury burdens in furbearers in Wisconsin, *Jour. Wildl. Manage.*, 46, 1117–1120.

Siegel, B.Z., S.M. Siegel, T. Correa, C. Dagan, G. Galvez, L. Leeloy, A. Padua, and E. Yaeger. 1991. The protection of invertebrates, fish, and vascular plants against inorganic mercury poisoning by sulfur and selenium derivatives, *Arch. Environ. Contam. Toxicol.*, 20, 241–246.

Silva, P., F.H. Epstein, and R.J. Solomon. 1992. The effect of mercury on chloride secretion in the shark (*Squalus acanthias*) rectal gland, *Comp. Biochem. Physiol.*, 103C, 569–575.

Solonen, T. and M. Lodenius. 1984. Mercury in Finnish sparrowhawks *Accipiter nisus*, *Ornis Fennica*, 61, 58–63.

Spann, J.W., R.G. Heath, J.F. Kreitzer, and L.N. Locke. 1972. Ethyl mercury p-toluene sulfonanilide: lethal and reproductive effects on pheasants, *Science*, 175, 328–331.

Stokes, P.M., S.I. Dreier, N.V. Farkas, and R.A.N. McLean. 1981. Bioaccumulation of mercury by attached algae in acid stressed lakes, *Canad. Tech. Rep. Fish. Aquat. Sci.*, 1151, 136–148.

Suzuki, T. 1979. Dose-effect and dose-response relationships of mercury and its derivatives, in *The Biogeochemistry of Mercury in the Environment*, J.O. Nriagu, (Ed.), Elsevier/North-Holland, New York, 399–431.

Thain, J.E. 1984. Effects of mercury on the prosobranch mollusc *Crepidula fornicata*: acute lethal toxicity and effects on growth and reproduction of chronic exposure, *Mar. Environ. Res.*, 12, 285–309.

Thompson, D.R. 1996. Mercury in birds and terrestrial mammals, in *Environmental Contaminants in Wildlife: Interpreting Tissue Concentrations*, W.N. Beyer, G.H. Heinz, and A.W. Redmon-Norwood, (Eds.), CRC Press, Boca Raton, FL, 341–356.

Tupyakov, A.V., T.G. Laperdina, A.I. Egerov, V.A. Banshchikov, M.V. Mel'nikova, and O.B. Askarova. 1995. Mercury concentration in the abiogenic environmental components near gold and tungsten mining and concentration complexes in Eastern Transbaikalia, *Water Resour.*, 22, 163–169.

Ullrich, S.M., T.W. Tanton, and S.A. Abdrashitova. 2001. Mercury in the aquatic environment: a review of factors affecting methylation, *Crit. Rev. Environ. Sci. Technol.*, 31, 241–293.

U.S. Environmental Protection Agency (USEPA). 1985. *Ambient Water Quality Criteria for Mercury — 1984*. USEPA, Washington, D.C. Rep. 440/5-84-026, 136 pp.

U.S. Public Health Service (USPHS). 1994. *Toxicological Profile for Mercury*. Toxic Substances and Disease Registry, Atlanta, GA. TP-93/10.

Van der Molen, E.J., A.A. Blok, and G.J. De Graaf. 1982. Winter starvation and mercury intoxication in grey herons (*Ardea cinerea*) in the Netherlands, *Ardea*, 70, 173–184.

van Straaten, P. 2000. Human exposure to mercury due to small scale gold mining in northern Tanzania, *Sci. Total Environ.*, 259, 45–53.

Veiga, M.M., J.A. Meech, and R. Hypolito. 1995. Educational measures to address mercury pollution from gold-mining activities in the Amazon, *Ambio*, 24, 216–220.

Verta, M. 1990. Changes in fish mercury concentrations in an intensively fished lake, Canada. *Jour. Fish. Aquat. Sci.*, 47, 1888–1897.

Vieira, L.M., C.J.R. Alho, and G.A.L. Rerreira. 1995. Mercury contamination in sediment and in molluscs of Pantanel, Mato Grosso, Brazil. *Revta. Brasil, Zool.*, 12, 663–670.

Voccia, I.,K. Krzystniak, M. Dunier, D. Flipo, and M. Fournier. 1994. *In vitro* mercury-related cytotoxicity and functional impairment of the immune cells of rainbow trout (*Oncorhynchus mykiss*), *Aquat. Toxicol.*, 29, 37–48.

von Tumpling Jr., W., P. Zeilhofer, U. Ammer, J. Einax, and R.D. Wilken. 1995. Estimation of mercury content in tailings of the gold mine area of Pocone, Mato Grosso, Brazil, *Environ. Sci. Pollut. Res. Int.*, 2, 225–228.

Walter, C.M., F.C. June, and H.G. Brown. 1973. Mercury in fish, sediments, and water in Lake Oahe, South Dakota, *Jour. Water Pollut. Control Feder.*, 45, 2203–2210

Watras, C.J. and J.W. Huckabee (Eds.). 1994. *Mercury Pollution: Integration and Synthesis.* Lewis Publishers, CRC Press, Boca Raton, FL, 727 pp.

Wayne, D.M., J.J. Warwick, P.J. Lechler, G.A. Gill, and W.B. Lyons. 1996. Mercury contamination in the Carson River, Nevada: a preliminary study of the impact of mining wastes, *Water Air Soil Pollut.*, 92, 391–408.

Weis, J.S. 1976. Effects of mercury, cadmium, and lead salts on regeneration and ecdysis in the fiddler crab, *Uca pugilator*, U.S. Natl. Mar. Fish. Serv., Fish. Bull., 74, 464–467.

Weis, J.S. 1984. Metallothionein and mercury tolerance in the killifish, *Fundulus heteroclitus*, *Mar. Environ. Res.*, 14, 153–166.

Weis, J.S. and P. Weis 1995. Effects of embryonic exposure to methylmercury on larval prey-capture ability in the mummichog, *Fundulus heteroclitus*, *Environ. Toxicol. Chem.*, 14, 153–156.

West, J.M. 1971. *How to Mine and Prospect for Placer Gold.* U.S. Dept. Interior, Bur. Mines Inform. Circ. 8517, 43 pp.

Wheatley, B. and S. Paradis. 1995. Exposure of Canadian aboriginal peoples to methyl-mercury, *Water Air Soil Pollut.*, 80, 3–11.

Wiener, J.G., D.P. Krabbenhoft, G.H. Heinz, and A.M. Scheuhammer, 2002. Ecotoxicology of mercury, in *Handbook of Ecotoxicology, 2nd ed.,* D.J. Hoffman, B.A. Rattner, G.A. Burton, Jr., and J. Cairns, Jr., (Eds.), Lewis Publishers, Boca Raton, FL, 409–463.

Wiener, J.G. and D.J. Spry. 1996. Toxicological significance of mercury in freshwater fish, in *Environmental Contaminants in Wildlife: Interpreting Tissue Concentrations*, W.N. Beyer, G.H. Heinz, and A.W. Redmon-Norwood, (Eds.), CRC Press, Boca Raton, FL, 297–339.

Wolfe, M. and S. Norman. 1998. Effects of waterborne mercury on terrestrial wildlife at Clear Lake: evaluation and testing of a predictive model, *Environ. Toxicol. Chem.*, 17, 214–217.

Wolfe, M.E., S. Schwarzbach, and R.A. Sulaiman. 1998. Effects of mercury on wildlife: a comprehensive review, *Environ. Toxicol. Chem.*, 17, 146–160.

Wood, P.B., J.H. White, A. Steffer, J.M. Wood, C.F. Facemire, and H.F. Percival. 1996. Mercury concentrations in tissues of Florida bald eagles, *Jour. Wildl. Manage.*, 60, 178–185.

Wren, C.D. 1986. A review of metal accumulation and toxicity in wild mammals. I. Mercury, *Environ. Res.*, 40, 210–244.

Wright, D.A., P.M. Wellbourn, and A.V.M. Martin. 1991. Inorganic and organic mercury uptake and loss by the crayfish *Orconectes propinquus*, *Water Air Soil Pollut.*, 56, 697–707.

Yeardley, R.B., J.M. Lazorchak, and S.G. Paulsen. 1998. Elemental fish tissue contamination in northeastern U.S. lakes: evaluation of an approach to regional assessment, *Environ. Toxicol. Chem.*, 17, 1875–1884.

Zilincar, V.J., B. Bystrica, P. Zvada, D. Kubin, and P. Hell. 1992. Die Schwermeallbelastung bei den Braunbaren in den Westkarpten, *Z. Jagdwiss.*, 38, 235–243.

Zillioux, E.J., D.B. Porcella, and J.M. Benoit. 1993. Mercury cycling and effects in freshwater wetland ecosystems, *Environ. Toxicol. Chem.*, 12, 2245–2264.

Zumbroich, T. 1997. Heavy metal pollution of vegetables in a former ore-mining region, in *Ecological Issues and Environmental Impact Assessment*, P.N. Cheremisinoff, (Ed.), Gulf Publishing Co., Houston, 207–215.

CHAPTER 14

Abandoned Underground Gold Mines

Gold has been mined since antiquity at numerous locations throughout the world (Kirkemo et al. 2001). The Republic of South Africa is the major commercial global producer of gold; secondary producers include the United States, the former Soviet Union, Canada, Australia, the People's Republic of China, Brazil, the Philippines, the Dominican Republic, Papua New Guinea, Ghana, Tanzania, and Ecuador (Elevatorski 1981; Gasparrini 1993; Greer 1993). Underground gold mines are usually abandoned or closed owing to poor yields or adverse economic conditions. At present, the most environmentally responsible gold mining companies spend millions of dollars restoring the sites of closed mines and developing technologies to minimize the impact of active mines, although many attempts are ineffective (Fields 2001). Data are currently scarce or incomplete on the influence of inactive underground gold mines on the surrounding biosphere.

This chapter briefly synthesizes available information on abandoned underground gold mines as habitats for animals and plants, as deterrents to land development, as sources of drainage water toxic to natural resources, and as science sites. These findings may have application to other inactive or abandoned underground mines.

14.1   HABITAT FOR BIOTA

Abandoned underground gold mines constitute unique habitats for recently identified species of microorganisms. In South Korea, for example, three new species of fungi were isolated from the soils of gold mines: *Catellatospora koreensis* (Lee et al. 2000a), *Saccharothrix violacea*, and *S. albidocapillata* (Lee et al. 2000b). Because gold mine ores usually contain high concentrations of arsenic (Kirkemo et al. 2001), many species of arsenic-resistant bacteria are found there. In one case, a new species of unique anaerobic bacterium, *Chrysiogenes arsenatis*, was isolated from wastewater of an Australian gold mine. This organism grew with As^{+3} as the electron donor, and CO_2 or HCO_3^- as the carbon source; growth was rapid with a doubling time of 7.6 hours (Macy et al. 1996; Santini et al. 2000). More research seems needed on physiological mechanisms of action of microorganisms in coping with chemically hostile conditions, and their possible role in detoxification of mine drainage waters containing arsenic and other potentially harmful substances.

Abandoned underground gold mines are also used as habitat by representative species of local vertebrates, although much of the evidence available is tenuous. In Arkansas, for example, between 1870 and 1890, subterranean habitat from mining was formed during the gold rush (Saugey et al. 1988). During the period of greatest activity, 1885 to 1888, over a dozen gold mines were in operation, some extending over 150 meters into the surrounding mountains. The gold and silver boom ended in 1888 with the issuance of a federal report stating that there were no precious metals in paying quantities to be found within the area. Soon thereafter, the mines were abandoned as proprietors moved west. The legacy has been the creation of unusual and unique wildlife habitat for six species of salamanders and nine species of bats, including some that were considered threatened. Salamanders found in abandoned Arkansas mines included the Ouachita dusky salamander (*Desmognathes brimleyorum*), the many ribbed salamander (*Eurycea multiplicata*), and four species of *Plethodon*; bat species identified included the eastern pipistrelle (*Pipistrellus subflavus*), the big brown bat (*Eptesicus fuscus*), five species of *Myotis*, the red bat (*Lasiurus borealis*), and the silver-haired bat (*Lasionycteris noctivagans)*. A total of 27 vertebrate species were identified in these abandoned mines (Saugey et al. 1988).

Other abandoned underground mines in Arkansas — originally constructed for gold extraction, but eventually used commercially for lead and zinc production — were examined between December 1991 and March 1995 (McAllister et al. 1995). These mines were considered home to 16 species of vertebrates, including eight species of amphibians (two species of salamanders, one species of toad, and five species of frogs), three species of reptiles (northern fence lizard, *Sceloporus undulatus hyacinthinus*; southern copperhead, *Agkistrodon contortix*; and broadhead skink, *Eumeces laticeps*), house wren (*Troglodytes aedon*), two species of bats (*Myotis* spp.), and two species of mice (*Peromyscus* spp.). Another four species of mammals (coyote, *Canis latrans*; opossum, *Didelphis virginiana*; raccoon, *Procyon lotor*; and striped skunk, *Mephitis mephitis*) were not collected but presumably present because their tracks were identified in the mine entrance (McAllister et al. 1995).

Abandoned mines play an important role in the ecology of many species, serving as permanent or temporary habitats (Heath et al. 1986), especially for bats (Taylor 1995). About 30 species of North American bats — including six endangered species — use abandoned underground mines for rearing young and for hibernation (Taylor 1995). In fact, nearly half of all species of bats in North America live in regions where abandoned underground mines provide suitable temperatures for year-round use; of more than 6000 underground mines surveyed in Arizona, California, Colorado, and New Mexico, about half showed signs of use by bats, and 10% contained important colonies. The closure of abandoned mines without first evaluating their importance to bats is considered a major threat to North American bat populations (Taylor 1995).

The Millionaire mine in Beaverhead National Forest, Montana, is now home to feral rock doves, *Columba livia*. These pigeons nested on the ground 91 meters below the surface in near or total darkness at 6°C when the outside temperature was 22°C, presumably to protect against predators and fluctuating temperatures (Hendricks 1997). The Millionaire mine was active between 1911 and 1921, but produced only 11 troy ounces of gold during that period, as well as 2556 troy ounces of silver and at least 24 metric tons of lead. Nesting in abandoned underground mines is also reported for house sparrows, *Passer domesticus* (Hendricks 1997). Hibernating bats (little brown bat, *Myotis lucifugus*; northern myotis, *Myotis septentrionalis*) frequented a 150-meter adit of an abandoned underground mine near Windsor, Quebec (Thomas 1995). Of 676 abandoned mine sites examined in northern Utah, 196 (24%) were occupied by day-roosting Townsend's big-eared bat, *Plecotus townsendii* (Sherwin et al. 2000). Similar studies conducted throughout the western United States indicated a trend toward use of abandoned gold mines by *P. townsendii*, with up to 40% occupancy of known bat roosts in California, Oregon, and western central Nevada. It was suggested that abandoned mines may be colonized by pioneering individuals or groups of bats that have not had sufficient time to build large colonies relative to groups in caves (Pierson 1989, Sherwin et al. 2000). The U.S. Forest Service (USFS) considers abandoned mines unique subterranean habitat and is actively acquiring and managing lands containing abandoned mines, including abandoned gold mines, as protected wildlife areas (Saugey et al. 1988). Further, the USFS has prohibited additional mining from these abandoned mines, has designated key areas immune from logging, and is closing the protected areas and associated aboveground habitat during critical parts of the year to offer additional protection to species of concern (Saugey et al. 1988).

Monitoring and research efforts now seem warranted on the suitability of abandoned underground gold mines and environs as habitat for macrofauna, with emphasis on suitability for sensitive species now classified as threatened or endangered, and on changes in species abundance and diversity due to seasonal food availability and migratory patterns.

14.2 LAND DEVELOPMENT

Abandoned mines at shallow depth represent a serious problem in areas that are being developed or redeveloped (Bell et al. 2000) throughout the world. In many areas where gold deposits have been worked for more than 100 years, abandoned mine sites are frequently unrecorded. In England, the first statutory obligation to keep mine records dates from 1850, and it was not until 1872 that the production and retention of mine plans became compulsory. And if old records exist, they may be inaccurate. For example, in Johannesburg, Republic of South Africa, early underground gold mines, circa 1886, now abandoned, were at shallow depths and their presence has resulted in subsidence, imposing limitations on development. In fact, the erection of buildings on the honey-combed land is now controlled by the Government Mining Engineer, who determines whether building is permissible as well as the permissible heights of buildings in relation to the depth at which mining occurred. Existing building regulations in Johannesburg prohibit development or construction if mining has occurred 0 to 90 meters underground, limit building height to a single story if mining has occurred 90 to 120 meters below ground, two stories if 120 to 150 meters, three at 150 to 180 meters, four stories at 180 to 210 meters, and no restrictions over 240 meters unless mining circumstances are unusual (Bell et al. 2000).

14.3 EFFECTS ON WATER QUALITY

Abandoned underground gold mines in the Black Hills of South Dakota contributed acid, metals, metalloids, and cyanides to streams (Rahn et al. 1996). In some areas of sulfide mineralization, local impacts were severe; however, in most areas the impacts were negligible because most ore deposits consisted of small quartz veins with few sulfides. The maximum daily discharge into nearby creeks from 11 abandoned underground gold mines in the Black Hills was 2.5 million kg of tailings containing 15 kg of mercury, 140 kg of cyanide, 100 kg of zinc, and 10,000 kg of arsenopyrites. The pH values for surface waters at these sites ranged from 1.6 to 9.7. The most acidic waters were associated with low discharges from tailings dumps with sulfide-rich ores. Metal concentrations, with the exception of mercury and iron, were usually low. The highest concentration recorded of iron was 498.0 mg/L, and for mercury 5.48 mg/L (Rahn et al. 1996).

The Serengeti National Park (SNP) in northern Tanzania supports more than 2 million large mammals (Bowell et al. 1995). This area is also part of the Lake Victoria gold fields, which produced 8810 kg of gold between 1933 and 1966. Flooding of tailings from a gold mine impacted the Orangi River, an important year-round source of water for wildlife in the northern part of SNP. Drainage water from the tailings was characterized by low pH of 2.3 and elevated concentrations of sulfate (3280 mg/L), aluminum (275 mg/L), arsenic (324 mg/L), copper (125 mg/L), iron (622 mg/L), lead (21 mg/L), manganese (65 mg/L), and zinc (126 mg/L). Mixing of these acidic waters with the alkaline river resulted in rapid precipitation as iron-ocher coatings on clastic sediments. Buffering of the mine drainage waters confined damage effects to within 1 km of mine workings. Protozoan bioassays indicated that growth was inhibited, presumably by metals and metalloids, from all locations tested, and that protozoan mortality was common at most sampling locations. The species considered most at risk in the SNP are mature bull African elephants (*Loxodonta africana*) which forage over a small part of the impacted area and have high bulk requirements of vegetation and water (Bowell et al. 1995).

Tailings from an abandoned Au-Ag-Mo mine in Korea was the main contamination source for cadmium (6 mg/kg DW tailings), copper (111 mg/kg), zinc (2010 mg/kg), lead (3250 mg/kg), and arsenic (20,140 mg/kg) in the soil–water system near the Songcheon mine (Lim et al. 2003). Similar findings were documented for an abandoned Au-Ag-Cu-Zn mine near Dongil, Korea (Table 14.1).

Elevated concentrations of arsenic, copper, cyanide, lead, mercury, and zinc in drainage waters from abandoned gold mines in South Dakota (Rahn et al. 1996) and Tanzania (Bowell et al. 1995) exceeded recommended concentrations for the protection of human health, plants, and animals (Eisler 2000a, 2000b, 2000c). For

Table 14.1   Metals and Arsenic in Tailings,
              Soils, Rice, and Groundwater
              near an Abandoned Gold-
              Silver-Copper-Zinc Mine,
              Dongil, Korea, 2000–2001

Sample and Element	Concentration
Tailings (mg/kg dry weight)	
Arsenic	8720
Lead	5850
Copper	3610
Zinc	630
Cadmium	6
Farm soils (mg/kg DW) vs. paddy soils (mg/kg DW)	
Arsenic	40 vs. 31
Lead	39 vs. 27
Copper	139 vs. 31
Rice grains (µg/kg DW) vs. groundwater (µg/L)[a]	
Arsenic	150 vs. 24
Cadmium	300 vs. ND[b]
Copper	6900 vs. 7
Lead	2160 vs. 4
Zinc	38,200 vs. 15

[a] Groundwater used as drinking water
[b] Not detectable

Source: Lee and Chon 2003b.

arsenic, sensitive aquatic species were damaged at water concentrations between 19 and 48 µg/L (Eisler 2000c). Inorganic arsenic levels recommended for human health protection include <10 µg As/m^3 air, and <10 µg As/L in drinking water. Copper was lethal to representative species of freshwater plants and animals at 5.0 to 9.8 µg Cu/L, lethal to sensitive species of freshwater biota at 0.23 to 0.91 µg Cu/L, toxic to terrestrial plants at >40 µg Cu/L irrigation water, and harmful to human infants at >3.0 mg Cu/L drinking water (Eisler 2000a). Cyanide was lethal to freshwater biota at water concentrations from 20 to 76 µg/L as HCN and produced adverse effects on swimming and reproduction of fishes at 5.0 to 7.2 µg HCN/L. The recommended cyanide drinking water criterion for human health protection is <10 µg HCN/L (Eisler 2000b). For lead, concentrations >25 µg total Pb/L at acidic pH were associated with adverse effects on fish embryo survival and growth, and for crops it was recommended that irrigation waters not exceed 5 mg Pb/L (Eisler 2000a). Mercury was lethal to representative aquatic organisms at 0.1 to 2.0 µg total Hg/L and produced significant adverse effects to sensitive aquatic organisms at 0.03 to 0.1 µg total Hg/L. Recommended safe levels of total mercury in drinking water for human health protection ranged from 0.2 µg/L in Brazil to 1.0 to 2.0 elsewhere (Eisler 2000a). Zinc was lethal to sensitive species of aquatic organisms at concentrations from 32 to 66 µg Zn/L; significant adverse effects on growth and reproduction were documented from 10 to 25 µg Zn/L among sensitive species of aquatic plants and invertebrates, fishes, and amphibians (Eisler 2000a).

To protect sensitive species of fish and other wildlife from toxic components in drainage waters from abandoned underground gold mines, it seems necessary to reduce discharges and to detoxify the waste stream. More studies are recommended on evaluation of detoxifying properties of metals-resistant strains of microorganisms isolated from underground gold mines, with emphasis on mercury- and iron-resistant strains.

14.4   SCIENCE SITE POTENTIAL

An abandoned underground gold mine in Lead, South Dakota, with a 2500-meter-deep shaft, is being considered by physicists as a facility to shelter sensitive instruments from cosmic radiation and to serve various educational and visitor needs (Malakoff 2001). This facility has more than 1000 km of tunnels equipped with electrical wiring and ventilation (Malakoff 2001). Abandoned underground gold mines and other abandoned underground mines should be reexamined for research and recreational potential.

14.5   SUMMARY

Abandoned underground gold mines provide habitat and unique environments for certain microorganisms and local vertebrate fauna, and show potential as science education centers. Uncontrolled mine drainage waters, however, may be toxic to aquatic life and other wildlife that depend on clean water. Land development is curtailed in certain heavily mined areas.

Additional studies seem warranted on the suitability of abandoned underground gold mines as habitat for local biota, on physiological mechanisms of resistance to metals by bacterial strains found in abandoned gold mines, and on biological detoxification processes of acidic gold mine drainage waters.

LITERATURE CITED

Bell, F.G., T.R. Stacey, and D.D. Genske. 2000. Mining subsidence and its effect on the environment: some differing examples, *Environ. Geol.*, 40,135–152.

Bowell, R.J., A. Warren, H.A. Minjera, and N. Kimaro. 1995. Environmental impact of former gold mining on the Orangi River, Serengeti N.P., Tanzania, *Biogeochem.*, 28, 131–160.

Eisler, R. 2000a. *Handbook of Chemical Risk Assessment: Health Hazards to Humans, Plants, and Animals. Volume 1. Metals.* Lewis Publishers, Boca Raton, FL, 93–409, 605–714.

Eisler, R. 2000b. *Handbook of Chemical Risk Assessment: Health Hazards to Humans, Plants, and Animals. Volume 2. Organics.* Lewis Publishers, Boca Raton, FL, 903–959.

Eisler, R. 2000c. *Handbook of Chemical Risk Assessment: Health Hazards to Humans, Plants, and Animals. Volume 3. Metalloids, Radiation, Cumulative Index to Chemicals and Species.* Lewis Publishers, Boca Raton, FL, 1501–1566.

Elevatorski, E.A. 1981. *Gold Mines of the World*. Minobras, Dana Point, CA, 107 pp.

Fields, S. 2001. Tarnishing the earth: gold mining's dirty secret, *Environ. Health Perspec.*, 109, A4741501–1566 A482.

Gasparrini, C. 1993. *Gold and Other Precious Metals. From Ore to Market.* Springer-Verlag, Berlin, 336 pp.

Greer, J. 1993. The price of gold: environmental costs of the new gold rush, *The Ecologist*, 23(3), 91–96.

Heath, D.E., D.A. Saugey, and G.A. Heidt. 1986. Abandoned mine fauna of the Ouachita Mountains, Arkansas: vertebrate taxa, *Proc. Arkansas Acad. Sci.*, 40, 33–36.

Hendricks, P. 1997. Feral pigeons nesting underground in an abandoned mine, *Northwest. Natural.*, 78, 74–76.

Kirkemo, H., W.L. Newman, and R.P. Ashley. 2001. *Gold.* U.S. Geological Survey, Federal Center, Denver, 23 pp.

Lee, J.S. and H.T. Chon. 2003a. Toxic risk assessment of heavy metals on abandoned metal mine areas with various exposure pathways. 6th International Symposium on Environmental Geochemistry, Edinburgh, Scotland, 7–11 Sept. 2003, Book of Abstracts, 191.

Lee, J.S. and H.T. Chon. 2003b. Exposure assessment of heavy metals on abandoned metal mine areas by ingestion of soil, crop plant and groundwater, *Jour. de physique. IV, Proceedings*, 107, 757–760.

Lee, S., S.O. Kang, and Y.C. Hah. 2000a. *Catellatospora koreensis* sp. nov., a novel actinomycete isolated from a gold-mine cave, *Int. Jour. System. Evolut. Microbiol.*, 50, 1103–1111.

Lee, S.D., E.S. Kim, J.H. Roe, J.H. Kim, S.O. Kang, and Y.C. Hah. 2000b. *Saccharothrix violacea* sp. nov., isolated from a gold mine cave, and *Saccharothrix albidocapillata* comb. nov., *Int. Jour. System. Evolut. Microbiol.*, 50, 1315–1323.

Lim, H-S., J.S. Lee, and H.T. Chon. 2003. Arsenic and heavy metal contamination in the vicinity of abandoned Songcheon Au-Ag-Mo mine, Korea. 6th International Symposium on Environmental Geochemistry, Edinburgh, Scotland, 7–11 Sept. 2003, Book of Abstracts, 157.

Macy, J.M., K. Nunan, K.D. Hagen, D.R. Dixon, P.J. Harbour, M. Cahill, and L.I. Sly. 1996. *Chrysiogenes arsenatis* gen. nov., sp. nov., a new arsenate-respiring bacterium isolated from gold mine wastewater, *Int. Jour. System. Bacteriol.*, 46, 1153–1157.

Malakoff, D. 2001. U.S. researchers go for scientific gold mine, *Science*, 292, 1979.

McAllister, C.T., S.E. Trauth, and L.D. Gage. 1995. Vertebrate fauna of abandoned mines at Gold Mine Springs, Independence County, Arkansas, *Proc. Arkansas Acad. Sci.*, 49, 184–187.

Pierson, E.D. 1989. Help for Townsend's big-eared bats in California, *Bats*, 7(1), 5–8.

Rahn, P.H., A.D. Davis, C.J. Webb, and A.D. Nichols. 1996. Water quality impacts from mining in the Black Hills, South Dakota, USA, *Environ. Geol.*, 27, 38–53.

Santini, J.M., L.I. Sly, R.D. Schnagl, and J.M. Macy. 2000. A new chemolithoautotrophic arsenite-oxidizing bacterium isolated from a gold mine: phylogenetic, physiological and preliminary biochemical studies, *Appl. Environ. Microbiol.*, 66, 92–97.

Saugey, D.A., G.A. Heidt, and D.R. Heath. 1988. Utilization of abandoned mine drifts and fracture caves by bats and salamanders: unique subterranean habitat in the Ouachita Mountains, in *Management of Amphibians, Reptiles, and Small Mammals in North America*, R.C. Szaro, K.E. Severson, and D.R. Patton, (Eds.), USDA Forest Serv., Gen. Tech. Rep. RM-166, 64–71.

Sherwin, R.E., D. Stricklan, and D.S. Rogers. 2000. Roosting affinities of Townsend's big-eared bat (*Corynorhinus townsendii*) in northern Utah, *Jour. Mammal.*, 81, 939–947.

Taylor, D.A.R. 1995. The North American bats and mine project: a cooperative approach for integrating wildlife, ecosystem management, and mine land reclamation. Sudbury 95, Conference on Mining and the Environment, Sudbury, Ontario, May 28–June 1, 1995. *Confer. Proceed.*, 319–327.

Thomas, D.W. 1995. Hibernating bats are sensitive to nontactile human disturbance, *Jour. Mammal.*, 76, 940–946.

PART 4

Mining Legislation, Concluding Remarks, and Indices

CHAPTER 15

Selected Mining Legislation

Major laws and regulations governing gold mining operations in the United States are discussed, including the Clean Water Act, CERCLA (the Comprehensive Environmental Response, Compensation and Liability Act), the National Environmental Policy Act, the Federal Land Policy and Management Act of 1976, the National Forest Management Act of 1976, the Wilderness Act of 1964, the Resource Conservation and Recovery Act, the General Mining Law of 1872, the Endangered Species Act, the Surface Mining Control and Reclamation Act, the Migratory Bird Treaty Act, the Clean Air Act, and the National Historic Preservation Act. Proposed reforms of current mining laws are listed. Selected regulatory aspects of gold mining on public and state lands in the United States, Australia, Brazil, Chile, Guyana, Ghana, Peru, and Papua New Guinea are presented.

15.1 UNITED STATES

Federal and state regulations, as currently practiced, are less than satisfactory for the protection of national water and other resources from hardrock mining activities (Galloway and Perry 1997). Deficiencies include a lack of consistent and integrated regulation at the state and federal level; inadequate funding and staffing; less than effective monitoring and evaluation systems; vague reclamation standards; and exemptions that undermine the effectiveness of existing laws (Galloway and Perry 1997).

15.1.1 Federal Laws

A number of federal laws address aspects of hardrock mining. These laws include: the Clean Water Act; the Comprehensive Environmental Response, Compensation, and Liability Act (CERCLA), better known as Superfund; the Federal Land Policy and Management Act of 1976 (FLPMA); the National Environmental Policy Act (NEPA); the National Forest Management Act of 1976; the Resource Conservation and Recovery Act (RCRA); the Wilderness Act of 1964; and the General

Mining Law of 1872 (Da Rosa and Lyon 1997; Galloway and Perry 1997). These, and others, are summarized below.

Clean Water Act (CWA)

The CWA sets limits on pollutants that can be discharged to surface waters from fixed point sources, such as pipes and other outlets. However, it fails to directly regulate discharges to groundwater — though groundwater contamination is a problem at many mine sites — and does not set any operational or reclamation standards for contaminated discharges from abandoned mines that may affect water sources (Da Rosa and Lyon 1997).

The CWA established the National Pollutant Discharge Elimination System (NPDES) as a method of allocating and regulating the amounts of wastes that can be discharged into the waters of the United States (Galloway and Perry 1997). However, the basic NPDES permit program is of limited usefulness in controlling water pollution from hardrock mining because NPDES applies only to point source discharges. Many mining-related pollutant discharges are from nonpoint sources, that is, pollution that typically results from rainfall or snowmelt flowing from a specific site and carrying with it mining-related contaminants, usually in the form of acid mine drainage. The shortcomings of the NPDES program have allowed many hardrock mine sites to operate without permits under the CWA. In addition to unpermitted active mines, there are many abandoned hardrock mines that are polluting water resources, recently estimated at 557,650 mines — including many gold mines — in 32 states. The number of abandoned mine sites is growing, and many are capable of future contamination of surface and groundwater (Galloway and Perry 1997).

The NPDES requires permits of stormwater discharges associated with hardrock mining. But under the stormwater rules, nonpoint sources at mine sites can escape regulation, as they could under the basic NPDES program, because stormwater discharge permits are required only if runoff drains through a defined point source. The NPDES program also requires states to identify water bodies that fail to meet current water quality standards and to determine the Total Maximum Daily Load (TMDL), or amount each day of each contaminant that the water body can assimilate without violating water quality standards. The TMDL is calculated according to a three-part formula: point sources are assigned to waste-load allocations; nonpoint sources are assigned load allocations; and a margin-of-safety factor is incorporated to account for the uncertainty. The process is delayed significantly by slow efforts by the states and failure by regulatory agencies to prosecute (Galloway and Perry 1997).

Section 404 of the CWA protects wetlands and domestic waters by regulating the discharge of dredged and fill materials into these ecosystems. It is administered by the U.S. Army Corps of Engineers and the U.S. Environmental Protection Agency, after consultations with the U.S. Fish and Wildlife Service, the U.S. National Marine Fisheries Service, and state resource agencies (<http://www.epa.gov/owow/wetlands/facts/fact10.html>; <http://www.wetlands.com//regs/sec.404fc.htm>).

Comprehensive Environmental Response, Compensation and Liability Act (CERCLA)

CERCLA, or Superfund, affixes liability to responsible parties for severe environmental pollution that threatens public health and safety after the pollution occurs, and provides monetary resources to restore these affected sites. As was the case for the Clean Water Act, CERCLA does not address aspects of mine openings, operations, or closings, and is an after-the-fact program, not a preventive one (Da Rosa and Lyon 1997).

Only the largest and most severely contaminated sites are placed on the National Priorities List (NPL) (Galloway and Perry 1997). Of the approximately 1200 sites on the NPL, 66 are mining-related sites. Many of these are old mines that operated before the provisions of the Clean Water Act or the Resource Conservation and Recovery Act (= RCRA, see later). A major problem in applying CERCLA to hazardous mine sites is lack of official records of a mine's existence; thus, no responsible parties can be found for such sites. Many thousands of smaller, potentially hazardous, inactive and abandoned hardrock mines now exist across the nation; these mines continue to discharge wastes into streams with little hope for reclamation under CERCLA (Galloway and Perry 1997).

The Natural Resource Damage Assessment and Restoration Program (NRDAR) portion of CERCLA addresses the restoration of natural resources — such as fish, wildlife, other living resources, water, lands, and protected areas — degraded by the release of hazardous materials at toxic waste sites. NRDAR is administered by federal, state, and tribal trustees (<http://contaminants.fws.gov/Documents/beyond_cleanup.pdf>; <http://njfieldoffice.fws.gov/Publications%20Holding/SEP%201999/fn999p6.html>).

National Environmental Policy Act (NEPA)

NEPA requires environmental studies of proposals of major federal actions affecting the environment. The law requires evaluation of a company's plan and assessment of environmental risk. NEPA is an important supplement to an actual regulatory framework for mining; however, evaluations are limited by being only supplements (Da Rosa and Lyon 1997).

Federal Land Policy and Management Act of 1976 (FLPMA)

FLPMA regulates aspects of natural resource management activities on federal public lands administered by the Bureau of Land Management (BLM), an agency within the U.S. Department of the Interior. FLPMA requires that all activities on public lands be conducted so as to prevent unnecessary or undue degradation of these lands, although Da Rosa and Lyon (1997) suggest that BLM is not yet satisfactorily positioned to cope with the many impacts of large, complicated, modern mining operations on federal lands.

Reclamation is required for all surface-disturbing activities on BLM lands (Galloway and Perry 1997). Hardrock operations need to be inspected at least twice a year by BLM, and producing operations using cyanide a minimum of four times annually; however, this has been accomplished with a success rate of only 47% in some parts of Nevada and 15% in California.

Additional funding and staffing are required to implement and enforce FLPMA (Galloway and Perry 1997).

National Forest Management Act of 1976

The regulatory program of the U.S. Forest Service (USFS) for hardrock mining operations is similar to that of the BLM (Galloway and Perry 1997). The National Forest Management Act of 1976 requires the USFS, an agency of the U.S. Department of Agriculture, to institute a comprehensive, interdisciplinary planning process for the 198 million acres (80.1 million ha) comprising the national forest system, of which about 140 million are public domain lands open to mineral exploration and development. The USFS requires a plan of operation for any operation where the district ranger determines significant disturbance of the surface resources is likely to occur, unlike BLM which waives this requirement for areas less than 5 acres. This will be discussed later.

Wilderness Act of 1964

Both BLM and USFS administer lands in the National Wilderness Preservation System, created by the Wilderness Act of 1964 (Galloway and Perry 1997). Although the law mandates that wilderness-designated lands be left unimpaired and protected for future use and enjoyment, hardrock mining is permitted in some wilderness areas. This exception to the Act of 1964 does not apply to existing mining claims as of December 31, 1983, the date wilderness areas were withdrawn from mineral development. These claims may continue to be mined, and the Chief of the USFS must allow any activity, including prospecting, for the purpose of gathering information about minerals in the wilderness areas of the national forest system. In general, however, mining operations in wilderness areas are more strictly regulated by both BLM and USFS than operations in non-wilderness areas. For example, in wilderness areas the construction of roads is limited and surface resources protected (Galloway and Perry 1997).

Prospecting and mining are prohibited in the national parks and other areas administered by the National Park Service (NPS), an agency of the U.S. Department of the Interior (Galloway and Perry 1997). But existing valid claims in areas subsequently brought under the jurisdiction of NPS may continue to be developed. The greatest mining-related problem in the national parks is not from active mining operations, but from abandoned mines. NPS estimates that there are more than 4000 abandoned mines in 45 states and that approximately 33,000 disturbed acres remain unclaimed (Galloway and Perry 1997).

Resource Conservation and Recovery Act (RCRA)

RCRA was intended to regulate all aspects of the nation's management of solid and hazardous waste. Although mining generates almost as much hazardous waste as all other industries combined (Da Rosa and Lyon 1997), Congress in 1980 passed an amendment to RCRA, the Bevill Amendment, which exempted much mining-related waste from RCRA Subtitle C, the federal program regulating the production, storage, transportation, and disposal of hazardous wastes (Da Rosa and Lyon 1997).

RCRA was enacted in 1976 as a set of amendments to the Solid Waste Disposal Act of 1965 in order to regulate hazardous and other solid wastes that are discharged onto land but ultimately create surface or groundwater contamination (Galloway and Perry 1997). To fall under RCRA, wastes need to be considered either toxic, ignitable, corrosive, or reactive. The wastes produced by mining and mineral processing largely have been exempted from the hazardous waste provisions of RCRA because they were not considered a substantial threat to human health when improperly handled, although cyanide and metals-laden tailings might conceivably fall under this provision. The Bevill Amendment covered low-hazard, high-volume wastes produced in the extraction, recovery, and processing stages of mining. These wastes were now covered under Subtitle D, RCRA's less stringent standards for solid waste. One difference between Subtitle C and Subtitle D is that the federal government has a direct role in the regulation of the former and individual states the latter. Subtitle D allows states to develop their own programs for solid wastes with minimum federal guidelines. Subtitle D requirements are not independently enforceable by the U.S. Environmental Protection Agency, except where violations pose an imminent and substantial endangerment to health or the environment (Galloway and Perry 1997).

General Mining Law of 1872

The 1872 Mining Law, which governs the extraction of hardrock minerals on public lands, makes mining the highest and best use of public lands. The 1872 Mining Law is flawed, according to Da Rosa and Lyon (1997), because it has no environmental protection provisions, requires that no royalties be paid to the federal government for mineral extraction from public lands, and allows mining companies to lease public lands for US$5 per acre or less. Under the law, rights to locatable minerals (including gold and other metals) on public lands can be claimed, and these claims may be registered with the federal government upon discovery of valuable mineral deposits (Galloway and Perry 1997). These claims can be patented — a process giving the patent holder full rights to both the minerals and the land on the surface — for as little as $2.50 to $5.00 per acre. As recently as 1970, with the passage of the Mining and Minerals Policy Act, federal policy remained heavily weighted toward promoting private enterprise in the development of national mineral resources. This act encouraged the development of reclamation and mining waste disposal methods, but the overriding federal policy in the statute was to promote economically sound and stable domestic mining and mineral reclamation activities (Galloway and Perry 1997).

Regulation of environmental impacts of hardrock mining is left primarily to the states, even if the impact is primarily to federal lands (Galloway and Perry 1997). In 1988, the U.S. General Accounting Office (GAO) reported that more than 424,000 acres of federal lands disturbed by mining had been left unreclaimed. Active, authorized mining operations accounted for about 150,000 acres of mining-disturbed lands, and abandoned, suspended, and unauthorized operations accounted for the remainder. GAO found that almost 25,000 acres of these unreclaimed lands required removal of mine wastes that threatened water resources, and another 74,000 acres needed reclamation to control erosion, landslides, and water runoff (Galloway and Perry 1997).

Surface Mining Control and Reclamation Act

This act attempts to minimize environmental impacts from coal mining and does not pertain to other minerals (Starnes and Gasper 1996). However, similar legislation and enforcement should be encouraged for mining operations involving gold and other metals. At present, gold, silver, copper, lead, zinc, and other metals are frequently mined and milled during the same mining operation, resulting in large quantities of powdered mill wastes or tailings. Trace metals in the tailings are potential water contaminants. Land restoration is difficult because tailings cannot be returned to the mine and ultimately contribute to erosion problems at the site (Starnes and Gasper 1996).

Also needed are legislation and enforcement based on: methods to predict biological perturbations from mining wastes; identification of areas unsuitable for mining; development of appropriate wetlands reclamation strategies; accelerated stream restoration methodologies; and reevaluation of water quality criteria for aquatic life protection in terms of mine discharges (Starnes and Gasper 1996).

Endangered Species Act

The Endangered Species Act (16 USC 1536) requires all federal agencies to consult with the U.S. Fish and Wildlife Service or the National Marine Fisheries Service (an agency in the U.S. Department of Commerce) or both when threatened or endangered species may be adversely affected by a proposed mining operation on federal lands (USNAS 1999).

Migratory Bird Treaty Act

The Migratory Bird Treaty Act of 1918 (16 U.S.C. 703-12) and amendments protects migratory birds from Canada, Mexico, Japan, and Russia. Specifically, it prohibits their capture, death, possession, sale, shipment, export, or any part, nest, or egg of such bird. The law is interpreted and enforced by the U.S. Department of the Interior (<http://laws.fws.gov//lawsdigest/migtrea.html>). The killing of birds, such as by cyanide poisoning at gold mines, as well as the destruction of active nests with eggs or young, are offenses under this act. The former prohibition has led to various efforts to avoid avian mortality in toxic ponds at gold mines. The latter may require surveys of land areas during the breeding season to determine the presence of active bird nests and, if found, would preclude vegetation clearing until after nesting (S.N. Wiemeyer, personal communication).

Clean Air Act of 1990 (P.L. 101-549)

This act amends the Clean Air Act of 1970 to "… provide for attainment and maintenance of health protective national ambient air quality standards … ." One criteria air pollutant is particulate matter, which includes soot, dust, and other tiny bits of solid material that are released into the atmosphere. Particulates emanate from many sources, including mining operations, and may cause irritation of the eye, nose, and throat, and other health problems. The Clean Air Act is enforced by the U.S. Environmental Protection Agency, state, and local governments (<http://www.epa.gov.oar/oaqps/peg_caa/pegcaa10.html>).

National Historic Preservation Act/American Indian Religious Freedom Act

The National Historic Preservation Act (16 USC 470 et seq.) and the American Indian Religious Freedom Act (42 USC 1996) require the U.S. Forest Service and the U.S. Bureau of Land Management to consult with other agencies and with tribes to consider and mitigate potential effects of mining and other human activities on protected resources or interests. These consultations are usually integrated with NEPA (National Environmental Policy Act) review (USNAS 1999).

15.1.2 Mining Public Lands in the Western United States

In 1999, the National Research Council of the U.S. National Academy of Sciences (USNAS) published a report (USNAS 1999) to assess the adequacy of the regulatory framework for hardrock (locatable minerals) mining on 350 million acres of federal lands in the western United States managed by two federal agencies: the Bureau of Land Management (BLM) in the Department of the Interior and the Forest Service in the Department of Agriculture. The BLM is responsible for 260 million acres of land in the western United States, including Alaska, of which about 90% are open to hardrock mining. About 160,000 acres of BLM lands are affected by active mining and mineral exploration. The Forest Service manages 163 million acres, of which 80% are open to hardrock mining. Together, these two land management agencies are responsible for 38% of the total area of the western United States. In addition to potential mineral wealth, the lands are important for timber, grazing purposes, clean water sources, recreational activities, wildlife habitat, scenic areas, and other resources (USNAS 1999).

Hardrock mining occurs where minerals are concentrated in economically viable deposits; these comprise less than 0.01% of the earth's crust. The mining process consists of exploration, mine development, extraction of gold and other minerals, mineral processing, and reclamation — including post-closure (USNAS 1999). Each step from exploration through post-closure has the potential to cause environmental impacts. In addition to the obvious disturbance of the land surface, mining may also affect groundwater, surface water, aquatic fauna, aquatic and terrestrial vegetation, wildlife, soils, air, and cultural resources. Actions based on environmental regulations may avoid or limit many of the potential impacts; however, mining will, to some degree, always alter landscapes and environmental resources. Regulations intended to control and manage these alterations are in place but are not always wholly successful, and are updated as new techniques are developed to improve mineral extraction, to reclaim mined lands and to limit environmental perturbations (USNAS 1999).

The basic statute for hardrock mining on federal lands is the General Mining Law of 1872. Land management direction is provided in the Federal Land Policy and Management Act of 1976 (FLPMA) for the BLM and in the 1897 Organic Act and the 1976 National Forest Management Act for the Forest Service (USNAS 1999). Proposed mining activities on federal lands trigger the application of BLM's Part 3809 regulations (43 CFR Part 3809) and the Forest Service's Part 228 regulations (36 CFR Part 228). BLM's Part 3809 regulations establish guidelines intended to assure compliance with the FLPMA prohibition of unnecessary or undue degradation of public lands. Part 228 of the Forest Service's regulations is intended to assure compliance with the Forest Service regulations requirement of minimizing adverse environmental impacts on national forest resources, as based on the Organic Act. The National Environmental Policy Act (NEPA) integrates BLM and Forest Service decision making on particular mining proposals with evaluation of other environmental concerns, as well as with state and federal permitting requirements. For mining operations on federal lands that may adversely affect the environment, an environmental impact statement (EIS) is prepared for use by the federal land manager in the decision-making process. For smaller operations, an environmental assessment often is produced instead of an EIS in order to assist the federal land management agency in deciding where environmental impacts will be significant (USNAS 1999).

Permission to mine for gold on public lands may be relatively simple to acquire. In the case of gold mining in national forests (USFS 1981), for example, no formal plan of operations needs to be submitted to the district ranger provided that the operation is limited to:

1. The use of vehicles on existing public roads or roads used and maintained for national forest purposes
2. Individuals who search for and occasionally remove small mineral samples or specimens
3. Prospecting and sampling which will not cause significant surface resource disturbance and will not involve removal of more than a reasonable amount of mineral deposit for analysis and study
4. Marking and monumenting a mining claim
5. Subsurface operations which will not cause significant surface resource disturbance (USFS 1981; see also <http://www.fs.fed.us/geology/36CFR228a.txt>)

No notice of intent needs to be filed if operators do not use mechanized equipment such as bulldozers or backhoes and do not cut down trees.

However, in all other cases, a notice of intent and plan of operations needs to be filed and approved by the district ranger before operations can commence (USFS 1981). These operators need to comply with:

1. Applicable federal and state air quality standards, including the requirements of the Clean Air Act, as amended (42 U.S.C. 1857 et seq.).
2. Applicable federal and state water quality standards, including regulations issued pursuant to the Federal Water Pollution Control Act, as amended (33 U.S.C. 1151 et seq.).
3. All applicable standards for the disposal and treatment of solid wastes (i.e., all garbage, refuse, or waste needs to be removed from national forest lands or disposed of or treated so as to minimize, so far as is practicable, its impact on the environment and the forest surface resources; all tailings, dumpage, deleterious materials, or substances and other waste produced by operations shall be deployed, arranged, disposed of or treated so as to minimize adverse impact upon the environment and forest surface resources).
4. Scenic values (operator shall, to the extent practicable, harmonize operations with scenic values through measures that include design and location of operating facilities, and construction of structures and improvements which blend with the landscape).
5. Fisheries and wildlife habitat (in addition to compliance with required water quality and solid waste disposal standards, operator shall take all practicable measures to maintain and protect affected fisheries and wildlife habitat).
6. Roads (all roads constructed and maintained by the operator shall have adequate drainage to minimize or eliminate damage to soil, water, and other resource values).
7. Reclamation (upon exhaustion of the mineral deposit and within 1 year of the conclusion of operations, unless a longer time is authorized, operators shall, where practicable, reclaim the surface disturbed in operations by taking such measures as will prevent or control on-site and off-site damage to the environment and forest surface resources, including: control of erosion and landslides; control of water runoff; toxic substances isolation, removal or control; reshape and revegetate disturbed areas, where reasonably practicable; and rehabilitation of fisheries and wildlife habitat. During all operations, the operator shall maintain his structures, equipment, and other facilities in a safe, neat, and workmanlike manner; shall remove all structures and equipment after cessation of operations; and shall comply with all applicable fire laws, take all reasonable measures to prevent and suppress fires on the area of operations; and shall require compliance with fire laws by his employees, contractors, and subcontractors) (USFS 1981).

The Forest Service and BLM have at least five mechanisms for protecting valuable resources and sensitive areas from mining (USNAS 1999). The first is to formally withdraw federal land from mining and mineral development in recognition of extraordinary natural, scenic, or cultural values that are of such significance that other competing resource uses must be excluded. The withdrawal authority is also used to protect administrative and other public facilities. The second mechanism is to prepare land management plans that identify natural and cultural resources and allocate competing uses accordingly. These plans specify limitations and mitigation that apply in particular locations and to particular uses, such as mining. These plans are intended to be based on thorough resource inventories and to be developed with broad public input. For BLM lands, the Federal Land Policy and Management Act of 1976 specifies that land use plans give priority to the designation and protection of areas of critical environmental concern. These are areas recognized as needing special management, including consideration of important historic, cultural, or scenic values, fish and wildlife resources and other natural systems or processes, or protection from natural hazards. Designated areas of critical environmental concerns are usually withdrawn from hardrock mining. The third mechanism for protecting valuable resources and sensitive areas is the use of advisory guidelines that identify categories of resources or lands that deserve special consideration. Use of such guidelines is consistent with the principle that regulatory guidelines should be based on site-specific evaluations and conditions. A fourth mechanism is through the use of BLM's Visual Resource Management System (VRMS), which recognizes that public lands have a variety of visual values. A VRMS classification is part of BLM's land use planning process, and applies to all activities on BLM lands, not just mine-related activities. The fifth mechanism by which BLM and the Forest Service routinely protect environmental resources not governed by specific law is the adoption of site-specific measures imposed on a plan of operations after the environmental assessment or EIS process. This is the most common way for the federal land management agencies to protect cultural values, riparian habitat, springs, seeps, and ephemeral streams that are not otherwise protected by specific laws (USNAS 1999).

Regulatory gaps perceived by the National Research Council (USNAS 1999) in mining public lands are stated in the following recommendations:

1. Financial assurance should be required for reclamation of environmental disturbances caused by all mining activities beyond those classified as casual use, even if the disturbed area is less than 5 acres.
2. Plans of operations should be required for mining and milling operations, other than those classified as casual use or exploration activities, even if the area disturbed is less than 5 acres.
3. BLM and the Forest Service should revise their regulations to provide more effective criteria for modifications to plans of operations, where necessary, to protect the federal lands.
4. BLM and the Forest Service should adopt consistent regulations that (a) define the conditions under which mines will be considered to be temporarily closed, (b) require that interim management plans be submitted for such periods, and (c) define the conditions under which temporary closure becomes permanent and all reclamation and closure requirements must be completed.
5. Existing environmental laws and regulations should be modified to allow and promote the cleanup of abandoned mine sites in or adjacent to new mine areas without causing mine operators to incur additional environmental liabilities.
6. BLM and the Forest Service should have the authority to issue administrative penalties for violations of their regulatory requirements, subject to due process, and also clear procedures for referring activities to other federal and state agencies for enforcement (USNAS 1999).

The National Research Council also recognized that successful environmental protection is based on sound science and recommended that the Congress should fund an aggressive and coordinated research program related to the environmental impacts of hardrock mining (USNAS 1999). The group concluded that conditions are changing for regulations and mining, that portions of the public and the mining industry have little confidence in the propriety or fairness of the regulatory and permitting system, and that the Bureau of Land Management and the Forest Service need not have identical regulations, but some changes are warranted (USNAS 1999).

15.1.3 State Laws

Various state and federal laws establish environmental requirements applicable to mining on federal lands, including state reclamation laws, state and federal water pollution laws, state and federal fish and wildlife laws, state and federal air quality laws, wetland laws, and others (USNAS 1999).

State mining regulations, in general, are difficult to interpret uniformly (Da Rosa and Lyon 1997). Examples of vague language appear in the mining laws of Arizona ("... unless this is not practical ..."; "if reasonable likelihood exists ..."), Nevada (The State Division of Environmental Protection is granted the right to approve "any appropriate method of reclamation" consistent with the regulations ...), and Colorado (..."regrading shall be appropriate to land use"; "... disturbances ... to groundwater shall be minimized ..."; "... reclamation shall be completed with all reasonable diligence ... to the extent practicable, taking into consideration ... economics, available equipment and material ...").

State enforcement of environmental regulations may be difficult, as judged by the Summitville Gold Mine situation in southern Colorado (Da Rosa and Lyon 1997). This cyanide heap leach gold mine, opened in 1986 by a Canadian-based corporation, experienced leakage of the protective liner shortly after operations were initiated, allowing cyanide to leak into the surface waters and groundwater. Waste rock and other mining materials unearthed and exposed by the mine began generating acid and heavy metals. The cyanide, acid, and metals flowing from the mine contaminated about 27 km of the nearby Alamosa River, an important source of irrigation water for local crops. The Canadian corporation declared bankruptcy and abandoned the site in 1992. Colorado requested assistance from the Superfund program, administered by the U.S. Environmental Protection Agency (EPA). By 1996, EPA had spent more than $100 million in efforts to clean up the site, with total remediation costs estimated at $220 million. Canadian courts have rebuffed efforts by the U.S. Department of Justice to recover these costs from the bankrupt Canadian corporation (Da Rosa and Lyon 1997).

The absence of a comprehensive national program for regulating hardrock mining leaves the primary responsibility for this task to the states (Galloway and Perry 1997). Within each mining state, the responsibility for regulating water quality effects of hardrock mining is usually shared between two agencies: a mining and reclamation agency, which permits and oversees mining activities, and a water quality agency, which assumes the primary enforcement role for implementing the Clean Water Act within the state. Frequently, there is little interaction between the agencies. In general, these agencies are understaffed, with the ratio of mine inspectors to active mining operations in some western states of less than 1 inspector per 100 mine sites (Galloway and Perry 1997).

15.1.4 Mining Law Reform

Many bills have been introduced into the U.S. Congress to reform existing mining laws, but none has been passed to date. Collectively, these bills address five key areas of reform:

1. Royalties
2. The price for patents
3. Maintenance fees
4. Uses of patented lands
5. Abandoned mine lands cleanup (Dobra 1997)

It is probable that these issues will continue to be debated by Congress in the future. Selected aspects of the failed legislation follow.

Both the House and Senate bills included provisions for a royalty (Dobra 1997). The Senate bill included a royalty of 2.5% of net smelter return. This would mean a 2.5% tax on production, since the gold industry's smelting and refining costs are minimal. The House bill carried a 3.5% royalty patterned after the Nevada Net Proceeds of Mines Tax. A compromise of a 5% net proceeds royalty was reached, which allows producers to deduct direct mining, processing, and selling expenses incurred on site. Second, both the House and Senate agreed that the current law that allows sale of patented mineral lands for $2.50 to $5.00 per acre should be modified to reflect fair market value with the reservation of the federal government's right to a royalty. Third, the annual claim maintenance or rental fee of $100 passed in 1993 would have been made permanent in legislation introduced in 1995 and would have required the fee to be doubled after the third year a claim is held to discourage claimstaking. Fourth, the post-mining use of patented lands was addressed in the Senate bill by providing that the Secretary of the Interior could revoke title to lands acquired for mining purposes that were used for non-mining related purposes. And finally, the legislation proposed that 40% of the royalties collected would be used to fund state reclamation of abandoned mine lands, with the balance going to the federal treasury. This provision would apply to lands disturbed by mining and abandoned by operators before the advent of reclamation laws in the 1980s (Dobra 1997).

15.2 FOREIGN

Selected examples of legislation impacting gold mining in other parts of the world are listed below.

Australia. The Alligator Rivers region of tropical northern Australia, east of Darwin, contains large economically important mineral reserves, including gold (Humphrey et al. 1990). The area also contains a diverse flora and fauna and an abundance of aboriginal rock art. To protect this unique environment, only certain areas can be mined, and the wastes monitored. A Supervisory Scientist position, appointed by the Commonwealth, is now the first-line legal and scientific authority to ensure protection, with special emphasis on wastewater releases (Humphrey et al. 1990).

Brazil. The Brazilian Mining Law of 1967 defines the garimpeiro as a professional who works the outcropping deposits manually (Pessoa et al. 1995). These individuals use elemental mercury extensively in recovering gold. Widespread contamination of the environment with mercury is attributed to this activity. The 1988 Constitution requires that states and municipalities legislate and supervise environmental matters and that class action may void any act harmful to the environment. In 1988, the federal government passed legislation to establish areas and conditions for the "garimpo" activity with the goal of encouraging the formation of cooperatives; specifically, passage of Law Number 7805/89, also known as the Garimpo Mining Permit. This legislation declares that the Amazon Forest, Pantanal, the Coastal Zone, the Mata Atlantica, and the Serra do Mar are protected national properties and that all economic activities, including mining, causing environmental degradation must be preceded by an environmental impact study, an obligation for the miners to restore degraded environments, and penal and monetary penalties (Pessoa et al. 1995). However, police enforcement of the existing mining code is almost impossible because of the long distances, total absence of infrastructure, severe climate, and inaccessibility of mining sites. Authorities now concede that systematic educational campaigns on environmental problems associated with the mercury amalgamation process in recovering gold may be more effective than police action (Pessoa et al. 1995).

Chile. In Chile, the Environmental Basis Law enacted in 1994 requires environmental impact assessments for large new projects, including mines (Da Rosa and Lyon 1997). It also provides the foundation for citizen lawsuits to enforce environmental laws and compensation for environmental damage. However, the requirements for the environmental impact statement are unclear, effluent standards for mining contaminants are not yet implemented, and enforcement is lax. Maintaining enforceable regulatory standards for mining wastes in developing countries is sometimes confounded by the high level of economic dependency upon foreign exchange earnings from mining (Da Rosa and Lyon 1997). For example, hardrock minerals accounted for more than 99% of the total value of Zambia's exports from 1988 to 1990. In Zaire, mineral exports accounted for 80%. For Papua New Guinea, mineral exports were 72% of total national export values; for Chile and Guyana, these values were 57% and 41%, respectively (Da Rosa and Lyon 1997).

Guyana. In some cases, the governments of developing countries become business partners in mining ventures (Da Rosa and Lyon 1997). For example, the Guyana government owned 5% of the Omai gold mine, the second largest gold mine in South America, together with a Canadian-based firm. The output from this mine alone comprised 20% of Guyana's gross national product. Guyana has no national environmental protection statute nor adequate mining regulations in place (Da Rosa and Lyon 1997). Persistent severe environmental problems associated with a tailings dam at the Omai mine — despite the claim of the Canadian partner that it met North American standards — did not result in mine shutdown, allegedly due to economic incentives by the government to keep the mine operating (Da Rosa and Lyon 1997).

Ghana. In December 1994, legislation was passed in Ghana establishing an Environmental Protection Agency, with authority to conduct environmental impact assessments and enforce environmental laws; however, the lack of effluent standards for water hampers the effectiveness of this agency (Da Rosa and Lyon 1997).

Peru. In January 1996, Peru adopted regulations establishing maximum allowable limits on concentrations of metals and cyanide in liquid effluents from mining operations, but enforcement is inadequate (Da Rosa and Lyon 1997).

Papua New Guinea (PNG). In PNG, an Australian-owned gold mine dumped more than 80,000 tons of untreated tailings into a major PNG river every day (Da Rosa and Lyon 1997). This mining operation altered the course of the navigable river; introduced potentially toxic concentrations of copper, cadmium, and zinc into the waterway; destroyed most of the aquatic life along the first 70 km of the river, including almost all species of fishes and all species of crocodiles, turtles, and crustaceans; and caused tailings-associated flooding that drowned community gardens and severely damaged trees and other vegetation. This area of dead and dying vegetation covered 30 km^2, according to the mine owners, and mining is expected to continue until 2011. The original agreement establishing the mining operation, signed in 1976 between the private investors and the PNG government — minority owners of the corporation — exempted the mining company from most of PNG's environmental laws. An Australian law firm, claiming to represent 30,000 PNG landowners, sued the mine owners for $4 billion (Australian) in damages due to loss of subsistence fisheries and fauna, compensation for harm to the traditional way of life and culture of villagers, and punitive damages. The suit was filed in an Australian court rather than in PNG because the PNG government is a 30% co-owner in the mine. The suit was settled for $110 million, plus additional benefits for the most severely affected residents. The final agreement did not specifically prohibit the management of tailings (Da Rosa and Lyon 1997).

15.3 SUMMARY

Legislation governing gold mining operations in the United States are discussed, including the Clean Water Act, the Comprehensive Environmental Response, Compensation and Liability Act, the National Environmental Policy Act, the Federal Land Policy and Management Act of 1976, the National Forest Management Act of 1976, the Wilderness Act of 1964, the Resource Conservation and Recovery Act, the General Mining Law of 1872, and others. Proposed reforms of current mining laws are listed.

Regulatory aspects of gold mining on foreign and domestic and state lands are presented. It is concluded that U.S. federal and state regulations, as currently practiced, are less than satisfactory for protection of water quality and wildlife resources from hardrock mining activities. Deficiencies include a lack of consistent and integrated regulation at the state and federal level, inadequate funding and staffing, less than effective monitoring and evaluation systems, vague reclamation standards, and exemptions which undermine the effectiveness of existing laws.

LITERATURE CITED

Da Rosa, C.D. and J.S. Lyon (Eds.). 1997. *Golden Dreams, Poisoned Streams*. Mineral Policy Center, Washington, D.C., 269 pp.

Dobra, J.L. 1997. *The U.S. Gold Industry 1996*. Nevada Bur. Mines Geol., Spec. Publ. 21, 32 pp.

Galloway, L.T., and K.L. Perry. 1997. Mining regulatory problems and fixes, in *Golden Dreams, Poisoned Streams*, C.D. Da Rosa and J.S. Lyon, (Eds.), Mineral Policy Center, Washington, D.C., 193–215.

Humphrey, C.L., K.A. Bishop, and V.M. Brown. 1990. Use of biological monitoring in the assessment of effects of mining wastes on aquatic ecosystems of the Alligator Rivers region, tropical northern Australia, *Environ. Monitor. Assess.*, 14, 139–181.

Pessoa, A., G.S. Albuquerque, and M.L. Barreto. 1995. The "garimpo" problem in the Amazon region, in *Chemistry of the Amazon: Biodiversity, Natural Products, and Environmental Issues*. American Chemical Society, Washington, D.C., ACS Sympos. Ser. 588, 281–294.

Starnes, L.B. and D.C. Gasper. 1996. Effects of surface mining on aquatic resources in North America, *Fisheries*, 21(5), 24–26.

U.S. Forest Service (USFS). 1981. Regulations for locatable minerals. USFS 74, 9 pp.

U.S. National Academy of Sciences (USNAS), National Research Council, Committee on Hardrock Mining on Federal Lands. 1999. *Hardrock Mining on Federal Lands*. National Academy Press, Washington, D.C., 247 pp.

CHAPTER 16

Concluding Remarks

Despite all evidence to the contrary and for reasons that seem neither logical nor rational, it is probable that society will continue to value gold as a commodity in the forseeable future and to provide continued employment to about 30 million individuals worldwide who derive a significant portion of their income from the mining, refining, and sale of raw and finished gold. Jewelry, coins, and bullion will doubtless continue to account for the great majority of all new gold mined — about 2000 tons annually — although new gold and gold products are constantly being developed for use in electronics, medicine, and other disciplines. As is true for other commodities, it is expected that the price of gold will continue to fluctuate over time based on demand and supply. As demand for gold decreases, the price for raw gold will decrease and marginally productive mines will be closed or abandoned, often with little or no regard for the environmental consequences. Gold miners and refiners will continue to suffer from increased illnesses and death rates when compared with the general population, although this is expected to lessen on exposure of miners to additional training and education, implementation of existing safety regulations, and installation of adequate mine safety equipment. Environmental degradation associated with the extraction of gold will continue or increase in developing nations, and at a reduced rate among nations with strong and enforceable environmental laws. New mining technologies now under development to produce gold from very low-grade ores are expected to be both cost-effective and environmentally friendly. Additional research efforts and information on various aspects of gold and gold mining are merited. Some of the more pressing needs are listed below.

1. Reliable production data

Major producers of gold include the Republic of South Africa, the United States, the former Soviet Union, Canada, Australia, the Peoples's Republic of China, and Brazil. However, official gold production data from many localities and nations are unreliable and are usually underreported. Accurate reporting of gold production is needed in order to establish a realistic price for this commodity.

2. Geologic characterization of gold deposits

Geologic characteristics of commercial gold-bearing strata are not known with certainty, although many deposits are related to proximity to volcanic settings, granitic magmas and fluids, pyrites of iron and other metals, and potassium-containing igneous rocks. Cost-effective technologies need to be developed to locate gold and other valuable mineral ore deposits.

3. Gold properties database

Gold is a complex and reactive element, with unique chemical, physical, and biochemical properties. Additional research is strongly recommended on expansion of knowledge concerning properties of gold and its salts, and to centralize all findings in a single accessible database.

4. Significance of gold concentrations in abiotic materials and living organisms

Gold concentrations are unusually high in certain abiotic materials, such as sewage sludge from a gold mining community (4.5 mg Au/kg DW), polymetallic sulfides from the ocean floor (28.7 mg/kg DW), and freshwater sediments near a gold mine tailings pile (256.0 mg/kg DW), suggesting that gold recovery is commercially feasible from these materials. A similar situation exists for gold accumulator plants, which may contain as much as 100.0 mg Au/kg DW. However, the significance of gold concentrations in various organisms and materials and mechanisms governing its uptake and retention are relatively unknown when compared with most metals. To more fully evaluate the role of gold in the environment, systematic monitoring of gold concentrations is recommended in abiotic materials and organisms comprising complex food chains; samples should also be analyzed for selected metals, metalloids, and other compounds known to modify gold uptake and retention.

5. Expanded use of gold and gold salts

The unique properties of gold have led to increasing use of the metal and its salts in the disciplines of electronics, dentistry, physiology, immunology, electron microscopy, and human medicine. Additional applications need to be explored.

6. Gold effects on biota

Recommended areas of research emphasis include: sublethal effects of Au^{+3} species to aquatic organisms because concentrations as low as 98 μg Au^{+3}/L adversely affect algal growth; mechanisms of accumulation of Au^0, Au^+, and Au^{+3} by microorganisms and algae; and effects of injected colloidal gold Au^0 and monovalent gold drugs (Au^+) in mammals on temperature regulation, brain chemistry, carcinogenicity, teratogenicity, and nephrotoxicity, with potential application to human medicine.

7. Biorecovery of gold

Gold recovery technologies based on bacteria, fungi, and other organisms are now commercially available; additional research is recommended on biorecovery of gold under different physicochemical conditions.

8. Health protection of gold miners

To protect the health of underground workers, continued intensive monitoring of atmospheric dust levels is recommended for conformance with safe occupational levels, implementation of regular and frequent medical examinations with emphasis on early detection and treatment of disease states, and continuation of educational programs on hazards of risky behaviors outside the mine environment. Miners who use elemental mercury to extract gold need to control mercury emissions in confined environments and limit consumption of larger carnivorous fishes. Intensive monitoring by physicians and toxicologists of populations at high risk for mercury poisoning is recommended in order to provide adherence to existing mercury criteria, as is critical examination of current mercury criteria to protect human health.

9. Gold sensitivity to humans

There is increasing documentation of allergic contact dermatitis and other effects to metallic gold from jewelry, dental restorations, and occupational exposure; these effects are most frequent in females wearing body-piercing gold objects. Research is needed to determine the extent of sensitivity (one estimate lists gold allergy at 13% worldwide), the mechanisms of action, and whether nursing infants are at risk. Similar studies are recommended for gold salts.

10. Gold drugs in medicine

Gold and gold drugs have been used in human medicine for centuries, and certain gold drugs have been used routinely for more than 75 years to treat rheumatoid arthritis. However, anti-inflammatory properties of gold metabolites and other mechanisms of action of gold drugs are not known with certainty and merit additional research, as does development of gold drugs with minimal side effects, alternate routes of gold drug administration for maximum efficacy, evaluation of gold drugs in combination with other drugs to enhance relief, and development of more sensitive and uniform indicators for evaluation of gold drug therapy.

11. Gold mine wastes

Acidic metal-rich water and tailings wastes from active gold mines devastate receiving aquatic ecosystems. More research is needed on prediction of extent of acid mine drainage and its prevention through physical, chemical, and biological remediation technologies, and on appropriate storage of tailings wastes.

12. Amalgamation contaminant problems

The use of mercury to recover gold has resulted in wholesale and persistent contamination of the biosphere, with direct — and frequently fatal — consequences to all members of the immediate biosphere, including humans. The use of mercury for this purpose must be abandoned, and improved remediation methodologies developed for mercury-contaminated environments using physical, chemical, and biological technologies.

13. Cyanide extraction and water management issues

The use of cyanide to extract gold from low-grade ores by major mine operators has caused a variety of lethal and significant sublethal effects on wildlife, especially in desert-like arid areas. But this situation has been steadily improving as wildlife are shielded from cyanide-containing solutions and new habitat has been created from waste rock, overburden, and tailings piles. Pit lakes, however, resulting from surface mining technologies, including cyanidation, may create a variety of problems, as yet poorly documented, to wildlife and human health. Additional research is merited on ecotoxicology of pit lakes.

14. Arsenic wastes

Because arsenic interferes with gold extraction and most gold-containing ores contain significant quantities of arsenic, these ores are roasted to remove the arsenic, with resultant arsenic contamination of air, water, and soil ecosystems. Cost-effective methods to remove the arsenic without release to the surrounding environment need to be refined, especially those involving microorganisms.

15. Disposition of abandoned underground mine sites

Abandoned underground gold mines are known habitats for species of concern, including threatened and endangered species of bats. Abandoned mine use by vertebrates needs to be quantified through monitoring, and their entrances structurally modified for greater protection of these species. Acid mine drainage from abandoned mines is one of the more environmentally devastating by-products of gold mining; additional research is needed to predict and control acidity and metals contamination from this source.

16. Mining legislation

All mining laws should be reexamined, especially the Surface Mining Act of 1872, in order to provide economic and other safeguards against environmental damage caused by current gold extraction practices.

General Index

A

Abandoned underground gold mines
  acid mine drainage 223
  arsenic contamination 222, 223, 225, 227
  barrier to land development 309, 310
  habitat for biota 307–309, 336
  potential science site 312
  water quality effects 310–312, 336
Acadian–Ponchartrain Basin 271
Acid–base accounting 171, 184
Acid mine drainage
  effects 163, 168–170, 184, 310
  metals content 168–172, 327
  mitigation 163, 170–172, 184, 335
  source 168, 169, 171, 223
  transport 168, 169, 184
Acquired Immune Deficiency Syndrome (AIDS)
  98, 100
Africa 91, 167, 253, 269, 292
Akkod 24
Alabama 16, 17
Alamosa River, Colorado 195, 327
Alaska 3, 7, 11, 13, 15–18, 25, 91, 92, 166, 167,
  173, 174, 177, 191, 323
Alberta, Canada 12
Alligator river region, Australia 328
Allochrysin 30
Allochrysine 77
Allocrysine 133
Alta Floresta, Brazil 259
Aluminum 24, 40, 66, 95, 168, 169, 310
Amalgam 32, 101, 103, 104, 124, 125, 253, 269
Amalgamation 251–293
Amazon Region, Brazil 14, 92, 104, 252, 253,
  255–259, 265, 329

Amazon River Basin 96
Amazonia 93, 104, 262
American College of Rheumatology 144
American Conference of Governmental Industrial
  Hygienists 104
American Contact Dermatitis Society 116
American Indian Religious Freedom Act 323
American River, California 166
p-amino-benzenearsonic acid 232
Angola 97
Antimony 9, 10, 39, 54, 66, 183
Appalachian region, USA 14
Aqua regia 42
Aquatic organisms
  arsenic
    concentrations 172, 179, 184, 226, 227,
     229, 230
    criteria 241–244, 311
    toxicity 227, 230–233, 244, 311
  cadmium concentrations 179, 184
  copper
    concentrations 179
    criteria 311
    toxicity 172, 311
  cyanide
    criteria 205, 311
    toxicity 192, 195–199, 201, 202, 214, 311
  gold
    concentrations 56–60
    toxicity 65, 66, 180–182
    uptake 68–72
  lead
    concentrations 172, 175, 179, 184
    criteria 311
    toxicity 172, 311

mercury
    concentrations 91–93, 101–105, 179, 255,
        263, 264, 266–270, 273, 291–293
    criteria 268, 286, 287, 289–293, 311
    toxicity 93, 264, 269, 276–281, 291, 292,
        311
nickel concentrations 179
suspended solids toxicity 207
tailings toxicity 230, 330
zinc
    concentrations 172, 179, 184
    criteria 311
    toxicity 311
Argen Corporation 32
Argentina 262
Argedent "52" 31
Arizona 16, 17, 199–202, 213, 289, 308, 327
Arkansas 308
Arsanilic acid 237, 244
Arsenates 223, 225, 227–229, 231–238, 240, 241,
        244
Arsenic
    as carcinogen 94–97, 225, 244
    concentrations in
        abiotic materials 9, 13, 54, 56, 57, 69, 163,
            165, 168, 170–180, 182–184, 191,
            192, 207, 208, 211, 214, 221–230,
            233, 244, 268, 269, 307, 310, 311
        biota 172, 177, 179, 181, 184, 222,
            224–230, 244, 311
        humans 60, 224–226, 244
    criteria to protect health and natural resources
        177, 181, 182, 184, 192, 208, 222,
        226, 229, 241–244, 310, 311
    health risks 224, 225, 238, 239
    interactions with cyanide 198, 221
    interactions with gold 39
    measurements 225
    metabolism 223, 240, 241
    mitigation 229
    persistence 241
    release from ore roasting 166, 229
    resistant microorganisms 307, 308
    risks to human health 94, 223, 224, 244
    sources 224, 225, 244
    teratogenicity 225
    toxicity to plants and animals 66, 223, 225,
        227, 230–241, 244
    uses 222, 223
    wastes from gold mining 221–243
    waste treatment 222, 223
Arsenic sulfide 68
Arsenic trioxide 222, 223, 235–237, 239
Arsenical pyrites 8
Arsinilic acid 241

Arsenites 223, 225, 227–240, 244
Arsenobetaine 225, 237, 241, 244
Arsenocholine 237
Arsenopyrites 68, 69, 166, 221–224, 229
Arthritis
    psoriatic 113, 114
    rheumatoid 118
    treatment with gold drugs 131–155
Asia 252, 253
Asia Minor 91
Aswan, Egypt 53, 57
Auranofin 30, 65, 78, 79, 80, 114, 121, 141, 143,
        154
Auric chloride 24, 28, 42, 66, 154
Auric iodide 42
Aurichloric acid 42
Auric oxide 42
Auriferous pyrite 9
Auriferous quartz 165
Aurocyanide 135, 136, 137
Aurosomes 44, 132, 138, 148
Aurous chloride 42
Aurous iodide 42
Australia 3, 7, 8, 11, 13, 19, 43, 53, 55, 91, 92,
        94, 101, 149, 196, 206, 207, 252,
        289, 307, 317, 328–330, 333
Austria 58

B

Bangladesh 224
Barium 67
Bastogne, Belgium 13
Belo Horizonte, Brazil 224
Bendigo, Australia 94
Benelux countries 289,
Bering Sea 179, 271
Berkeley Pit, Butte, Montana 212
Beryllium 66
Bevill Amendment 321
Big Belt Mountains, Montana 173
Bingham, Utah 16
Bingham Canyon, Utah 15
Biorecovery of gold
    animal fibrous proteins 73
    aquatic macrofauna 67, 72, 73
    fungi 68, 69
    higher plants 67, 68, 69, 71
    microorganisms 67, 68, 69
Birds
    arsenic
        concentrations 226, 229
        criteria 241, 243, 244
        toxicity 230, 234, 235, 239

cyanide
    concentrations 199
    criteria 205
    toxicity 192, 196, 199–202, 204, 214, 322
lead concentrations 175
mercury
    concentrations 263, 271, 273–275,
        282–284, 291–293
    criteria 275, 283, 284, 287, 288, 291–293
    toxicity 277, 278, 280–283, 291, 292
Bismuth 24, 26, 39, 40
Bis(thiosulfate) gold 30
Bitter crab disease 182
Black Hills, South Dakota 170, 172, 199, 222, 310
Black Sea 196
Bo Hai Sea, People's Republic of China 12
Bolivia 11, 92, 103, 105, 206, 254, 262
Boron 208, 214
Botswana 97, 100
Brazil
    arsenic contamination from gold mining 226
    bacterial degradation of cyanide 194
    Brazilian Mining Law of 1967 329
    Garimpo Mining Permit 329
    gold
        mining-related population shifts 7, 92,
            252–255, 257
        production 11, 12, 14, 19, 91–93, 224,
            253, 255, 307, 333
        concentrations in biota and abiotic
            materials 57
        extraction from cyanide leach liquor 70
        mining, military intervention 101
    health risks to gold miners 91, 93, 96,
        102–105
    mercury
        contamination from gold mining 11, 92,
            93, 100–105, 251, 254, 258–264,
            292
        criteria for protection of health and natural
            resources 262–264, 287, 289, 292,
            311
        ecotoxicological aspects of amalgamation
            255–265
        mitigation 264, 265
        poisoning, case histories 101–103
        release rates 256–258, 262–264
        sources 256–258
    mining legislation 329
Brazilian Mining Law of 1967 329
Breton Woods Agreement 11
Bristol, England 119
British Columbia, Canada 12, 13, 225
Burundi 14

C

Cacodylic acid 233–239
Cadmium
    concentrations in
        abiotic materials 32, 168–170, 172, 173,
            175–180, 182–184, 191, 192, 207,
            212, 225, 269, 280, 310, 311, 330
        biota 60, 176, 177, 179, 181, 184, 224, 311
    criteria for protection of health and natural
        resources 176, 181, 182, 184, 192
    interactions with gold 39, 43
    metabolism 45, 80
    toxicity to plants and animals 66
Calavarite 8, 39
Calcium arsenate 236
Calcium cyanide 43
California 3, 7, 8, 13–18, 92, 167, 168, 192, 201,
    202, 204, 251, 287, 289, 308, 309,
    320
California Air Pollution Control District 192
California Debris Commission 167
California Department of Health Services 192
California Regional Water Quality Board 192
California Water Board 192
Californium 27
Cameroon 14
Caminetti Act 167
Canada 7–9, 11, 13, 19, 91, 92, 94, 165, 168, 169,
    174, 189, 193, 195, 201, 221, 222,
    224, 225, 269, 287, 289, 292, 307,
    322, 327, 329, 333
Cape Yakataga, Alaska 17
Carlin gold deposit, Nevada 9
Carmos stream, Minas Gerais, Brazil 262
Carson River, Nevada 254, 271, 273, 274
Carson Sink 274
Carthaginians 92, 251
Casper, Wyoming 226
Cassiterite 8
Central African Republic 14
Cesium 42
Cesium–137 270
Chalcopyrite 69
Charles II 25
Cheyenne River, South Dakota 276
Chile 7, 69, 92, 317, 329
China 19, 24, 28
Chloroauric acid 42
Chlorobischolylglycinatogold 80
Chromate 121
Chromite 8
Chromium 60, 66, 94, 98, 168, 175–181, 183
Chryseis 30
Chrysiasis 132, 134, 149, 150, 155

Chrysotherapy 30, 31, 44, 45, 46, 58, 60, 118,
        126, 131–155
Circle Mining District, Alaska 10, 18
Clean Air Act 317, 323, 325
Clean Water Act 172, 317–319, 327, 330
Cobalt 54, 56, 60, 66, 117–119, 121, 177
Cobaltite 221
Colorado 7, 16, 17, 92, 171, 308, 327
Columbia, South America 11, 92, 95, 100, 103,
        105, 177, 221, 226, 254, 266
Comprehensive Environmental Response,
        Compensation and Liability Act
        (CERCLA, Superfund) 274,
        317–319, 327, 330
Comstock Lode, Nevada 270, 271, 273
Congo 14
Connecticut 289
Cook Inlet, Alaska 168
Copper
    as contact allergen 125
    as catalyst in cyanide mitigation 193
    concentrations in
        abiotic materials 7, 10, 11, 13, 39, 53, 69,
            70, 91, 165, 168, 169, 172, 173–180,
            182–184, 191, 192, 207, 208, 212,
            214, 228, 269, 310, 311, 330
        alloys 23, 24, 31, 40
        biota 60, 172, 176, 177, 179, 181, 184, 311
        bullion 25, 26
        gold prostheses 117, 124
    criteria to protect health and natural resources
        176, 177, 192, 310, 311
    interactions with cyanide 200
    interactions with gold 78, 79, 81
    metabolism 45
    toxicity to plants and animals 66, 172, 322
Copper pyrites 8
Cripple Creek, Colorado 16
Croesus 7, 24, 91
Cuiba River, Brazil 263
Cuyuni River, Venezuela 267
Cyanide
    concentrations in abiotic materials 163, 165,
        173, 192, 194, 196, 199, 204, 310
    criteria for protection of health and natural
        resources 192, 199, 205, 310, 311
    degradation 194
    leaching from dry tailings impoundments 183
    metabolism 195
    mitigation 189, 193–196, 198, 199, 203–206
    mode of action 201
    persistence 191, 193–196, 198
    toxicity 189, 190, 192, 193, 195–204, 214,
        336

    use in gold recovery 189–193, 195, 202, 214,
        336
    wastes from gold mining 193, 195, 196, 199,
        201–203, 205–208, 214
    water management issues in gold extraction
        206–214, 336
    weak acid dissociable cyanide 193, 200, 201,
        295
Cyanide hydratase 193
Cyanide nitrilase 193
Cyanide oxygenase 193
Cyanogen chloride 193
Cyprus 56
Czech Republic 52, 56, 57, 193, 227

                            D

Danube River, Yugoslavia 196
Darwin, Australia 328
Delaware 289
Dental gold
    allergic reactions 113, 116, 117, 123–126
    alloys 124–126
    case histories 125, 126
    chrysiasis 125
    gingivitis 125
    hypersensitivity 113, 119–124
    interleukin inhibition 124
Dexing County, Jiangxi Province, People's
            Republic of China 104, 268
Dicyanoaurate 46
Dimethyl arsinic acid 232, 237, 238
Disodium arsenate heptahydrate 232
Disodium gold thiomalate 81, 137, 138
Disodium methylarsenate 231–233, 235, 237
Dodecylamine p-chlorophenylarsonate 235
Dominican Republic 11, 307
Dongil, Korea 228, 310, 311

                            E

Ecuador 11, 12, 226, 307
Edmonton, Alberta, Canada 31
Edward III 25
Egypt 3, 7, 10, 23, 24, 25, 56, 91, 152
Electrum 8
Elko, Nevada 171
Empire Rheumatism Council 30
Enargite 221
Endangered Species Act 317, 322
England 167, 252, 256, 309

Environment Canada National Contaminated Site
        Program 270
Environment Canada Sediment Mercury Criterion
        268
Environmental Basis Law of 1994 (Chile) 329
Essequibo River, Guyana 196
Ethiopia 14
Eureka County, Nevada 211
Europe 7, 13, 25, 91, 94, 116, 118, 229
European Union 184

F

Fairbanks, Alaska 166, 174, 194, 222
Fazenda Ipiranga Lake, Brazil 263
Federal Land Policy and Management Act of 1976
        317, 319, 320, 324, 330
Federal Reserve Bank of New York 26
Federal Water Pollution Control Act 325
Ferric arsenate 222, 223
Ferric sulfate 222
Fiji 54
Florida 289
Folsom-Natomas Region, California 252
Fort Knox Bullion Depository 19
Former Soviet Union 8, 11, 12, 19, 307, 333
France 53, 55, 224
Frankfurt/Main, Germany 52
French Guiana 14, 103, 105
Fulminating gold 42

G

Gabon 14, 96, 97
Gallium 32
Garimpo Mining Permit 329
Garimpos 256, 257
Gastineau Channel, Alaska 179, 180
Gaul 10
Gegogen Harbor, Nova Scotia 175
General Mining Law of 1872 317, 321, 322, 324,
        330, 336
Georgia 15–17, 271, 276
Germany 51, 53, 57, 184, 256, 287
Ghana 11, 12, 14, 96, 97, 222, 227, 229, 307, 317,
        330
Giauque Lake, Yellowknife, Canada 270
Gold
    adverse reactions by humans
        as carcinogen 115
        as contact allergen 4, 114–118, 120, 123,
            125, 126
    gold drug reactions 148–154
    Goldschlager syndrome 113, 121, 122
    hypersensitivity 113, 115–122, 124–126,
        148–155
    implant rejection 123
    respiratory complaints 114
    teratogenicity 115
    toxicity 118, 126
    alloys 23, 24, 26, 31, 40, 124
    biorecovery 67–73, 81, 223, 335
    caratage 23, 24, 25, 40
    coins
        escudo d'oro 25
        florin 25
        mina 24
        noble 25
        shat 24
        shekel 24
        talent 24
        tower pound 25
    colloids 43
    concentrations in
        air 51, 52
        atmospheric dust 60
        earth's crust 51, 52, 60
        fish 58, 59, 60
        freshwater 51, 52, 60
        humans 44, 45, 58–60, 122, 124, 126, 132,
            133, 135, 144, 145, 148, 150–153
        invertebrates 59, 60
        mammals 58, 73
        ores 69, 177, 222, 252, 256
        plants 56, 57, 59, 60, 334
        rainwater 51, 52, 60
        seawater 51, 52, 54, 60
        sediments 51–55, 57, 60, 178, 334
        sewage sludge 51, 53, 55, 60, 334
        snow 51, 53, 55, 60
        soil 51, 53, 54, 55, 60
        stream water 176
        tailings 164, 274
        volcanic rock 51, 54
    drugs
        allocrysine 133
        allochrysin 30
        allochrysine 77
        auranofin 30, 65, 78–80, 114, 121, 133,
            134, 136, 141, 143, 154
        aurothion 133
        bis(thiosulfate) gold 30
        crisalbine 133
        disodium gold thiomalate 137–139
        gold chloroquinine 29
        gold sodium thiomalate 30, 66, 67, 76–78,
            80, 81, 116–121, 123–125, 147, 151

gold sodium thiosulfate 116–121, 125
gold thioglucose 132, 134, 148, 151, 154
gold thiomalate 132
gold thiosulfate 132
krysolgan 133
myochrisin 30
myochrisine 76, 133
myochrisis 133
myocrisin 133
sanochrysin 30
sanocrysin 133
sanocrysis 133
sodium gold 4-amino-2-mercaptobenzoic
    acid 133
sodium gold thioglucose 46, 133, 150–152
sodium gold thiomalate 30, 46, 73, 78,
    114, 115, 133, 134–137, 140–150,
    152–154
sodium gold thiosulfate 133
sodium gold thiopropanosulfonate 133
solganol 30, 133
2,3,4,6-tetra-O-acetyl-1 thio-beta-D-
    pyranosato-S-(triethylphosphine)-
    Au+ 133
thiomalatogold 30
thioglucose gold 30
thiopropanosulfonate gold 30
3-triethylphosphine gold-2,3,4,6-tetra-O-
    acetyl-1-thio-beta-D-
    glucopyranoside 133
trisodium bis (thiosulfato) aurate (I) 132
economics 15, 25, 190, 191, 252–254, 333
extraction
    by amalgamation 91, 92, 93, 101, 102,
        103–105, 167, 173, 175, 251–293
    by cyanidation 54, 67, 68, 69, 70, 91, 166,
        173, 189–197, 199, 201, 214, 257
    by thiourea 67, 68, 206
fineness 23–26, 40
geology 8–10
measurement 40, 41, 71, 73, 74
medical aspects
    adverse reactions 73, 148–154
    contact dermatitis 4, 114–118, 120, 123,
        125, 126
    drugs used medicinally 30, 114
    risks to miners 91–106
    toxicity 78
metabolism 71, 73, 78–80, 122, 124, 132–138,
    143, 145, 146, 148, 154
mine wastes
    acid mine drainage 163, 167, 169–172,
        335
    arsenic 166, 221–242
    cyanide 189, 193–214

history 163–168
mercury 251–293, 336
research recommendations 333–336
risks to human health 91–105, 224
slimes 183
suspended solids 164, 165, 167
tailings 163–165, 170, 172–183, 192, 201,
    207, 230, 252, 335
waste rock 163–165, 170, 171, 183, 192
miners: health risks
    historical background 91–93
    underground miners 91–100, 105
    surface miners who use mercury 91–93,
        100–105, 258
mining-related population shifts 92, 252–255,
    257, 267
oxidation states 41
panning and sluicing 166
placers 7, 8, 11, 166
production
    Africa except RSA 253, 310
    Asia and environs 12
    Canada 12–13, 269
    Europe 13–14
    Global 25
    Papua New Guinea 179
    Philippines 268
    Republic of South Africa 14
    South America 14, 91, 224, 253, 255, 256
    United States 14–19, 168, 191, 270, 310
properties
    biological 43, 44, 45
    biochemical 43–47
    chemical 26, 41–43
    physical 26, 39–41
prostheses 29, 30, 117, 122
protective effects 123
research recommendations 333–336
refining 43, 190
solubility 44, 45, 46
sources 3, 7, 10–18, 91, 92, 166, 171, 172,
    179, 196, 257, 263, 266–270, 275,
    276, 307, 309, 333, 334
tenor 8, 165
toxicity
    aquatic organisms 65–67, 81, 179,
        180–182
    mammals 73–78, 81
uptake stimulating factor 45
uses
    in alcoholic beverages 33
    coinage 24–26, 40, 333
    decorations 7, 23, 24, 33, 91
    delivery vehicle 32
    dentistry 31, 32, 59, 334

electron microscopy 33
electronics 26, 32, 333, 334
jewelry 3, 7, 23, 24, 40, 59, 91, 116, 117, 333
medicine 28–31, 46, 79, 113–118, 123, 131–155, 333–335
physiology, obese mouse model 31, 74, 75, 334
radiogold isotopes 26, 27, 78
Gold accumulator plants 56, 334
Gold (1,2-bis(diphenylphosphino)ethane 77
Gold bromide 42
Gold chloride 33, 42, 46, 65
Gold coast 25
Gold complexes 59
Gold cyanide 28, 33, 42, 46, 47, 55, 69, 72, 73, 135, 143
Gold disodium thiomalate 45
Gold halogens 51
Gold hydroxide 55
Gold-induced obesity in mice 31, 74, 75
Gold iodide 33, 47
Gold isotopes 40
Gold L-cysteine 78
Gold miners: health risks
historical background 91–93
research recommendations 333
surface miners who use mercury
case histories 101, 102
mercury in air and in fish diet 104–105
mercury in tissues 102–104
underground miners
Africa 91, 96–100, 105
Australia 91, 93, 94, 105
Europe 91, 96, 105
North America 91, 94, 95, 105
South America 91, 95, 96, 105
Gold prostheses 117, 122, 123
Gold rings 123
Gold selenate 42
Gold sodium aurothiomalate 66
Gold sodium thioglucose 148
Gold sodium thiomalate 30, 66, 67, 76–78, 80, 81, 116–125, 151
Gold sodium thiopropanosulfonate 77
Gold sodium thiosulfate 116–121, 125, 147
Gold sulfide 52, 54, 55
Gold thiocaproate 74
Gold thiocyanate 29, 136
Gold thiogalactose 74
Gold thioglucose 29, 31, 74, 75, 82, 132, 151
Gold thioglycoanilide 74
Gold thiolate 136
Gold thiomalate 74, 132
Gold thiosorbitol 74

Gold thiosulphate 74, 132
Gold trichloride 118, 119
Golden fleece 3
Goldenville, Nova Scotia 175, 269
Goldenville gold mine, Nova Scotia 176, 226
Goldschlager syndrome 113, 121, 122
Goldsmiths 58–60
Grass Valley, California 16
Greece 193
Grosz, Austria 58
Gulf of Papua 207
Guyana 14, 190, 196, 317, 329

H

Heap leach gold mining
cyanide hazards 91, 189–193, 199, 204
gold recovery 189
history 18, 189–193
water management issues 189, 206–214
Hematite 69
Homestake Mine, Lead, South Dakota 16
Hong Kong 3
Human health
arsenic
criteria 242, 244
toxicity 239, 240
cyanide toxicity 196, 197
gold toxicity 78, 142, 151
mercury
concentrations 255, 259, 291
criteria 286, 289–292
toxicity 268, 284, 291, 292
risks to gold miners 93–105
Humboldt River Basin, Nevada 207, 208, 210, 211, 213, 271
Humboldt Sink, Nevada 208, 210, 222
Hungary 24, 196
Hydraulic gold mining 173
Hydraulicking 167

I

Icy Bay, Alaska 17
Idaho 7, 16–18, 92
Illinois 289
Ilmenite 8
Imperial Rome 10, 251
India 12, 24, 28, 174, 287
Indium 32, 39, 60
Indonesia 11, 92, 172, 253, 254

Interleukin 114,116, 124, 139, 143, 154
International Agency for Research on Cancer 94
International Gold Council 23
Inuit people 285
Iodine 27
Iowa 287
Iran 25
Iraq 25
Iridium 39
Iron 8–10, 26, 39, 42, 54–56, 66, 68–70, 117, 168, 169, 174, 176, 223, 225, 310
Iron pyrite 8, 9, 168
Iron sulfide 170, 193
Israel 120
Italy 55, 58, 59, 251

J

Jaba River, Papua New Guinea 207
Jack of Clubs Lake, British Columbia, Canada 226
Japan 3, 24, 53, 54, 57, 116, 119, 196, 284, 289, 322
Jefferson City, Montana 195
Johannesburg, RSA 309
Juneau, Alaska 179, 180, 182

K

Kawerjong/Jaba River system, Papua New Guinea 207
Kazakhstan 12
Kenya 14, 96, 97, 100, 268
Kinetic testing of acid mine wastes 171, 184
King Solomon 10
Klondike region, Canada 166
Klondike River, Yukon Territory, Canada 13
Kola Peninsula, Russia 53
Korea 3, 175, 176, 222, 310
Kyrgyzstan 190

L

La Oraya, Peru 225
Lahontan Reservoir, Nevada 254, 271–274
Lake Gegogen, Nova Scotia 175
Lake Oahe, South Dakota 273
Lake Victoria, Serengeti National Park, Tanzania 227, 267, 310
Lanthanum 60, 66
Latvia 190

Lead
    as carcinogen 76
    concentrations in
        abiotic materials 26, 31, 39, 40, 54, 163, 168–170, 172, 174–180, 182, 184, 191, 192, 207, 225, 227, 230, 268, 269, 310, 311
        biota 60, 172, 175–177, 179, 181, 311
    criteria for protection of health and natural resources 176, 181, 182, 192, 310, 311
    interactions with gold 39
    production 308, 309
    toxicity 172
Lead-210 270
Lead arsenate 236
Lead oxide 41
Lead, South Dakota 16, 312
Lesotho 100
Liberia 14
Liadong-Koren peninsula, People's Republic of China 12
Liquid gold 24
Limonote 8
Lithium 42, 66, 67
Lode gold mining 11, 13, 15, 16, 18, 165, 166, 172–174, 179
London Bullion Market 25
Long Island, New York 270
Lynches River, South Carolina 195

M

Madeira River Basin, Brazil 255, 259, 260, 262
Magnesium 54, 56, 66
Magnetite 8
Maine 287
Malagasy Republic 14
Malawi 97
Malaysia 177, 228, 229
Malaysian Food Act of 1983 177, 229
Mali 14
Mammals
    arsenic
        concentrations 224, 225
        criteria 241, 243, 244
        toxicity 225, 235–241
    copper
        criteria 311
        toxicity 311
    cyanide
        criteria 205, 311
        toxicity 192, 196, 199–204, 311

gold
    accumulations in tissues 44, 45
    as anti-cancer agent 44, 46,47
    as anti-HIV agent 47
    as anti-inflammatory agent 44, 46
    circulation 44
    distribution in body 43
    essentiality 43
    -induced obesity in mice 31, 74, 75
    metabolism 43, 45, 80, 81
    radiogold distribution in rats 79
    toxicity 73–78, 80
mercury
    concentrations 91, 92, 93, 101–103, 259,
        264, 266, 284, 285, 291, 293
    criteria 266, 284, 286, 288, 291–293
    toxicity 91, 92, 93, 101–103, 264, 266,
        277–281, 284, 285, 291, 293
silver
    criteria 311
    toxicity 311
Manganese 51, 56, 67, 168, 175–177
Manitoba, Canada 12, 13
Marcasites 168
Maryland 32
Marysville, California 167
Mata Atlantica, Brazil 329
Mato Grosso, Brazil 260, 264
Mercuric chloride 279
Mercuric iodide 103
Mercury
    as contact allergen 125
    amalgamation 252
    biological function 254
    biomagnification 254
    carcinogenicity 254, 276, 284
    concentrations
        in abiotic materials 9, 39, 91, 93, 100, 104,
            105, 168, 171–173, 175–179, 183,
            191, 192, 207, 228, 253, 255,
            257–263, 265–276, 310
        in biota 32, 60, 91, 93, 100, 101–105, 177,
            179, 224, 253–255, 258–276, 280,
            282, 283, 285
        in humans 255, 258, 259, 265
        near gold mining sites
            Brazil (active) 92, 93, 100, 103, 104,
                259–262, 292
            United States (historical) 270–276, 292
    contamination from gold mining 92, 93,
        100–106, 251, 252, 254, 292, 336
    criteria for protection of health and natural
        resources 91, 181, 184, 192, 251,
        258, 262–264, 266, 268, 270, 279,
        283, 284, 286–293, 310, 311, 335

cycling 253
ecotoxicological aspects of amalgamation
    Africa 267–269
    Brazil 255–265
    Canada 269, 270
    People's Republic of China 268
    Philippines 268, 269
    Siberia 269
    South America except Brazil 265–267
    United States of America 270–276
health advisories 274
health risks to gold miners 101–106, 251, 258
history in gold mining 8, 15, 43, 91, 92,
    100–103, 251–255
interactions with other metals 283, 285
lethal and sublethal effects 66, 124, 251, 254,
    276–286
measurement 270
medical aspects 251
metabolism 45, 251, 281, 284, 285
mitigation 264, 265, 292, 336
mobilization 252
mutagenicity 254, 276, 284
persistence 285
release rates 251, 253, 254, 256, 262–270,
    273, 274, 276
sources 100, 123, 251, 253, 256, 274
teratogenicity 254, 276, 284
transformations 254, 257, 262, 263, 270, 274,
    275, 277, 284, 292
transport 254, 255, 257, 262, 265, 270, 274,
    276, 292
Metallic gold (Au^0)
    effects
        humans 113, 115–123, 125, 126
        mammals 81, 334
    mechanisms of accumulation 334
    metabolism 135, 138, 139, 154
    toxicity 118, 126
    use in patch tests 118
Metallothioneins 45, 79, 80, 81
Methanearsonic acid 236
Methotrexate 31
Methylmercury 280
Mexico 7, 252, 265, 322
Michigan 16
Migratory Bird Treaty Act 201, 212, 317, 322, 323
Millionaire mine, Beaverhead National Forest,
    Montana 308, 309
Minamata, Japan 283, 284
Minas Gerais, Brazil 224, 265
Mindanao, Philippines 104, 268
Mine dewatering operations: effects on natural
    resources 207–211
Mining and Minerals Policy Act 321

Mining Legislation
    Foreign 328–330
    United States 317–327
Mining law reform 328, 330
Mining Public Lands in western USA 323–327
Misima Island, Papua New Guinea 179
Molybdenum 54, 67, 229
Monazite 8
Monomethylarsonic acid 238
Monosodium methanearsonate 231, 235–237
Monovalent gold
    adverse reactions in humans 113–121, 125
    as anti-tumor agent 29, 114
    as anti-inflammatory agent 30, 76
    concentrations in humans 132
    effects
        aquatic organisms 65, 66, 81
        mammals 74–82
    mechanisms of accumulation 334
    metabolism 132, 135–140, 154
    use in medicine 113–116, 118–121, 123, 125,
        132
Montana 7, 16–18, 92, 167, 202, 213, 289
Mont Blanc, Italy 53
Monte Rosa, Italy 53
Montreal River, Ontario, Canada 174
Mozambique 97, 100
Murray Brook, New Brunswick, Canada 52, 55
Myochrisin 30
Myochrisine 76
Myungbong, Korea 228

N

Nagyagite 39
Nashville, Tennessee 122
National Environmental Policy Act (NEPA) 317,
    319, 324, 330
National Forest Management Act of 1976 317,
    320, 324, 330
National Historic Preservation Act 317, 323
National Pollution Discharge Elimination System
    318
National Priorities List 319
National Research Council of US National
    Academy of Sciences 323, 326
Natural Resource Damage Assessment and
    Restoration Program 319
Negro River, Brazil 260
Nephelometric units 174, 178
Netherlands 25, 151, 256, 282

Nevada 7, 9, 11, 15–17, 19, 60, 92, 190, 192, 201,
    202, 204, 207, 208, 211, 212, 270,
    271, 292, 309, 320, 327
Nevada Administrative Code "502.460–502.499"
    192
Nevada Division of Wildlife 192
Nevada Net Proceeds of Mines Tax 328
New Brunswick, Canada 55
New Guinea 54
New Jersey 270, 289
New Mexico 16, 17, 166, 308
New York 54, 289
New Zealand 7, 53, 54, 92, 190
Niccolite 221
Nickel
    as contact allergen 125, 151
    concentrations in
        abiotic materials 39, 53, 165, 173, 175,
            176–180, 207
        alloys 23, 24, 40, 118
        biota 60, 179, 181
        in gold prostheses 117
    contaminant in gold drugs 151
    criteria for protection of health and natural
        resources 181, 182
    interactions with cyanide 198
    production 59
    toxicity 66, 116, 119–121, 125
Nickel sulfate 124, 198
Nigeria 58, 60
Niigata Prefecture, Japan 52
3-nitro-4-hydroxyphenylarsonic acid 238, 241,
    244
Nome, Alaska 17, 166, 168, 178, 276
North America 91, 116, 118, 167, 168, 189, 251
North Carolina 7, 14–17, 92, 191, 273
North Dakota 289
North Island, New Zealand 57
Norton Sound, Alaska 178
Norwegian National Adverse Reaction Group 125
Nova Linda, Brazil 70
Nova Scotia, Canada 12, 13, 221, 269

O

Obuasi, Ghana 267
Oklahoma 289
Omai gold mine, Guyana 329
Ontario, Canada 13, 68, 94, 95, 175, 224
Ontario Provincial Quality Guidelines 175
Open pit gold mining 16, 164, 165, 169, 192, 206,
    207
Open pit mines, effects on resources 207–214

Orangi River, Tanzania 310
Oregon 16, 17, 309
Organic Act of 1897 324
Orpiment 221
Osmium 39

P

Pactolus River 7, 91
Palladium 26, 31, 32, 39, 40, 55, 59, 69, 117, 124,
          125
Panning and sluicing 166, 167
Pantanal, Brazil 260, 262, 263, 329
Papua New Guinea 11, 12, 53, 54, 180, 207, 307,
          317, 329, 330
Paraguay 262
Parana River, Brazil 260
Payapal, Venezuela 96
Pennsylvania 16
Pentachlorophenol 177
People's Republic of China 11, 12, 92, 103, 105,
          252, 254, 268, 292, 307, 333
Peru 7, 11, 92, 226, 254, 266, 317, 330
Philippines 11, 12, 54, 91, 92, 100, 101, 103–105,
          172, 254, 268, 269, 292, 307
Phoenicians 92, 251, 252
Phytoremediation 183
Pit lakes
    effects on wildlife 168, 206, 212–214
    formation 207, 211
    groundwater 206, 207, 212, 214
    hydrology 211, 212
    research needs 213, 214, 336
    water management issues 206, 207, 211–214
    water quality 170, 207, 211–214
Placer gold mining 13, 15–18, 43, 166–168, 173,
          174, 177–179
Platinum 8, 26, 29, 31, 39, 40, 43, 46, 55, 59, 60,
          124, 190
Pocone, Brazil 258, 260
Poland 52, 56, 57, 227
Port Clarence, Alaska 168
Porto Velho, Brazil 260
Portovela-Zaruma, Ecuador 192
Portugal 119
Potassium cyanide 201, 203
Purple of Cassius 43
Pyrites 68, 69
Pyrrhotites 168

Q

Quebec, Canada 12, 13, 175
Queen Charlotte Islands, British Columbia,
          Canada 13
Queensland, Australia 93, 101

R

Radiogold isotopes
    as cancer treatment 26, 27
    as chemical label 26
    distribution in rats 78, 79
    to measure radiation exposures 26
    to track animals 26, 27
    uptake by aquatic biota 72
Radon 95, 96
Realgar 221
Reef gold 9
Refractory ores 222, 223
Reptiles
    cyanide toxicity 192, 201, 202
    mercury
        concentrations 264
        toxicity 264
Republic of South Africa 7, 13, 19, 68, 71, 91,
          92, 96–100, 190, 202, 307, 333
Resource Conservation and Recovery Act 317,
          319, 321, 330
Rhenium 67
Rheumatoid arthritis: treatment with gold drugs
    adverse effects 45, 77, 78, 118, 121, 131,
          133–135, 140, 144–146, 148–155
    autoimmunity 46
    case histories 131, 144, 146–148, 155
    history 28, 31, 79, 131–134
    modes of action 45, 46, 131, 134–142, 144,
          154
    research recommendations 335
    treatment regimes 131, 139, 140, 144, 145,
          154, 155
Rhodium 24, 39, 55
Roberts Mountains, Nevada 9
Romania 3, 193
Rondonia, Brazil 255
Roosevelt, President Franklin Delano 15
Rubidium 42, 67
Ruby glass 24, 40
Russia 9, 14, 55, 60, 322
Ruthenium 32, 39
Rwanda 14

S

Sacramento River, California 170, 270, 271
Sado Island, Japan 52, 55
Sanochrysin 30
Santee River Basin, South Carolina 270, 271
Sardinia, Italy 13
Saudi Arabia 7, 10, 91
Sawyer, Judge Alonzo 167
Scotland 118, 189
Sea of Japan 55
Selenic acid 42
Selenium 40, 54, 78, 79, 168, 173, 183, 269, 275, 280, 285, 291
Seleno-DL-methionine 283
Seoul, Korea 176
Serengeti National Park, Tanzania 310
Serra do Mar, Brazil 329
Seward Peninsula, Alaska 271, 276
Shangdong Peninsula, People's Republic of China 12
Siberia 7, 24, 92, 269, 292
Sierra County, California 16
Sierra Nevada Mountains, California 8, 252
Silesia 24
Silica 95–100
Silicon 54, 67, 95
Silicosis 94, 95, 97–100
Silver
    concentrations in
        alloys 23, 24, 31, 32, 40, 125
        bullion 25, 26
        coinage 23, 24
        cyanide leach liquor 70
        gold refining 43
        gold prostheses 117, 124
        human hair 59
        ores 7–10, 39, 54, 91, 179, 229
        sediments 54
        tailings 274
        water 176
    metabolism 45
    production 18, 251, 265, 309
    recovery by cyanide 190
    toxicity 66, 67, 120
    use in fire assay 41
Singapore 3, 120
Sixmile Canyon, Nevada 53
Slovak Republic 288
Sodium arsenate 235–237, 238, 240, 243
Sodium arsenite 236, 237, 240
Sodium arsinilate 241, 244

Sodium aurothiomalate 29, 65
Sodium cacodylate 232, 235–238
Sodium cyanide 43, 189–191, 193, 195, 198, 200, 201
Sodium dimethylarsenate 231, 232
Sodium-gold chloride 42
Sodium gold thioglucose 46, 150–152
Sodium gold thiomalate 30, 46, 58, 73, 78, 114, 115, 135, 137, 140–150, 152–154
Sodium tetrachloroaurate 80
Sodium tetrachloroaurate dihydrate 80
Solganol 30
Solid Waste Disposal Act of 1965 321
Solsigne, France 96
Somes River, Romania 195
Songcheon, Korea 227, 229
South Carolina 16, 17, 191, 202
South Dakota 11, 16–18, 95, 202, 204, 205, 310
South Florida Basin, Florida 270, 271
Spain 92, 251, 252
Spangold 24
Sphalerite 53, 54
Spirit Mountain, Montana 170
Sri Lanka 56, 57
Stannous chloride 23
Strip gold mining 165
Strontium
Suction dredging 168
Sudan 14
Sudbury, Ontario, Canada 59
Summitville Gold Mine, Colorado 327
Superfund 274, 317–319, 327, 330
Surface Mining Control and Reclamation Act 317, 322
Suriname 14, 266
Suspended solids
    effects on
        agriculture 167
        avoidance by aquatic fauna 184
        channel alteration 164
        dissolved oxygen 164
        fish physiology 164
        fisheries 167, 168, 177
        habitat restoration 165
        light penetration 164, 173
        reproduction of benthic fauna 184
        species diversity 164, 184
        stream flow 164
        turbidity 174, 178–180, 184, 207
        water quality 165, 167, 173, 174, 207
Sweden 116, 119, 149
Sylvanite 8, 39

T

Tailings disposal
  effects
    in freshwater 172–178, 184, 230
    in seawater 172, 178–184, 330
    on land 183, 184
  habitat restoration 174
  mitigation 222
Tailings fields 183
Tailings ponds 183
Taiwan 3
Tanzania 14, 92, 100, 102, 103, 253, 254, 267, 310
Tapajos River Basin, Brazil 257, 260, 264
Teles River, Brazil 261
Tellurium 8, 39, 40, 42, 291
Tennantite 221
Tennessee 16, 289
Terrestrial invertebrates
  arsenic
    concentrations
    toxicity 230, 234
  gold concentrations 59
  mercury
    criteria 286
    toxicity 286
Terrestrial plants
  arsenic
    criteria 242–244
    concentrations 227, 228, 311
    toxicity 229, 230, 233, 234
  cadmium concentrations 311
  copper
    concentrations 311
    criteria 311
    toxicity 311
  cyanide toxicity 196, 203, 204
  gold concentrations 56, 57, 59
  lead
    concentrations 311
    criteria 311
    toxicity 311
  mercury
    concentrations 259, 263, 268, 273
    criteria 286, 287
    toxicity 286
  tailings toxicity 330
  zinc concentrations 311
Tetrachloroauric acid 42, 43
Tetramethylarsonium iodide 237
Texas 289
Thailand 254
Thallium 9, 66
Thiamin 291

Thiomalatogold 30
Thiomalic acid 67
Thiopropanosulfonate gold 30
Thiorin 70
Thiourea 70, 206
Tin 24, 32, 39, 43, 120
Tisza River, Hungary 196
Titanium 13, 24, 31, 43, 119, 125, 126
Tokaimura, Japan 27
Tonga 54
Torres Strait 207
Transylvania 7, 91
Trivalent gold
  as anti-tumor agent 29, 115
  as anti-arthritis agent 29
  effects
    aquatic organisms 66, 67, 69, 70–72, 81
    human health 117–119, 124
    mammals 80, 81, 334
  mechanisms of accumulation 334
  metabolism 135–139, 154
  toxicity 118
  use in patch tests 118
Troy system 25, 40
Troyes, France 25
Tucurui Reservoir and vicinity, Brazil 261, 263, 264
Tuksut Channel, Alaska 168
Tungsten 166
Turkey 193
Turkish Supreme Court 193

U

United Nations Environment Program 101
United States of America 7, 11, 15, 18, 19, 25, 91, 92, 94, 95, 116, 163, 165, 166, 168–170, 190, 207, 252, 254, 289, 292, 307, 333
U.S. Army corps of Engineers 318
U.S. Bureau of Land Management 164, 192, 199, 319, 320, 323–327
U.S. Bureau of Mines 189
U.S. Department of Justice 327
U.S. Environmental Protection Agency 205, 206, 268, 270, 318, 321, 323, 327
U.S. Fish and Wildlife Service 212, 318, 322
U.S. Food and Drug Administration 181
U.S. Forest Service 309, 320, 323–327
U.S. General Accounting Office 322
U.S. Minerals Management Service 168
U.S. National Institute of Occupational Safety and Health 104
U.S. National Marine Fisheries Service 318, 322

U.S. National Park Service 320
U.S. Occupational Safety and Health
         Administration 95
Uranium 95
Utah 11, 16–18, 309
Uzbekistan 52, 58, 60

V

Val d'Or, Quebec, Canada 269
Vanadium 176
Vein gold 9
Venezuela 11, 14, 91, 92, 95, 101, 102, 104, 254,
         267
Vermont 287
Vietnam 92, 253
Virginia 15–17, 289
Virginia City, Nevada 254, 270
Visual Resource Management System 326

W

Wales, UK 56, 57
War Production Board Order L-208 15
Washington 16, 17
Weak acid dissociable cyanide 193, 200, 201, 205
White River Glacier Basin, Alaska 17
Whites Creek, Montana 173
Whitewood Creek, South Dakota 173, 226, 229
Wilderness Act of 1964 317, 320, 330
Windsor, Quebec, Canada 309
Wisconsin 287–289
Wisconsin River, Wisconsin 285
Witwatersrand gold fields, Republic of South
         Africa 14, 97, 189
World Health Organization 98, 102–105, 177,
         183, 258, 263, 268
Wyoming 16, 17, 213, 229

Y

Yakataga Glacier Basin, Alaska 17
Yakutat, Alaska 168
Yellowknife, NWT, Canada 165, 198, 225, 226,
         228, 270
Yellowstone River Basin 271
Yttrium 66
Yugoslavia 196
Yukatat, Alaska
Yukon–Kuskokwim Coast, Alaska 285
Yukon region, Canada 166
Yukon River, Canada 177
Yukon Territory, Canada 12, 13, 164

Z

Zaire 14, 329
Zambia 97
Zimbabwe 11, 12, 14, 96, 97, 176, 190, 196, 224
Zinc
    concentrations in
        abiotic materials 13, 54, 70, 165, 168, 171,
             172–180, 182, 184, 207, 208, 212,
             214, 227, 280, 310, 311, 330
        alloys 23, 32, 40
        biota 172, 176, 177, 179, 181, 184, 311
        gold prostheses 117, 124
    criteria for protection of health and natural
             resources 176, 177, 184, 310, 311
    interactions with gold 39, 67, 78, 79, 81, 166,
             190
    metabolism 45, 80
    production 308
    toxicity 66
Zirconium 56

Species Index

ALGAE AND HIGHER PLANTS

Alfalfa, *Medicago sativa* 71
Algae
    Brown alga, *Sargassum natans* 71
    Chlorella spp. 71
    Freshwater alga, *Scenedesmus obliquis* 230
    Red alga, *Champia parvula* 232
    Scripsiella faeroense 279
    Skeletonema costatum 233
Bean
    fava bean, *Vicia* sp. 56, 57
    green, *Phaseolus vulgaris* 203
Blueberry, lowbush, *Vaccinum angustifolium* 59
Bromeliad epiphyte, *Tillandsia usenoides* 261
Bryophytes, aquatic
    Chiloscyphus pallescens 56
    Fontinalis antypyretica 56
    Various 57
Cassava, *Manihot esculenta* 227, 268
Cattail, *Typha latifolia* 172
Diatom, *Amphora coffeaeformis* 66
Fern
    Red water fern, *Azolla filiarloides* 71
    Water fern, *Ceratopterus* spp. 227, 268
Fruit, oil palm, *Elaaeis guineensis* 227
Gold accumulator plants
    Artemisia persia 56, 57
    Prangos popularia 56, 57
    Stripa spp. 56, 57
Grasses
    Bermudagrass, common, *Cynodon dactylon* 233
    Elephant grass, *Pennisetum* spp. 227, 268
    Stargrass, *Eleusine indica* 227
Horsetails, *Equisetum* spp. 183, 229

Hyacinth, water, *Eichornia crassipes* 204
Macrophytes, aquatic
    Egeria densa 57
    Eichincloa sp. 260
    Eichornia spp. 260, 263, 266
    Salvina sp.263
    Scurpus cubensis 263
    Victoria amazonica 259
Pea, *Pisum sativum* 230, 233
Pines
    Corsican pine tree, *Pinus larico* 56
    Scots pine, *Pinus sylvestris* 233
Plantain, *Musa paradisiaca* 227, 268
Peat, *Muther agawela* 57
Poplar, *Populus* sp. 57
Rice, *Oryza sativa* 176, 233, 286, 311
Seaweeds
    Eisenia spp. 71
    Gracilaria spp. 71
    Porphyra sp. 57
    Sargassum spp. 69, 71
    Ulva spp. 57, 71
Sesame, *Sesamum indicum* 176
Soybean, *Glycine max* 230, 233
Spruce, *Picea abies* 286
Sugarcane, *Saccharum officinarum* 56

AMPHIBIANS

Frogs
    Bullfrog, *Rana catesbeiana* 66, 67
    Columbia spotted frog, *Rana luteiventris* 210
    Leopard frog, *Rana pipiens* 279

Salamanders
  Many ribbed salamander, *Eurycea multiplacata* 308
  Marbled salamander, *Ambystoma opacum* 232
  Ouachita dusky salamander, *Desmognathes brimleyorum* 308
  Plethodon spp. 308
Toad, narrow-mouthed, *Gastrophryne carolinensis* 230, 232, 278

BACTERIA

Acinetobacter spp. 193
Alcaligenes spp. 193
Archaea spp. 68
Bacillus spp. 29, 56, 65, 69, 193
Bacteria spp. 68
Burkholderia spp. 70
Chrysiogenes spp. 307
Escherichia coli 194
Helicobacter pylori 29, 145
Leptospirillium spp. 68, 223
Mycobacterium spp. 77, 99
Plectonoma spp. 69
Pseudomonas spp. 69, 70, 193, 194, 205
Rickettsia spp. 97
Spirulina spp. 70
Staphylococcus spp. 29, 65
Streptomyces spp. 70
Sulfolobus spp. 68, 223
Thiobacillus spp. 68, 69, 168, 172, 223

BIRDS

Anhinga, *Anhinga anhinga* 263
Blackbird, red-winged, *Agelaius phoeniceus* 282
Bobwhite, common, *Colinus virginianus* 235, 278
Chicken
  Domestic chicken, *Gallus* spp. 200, 235, 282, 291
  Prairie chicken, *Tympanuchus cupido* 278
Cormorant, double crested, *Phalacrocorax auritus* 271, 272, 274, 275
Coturnix, *Coturnix coturnix coturnix* 278, 281
Cowbird, *Molothrus ater* 282
Crossbill, red, *Loxia curvirostrata* 199
Dove, rock, *Columba livia* 278, 283, 308
Ducks
  American black duck, *Anas rubripes* 175, 283
  Mallard, *Anas platyrhynchos* 175, 200, 201, 239, 243, 278, 281–283, 288
  Ring-necked duck *Aythya collaris* 175
Eagle
  bald eagle, *Haliaeetus leucocephalus* 210
  white-tailed sea eagle, *Haliaeetus albicilla* 196, 283
Egret
  Great egret, *Casmerodius albus* 263
  Snowy egret, *Egretta thula* 271, 272, 275
Finch, zebra, *Poephila guttata* 278, 281
Goose, lesser snow, *Chen caerulescens caerulescens* 212
Grackle, *Quiscalus quiscula* 282
Hawk
  Red-tailed hawk, *Buto jamaicensis* 282
  Sparrow hawk, *Accipiter nisus* 284
Heron
  Black-crowned night heron, *Nycticorax nycticorax* 272, 275
  Grey heron, *Ardea cinerea* 282
Jackdaw, *Corvus monedula* 282
Kestrel
  American kestrel, *Falco sparverius* 200
  Kestrel, *Falco tinnunculus* 282
Kite, snail, *Rostrhamus sociabilis* 258, 263
Limpkin, *Aramus guarauna* 263
Loon, common, *Gavia imer* 283, 288
Magpie, *Pica pica* 282
Merganser, red-breasted, *Mergus serrator* 200
Osprey, *Pandion haliaetus* 210
Owl, eastern screech, *Otus asio* 200
Partridge, gray, *Perdix perdix* 278, 283
Pheasant, ring-necked, *Phasianus colchicus* 235, 278, 282, 283, 288
Quail
  California quail, *Callipepla californica* 234
  Japanese quail, *Coturnix japonica* 200, 283
Siskin, pine, *Carduelis pinus* 199
Sparrow, house, *Passer domesticus* 278, 309
Starling, European, *Sturnus vulgaris* 200, 203, 282, 283
Tern, common, *Sterna hirundo* 288
Vulture, black, *Coragyps atratus* 200
Wren, house, *Troglodytes aedon* 226, 229, 308

FISHES

Aimara, *Hoplias macrophthalmus* 267
Bluegill, *Lepomis macrochirus* 196, 243
Catfish
  Channel catfish, *Ictalurus punctatus* 278
  Clarias sp. 267
Flagfish, *Jordanella floridae* 232

Goldfish, *Carassius auratus* 72, 231, 232, 241, 281

Goodeid fish, *Ameca splendens* 32

Guppy, *Poecilia reticulatus* 281

Grayling, Arctic, *Thymallus arcticus* 164, 177, 179

Herring, Pacific, *Clupea harengus pallasi* 180, 181

Lungfish, marbled, *Protopterus aethiopicus* 267

Longjaw mudsucker, *Gillichthys mirabilis* 72

Mackerel, *Pneumatophorous japonica* 58

Minnow

    Fathead minnow, *Pimephales promelas* 198, 232, 279

    Sheepshead minnow, *Cyprinodon variegatus* 72, 180, 181

Mudfish, *Heterobranchus bidorsalis* 227, 268

Mudskipper, *Boleophthalmus boddaerti* 197

Mummichog, *Fundulus heteroclitus* 66

Pavon, speckled, *Cichla temensis* 264

Perch, yellow, *Perca flavescens*175

Pike, northern, *Esox lucius* 270, 273, 281

Salmon

    Atlantic salmon, *Salmo salar* 196, 198

    Coho salmon, *Oncorhynchus kisutch* 168

    Pink salmon, *Oncorhynchus gorbuscha* 233

Sole, yellowfin, *Pleuronectes asper* 182, 183, 230

Steelhead, *Oncorhynchus mykiss* 168, 195

Trout

    Brook trout, *Salvelinus fontinalis* 278–281, 291

    Cutthroat trout, *Oncorhynchus clarki* 173

    Lahontan cutthroat trout, *Oncorhynchus clarki henshawi* 209, 210

    Lake trout, *Salvelinus namaycush* 270

    Rainbow trout, *Oncorhynchus mykiss* 197, 198, 232, 278–281, 287

    Steelhead trout, *Oncorhynchus mykiss*

Walleye *Stizostedion vitreum* 175

Whitefish, round, *Prosopium cylindraceum* 270

Zebrafish, *Brachydanio rerio* 277

FUNGI

Aspergillus spp. 70

Catellatospora spp. 307

Cladosporium spp. 70

Penicillium spp. 69

Saccharothrix spp. 307

INVERTEBRATES, AQUATIC

Amphipod

    Gammarus pseudolimnaeus 197, 230

    Gammarus pulex 231

Cladocerans

    Bosmina longirostris 231

    Daphnia magna 231, 278

    Daphnia pulex 231

    Simocephalus serrulatus 231

Copepods

    Acartia clausi 232, 281

    Eurytemora affinis 232

    Mesocyclops aspericornis 93, 94

Crabs

    Alaskan king crab, *Paralithodes camtschatica* 178–181

    Blue crab, *Callinectes sapidus* 72

    Dungeness crab, *Cancer magister* 233

    Red crab, *Cancer productus* 72

    Swamp ghost crab, *Ucides cordatus* 72

    Tanner crab, *Chionoecetes bairdi* 179–182, 230

Crayfish

    Astacus spp. 72

    Red crayfish, *Procambarus clarki* 231

    Orconectes sp. 278

Mosquito, *Aedes* spp. 93, 94

Molluscs, bivalves

    American oyster, *Crassostrea virginica* 66, 72

    Ampullaria sp. 258, 260

    Brotia costula 177

    Butter clam, *Saxidomus giganteus* 72

    California floater, *Anodonata californiensis* 210

    Clam, *Anodontitis* sp. 260

    Clam, *Castalia* sp. 260

    Clam, *Tapes* sp. 57

    Clithon sp. 177

    Common mussel, *Mytilus edulis* 57, 72

    Marisa sp.260

    Melanoides tuberculata 177

    Pacific oyster, *Crassostrea gigas* 233

    Quahog clam, *Mercenaria mercenaria* 72

    Winged floater clam, *Anodonta nuttalliana* 72

Molluscs, gastropods

    Slipper limpet, *Crepidula fornicata* 277, 279

    Snail, *Helisoma campanulata* 231

Planarian, *Dugesia dorotocephala* 277

Shrimps

    Mysid shrimp, *Mysidopsis bahia* 180, 181, 233, 279

    Northern shrimp, *Pandalus borealis* 180, 181

    Various, *Pandulus* sp. 57

Starfish, *Asterias amurensis* 33

Stonefly, *Pteronarcys californica*
Tunicates
   Diplosoma macdonaldi 66
   Molgula occidentalis 66
   Styela plicata 66

INVERTEBRATES, TERRESTRIAL

Budworm, western spruce, *Christoneura*
   occidentalis 233, 234
Earthworms
   Eisenia foetida 286
   Lumbricus terrestris 230, 234
   Octochaetus pattoni 286
Flea, *Pulex* spp. 97
Fruitfly, *Drosophila melanogaster* 286
Louse, *Pediculus* spp. 97
Stoneflys
   Pteronarcys californica 231
   Pteronarcys dorsata 231
Wasp, chalcid, *Hemadas nubilpennis* 59

MAMMALS

Anteater, giant, *Myercophaga tridactyla* 263
Badger, *Taxidea taxus* 201, 202
Bats
   Big brown bat, *Eptesicus fuscus* 308
   Eastern pipistrelle, *Pipistrellus subflorus* 308
   Little brown bat, *Myotis lucifugus* 200, 309
   Myotis spp. 308
   Northern myotis, *Myotis septentrionalis* 309
   Red bat, *Lasiurus borealis* 308
   Silver-haired bat, *Lasionycterius noctirogans*
     308
   Townsend's big-eared bat, *Plecotus*
     townsendii 202, 309
   Vampire bat, *Desmodus rotundus* 95, 96
Beaver, common, *Castor canadensis* 201
Capybara, *Hydrochoerus hydrochaeris* 261
Caribou, *Rangifer tarandus* 285
Cat, domestic, *Felis domesticus* 201, 236, 241,
   279, 284, 285, 291
Cattle, *Bos* spp. 58, 79, 96, 201, 202, 235, 241,
   259
Coyote, *Canis latrans* 200, 201, 308
Deer
   Mule deer, *Odocoileus hemionus* 201, 202,
     204, 278, 284, 285

Swamp deer, *Cervus duvauceli* 263
Dog, domestic, *Canis familiaris* 45, 77, 79, 80,
   201, 223, 235, 241, 278, 284
Elk, *Cervus elaphus* 204
Elephant, African, *Loxodonta africana* 310
Ferret, *Mustela* sp. 32
Fox
   Kit fox, *Vulpes macrotis* 201
   Red fox, *Vulpes vulpes* 285
Hamster, *Cricetus* spp. 236
Horse, *Equus caballus* 96, 236, 241
Jackrabbit, blacktail, *Lepus californicus* 201
Jaguar, *Panthera onca* 263, 264
Marten, *Martes martes* 285
Mink, *Mustela vison* 169, 279, 284, 285, 291
Monkey, rhesus, *Macaca mulatta* 241, 284
Moose, *Alces alces* 285
Mouse
   Eastern harvest mouse, *Reithrodontomys*
     humulis 27
   House mouse, *Mus musculus* 200
   Mus spp. 45, 74, 75, 76, 77, 81, 82, 138, 236,
     237, 241
   Peromyscus spp. 308
   White-footed mouse, *Peromyscus leucopus*
     200
Opossum, *Didelphis virginiana* 308
Otter
   European otter, *Lutra lutra* 196, 266, 288
   Giant otter, *Pteronura brasiliensis* 264, 266
   River otter, *Lutra canadensis* 169, 279, 284,
     285
Panther, Florida, *Felis concolor coryi* 288
Pig
   Domestic pig, *Sus* spp. 241, 259, 278, 284
   Guinea pig, *Cavia* spp. 73, 236, 238
Polecat, *Mustela putorius* 285
Rabbit, *Oryctolagus* sp. 73, 76, 237, 241, 285
Raccoon, *Procyon lotor* 308
Rat
   Cotton rat, *Sigmodon hispidus* 238
   Norway rat, *Rattus norvegicus* 77, 78, 200
   Rattus spp. 73, 76, 77, 79–81, 203, 237, 238,
     241
Seal
   Harp seal, *Phoca groenlandica* 278, 284, 291
   Ringed seal, *Phoca hispida* 285
Sheep
   Bighorn sheep, *Ovis canadensis* 202
   Domestic sheep, *Ovis aries,* 3, 237, 259
Skunk, striped, *Mephitis mephitis* 308

PROTOZOANS

Leishmania spp. 28
Plasmodium spp. 28, 29, 96
Tetrahymena spp. 65, 66

REPTILES

Copperhead, southern, *Agkistrodon contortix*
         308
Crocodile
    Caiman spp. 264
    Crocodilus crocodilus 264
    Melanosuchus niger 264
    Paleosuchus sp. 261, 264
Lizard, northern fence, *Sceloporus undulatus*
         hyacinthinus 27, 308

Skink, broadhead, *Eumeces laticeps* 308
Turtle, *Podocnemis unifilis* 261

VIRUSES

Ebola 96, 97
Epstein–Barr virus 144
HIV 29, 44, 47, 98–100, 105, 113, 114
Semiliki Forest Virus 76
West Nile virus 76
Yellow fever virus 76

YEASTS

Candida spp. 29, 65
Cryptococcus spp. 194
Rhodotorula spp. 194
Saccharomyces spp. 66, 70